Textbook of
Engineering Drawing

(as per BIS:SP46:2003)

with AutoCAD

Fourth Edition

Foreword

I have gone through the book **"Engineering Drawing"** written by **prof. K.Venkata Reddy,** a long time colleague at S.V.University College of Engineering, Tirupati. I really felt amazed at the treatment. Most books on the subject have explanations which are either difficult to understand or lengthy, in either case, students require others help to delineate the meaning. Explanation in this book is brief and to the point, and students can easily proceed with the drawing by themselves.

A noteworthy feature is introducing the students to the drawing practices as recommended by the **Bureau of Indian Standards (BIS) SP:46:2003**. These include drawing sheet sizes and layout, title block, types of lines, lettering and principles of dimensioning. Some common violations and their corrected versions are included as a guidance to the students.

The chapter on Freehand Sketching is an important feature of this book. Irrespective of the branch of study, effective means of communication of ideas is through sketching and freehand sketching is thus very important, specially, for Engineers.

Engineers always present their ideas in the form of a pictorial view, in which the object is shown in one view, approximately as it appears to the eye. But in the execution of the idea, orthographic views are used. All engineers must therefore, be conversant in constructing orthographic views from pictorial views and vice versa. To strengthen the skills in this field worksheets with exercises for practice are provided on orhographic views, missing lines and views at the end of the book.

This book has certain features not commonly found in other books and I congratulate the author for his effort in this direction.

(Prof. K. Narayana Reddy)
Professor of Mechanical Engg. (Retd.)
S.V.U.College of Engg., Tirupati

Preface to Fourth Edition

Engineering Drawing is said to be the universal language of Engineers. As such, Engineering Graphics or Drawing subject is made common to all branches in first year engineering diploma and degree professional courses.

Bureau of Indian Standards (BIS) has prescribed standard Code Book BIS : SP 46: 2003 "Engineering drawing for schools and colleges" so that we follow ISO standards of principles of drawing.

This book follows the above standards throughout the book in first angle projection as recommended by BIS and uniform aligned system of dimensioning. It covers all units systematically and step by step with numerous examples worked out in stages. This enables the student and staff to understand the subject with ease and enthusiasm by self study.

Students invariably find it difficult to follow the chapters on projection of points and lines especially line inclined to both the planes. Special care is taken to explain this in 4 stages of drawing for easy understanding. All other chapters are also enlarged with more worked examples.

The chapter on AutoCAD is revised to cover more details in making the drawing through AutoCAD.

-Author

Preface to First Edition

Drawing as an Art, is the picturisation of the imagination of the scene by an individual - the Artist. It has no standard guidelines and boundaries. Engineering Drawing on the other hand is the scientific representation of an object, according to certain national and international standards of practice. It can be drawn and understood by all with the knowledge of basic principles of drawing.

Drawing is said to be the universal language of all engineers. What can't be explained by any length of description in words can be made clear by a simple line diagram; be it a house plan in civil engg., circuit diagrams in electrical engg., plant layout in chemical engg, production drawing in mechanical engg. and graphics by computer science engineers and so on.

Whatever be the application, the language, in this case, the drawing should be simple, easily understood and liked by students of all branches. This book attempts to make the subject interesting and easy by limiting the topics to the extent necessary instead of too much of brain storming the learners. This has a practical approach with step by step introductions as per the syllabus of most the Universities.

The introduction of worksheets is unique. It follows the principles of drawing as recommended by Bureau of Indian Standards (BIS), in their standards "Engg., drawing practice for schools and colleges". BIS : SP46 : 1988.

Any suggestions to improve the contents are most welcome.

- Author

Contents

CHAPTER - 4

Geometrical Constructions 49-105

CHAPTER - 5

Orthographic Projections 107-144

Chapter - 6

Projection of Solids 157-226

Chapter - 7

Development of Surfaces 224-246

CHAPTER - 8

Intersection of Surfaces 247-260

CHAPTER - 9

Isometric Projection 261-287

CHAPTER - 13

Freehand Sketching 356-361

CHAPTER - 14

Computer Aided Design and Drawing (CADD) 362-413

CHAPTER 1

Drawing Instruments and Accessories

1.1 Introduction

Engineering drawing is a two dimensional representation of three dimensional objects. In general, it provides necessary information about the shape, size, surface quality, material, manufacturing process, etc., of the object. It is the graphic language from which a trained person can visualise objects.

Drawings prepared in one country may be utilised in any other country irrespective of the language spoken. Hence, engineering drawing is called the universal language of engineers. Any language to be communicative, should follow certain rules so that it conveys the same meaning to every one. Similarly, drawing practice must follow certain rules, if it is to serve as a means of communication. For this purpose, Bureau of Indian Standards (BIS) adapted the International Standards on code of practice for drawing. The other foreign standards are : DIN of Germany, BS of Britain and ANSI of America.

1.2 Role of Engineering Drawing

The ability to read drawing is the most important requirement of all technical people in any profession. As compared to verbal or written description, this method is brief and more clear. Some of the applications are : building drawing for civil engineers, machine drawing for mechanical engineers, circuit diagrams for electrical and electronics engineers, computer graphics for one and all.

The subject in general is designed to impart the following skills:

1. Ability to read and prepare engineering drawings.
2. Ability to make free-hand sketching of objects.
3. Power to imagine, analyze and communicate, and
4. Capacity to understand other subjects.

1.3 Drawing Instruments and Aids

The Instruments and other aids used in draughting work are listed below :

1. Drawing board
2. Mini draughter
3. Instrument box
4. Set squares
5. Protractor
6. Set of scales
7. French curves
8. Drawing sheets
9. Pencils
10. Templates

1.3.1 Drawing Board

Until recently drawing boards used are made of well seasoned softwood of about 25 mm thick with a working edge for T-square. Now a days mini-draughters are used instead of T-squares which can be fixed on any board. The standard size of board depends on the size of drawing sheet size required.

1.3.2 Mini-Draughter

Mini-draughter consists of an angle formed by two arms with scales marked and rigidly hinged to each other (Fig. 1.1). It combines the functions of T-square, set-squares, scales and protractor. It is used for drawing horizontal, vertical and inclined lines, parallel and perpendicular lines and for measuring lines and angles.

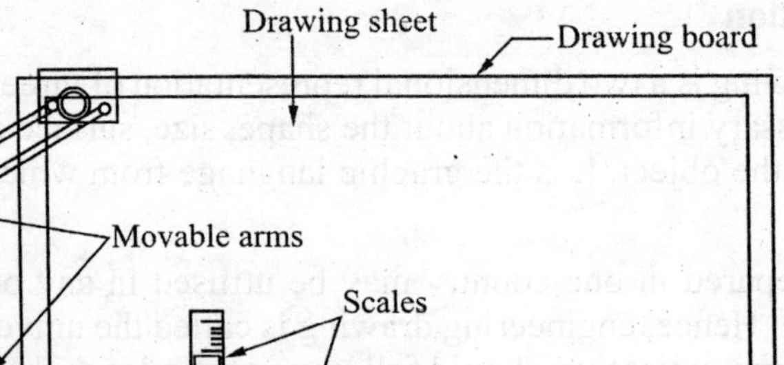

Fig. 1.1 Mini-draughter.

1.3.3 Instruments Box

Instruments box contains 1. Compasses, 2. Dividers and 3. Inking pens. What is important is the position of the pencil lead with respect to the tip of the compass. It should be atleast 1 mm above as shown in Fig.1.2 because the tip goes into the board for grip by 1 mm.

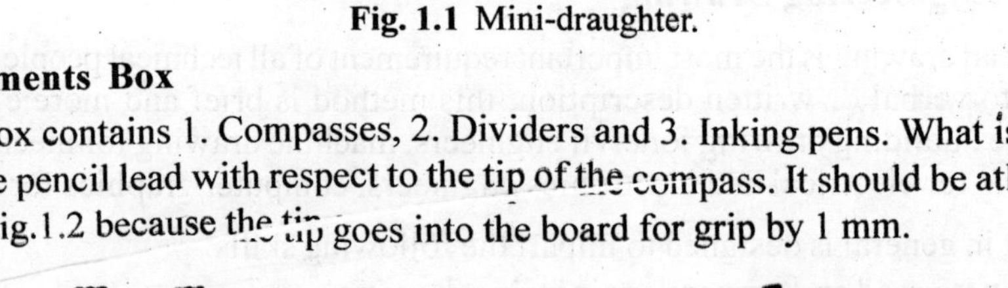

(a) Sharpening and position of compass lead

(b) Position of the lead leg to draw larger circles

Fig. 1.2

1.3.4 Set of Scales

Scales are used to make drawing of the objects to proportionate size desired. These are made of wood, steel or plastic (Fig.1.3). BIS recommends eight set-scales in plastic/cardboard with designations M1, M2 and so on as shown in Table 1.1 and Table 1.2.

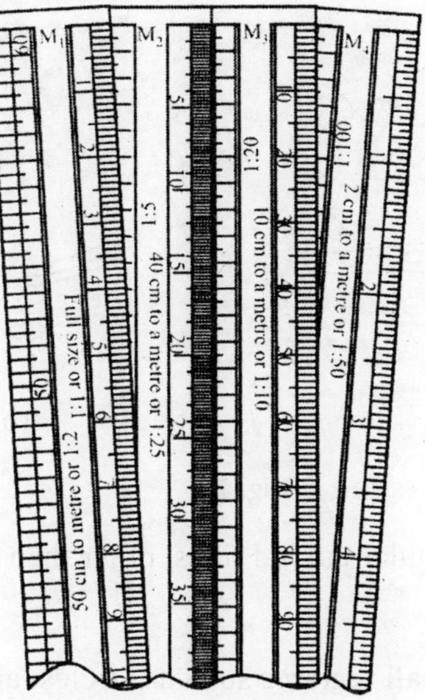

Fig. 1.3 Set of Scales.

Table 1.1 Designation of Scales.

	M1	M2	M3	M4	M5	M6	M7	M8
Scale on one edge	1:1	1:2.5	1:10	1:50	1:200	1:300	1:400	1:1000
Scale on other edge	1:2	1:5	1:20	1:100	1:500	1:600	1:800	1:2000

Note : Do not use the scales as a straight edge for drawing straight lines.

Table 1.2

Scales for use on technical drawings (IS : 46-1988)			
Category	**Recommended scales**		
Enlargement scales	50 : 1 5 : 1	20 : 1 2 : 1	10 : 1
Full size	1 : 1		
Reduction scales	1 : 2 1 : 20 1 : 200 1 : 2000	1 : 5 1 : 50 1 : 500 1 : 5000	1 : 10 1 : 100 1 : 1000 1 : 10000

1.3.5 French Curves

French curves are available in different shapes (Fig.1.4). First a series of points are plotted along the desired path and then the most suitable curve is made along the edge of the curve. A flexible curve consists of a lead bar inside rubber which bends conveniently to draw a smooth curve through any set of points.

(a) French curves (b) Flexible curve

Fig. 1.4

These are used for drawing irregular curved lines, other than circles or arcs of circles.

1.3.6 Templates

These are aids used for drawing small features such as circles, arcs, triangular, square and other shapes and symbols used in various science and engineering fields (Fig.1.5).

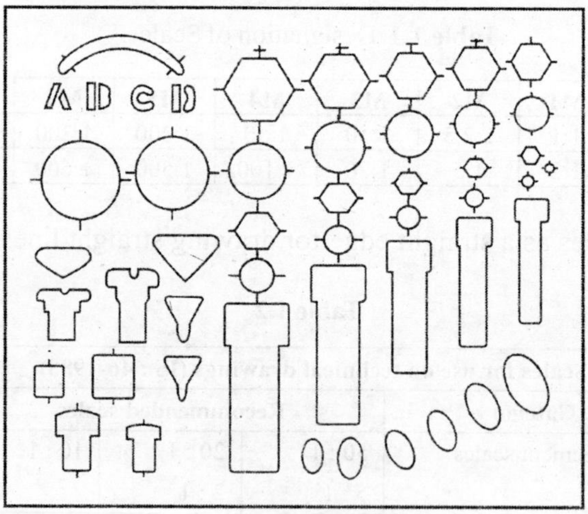

Fig. 1.5 Template.

1.3.7 Pencils

Pencils with leads of different degrees of hardness or grades are available in the market. The hardness or softness of the lead is indicated by 3H, 2H, H, HB, B, 2B, 3B, etc. The grade HB denotes medium hardness of lead used for general purpose. The hardness increases as the value of the numeral before the letter H increases. The lead becomes softer, as the value of the numeral before B increases (Fig. 1.6).

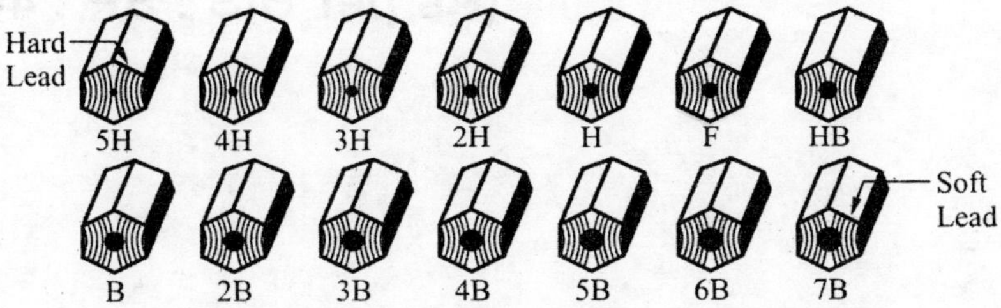

Fig. 1.6 Pencil Leads.

The selection of the grade depends on the line quality desired for the drawing. Pencils of grades H or 2H may be used for finishing a pencil drawing as these give a sharp black line. Softer grade pencils are used for sketching work. HB grade is recommended for lettering and dimensioning.

Now a days mechanical pencils are widely used in place of wooden pencils. When these are used, much of the sharpening time can be saved. The number 0.5, 0.70 of the mechanical pencil (Fig. 1.7) indicates the thickness of the line obtained with the lead and the size of the lead diameter.

Fig. 1.7 Mechanical Pencil.

Micro-tip pencils with 0.5 mm leads with the following grades are recommended.

HB Soft grade for border lines, lettering and free hand sketching

H Medium grade for visible outlines, visible edges and boundary lines

2H Hard grade for construction lines, dimension lines, leader lines, extension lines, centre lines, hatching lines and hidden lines.

Chapter 2

Lettering and Dimensioning Practices

(As per BIS : SP : 46 : 2003)

2.1 Introduction

Engineering drawings are prepared on standard size drawing sheets. The correct shape and size of the object can be visualised from the understanding of not only its views but also from the various types of lines used, dimensions, notes, scale etc. For uniformity, the drawings must be drawn as per certain standard practice. This chapter deals with the drawing practices as recommended by Bureau of Indian Standards (BIS) SP: 46:2003. These are adapted from what is followed by International Standards Organisation (ISO).

2.2 Drawing Sheet

The standard drawing sheet sizes are arrived at on the basic Principle of $X : Y = 1 : \sqrt{2}$ and $XY = 1$ where X and Y are the sides of the sheet. For example A0, having a surface area of 1 sq.m; X = 841 mm and Y = 1189 mm. The successive sizes are obtained by either by halving along the length or doubling the width, the area being in the ratio 1 : 2. Designation of sizes is given in Fig.2.1 and their sizes are given in Table 2.1. For class work A2 size drawing sheet is preferred.

Table 2.1

Designation	Dimension, mm Trimmed size
A0	841 × 1189
A1	594 × 841
A2	420 × 594
A3	297 × 420
A4	210 × 297

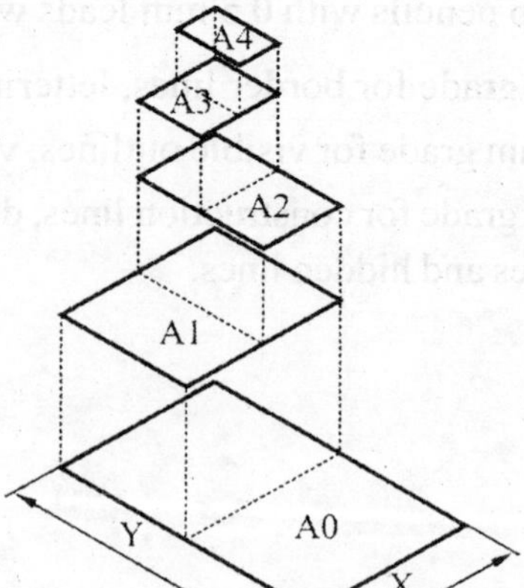

Fig. 2.1 Drawing Sheet Formats.

2.2.1 Title Block

The title block should lie within the drawing space at the bottom right hand corner of the sheet. The title block can have a maximum length of 170 mm providing the following information.

1. Title of the drawing.
2. Drawing number.
3. Scale.
4. Symbol denoting the method of projection.
5. Name of the firm, and
6. Initials of staff who have designed, checked and approved.

The title block used in industries and one suggested for students class work are shown in Fig. 2.2.

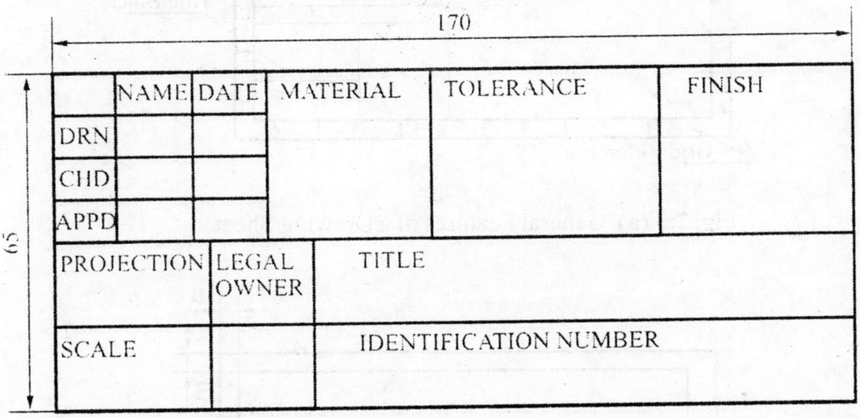

Fig. 2.2(a) Title Block for Industry.

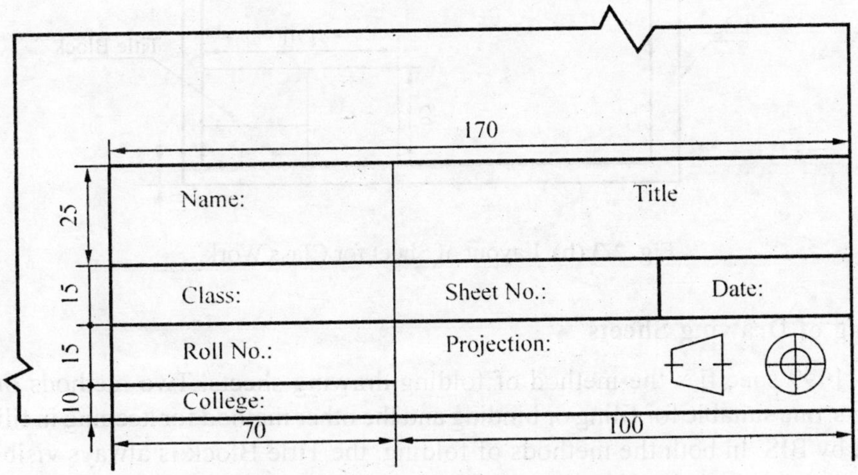

Fig. 2.2(b) Title Block for Class Work.

2.2.2 Drawing Sheet Layout (IS 10711 : 2001)

The layout of a drawing sheet used on the shop floor is shown in Fig. 2.3(a), The layout suggested to students is shown in Fig. 2.3(b).

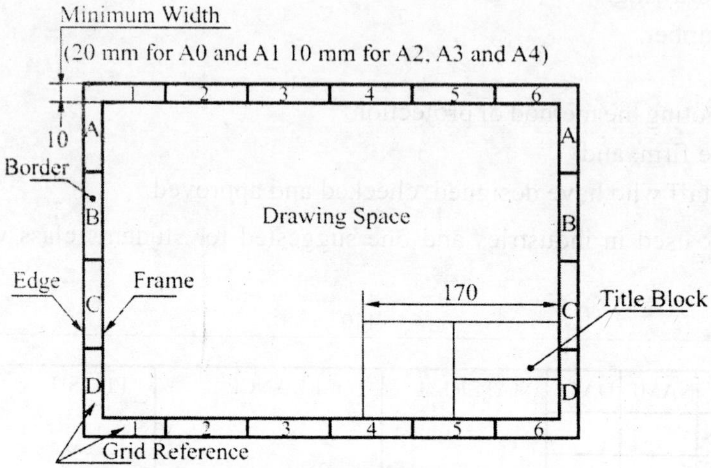

Fig. 2.3 (a) General Features of a Drawing Sheet.

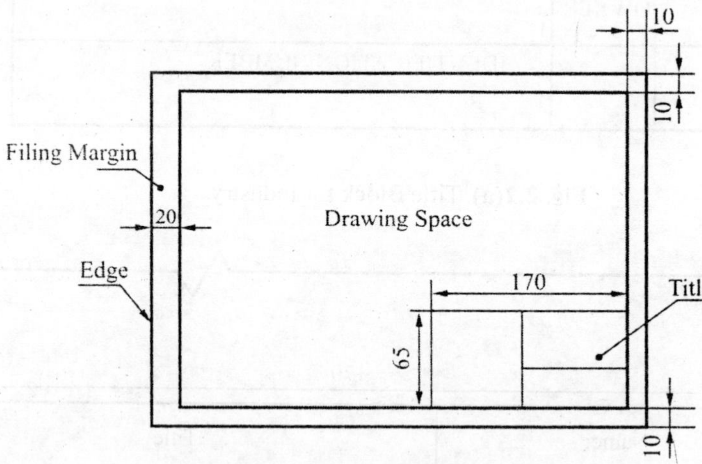

Fig. 2.3 (b) Layout of Sheet for Class Work.

2.2.3 Folding of Drawing Sheets

IS : 11664 – 1999 specifies the method of folding drawing sheets. Two methods of folding of drawing sheets, one suitable for filing or binding and the other method for keeping in filing cabinets are specified by BIS. In both the methods of folding, the Title Block is always visible. Fig. 2.4. shows the method in which drawing sheets may be unfolded and refolded, without the necessity of removal from the file.

Sheet Designation	Folding Diagram	Lengthwise Folding	Crosswise Folding
A1 594 × 841	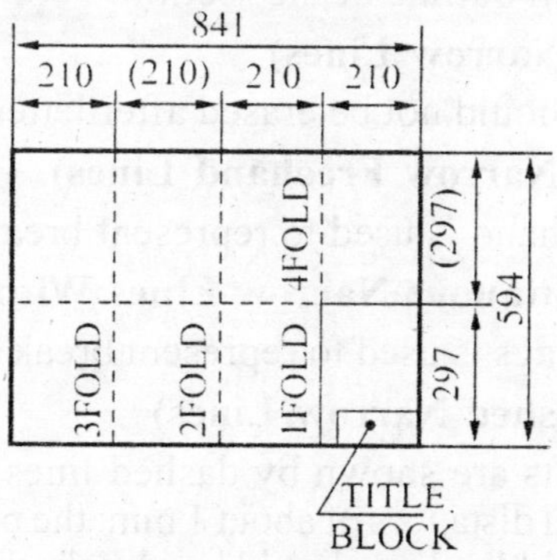	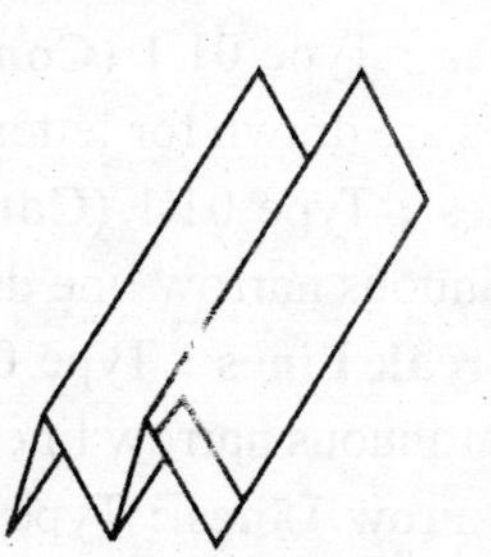	
A2 420 × 594	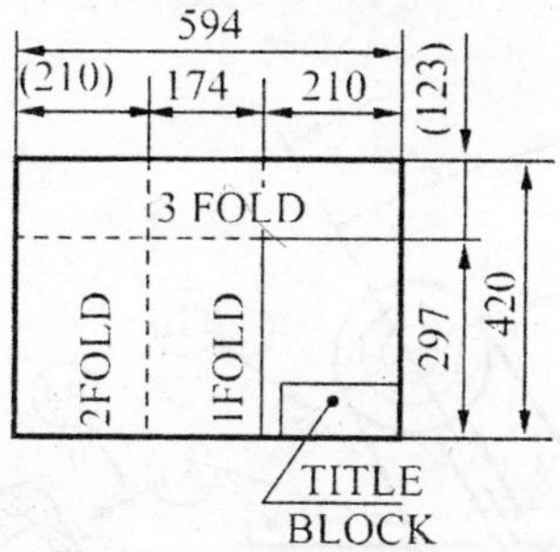		

Fig. 2.4 Folding of Drawing Sheets.

2.2.4 Types of Lines (IS 10714 (Part 20) : 2001 and BIS : SP 46 : 2003) (Fig. 2.5)

Just as in English textbook the correct words are used for making correct sentences; in Engineering Graphics, the details of various objects are drawn by different types of lines. Each line has a definite meaning and sense to convey.

IS 10714 (Part 20) : 2001 (General principles of presentation on technical drawings) and BIS : SP 46:2003 specify the following types of lines and their applications (Fig. 2.5 and Table 2.2).

- **Visible Outlines, Visible Edges :** Type 01.2 **(Continuous wide lines)**

  The lines drawn to represent the visible outlines/ visible edges / surface boundary lines of objects should be outstanding in appearance.

- **Dimension Lines :** Type 01.1 **(Continuous narrow Lines)**

  Dimension Lines are drawn to mark dimension.

- **Extension Lines :** Type 01.1 **(Continuous narrow Lines)**

  These are extended from out lines upto slightly beyond the respective dimension lines.

- **Construction Lines:** Type 01.1 **(Continuous narrow Lines)**

  Construction Lines are drawn for constructing drawings and should not be erased after completion of the drawing.

- **Hatching / Section Lines :** Type 01.1 **(Continuous Narrow Lines)**
  Hatching Lines are drawn for the sectioned portion of an object. These are drawn inclined at an angle of 45° to the axis or to the main outline of the section.
- **Guide Lines :** Type 01.1 **(Continuous Narrow Lines)**
  Guide Lines are drawn for lettering and should not be erased after lettering.
- **Break Lines :** Type 01.1 **(Continuous Narrow Freehand Lines)**
  Wavy continuous narrow line drawn freehand is used to represent break of an object.
- **Zig Zag Break Lines :** Type 01.1 **(Continuous Narrow Lines With Zigzags)**
  Straight continuous narrow line with zigzags is used to represent break of an object.
- **Dashed Narrow Lines :** Type 02.1 **(Dashed Narrow Lines)**
  Hidden edges / Hidden outlines of objects are shown by dashed lines of short dashes of equal lengths of about 3 mm, spaced at equal distances of about 1 mm; the points of intersection of these lines with the outlines / another hidden line should be clearly shown.

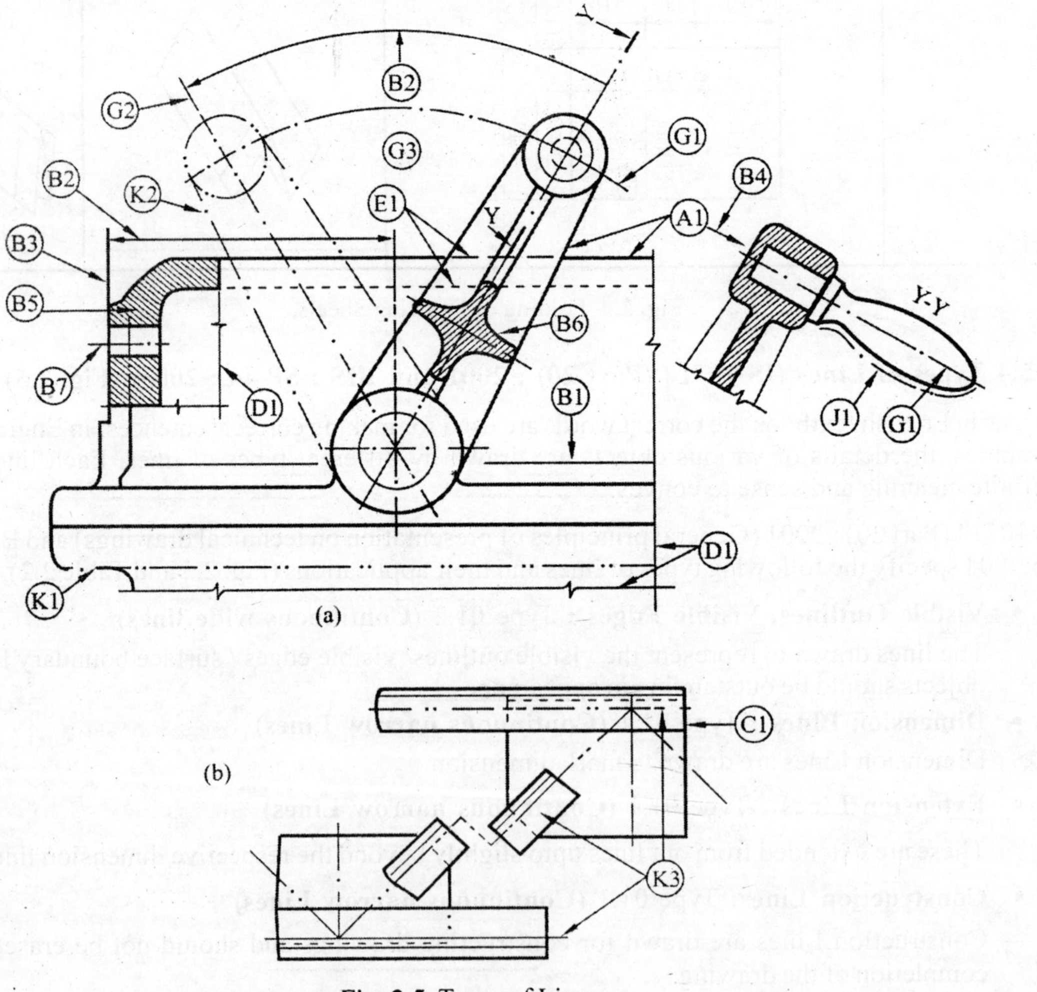

Fig. 2.5 Types of Lines.

- **Center Lines :** Type 04.1 **(Long-Dashed Dotted Narrow Lines)**
  Center Lines are drawn at the center of the drawings symmetrical about an axis or both the axes. These are extended by a short distance beyond the outline of the drawing.
- **Cutting Plane Lines :** Type 04.1 and Type 04.2
  Cutting Plane Line is drawn to show the location of a cutting plane. It is long-dashed dotted narrow line, made wide at the ends, bends and change of direction. The direction of viewing is shown by means of arrows resting on the cutting plane line.
- **Border Lines**
  Border Lines are continuous wide lines of minimum thickness 0.7 mm

Understanding the various types of lines used in drawing (i.e.,) their thickness, style of construction and appearance as per BIS and following them meticulously may be considered as the foundation of good drawing skills. Table 2.2 shows various types of lines with the recommended applications.

Table 2.2 Types of Lines and their applications (IS 10714 (Part 20) : 2001) and BIS: SP46 : 2003.

No.	Line Description and Representation	Application
01.1	Continuous narrow line B	Dimension lines, Extension lines (B2) Leader lines, Reference lines (B4) Short centre lines (B7) Projection lines (B3) Hatching (B5) Construction lines, Guide lines Outlines of revolved sections (B6) Imaginary lines of intersecton (B1)
01.1	Continuous narrow freehand line C	Preferably manually represented termination of partial or interupped views, cuts and sections, if the limit is not a line of symmetry or a center line[a]. (C1)
01.1	Continuous narrow lines eith zigzags A	Perferably mechanically represented termination of partial or interrupted views, cuts and sections, if the limit is not a line of symmetry or a centre line[a]. (D1)
01.2	Continuous wide line	Visible edges, visible outlines (A1) Main representations in diagrams, maps, flow charts
02.1	Dashed narrow line D	Hidden edges Hidden outlines (E1)
04.1	Long-dashed dotted narrow line E	Center lines / Axes, Lines if symmetry (G1) Cutting lines / Axes, Lines 04.2 at ends and changes o direction)
04.2	Long-dashed dotted wide lines F	Cutting planes at the ends and changes of direction outlines of visible parts situated infront of cutting plane (Y)
05.1	K	Outlines of adjacent parts Alternate and extreme position of movable parts (K,)

Line widths (IS 10714 : 2001)

Line width means line thickness.

Choose line widths according to the size of the drawing from the following range : **0.13, 0.18, 0.25, 0.35, 0.5, 0.7 and 1 mm.**

BIS recommends two line widths on a drawing. Ratio between the thin and thick lines on a drawing shall not be less than **1 : $\sqrt{2}$** .

Precedence of Lines

1. When a Visible Line coincide with a Hidden Line or Center Line, draw the Visible Line. Also, extend the Center Line beyond the outlines of the view.
2. When a Hidden Line coincides with a Center Line, draw the Hidden Line.
3. When a Visible Line coincides with a Cutting Plane, draw the Visible Line.
4. When a Center line coincides with a Cutting Plane, draw the Center Line and show the Cutting Plane line outside the outlines of the view at the ends of the Center Line by thick dashes.

2.3 Lettering [IS 9609 : 2001 AND BIS : SP 46 : 2003]

Lettering is defined as writing of titles, sub-titles, dimensions, etc., on a drawing.

2.3.1 Importance of Lettering

To undertake production work of engineering components as per the drawing, the size and other details are indicated on the drawing. This is done in the form of notes and dimensions.

Main Features of Lettering are legibility, uniformity and rapidity of execution. Use of drawing instruments for lettering consumes more time. **Lettering should be done freehand with speed**. Practice accompanied by continuous efforts would improve the lettering skill and style. Poor lettering mars the appearance of an otherwise good drawing.

BIS and ISO Conventions

IS 9609 : 2001 and SP 46 : 2003 (Lettering for technical drawings) specify lettering in technical product documentation. This BIS standard is based on ISO 3098 : 1997.

2.3.2 Single Stroke Letters

The word single-stroke should not be taken to mean that the lettering should be made in one stroke without lifting the pencil. It means that the thickness of the letter should be uniform as if it is obtained in one stroke of the pencil.

2.3.3 Types of Single Stroke Letters (Fig. 2.6 and 2.7)

1. Lettering Type A : (i) Vertical and (ii) Sloped (at 75° to the horizontal)
2. Lettering Type B : (i) Vertical and (ii) Sloped (at 75° to the horizontal)

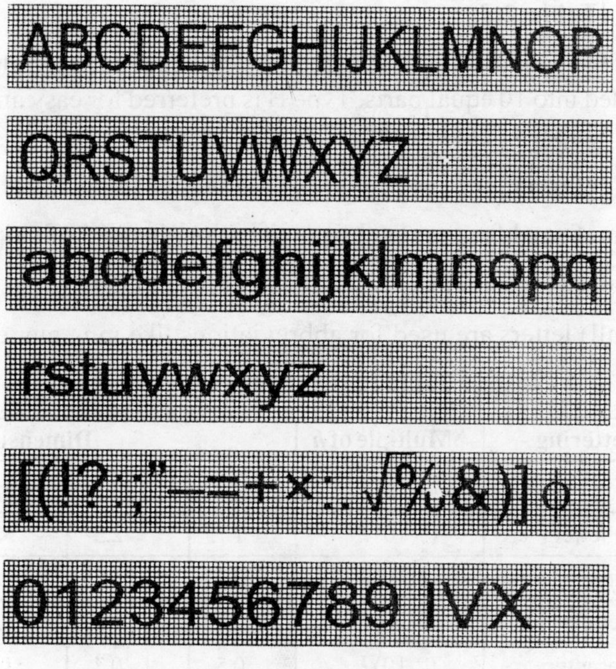

Fig. 2.6 Vertical Lettering

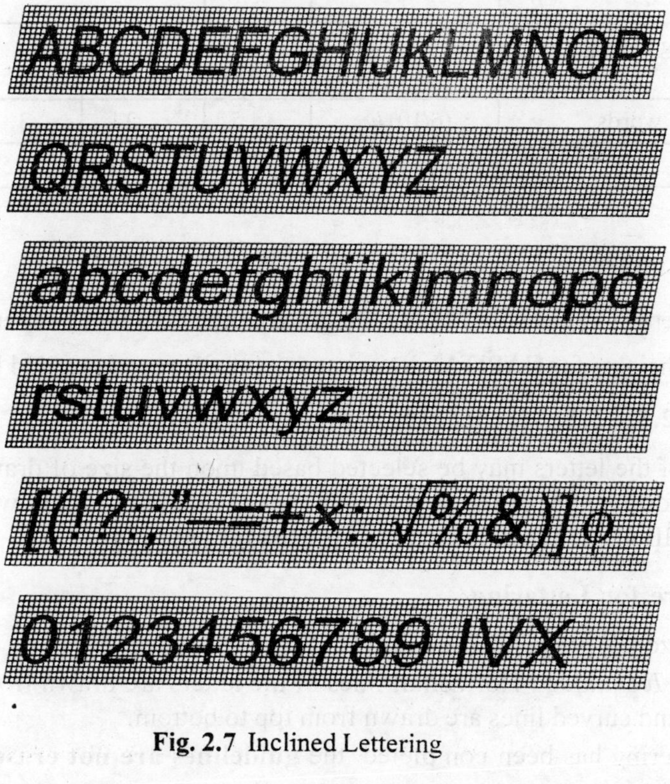

Fig. 2.7 Inclined Lettering

Type B Preferred (Table 2.3)

In Type A, height of the capital letter is divided into 14 equal parts, while in Type B, height of the capital letter is divided into 10 equal parts. Type B is preferred for easy and fast execution, because of the division of height into 10 equal parts.

Vertical Letters Preferred

Vertical letters are preferred for easy and fast execution, instead of sloped letters.

Note : Lettering in drawing should be in CAPITALS (i.e., Upper-case letters).

Lower-case (small) letters are used for abbreviations like mm, cm, etc.

Table 2.3 Lettering Proportions.

Characteristics of lettering		Multiple of h	Dimensions (mm)				
Lettering height	h	$(10/10)\,h$	2.5	3.5	5	7	10
Height of lower-case letters	c_1	$(7/10)\,h$	1.75	2.5	3.5	5	7
Tail of lower-case letters	c_2	$(3/10)\,h$	0.75	1.05	1.5	2.1	3
Stem of lower-case letters	c_3	$(3/10)\,h$	0.75	1.05	1.5	2.1	3
Spacing between characters	a	$(2/10)\,h$	0.5	0.7	1	1.4	2
Minimum, spacing between baselines)	b_1	$(15/10)\,h$	3.75	5.25	7.5	10.5	15
Minimum spacing between baselines)	b_2	$(13/10)\,h$	3.25	4.55	6.5	9.1	13
Spacing between words	e	$(6/10)\,h$	1.5	2.1	3	4.2	6
Line width	d	$(1/10)\,h$	0.25	0.35	0.5	0.7	1

2.3.4 Size of Letters

- Size of Letters is measured by the height **h** of the CAPITAL letters as well as numerals.
- Standard heights for CAPITAL letters and numerals recommended by BIS are given below :

1.8, 2.5, 3.5, 5, 6, 10, 14 and **20 mm**

Note: Size of the letters may be selected based upon the size of drawing. In order to obtain correct and uniform height of letters and numerals, guide lines are drawn, using 2H pencil with light pressure. HB grade conical end pencil is used for lettering.

2.3.5 Procedure for Lettering

1. Thin horizontal guide lines are drawn first at a distance '**h**' apart.
2. *Lettering Technique* : Horizontal lines of the letters are drawn from left to right. Vertical. inclined and curved lines are drawn from top to bottom.
3. After lettering has been completed, the **guidelines are not erased**.

2.3.6 Dimensioning of Type B Letters (Figs 2.8 and 2.9)

BIS denotes the characteristics of lettering as :

- h (height of capital letters),
- c_1 (height of lower-case letters),
- c_2 (tail of lower-case letters),
- c_3 (stem of lower-case letters),
- a (spacing between characters),
- b_1 & b_2 (spacing between baselines),
- e (spacing between words) and
- d (line thickness),

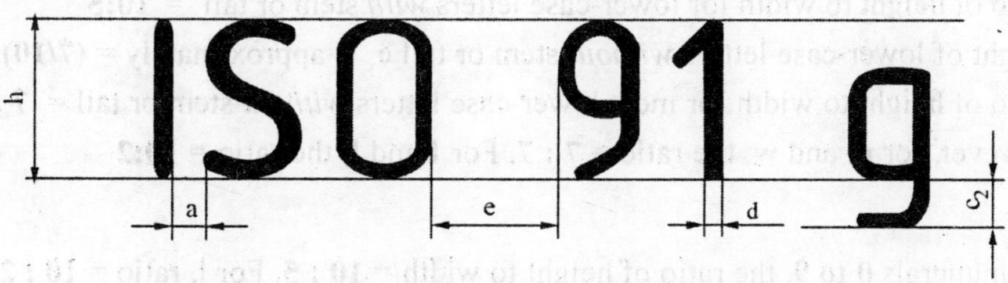

Fig. 2.8 Dimensions and Spacing of Letters.

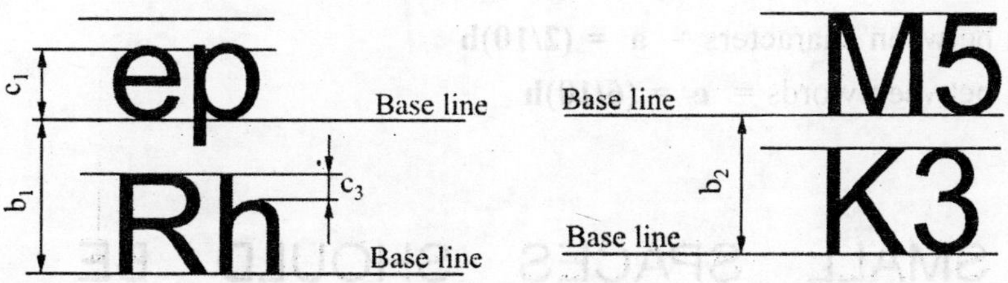

Fig. 2.9 Lettering.

2.3.7 Lettering practice

Practice of lettering capital and lower case letters and numerals of type B are shown in Figs.2.8 and 2.9.

The following are some of the guide lines for lettering (Fig 2.10 & 2.11)

1. Drawing numbers, title block and letters denoting cutting planes, sections are written in 10 mm size.
2. Drawing title is written in 7 mm size.
3. Hatching, sub-titles, materials, dimensions, notes, etc., are written in 3.5 mm size.
4. Space between lines = (3/4)h.
5. Space between words may be equal to the width of alphabet M or (3/5) h.
6. Space between letters should be approximately equal to (1/5) h. Poor spacing will affect the visual effect.
7. The spacing between two characters may be reduced by half if this gives a better visual effect, as for example LA, TV; over lapped in case of say LT, TA etc, and the space is increased for letters with adjoining stems.

CAPITAL Letters

- Ratio of height to width for most of the CAPITAL letters is approximately = **10:6**
- However, for **M** and **W**, the ratio = **10:8** for I the ratio = **10:2**

Lower-case Letters

- Height of lower-case letters *with* stem / tail (b, d, f, g, h, j, k, l, p, q, t, y) = $c_2 = c_3$ = **h**
- Ratio of height to width for lower-case letters *with* stem or tail = **10:5**
- Height of lower-case letters *without* stem or tail c_1 is approximately = **(7/10) h**
- Ratio of height to width for most lower-case letters *without* stem or tail = **7 : 5**
- However, for **m** and **w**, the ratio = **7 : 7**. For **I** and **l**, the ratio = **10:2**

Numerals

- For numerals **0 to 9**, the ratio of height to width = **10 : 5**. For l, ratio = **10 : 2**

Spacing

- Spacing between characters = **a = (2/10)h**
- Spacing between words = **e = (6/10)h**

SMALL SPACES SHOULD BE
USED FOR GOOD LETTER
SPACING

Correct

P O O R L E T T E R S P A C I N G
R E S U L T S F R O M S P A C E S
B E I N G T O O B I G

In correct

(a)

N I G H T N U M B E R S

Letters with adjoining stems require more spacing

VITAL ALTAR

(b)

Letter combinations with overlapping letters

Fig. 2.10 Guide Lines for Lettering.

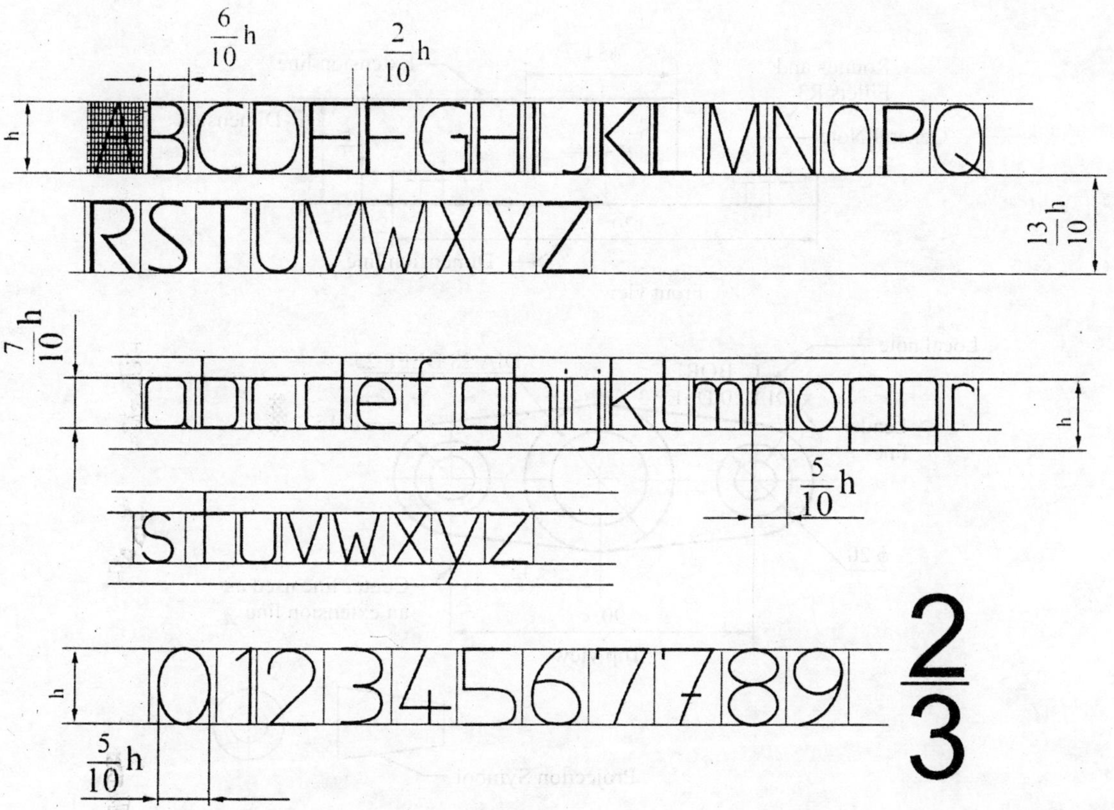

Fig. 2.11 Vertical Capital and Lowercase Letters and Numerals of Type B.

Example in Lettering Practice (Fig. 2.12)

Write freehand the following, using single stroke vertical CAPITAL letters of 5 mm (h) size

Fig. 2.12

2.4 Dimensioning

Drawing of a component, in addition to prividing complete shape description, must also furnish information regarding the size description (Fig. 2.13). These are provided through the distances between the surfaces, location of holes, nature of surface finish, type of material, etc. The expression of these features on a drawing, using lines, symbols, figures and notes is called dimensioning.

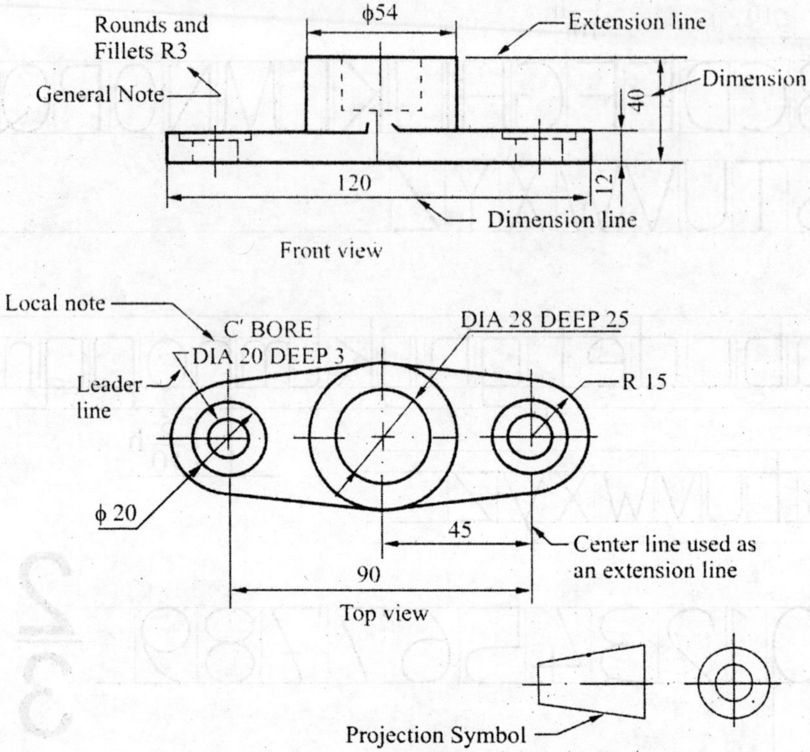

Fig. 2.13 Elements of Dimensioning.

2.4.1 Principles of Dimensioning

Some of the basic principles of dimensioning are given below.

1. All dimensional information necessary to describe a component clearly and completely shall be written directly on a drawing.
2. Each feature shall be dimensioned once only on a drawing, i.e., dimension marked in one view need not be repeated in another view.
3. Dimension should be placed on the view where the shape is best seen (Fig. 2.14).
4. As far as possible, dimensions should be expressed in one unit only preferably in millimeters, without showing the unit symbol (mm).
5. As far as possible dimensions should be placed outside the view (Fig. 2.15).
6. Dimensions should be taken from visible outlines rather than from hidden lines (Fig. 2.16).
7. No gap should be left between the feature and the start of the extension line (Fig. 2.17).
8. Crossing of centre lines should be done by a long dash and not at a dot or short dash (Fig. 2.18).

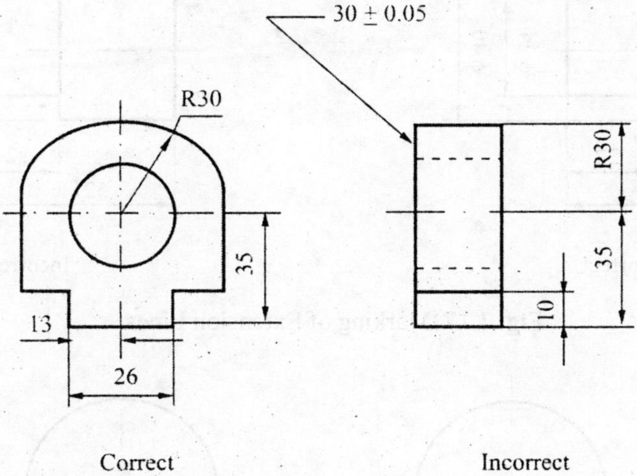

Fig. 2.14 Placing the Dimensions where the Shape is Best Shown.

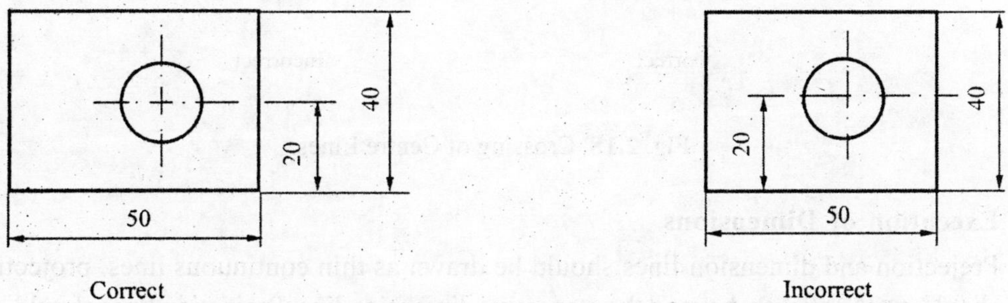

Fig. 2.15 Placing Dimensions Outside the View.

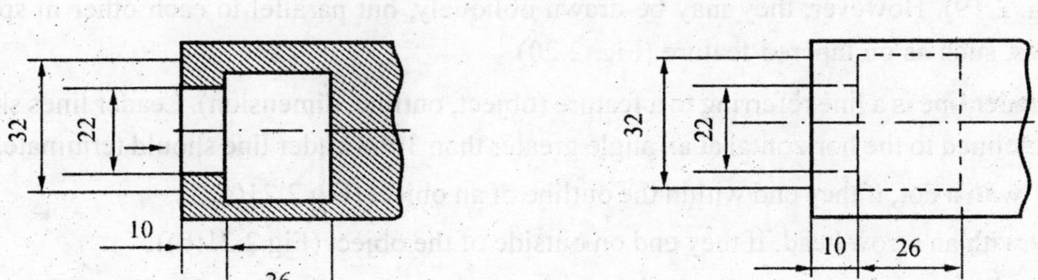

Fig. 2.16 Marking the Dimensions from the Visible Outlines.

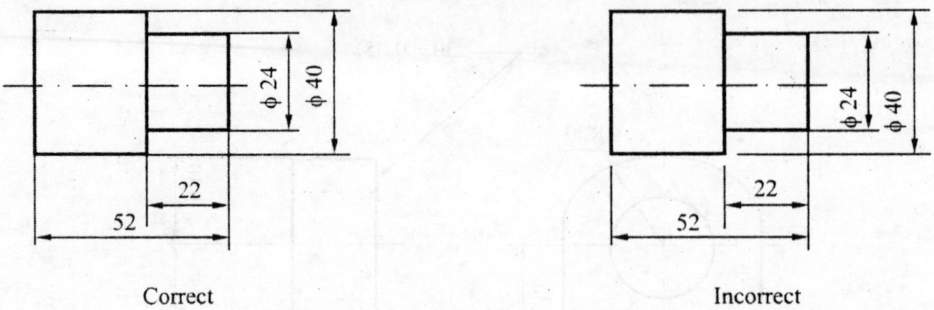

Fig. 2.17 Marking of Extension Lines.

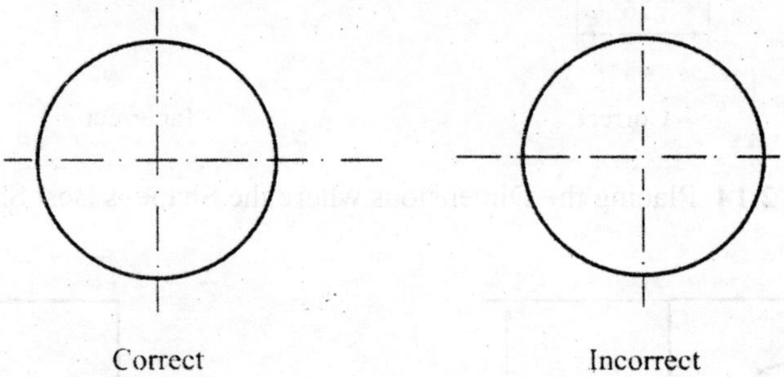

Fig. 2.18 Crossing of Centre Lines.

2.4.2 Execution of Dimensions

1. Prejection and dimension lines should be drawn as thin continuous lines. projection lines should extend slightly beyond the respective dimension line. Projection lines should be drawn perpendicular to the feature being dimensioned. If the space for dimensioning is insufficient, the arrow heads may be reversed and the adjacent arrow heads may be replaced by a dot (Fig. 2.19). However, they may be drawn obliquely, but parallel to each other in special cases, such as on tapered feature (Fig. 2.20).
2. A leader line is a line referring to a feature (object, outline, dimension). Leader lines should be inclined to the horizontal at an angle greater than 30^0. Leader line should terminate,
   (a) with a dot, if they end within the outline of an object (Fig.2.21(a)).
   (b) with an arrow head, if they end on outside of the object (Fig.2.21(b)).
   (c) without a dot or arrow head, if they end on dimension line (Fig.2.21(c)).

Dimension Termination and Origin Indication

Dimension lines should show distinct termination in the form of arrow heads or oblique strokes or where applicable an origin indication (Fig. 2.22). The arrow head included angle is 15^0. The origin indication is drawn as a small open circle of approximately 3 mm in diameter. The proportion lenght to depth 3 : 1 of arrow head is shown in Fig. 2.23.

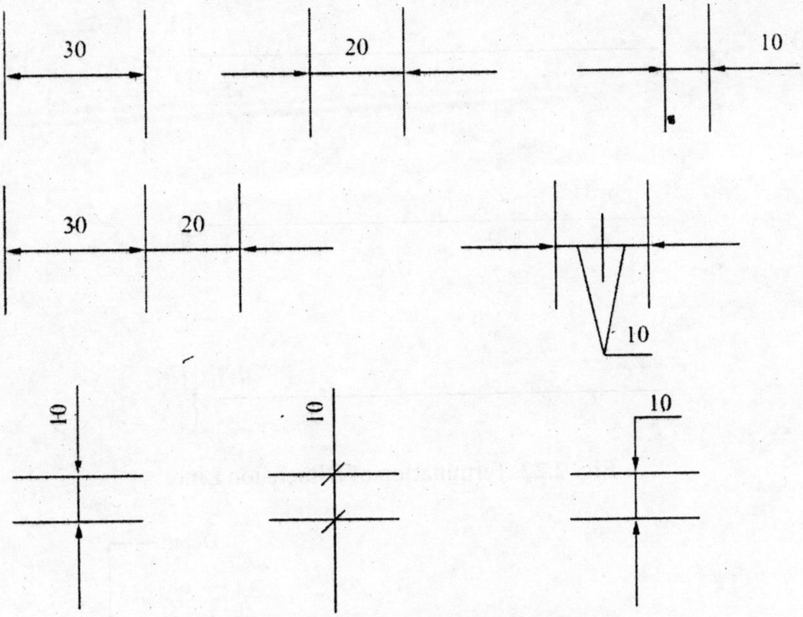

Fig. 2.19 Dimensioning in Narrow Spaces.

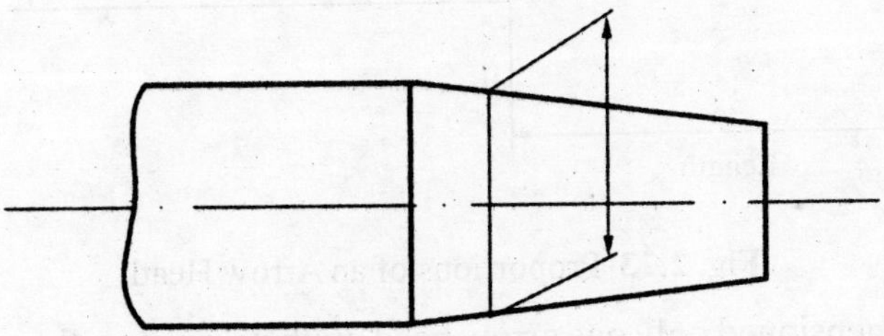

Fig. 2.20 Dimensioning a Tapered Feature.

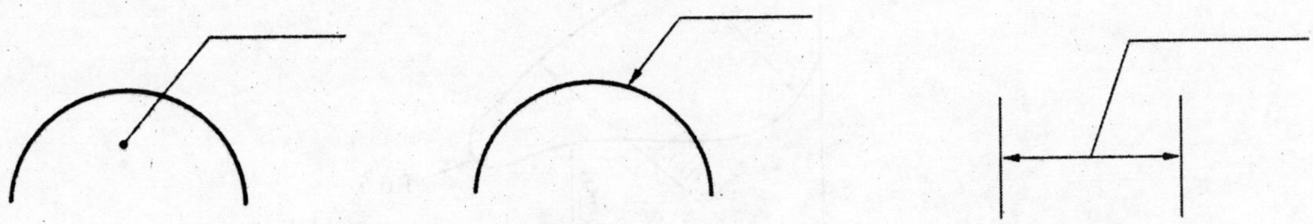

Fig. 2.21 Termination of Leader Lines.

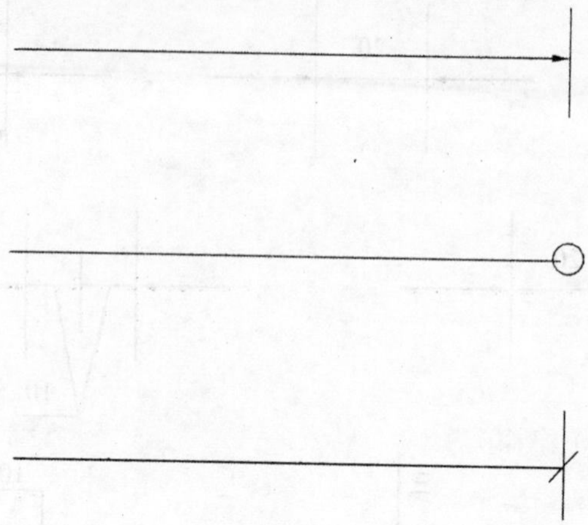

Fig. 2.22 Termination of Dimension Line.

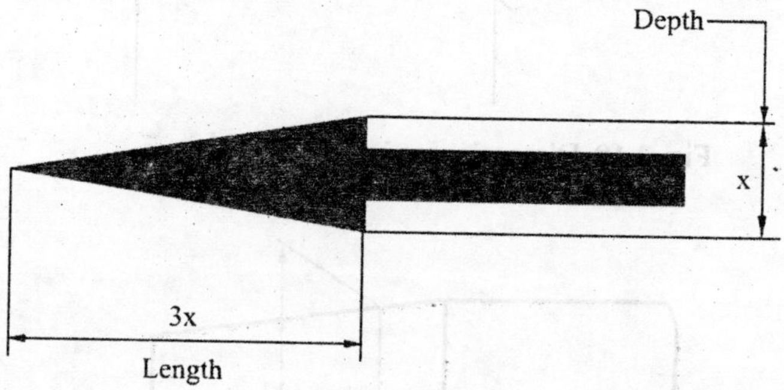

Fig. 2.23 Proportions of an Arrow Head.

When a radius is dimensioned only one arrow head, with its point on the arc end of the dimension line should be used (Fig. 2.24). The arrow head termination may be either on the inside or outside of the feature outline, depending on the size of the feature.

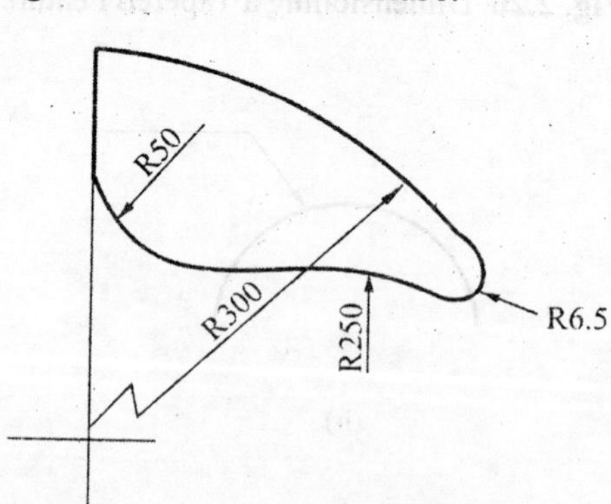

Fig. 2.24 Dimensioning of Radii.

2.4.3 Methods of Indicating Dimensions

The dimensions are indicated on the drawings according to one of the following two methods.

Method - 1 (Aligned method)

Dimensions should be placed parallel to and above their dimension lines and preferably at the middle, and clear of the line (Fig. 2.25).

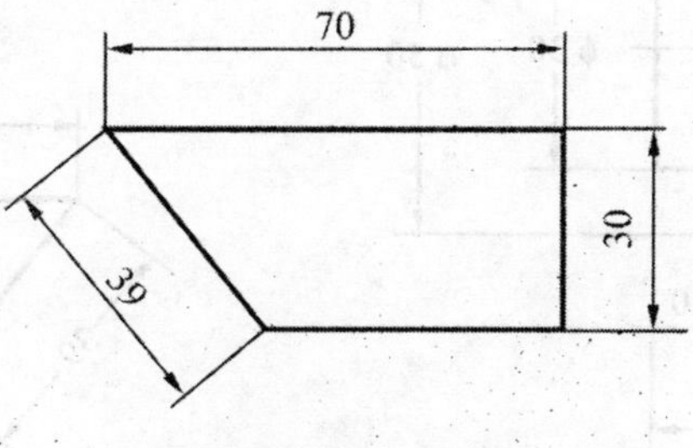

Fig. 2.25 Aligned Method

Dimensions may be written so that they can be read from the bottom or from the right side of the drawing. Dinensions on oblique dimension lines should be oriented as shown in Fig. 2.26(a) and except where unavoidable, they shall not be placed in the 30⁰ zone. Angular dimensions are oriented as shown in Fig. 2.26(b).

Method - 2 (uni-directional method)

Dimensions should be indicated so that they can be read from the bottom of the drawing only.

Non-horizontal dimension lines are interrupted, preferably in the middle for insertion of the dimension (Fig.2.27(a)).

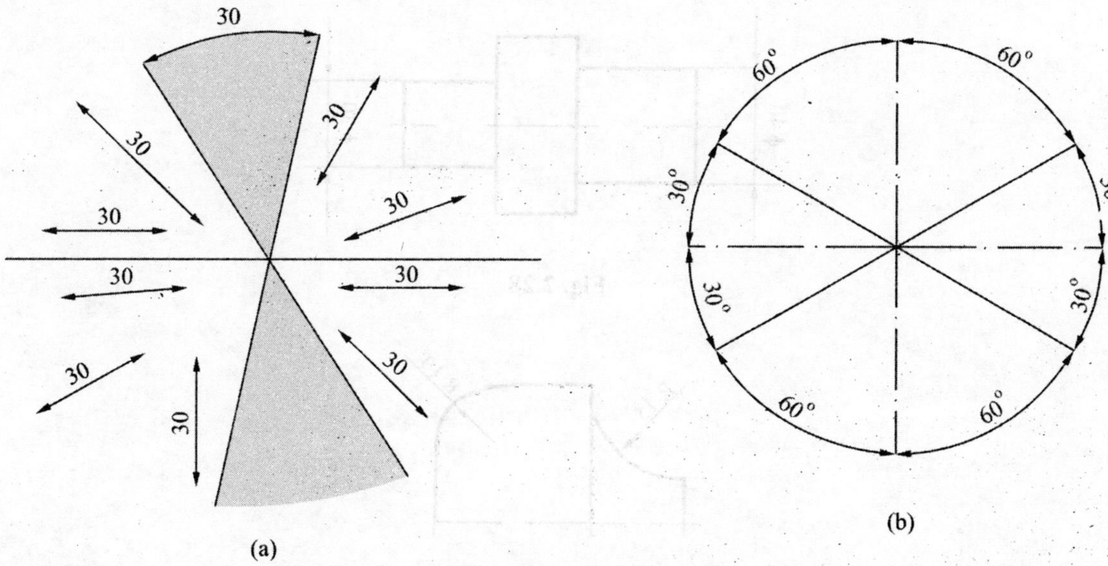

(a) (b)

Fig. 2.26 Angular Dimensioning.

Angular dimensions may be oriented as in Fig. 2.27(b).

Note : Horizontal dimensional lines are not broken to place the dimension in both cases.

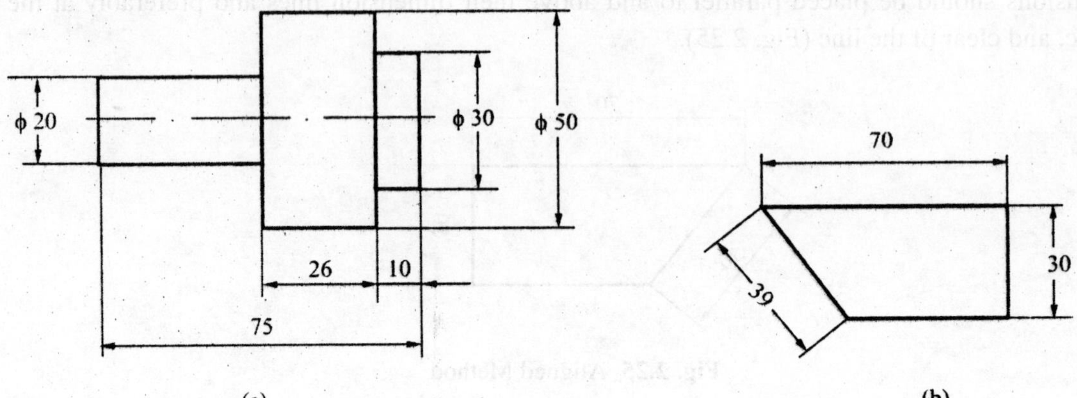

Fig. 2.27 Uni-directional Method.

2.4.4 Identification of Shapes

The following indications are used with dimensions to show applicable shape identification and to improve drawing interpretation. The diameter and square symbols may be omitted where the shape is clearly indicated. The applicable indication (symbol) shall precede the value for dimension (Figs. 2.28 to 2.32).

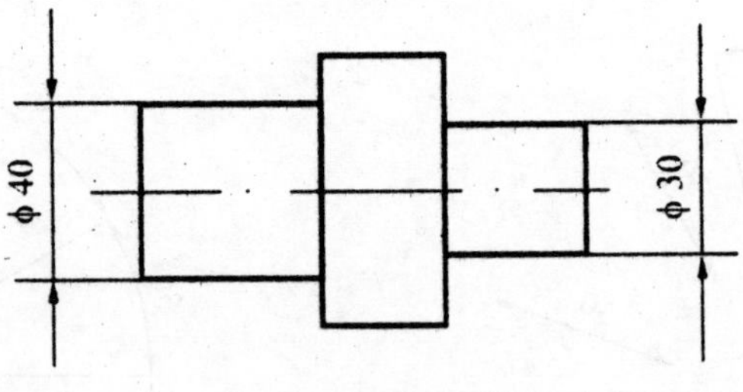

Fig. 2.28

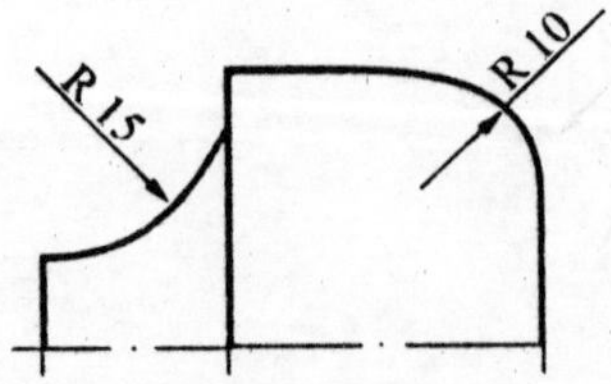

Fig. 2.29

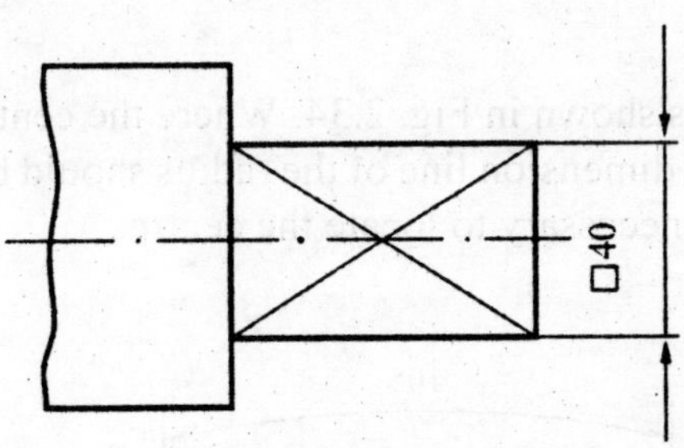

Fig. 2.30

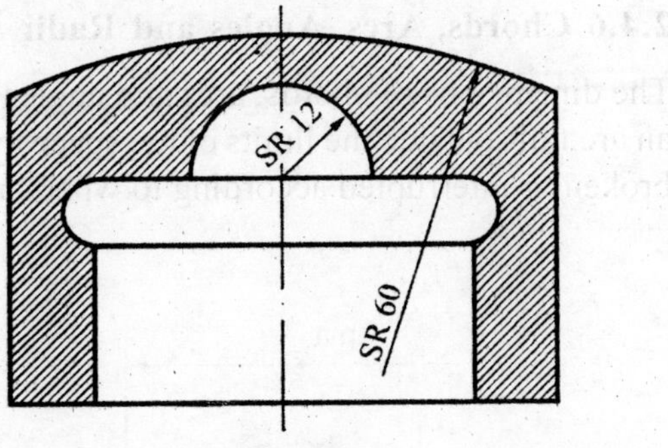

Fig. 2.31

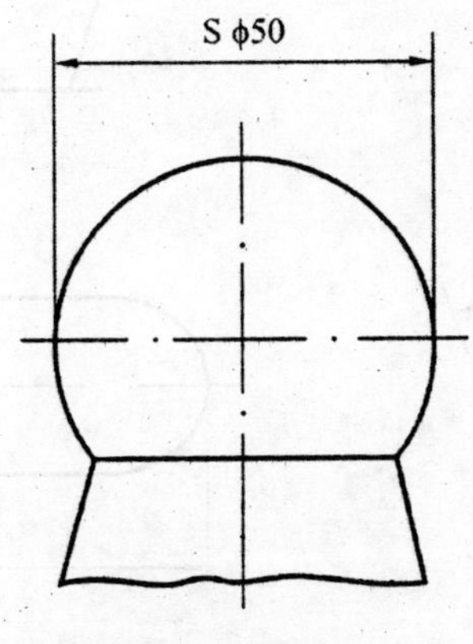

Fig. 2.32

2.4.5 Dimensioning of Diameters (Fig. 2.33)

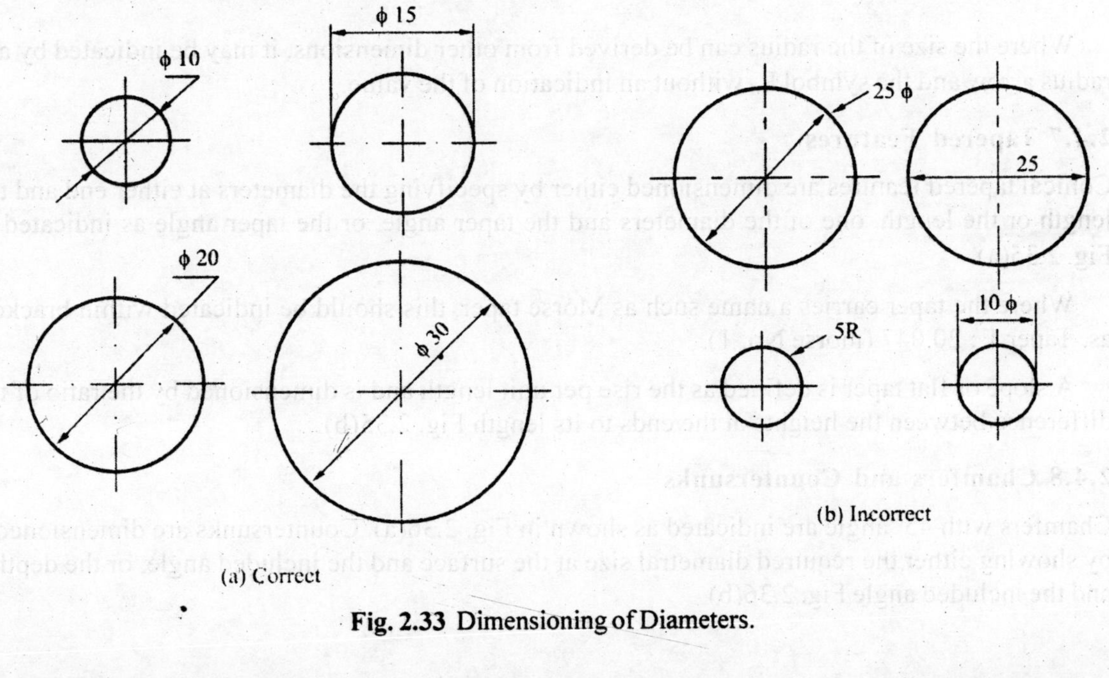

(b) Incorrect

(a) Correct

Fig. 2.33 Dimensioning of Diameters.

2.4.6 Chords, Arcs, Angles and Radii

The dimensions of chords, arcs and angles should be as shown in Fig. 2.34. Where the centre of an arc falls outside the limits of the space available, the dimension line of the radius should be broken or interrupted according to whether or not it is necessary to locate the centre.

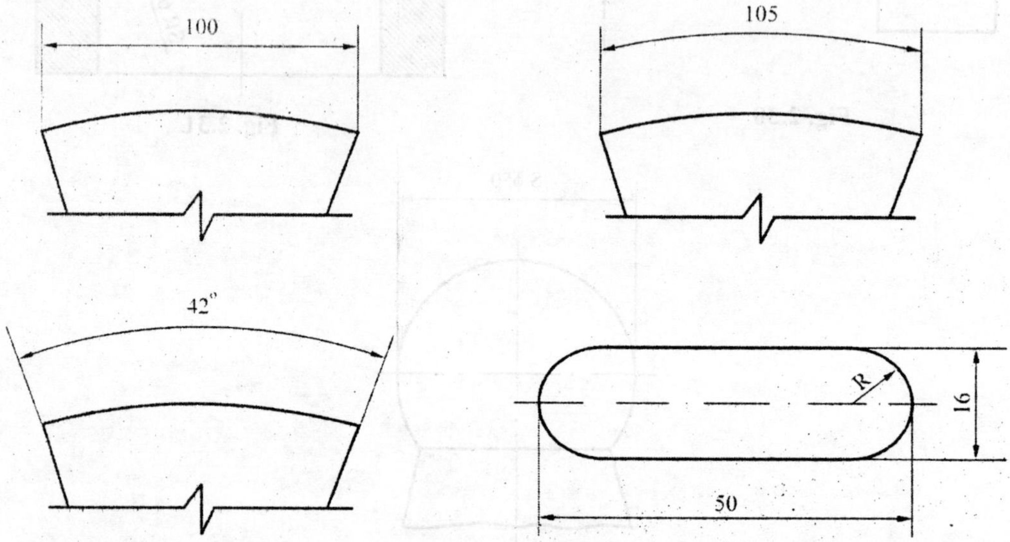

Fig. 2.34 Dimensioning of Chords, Arcs and Angles.

Where the size of the radius can be derived from other dimensions, it may be indicated by a radius arrow and the symbol R, without an indication of the value.

2.4.7 Tapered Features

Conical tapered features are dimensioned either by specifying the diameters at either end and the length or the length, one of the diameters and the taper angle, or the taper angle as indicated in Fig. 2.35(a).

Where the taper carries a name such as Morse taper, this should be indicated within brackets as, Taper 1 : 20.047 (morse No. 1).

A slope or flat taper is defined as the rise per unit length and is dimensioned by the ratio of the difference between the heights at the ends to its length Fig. 2.35(b).

2.4.8 Chamfers and Countersunks

Chamfers with 45° angle are indicated as shown in Fig. 2.36(a). Countersunks are dimensioned by showing either the required diametral size at the surface and the included angle, or the depth and the included angle Fig. 2.36(b).

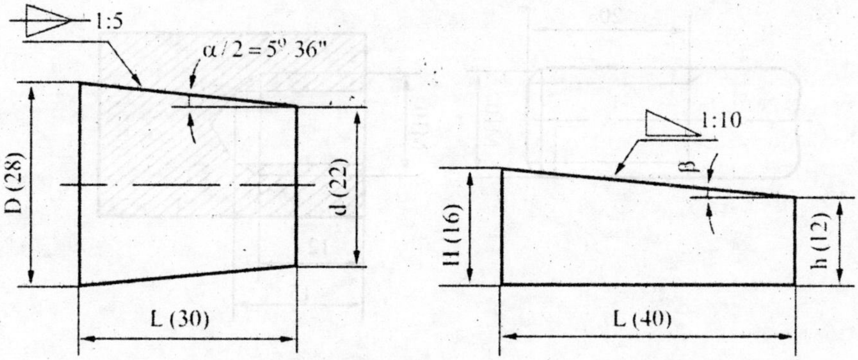

$$\text{Taper} = \frac{D-d}{L}\left[\frac{28-22}{30} = \frac{1}{5}\right] \qquad \text{Slope} = \frac{H-h}{L}\left[\frac{16-12}{40} = \frac{1}{10}\right]$$

(a) Conical taper (b) Flat taper

Fig. 2.35 Dimensioning Tapers.

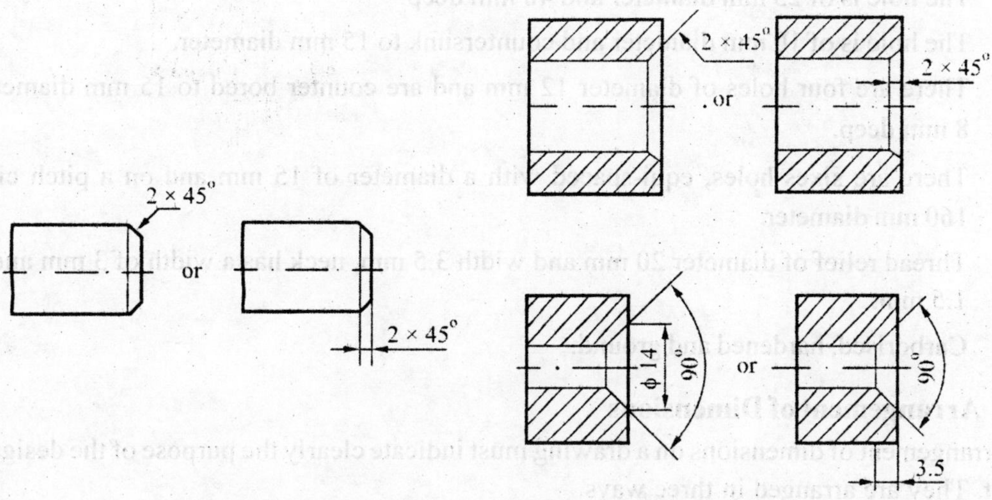

(a) (b)

Fig. 2.36 Dimensioning Chamfers and Countersunks.

2.4.9 Screw Threads

Screw threads are always specified with proper designation. The nomial diameter is preceded by the letter M. The useful length of the threaded portion only should be dimensioned (Fig. 2.37). While dimensioning the internal threads, the length of the drilled hole should also be dimensioned.

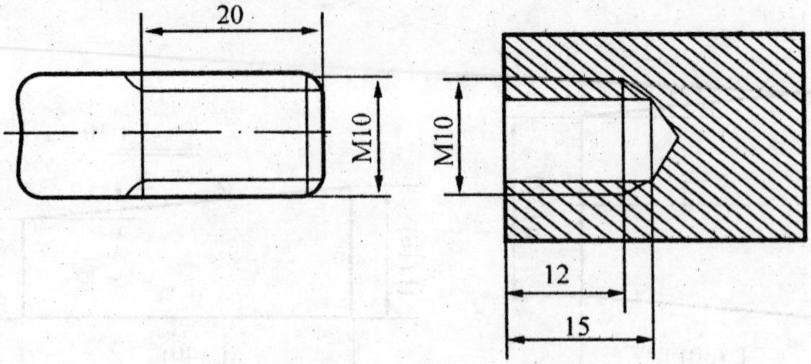

Fig. 2.37 Dimensioning Screw Threads.

2.4.10 Notes

Notes should always be lettered horizontally, in capitals and begin with the leader line. Further, notes should be brief and clear and the wording shall be standard in form. The meaning of the notes in Fig. 2.38 is given below.

(a) The hole is of 25 mm diameter and 40 mm deep.

(b) The hole is of 10 mm diameter and countersunk to 15 mm diameter.

(c) There are four holes of diameter 12 mm and are counter bored to 15 mm diameter and 8 mm deep.

(d) There are sixes holes, equi-spaced with a diameter of 15 mm and on a pitch circle of 160 mm diameter.

(e) Thread relief of diameter 20 mm and width 3.5 mm, neck has a width of 3 mm and depth 1.5 mm.

(f) Carborised, hardened and ground.

2.5 Arrangement of Dimensions

The arrangement of dimensions on a drawing must indicate clearly the purpose of the design of the object. They are arranged in three ways.

1. Chain dimensioning
2. Parallel dimensioning
3. Combined dimensioning.

1. Chain dimensioning

Chain of single dimensioning should be used only where the possible accumulation of tolerances does not endanger the fundamental requirement of the component (Fig. 2.39)

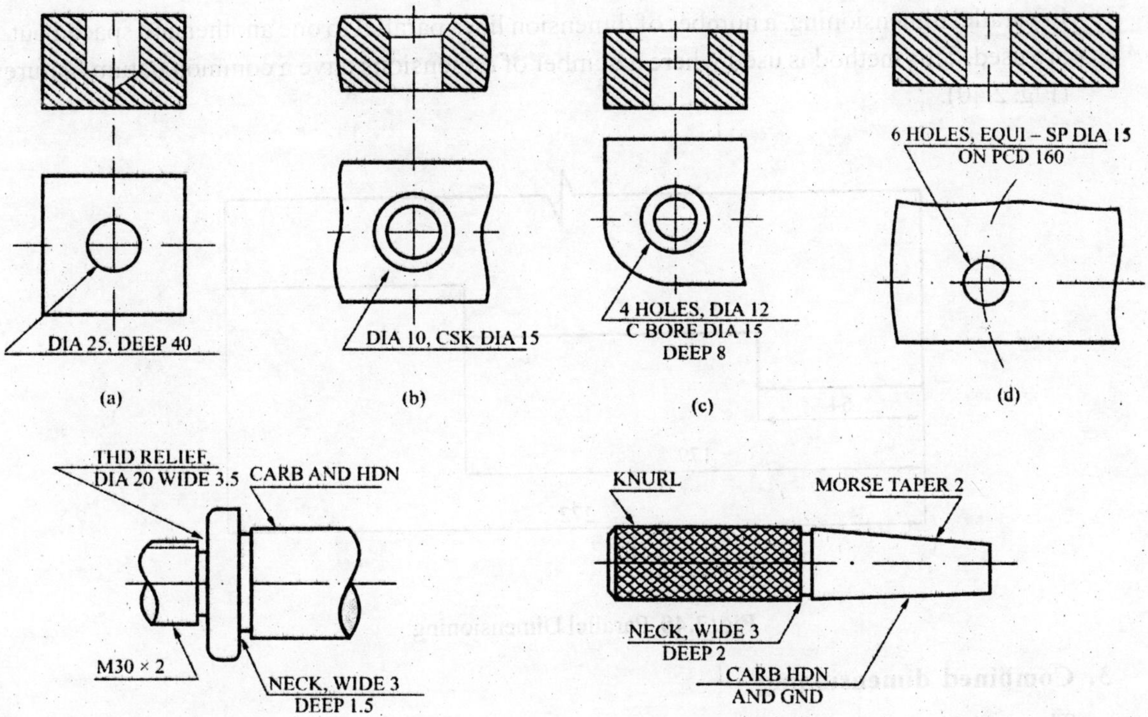

Fig. 2.38 Method of Indicating Notes.

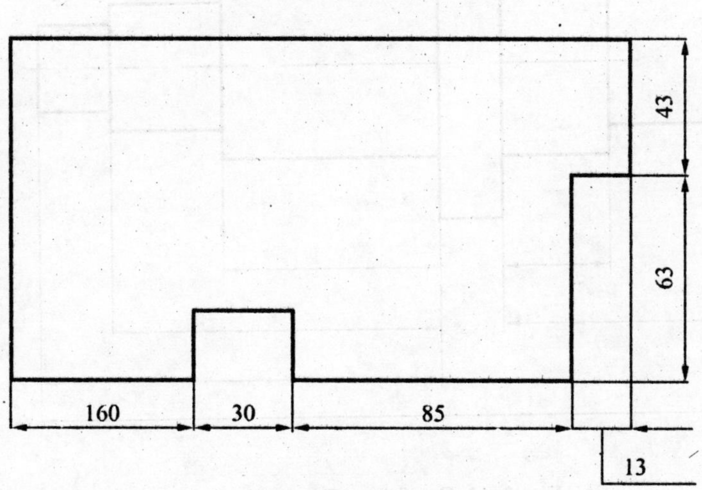

Fig. 2.39 Chain Dimensioning.

2. Parallel dimensioning

In parallel dimensioning, a number of dimension lines parallel to one another and spaced out. are used. This method is used where a number of dimensions have a common datum feature (Fig. 2.40).

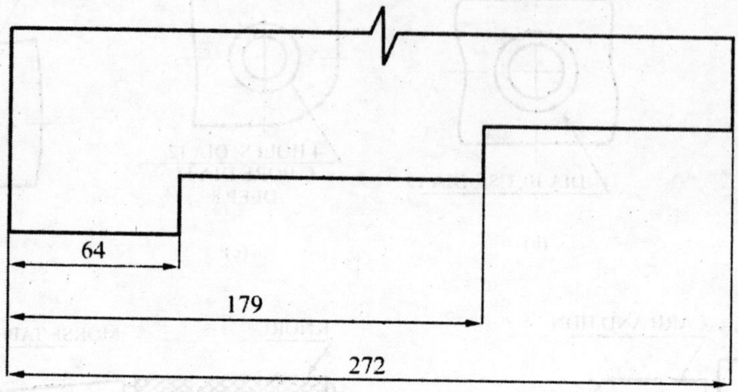

Fig. 2.40 Parallel Dimensioning.

3. Combined dimensioning

These are the result of simultaneous use of chain and parallel dimensioning (Fig. 2.41).

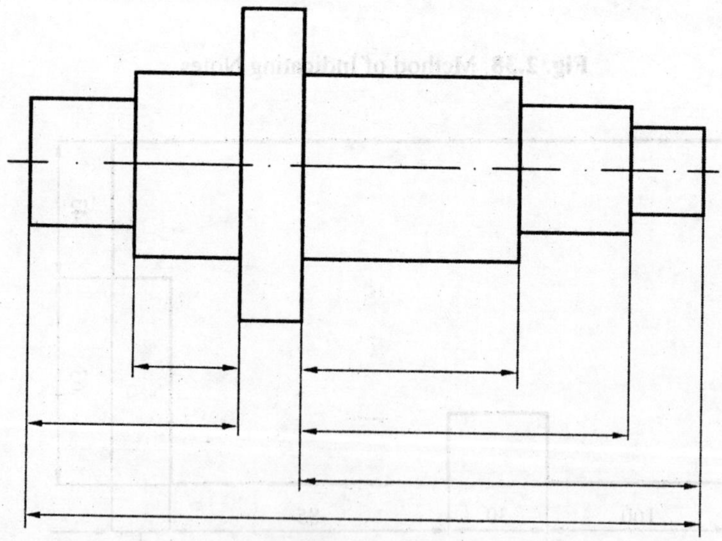

Fig. 2.41 Combined Dimensioning.

Example 1

Violation of some of the principles of drawing are indicated in Fig.2.42(a). The corrected version of the same as per BIS : SP : 46:2003 is given in Fig.2.42(b). The violations from 1 to 16 indicated in the figure are explained below.

1. Dimension should follow the shape symbol.
2. and 3. As far as possible, features should not be used as extension lines for dimensioning.
4. Extension line should touch the feature.
5. Extension line should project beyond the dimension line.
6. Writing the dimension is not as per aligned method.
7. Hidden lines should meet without a gap.
8. Centre line representation is wrong. Small dashes should be replaced by dots.
9. Horizontal dimension line should not be broken to insert the value of dimension in both aligned and uni-direction methods.
10. Dimension should be placed above the dimension line, not below or side.

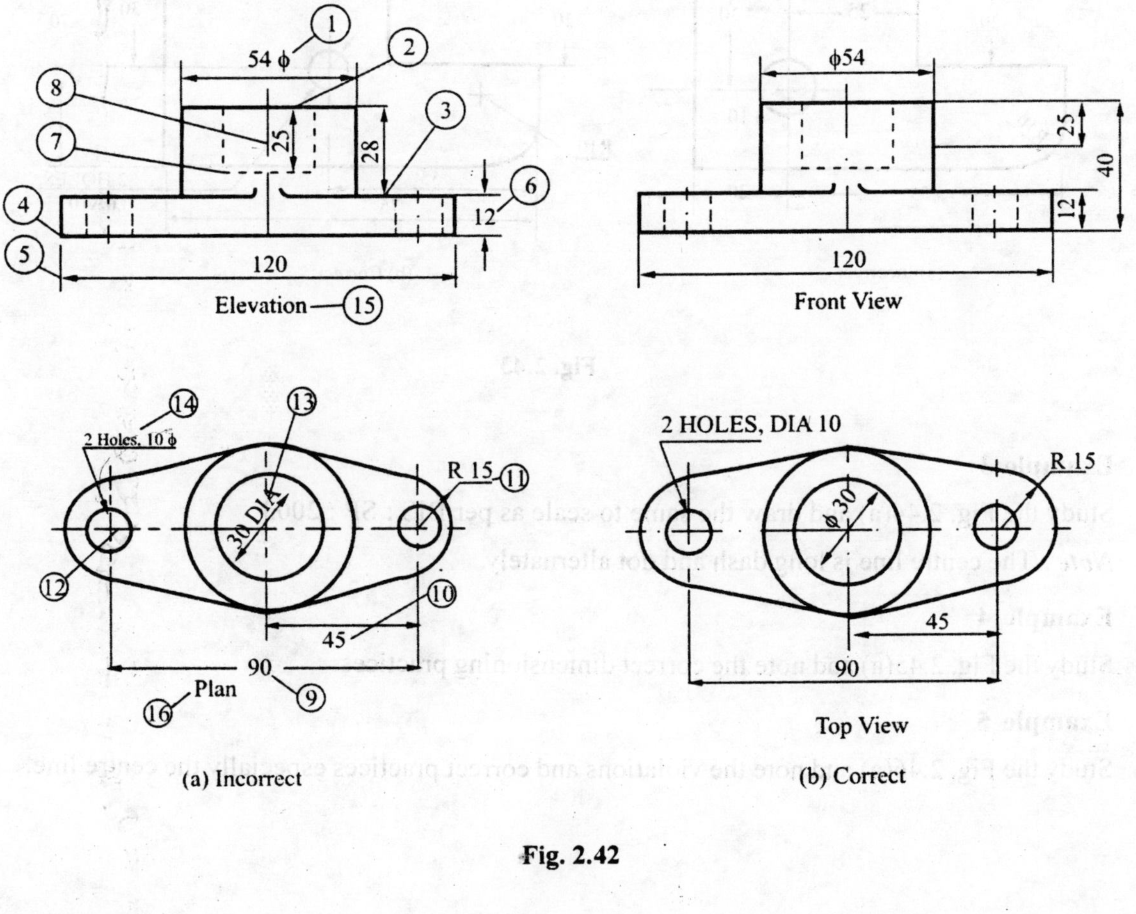

Fig. 2.42

11. Radius symbol should precede the dimension.
12. Centre line should cross with long dashes not dot.
13. Dimension should be written by symbol followed by its values and not abbreviation.
14. Note with dimensions should be written in capitals and not lower case letters.
15. Elevation is not correct usage.
16. Plan is obsolete in graphic language.

Example 2

Study the Fig. 2.43(a) and re-draw to scale using correct system of dimensioning by uni-directional method.

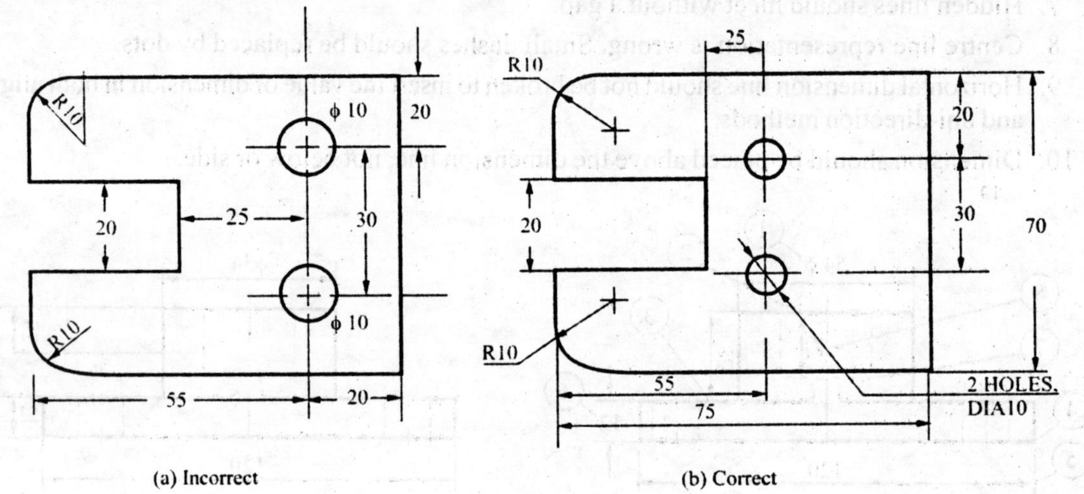

Fig. 2.43

Example 3

Study the Fig. 2.44(a) and draw the same to scale as per BIS : SP : 2003

Note : The centre line is long dash and dot alternately.

Example 4

Study the Fig. 2.45(a) and note the correct dimensioning practices.

Example 5

Study the Fig. 2.46(a) and note the violations and correct practices especially the centre line.

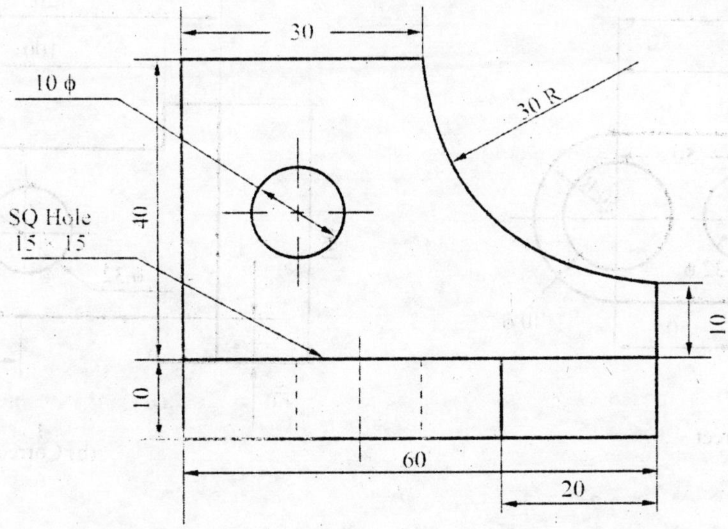

(a) Incorrect

Fig. 2.44(a)

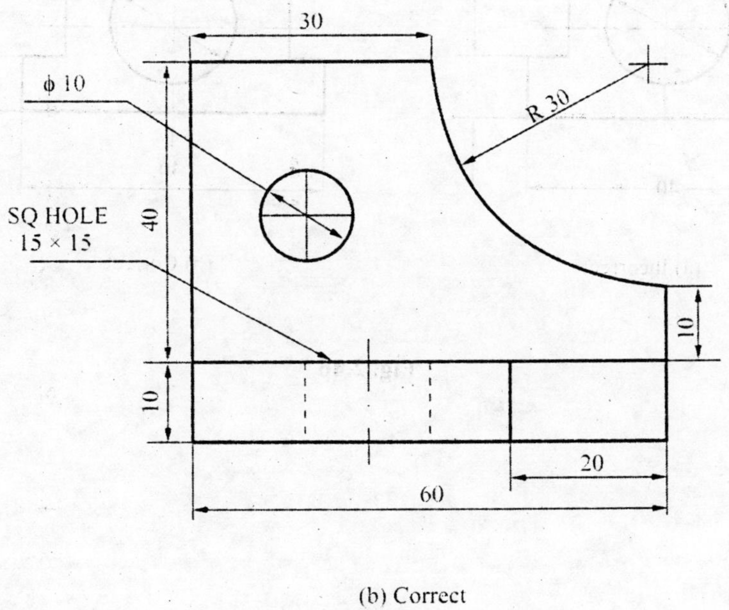

(b) Correct

Fig. 2.44(b)

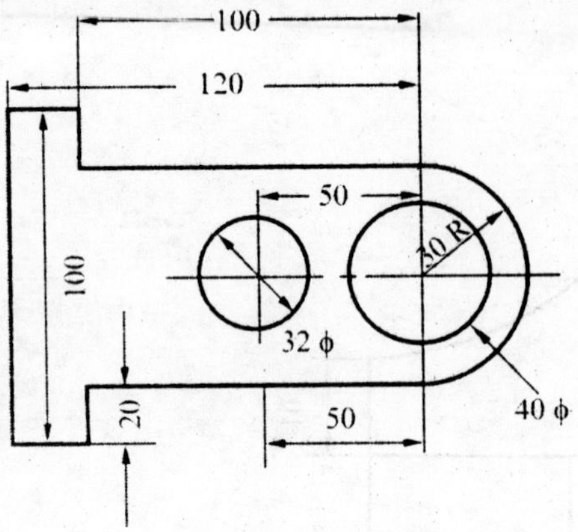

(a) Incorrect

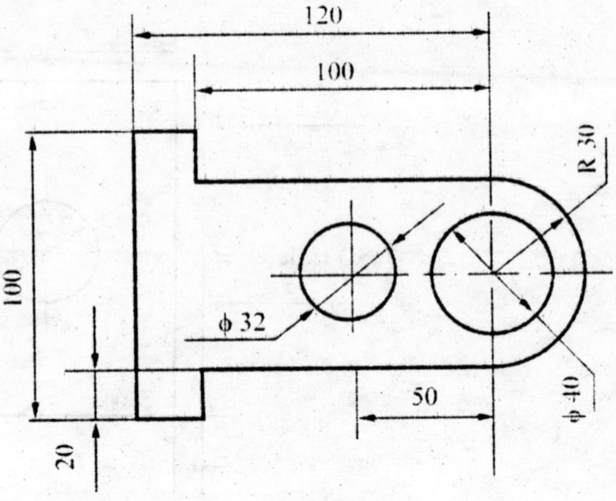

(b) Correct

Fig. 2.45

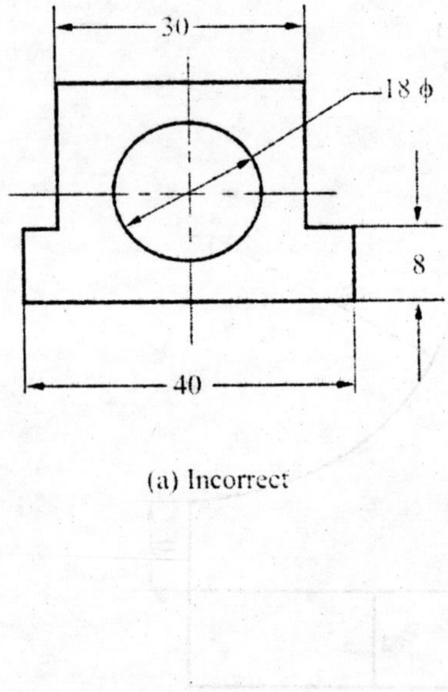

(a) Incorrect

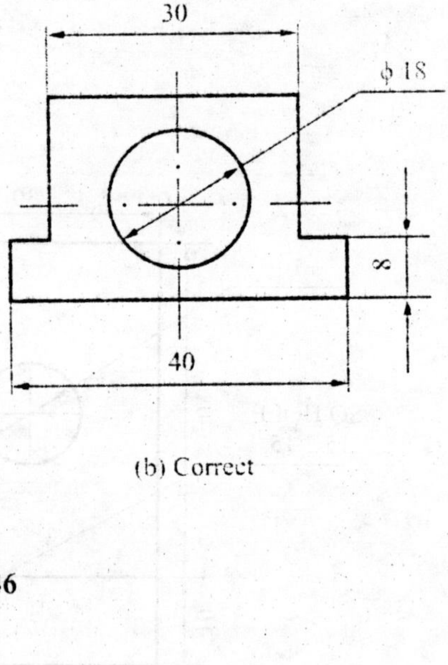

(b) Correct

Fig. 2.46

EXERCISES

1. Write with freehand the following, using single stroke vertical (CAPITAL and lower-case) letters:
   (a) Alphabets (Upper-case & Lower-case) and Numerals 0 to 9 (h = 5 and 7 mm)
   (b) PRACTICE MAKES A PERSON PERFECT (h = 3.5 and 5)
   (c) BE A LEADER NOT A FOLLOWER (h = 5)
   (d) LETTERING SHOULD BE DONE FREEHAND WITH SPEED (h = 5)
2. Study the Figs. 2.47 to 2.55 and re-draw the correct figures as per the principles of dimensioning as per BIS : SP 46 : 2003

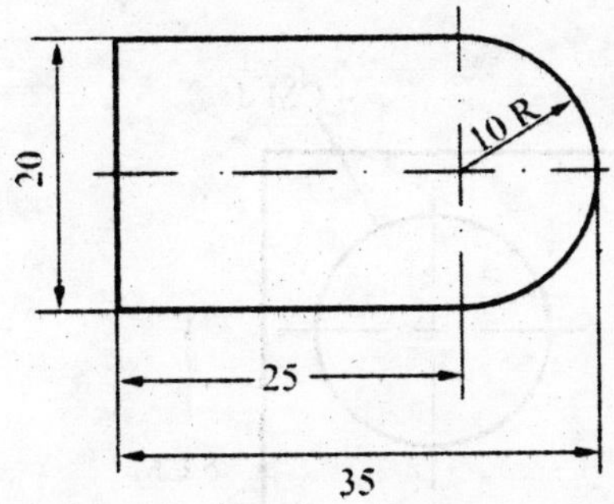

Fig. 2.47

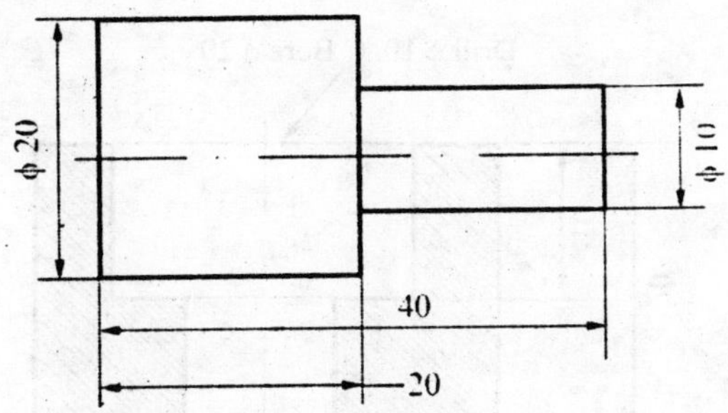

Fig. 2.48

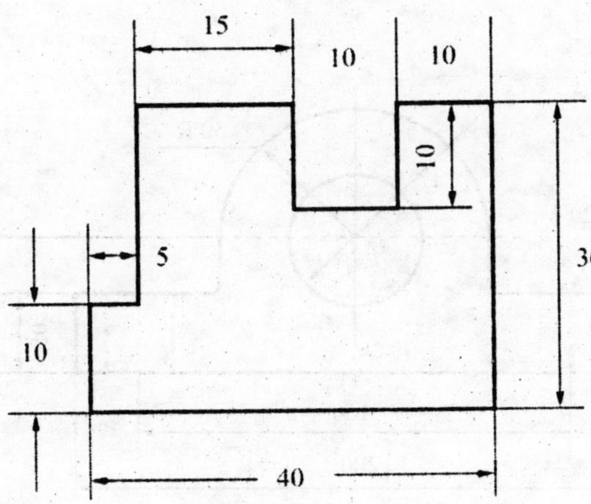

Fig. 2.49

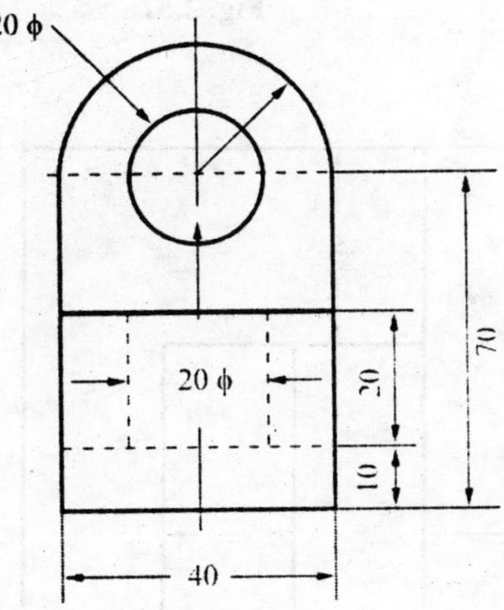

Fig. 2.50

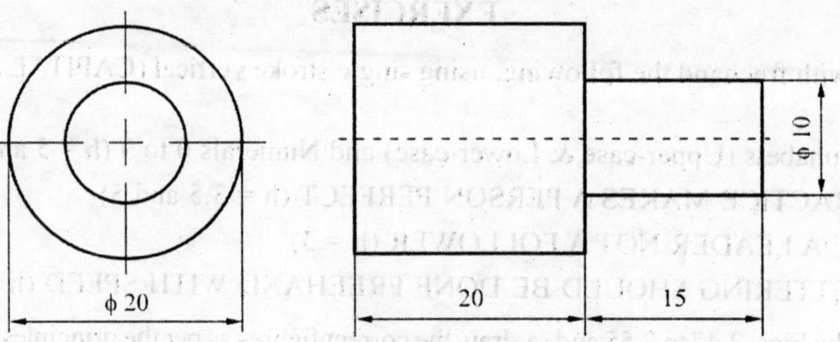

Fig. 2.51

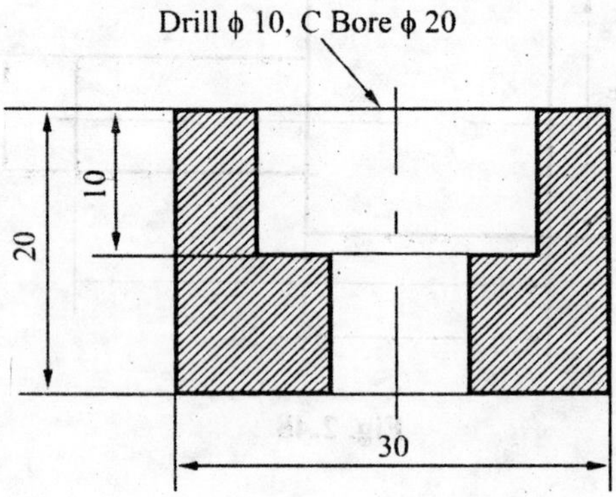

Fig. 2.52

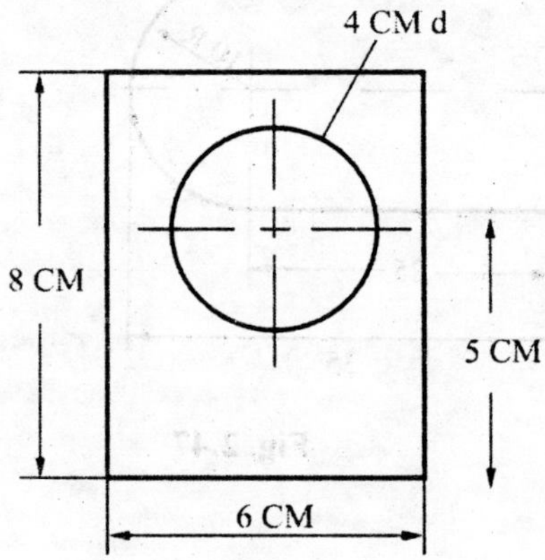

Fig. 2.53

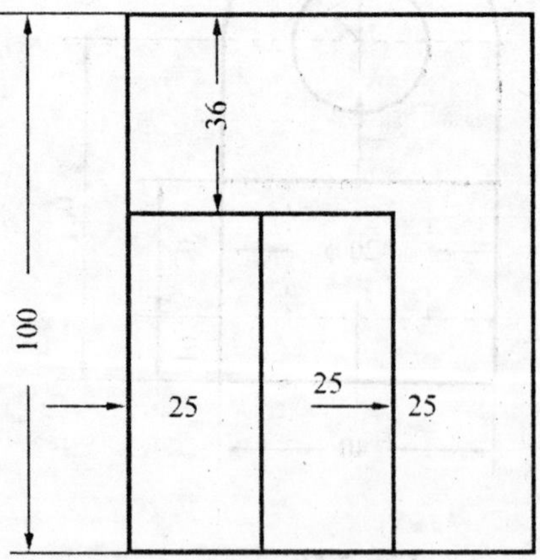

Fig. 2.54

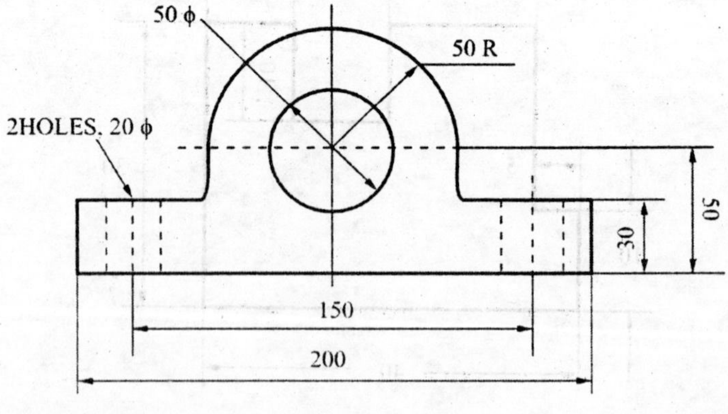

Fig. 2.55

CHAPTER 3

Scales

3.1 Introduction

It is not possible always to make drawings of an object to its actual size. If the actual linear dimensions of an object are shown in its drawing, the scale used is said to be a **full size scale**. Wherever possible, it is desirable to make drawings to full size.

3.2 Reducing and Enlarging Scales

Objects which are very big in size can not be represented in drawing to full size. In such cases the object is represented in reduced size by making use of reducing scales. Reducing scales are used to represent objects such as large machine parts, buildings, town plans etc. A reducing scale, say 1:10 means that 10 units length on the object is represented by 1 unit length on the drawing.

Similarly, for drawing small objects such as watch parts, instrument components etc., use of full scale may not be useful to represent the object clearly. In those cases enlarging scales are used. An enlarging scale, say 10:1 means one unit length on the object is represented by 10 units on the drawing.

The designation of a scale consists of the word. SCALE, followed by the indication of its ratio as follows (Standard scales are shown in Fig. 3.1).

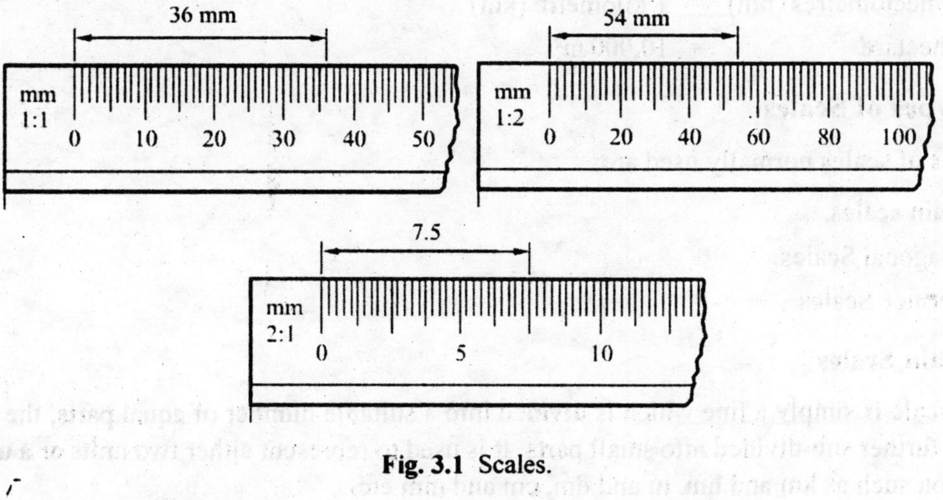

Fig. 3.1 Scales.

Scale 1:1 for full size scale

Scale 1: X for reducing scales (X = 10,20......etc.,)

Scale X:1 for enlarging scales.

Note : For all drawings the scale has to be mentioned without fail.

3.3 Representative Fraction

The ratio of the dimension of the object shown on the drawing to its actual size is called the Representative Fraction (RF).

$$RF = \frac{\text{Drawing size of an object}}{\text{Its actual size}} \text{ (in same units)}$$

For example, if an actual length of 3 metres of an object is represented by a line of 15mm length on the drawing

$$RF = \frac{15\text{mm}}{3\text{m}} = \frac{15\text{mm}}{(3 \times 1000)\text{mm}} = \frac{1}{200} \text{ or } 1:200$$

If the desired scale is not available in the set of scales it may be constructed and then used.

Metric Measurements

10 millimetres (mm)	=	1 centimetre(cm)
10 centimetres (cm)	=	1 decimetre(dm)
10 decimetre (dm)	=	1 metre(m)
10 metres (m)	=	1 decametre (dam)
10 decametre (dam)	=	1 hectometre (hm)
10 hectometres (hm)	=	1 kilometre (km)
1 hectare	=	10,000 m^2

3.4 Types of Scales

The types of scales normally used are:

1. Plain scales.
2. Diagonal Scales.
3. Vernier Scales.

3.4.1 Plain Scales

A plain scale is simply a line which is divided into a suitable number of equal parts, the first of which is further sub-divided into small parts. It is used to represent either two units or a unit and its fraction such as km and hm, m and dm, cm and mm etc.

Problem : *On a survey map the distance between two places 1km apart is 5 cm. Construct the scale to read 4.6 km.*

Solution : (Fig. 3.2)

$$RF = \frac{5\text{ cm}}{1 \times 1000 \times 100\text{ cm}} = \frac{1}{20000}$$

If **X** is the drawing size required $X = 5(1000)(100)\ X\ \frac{1}{20000}$

Therefore, X = 25 cm

Note : If 4.6 km itself were to be taken X = 23 cm. To get 1 km divisions this length has to be divided into 4.6 parts which is difficult. Therefore, the nearest round figure 5 km is considered. When this length is divided into 5 equal parts each part will be 1 km.

1. Draw a line of length 25 cm.
2. Divide this into 5 equal parts. Now each part is 1 km.
3. Divide the first part into 10 equal divisions. Each division is 0.1 km.
4. Mark on the scale the required distance 4.6 km.

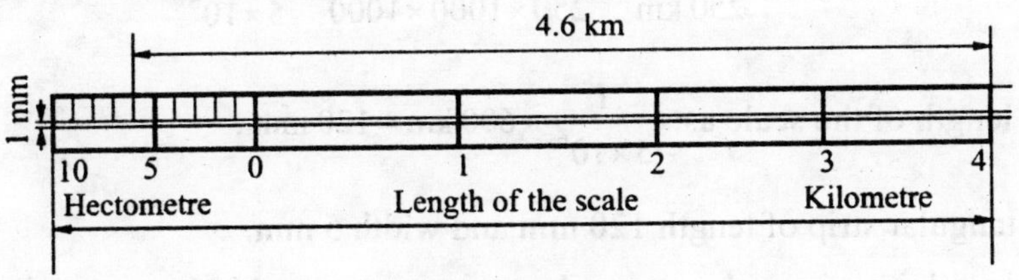

Fig. 3.2 Plain Scale.

Problem : *Construct a scale of 1:50 to read metres and decimetres and long enough to measure 6 m. Mark on it a distance of 5.5 m.*

Construction (Fig. 3.3)

1. Obtain the length of the scale as: RF × 6 m = $\frac{1}{50} \times 6 \times 100 = 12$ cm
2. Draw a rectangle strip of length 12 cm and width 0.5 cm.
3. Divide the length into 6 equal parts, by geometrical method each part representing 1m.
4. Mark 0(zero) after the first division and continue 1, 2, 3 etc., to the right of the scale.
5. Divide the first division into 10 equal parts (secondary divisions), each representing 1 cm.
6. Mark the above division points from right to left.
7. Write the units at the bottom of the scale in their respective positions.

8. Indicate RF at the bottom of the figure.
9. Mark the distance 5.5 m as shown.

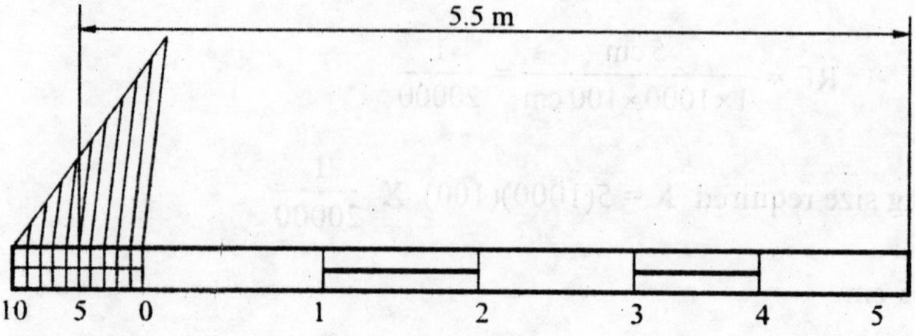

Fig. 3.3

Problem : *The distance between two towns is 250 km and is represented by a line of length 50 mm on a map. Construct a scale to read 600 km and indicate a distance of 530 km on it.*

Solution : (Fig. 3.4)

1. Determine the RF value as $\frac{50\text{ mm}}{250\text{ km}} = \frac{50}{250 \times 1000 \times 1000} = \frac{1}{5 \times 10^6}$

2. Obtain the length of the scale as: $\frac{1}{5 \times 10^6} \times 600\text{ km} = 120\text{ mm}$.

3. Draw a rectangular strip of length 120 mm and width 5 mm.
4. Divide the length into 6 equal parts, each part representing 100 km.
5. Repeat the steps 4 to 8 of construction in Fig. 3.3. suitably.
6. Mark the distance 530 km as shown.

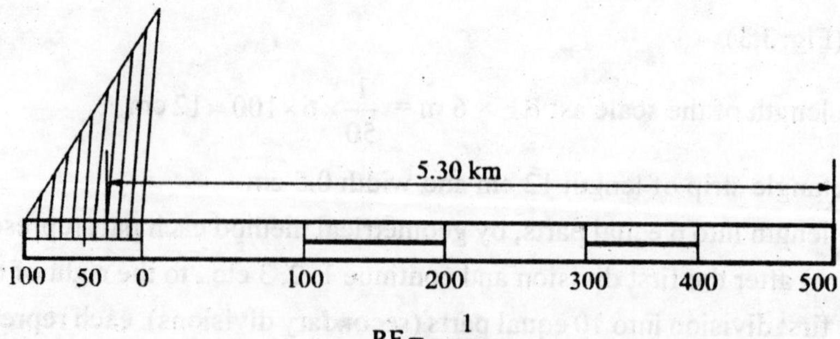

Fig. 3.4

Problem : *Construct a plain scale of convenient length to measure a distance of 1 cm and mark on it a distance of 0.94 cm.*

Solution : (Fig. 3.5)

This is a problem of enlarged scale.

1. Take the length of the scale as 10 cm
2. RF = 10/1, scale is 10:1
3. The construction is shown in Fig. 3.5

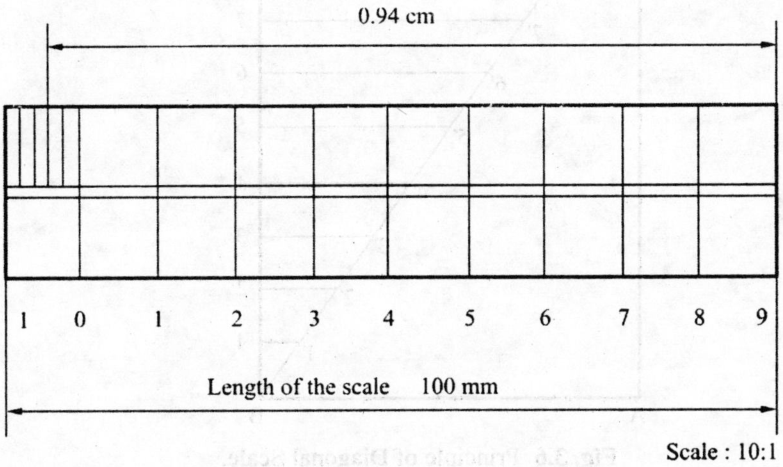

Fig. 3.5

3.4.2 Diagonal Scales

Plain scales are used to read lengths in two units such as metres and decimetres, centimetres and millimetres etc., or to read to the accuracy correct to first decimal.

Diagonal scales are used to represent either three units of measurements such as metres, decimetres, centimetres or to read to the accuracy correct to two decimals.

Principle of Diagonal Scale (Fig. 3.6)

1. Draw a line AB and errect a perperrdicular at B.
2. Mark 10 equi-distant points (1, 2, 3, etc) of any suitable length along this perpendicular and mark C.
3. Complete the rectangle ABCD
4. Draw the diagonal BD.
5. Draw horizontals through the division points to meet BD at 1′, 2′, 3′ etc.

Considering the similar triangles say BCD and B44′

$$\frac{B4'}{CD} = \frac{B4}{BC}; \; = \frac{4}{10} \times BC \times \frac{1}{BC} = \frac{4}{10};$$

$$44'' = 0.4CD$$

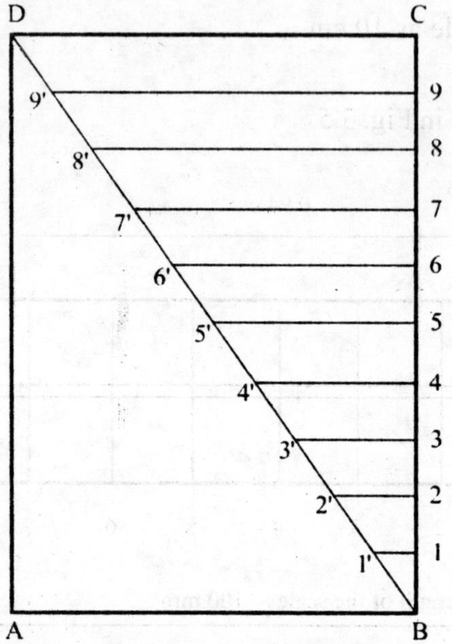

Fig. 3.6 Principle of Diagonal Scale.

Thus, the lines $1-1', 2-2', 3-3'$ etc., measure 0.1CD, 0.2CD, 0.3CD etc. respectively. The line CD is divided into 1/10 the divisions by the diagonal BD, i.e., each horizontal line is a multiple of 1/10 CD.

This principle is used in the construction of diagonal scales.

Note : B C must be divided into the same number of parts as there are units of the third dimension in one unit of the secondary division.

Problem : *On a plan, a line of 22 cm long represents a distance of 440 metres. Draw a diagonal scale for the plan to read up to a single metre. Measure and mark a distance of 187 m on the scale.*

Solution : (Fig. 3.7)

1. $RF = \frac{22}{440 \times 100} = \frac{1}{2000}$

2. As 187 m are required, consider 200 m.

   Therefore drawing size = R F × actual size = $\frac{1}{2000} \times 200 \times 100 = 10$ cm

   When a length of 10 cm representing 200 m is divided into 5 equal parts, each part represents 40 m as marked in the figure.

3. The first part is sub-divided into 4 divisions so that each division is 10 cm
4. On the diagonal portion 10 divisions are taken to get 1 m.
5. Mark on it 187 m as shown.

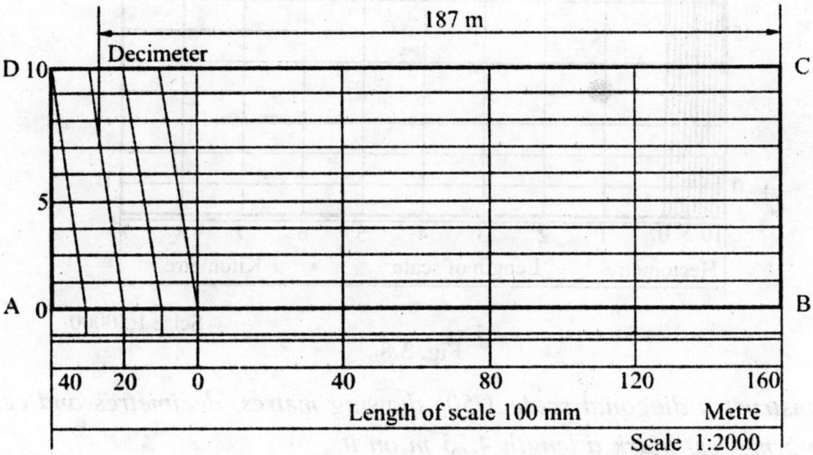

Fig. 3.7 Diagonal Scale.

Problem : *An area of 144 sq cm on a map represents an area of 36 sq km on the field. Find the RF of the scale of the map and draw a diagonal scale to show Km, hectometres and decametres and to measure upto 10 km. Indicate on the scale a distance 7 km, 5 hectometres and 6 decemetres.*

Solution : (Fig. 3.8)

1. 144 sq cm represents 36 sq km or 12 cm represent 6 km

$$R\,F = \frac{12}{6 \times 1000 \times 100} = \frac{1}{50000}$$

$$\text{Drawing size X} = \text{R F} \times \text{actual size} = \frac{10 \times 1000 \times 100}{50000} = 20 \text{ cm}$$

2. Draw a length of 20 cm and divide it into 10 equal parts. Each part represents 1 km.
3. Divide the first part into 10 equal subdivisions. Each secondary division represents 1 hecometre
4. On the diagonal scale portion take 10 equal divisions so that 1/10 of hectometre = 1 decametre is obtained.
5. Mark on it 7.56 km as shown.

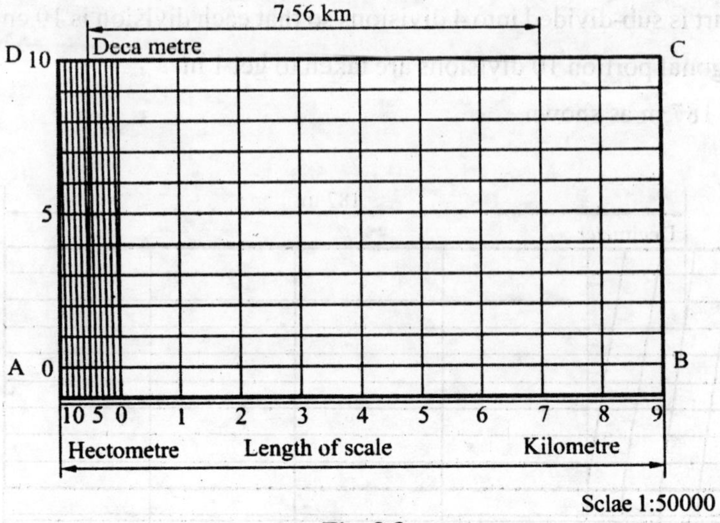

Fig. 3.8

Problem : *Construct a diagonal scale 1/50, showing metres, decimetres and centimetres, to measure up to 5 metres. Mark a length 4.75 m on it.*

Solution : (Fig. 3.9)

1. Obatin the length of the scale as $\frac{1}{50} \times 5 \times 100 = 10\,\text{cm}$
2. Draw a line A B, 10 cm long and divide it into 5 equal parts, each representing 1 m.
3. Divide the first part into 10 equal parts, to represent decimetres.
4. Choosing any convenient length, draw 10 equi-distant parallel lines above AB and complete the rectangle A B C D.

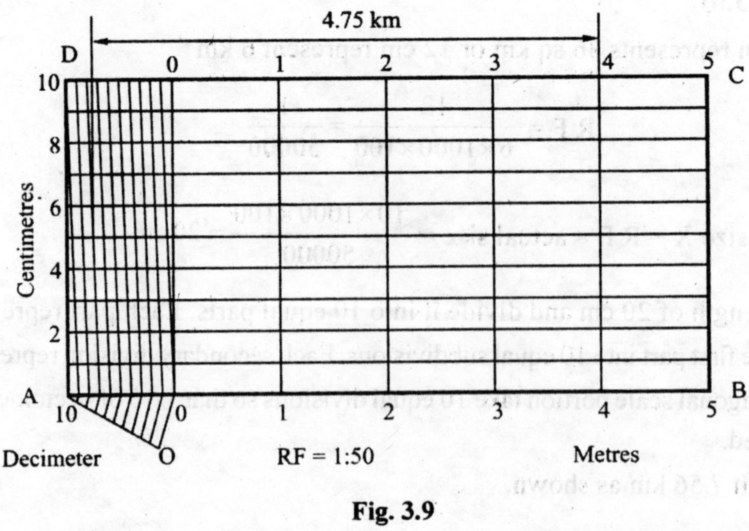

Fig. 3.9

5. Erect perpendiculars to the line A B, through 0, 1, 2, 3 etc., to meet the line C D.
6. Divide OA into 10 equal parts as shown and mark as 0, 1, 2, ... 10.
7. Similarly divide OD into 10 equal parts.
8. Join D to 9, the first sub-division from A on the main scale AB, forming the first diagonal.
9. Draw the remaining diagonals, parallel to the first. Thus, each decimetre is divided into 1/10th division by diagonals.
10. Mark the length 4.75m as shown.,

3.4.3 Vernier Scales

The vernier scale is a short auxiliary scale constructed along the plain or main scale, which can read up to two decimal places.

The smallest division on the main scale and vernier scale are 1 msd or 1 vsd repectively. Generally (n+1) or (n-1) divisions on the main scale is divided into **n** equal parts on the vernier scale.

Thus, $$1 \text{ vsd} = \frac{(n-1)}{n} \text{msd} \quad \text{or} \quad \left(1 - \frac{1}{n}\right) \text{msd}$$

When 1 vsd < 1 it is called forward or direct vernier. The vernier divisions are numbered in the same direction as those on the main scale.

When 1 vsd > 1 or (1+1/n), it is called backward or retrograde vernier. The vernier divisions are numbered in the opposite direction compared to those on the main scale.

The least count (LC) is the smallest dimension correct to which a measurement can be made with a vernier.

For forward vernier, L C = (1 msd – 1 vsd)

For backward viermier, LC = (1 vsd – 1 msd)

Problem : *Construct a forward reading vernier scale to read distance correct to decametre on a map in which the actual distances are reduced in the ratio of 1 : 40,000. The scale should be long enough to measure up to 6 km. Mark on the scale a length of 3.34 km and 0.59 km.*

Solution : (Fig. 3.10)

1. RF = 1/40000; length of drawing $= \frac{6 \times 1000 \times 100}{40000} = 15$ cm
2. 15 cm is divided into 6 parts and each part is 1 km
3. This is further divided into 10 divitions and each division is equal to 0.1 km = 1 hectometre.

   1ms d = 0.1 km = 1 hectometre

   L.C expressed in terms of msd = (1/10) m s d

   L C is 1 decametre = 1 msd – 1 vsd

1 vsd = 1 – 1/10 = 9/10 msd = 0.09 km

4. 9 msd are taken and divided into 10 divisions as shown. Thus 1 vsd = 9/10 = 0.09 km
5. Mark on it by taking 6 vsd = 6 × 0.9 = 0.54 km, 28 msd (27 + 1 on the LHS of 1) = 2.8 km and Total 2.8 + 0.54 = 3.34 km.
6. Mark on it 5 msd = 0.5 km and add to it one vsd = 0.09, total 0.59 km as marked.

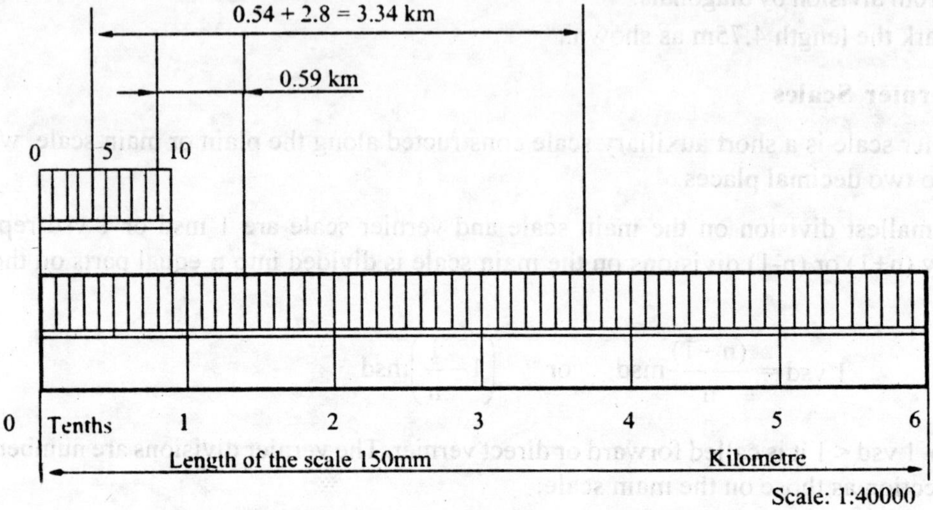

Fig. 3.10 Forward Reading Vernier Scale.

Problem : *Construct a vernier scale to read metres, decimetres and centimetres and long enough to measure up to 4 m. The RF of the scale in 1/20. Mark on it a distance of 2.28 m.*

Solution : (Fig. 3.11)

Backward or Retrograde Vernier scale

1. The smallest measurement in the scale is cm.
   Therefore LC = 0.01m
2. Length of the scale = RF × maximum distance to be measured

$$= \frac{1}{20} \times 4\text{ m} = \frac{1}{20} \times 400 = 20\text{ cm}$$

3. Draw a line of 20 cm length. Complete the rectangle of 20 cm × 0.5 cm and divide it into 4 equal parts each representing 1 metre. Sub divide all into 10 main scale divisions.
   1 msd = 1m/10 = 1dm.
4. Take 10 + 1 = 11 divisions on the main scale and divide it into 10 equal parts on the vernier scale by geometrical construction.
   Thus 1 vsd = 11 msd/10 = 1.1 dm = 11 cm
5. Mark 0, 55, 110 towards the left from 0 (zero) on the vernier scale as shown.

6. Name the units of the divisions as shown.
7. Distance 2.28 m = (8 × vsd) + 14 msd)

$$= (8 \times 0.11 \text{ m}) + (14 \times 0.1 \text{ m})$$

$$= 0.88 + 1.4 = 2.28 \text{ m.}$$

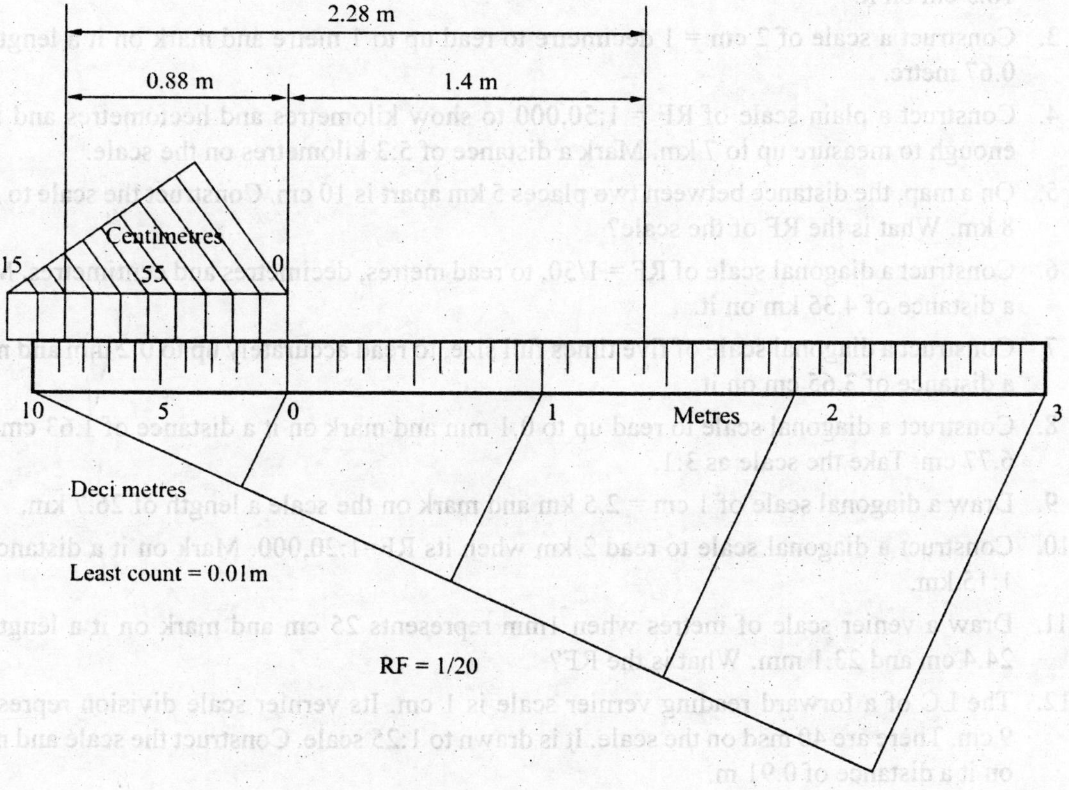

Fig. 3.11 Backward or Retrograde Vernier Scale.

EXERCISES

1. Construct a plain scale of 1:50 to measure a distance of 7 metres. Mark a distance of 3.6 metres on it.
2. The length of a scale with a RF of 2:3 is 20 cm. Construct this scale and mark a distance of 16.5 cm on it.
3. Construct a scale of 2 cm = 1 decimetre to read up to 1 metre and mark on it a length of 0.67 metre.
4. Construct a plain scale of RF = 1:50,000 to show kilometres and hectometres and long enough to measure up to 7 km. Mark a distance of 5:3 kilometres on the scale.
5. On a map, the distance between two places 5 km apart is 10 cm. Construct the scale to read 8 km. What is the RF of the scale?
6. Construct a diagonal scale of RF = 1/50, to read metres, decimetres and centimetres. Mark a distance of 4.35 km on it.
7. Construct a diagonal scale of five times full size, to read accurately up to 0.2 mm and mark a distance of 3.65 cm on it.
8. Construct a diagonal scale to read up to 0.1 mm and mark on it a distance of 1.63 cm and 6.77 cm. Take the scale as 3:1.
9. Draw a diagonal scale of 1 cm = 2.5 km and mark on the scale a length of 26.7 km.
10. Construct a diagonal scale to read 2 km when its RF=1:20,000. Mark on it a distance of 1:15 km.
11. Draw a venier scale of metres when 1mm represents 25 cm and mark on it a length of 24.4 cm and 23.1 mm. What is the RF?
12. The LC of a forward reading vernier scale is 1 cm. Its vernier scale division represents 9 cm. There are 40 msd on the scale. It is drawn to 1:25 scale. Construct the scale and mark on it a distance of 0.91 m.
13. 15 cm of a vernier scale represents 1cm. Construct a backward reading vernier scale of RF 1:4.8 to show decimetres, cm and mm. The scale should be capable of reading up to 12 decimetres. Mark on the scale 2.69 decimetres and 5.57 decimetres.

CHAPTER 4

Geometrical Constructions

4.1 Introduction

Engineering drawing consists of a number of geometrical constructions. A few methods are illustrated here without mathematical proofs.

1. To divide a straight line into a given number of equal parts say 5.

Construction (Fig. 4.1)

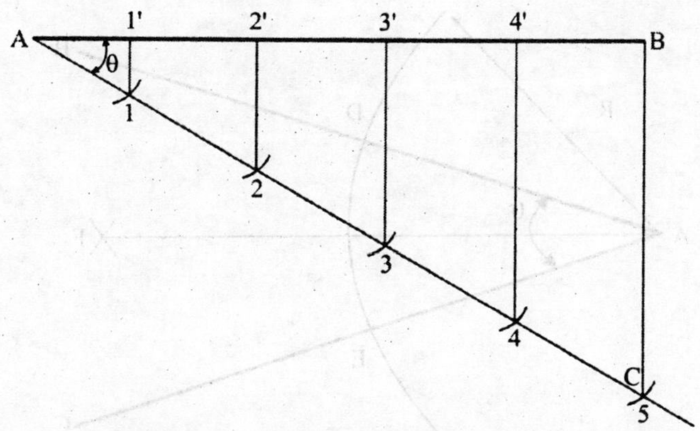

Fig. 4.1 Dividing a Line.

1. Draw AC at any angle θ to AB.
2. Construct the required number of equal parts of convenient length on AC like 1, 2, 3.
3. Join the last point 5 to B
4. Through 4, 3, 2, 1 draw lines parallel to 5B to intersect AB at 4', 3', 2' and 1'.

2. To divide a line in the ratio 1 : 3 : 4.

Construction (Fig. 4.2)

As the line is to be divided in the ratio 1:3:4 it has to be divided into 8 equal divisions. By following the previous example divide AC into 8 equal parts and obtain P and Q to divide the line AB in the ratio 1:3:4.

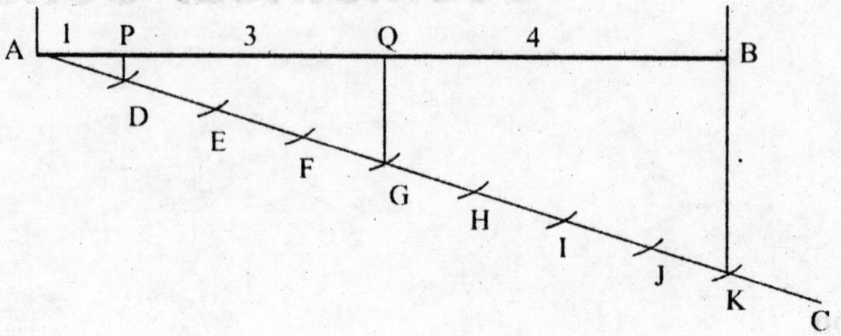

Fig. 4.2

3. To bisect a given angle.

Construction (Fig. 4.3)

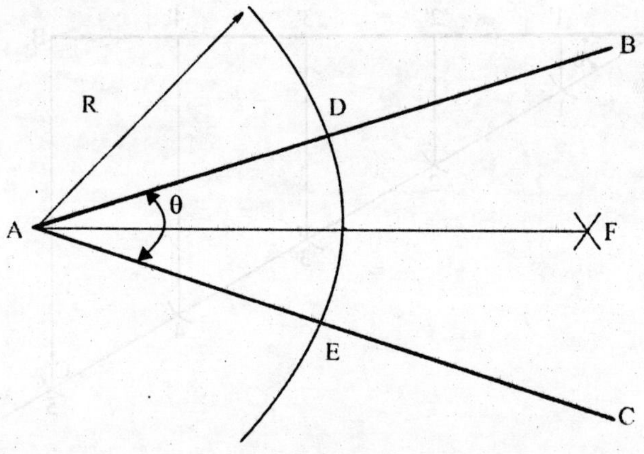

Fig. 4.3

1. Draw a line AB and AC making the given angle.
2. With centre A and any convenient radius R draw an arc intersecting the sides at D and E.
3. With centres D and E and radius larger than half the chord length DE, draw arcs intersecting at F
4. Join AF, <BAF = <FAC.

4. To inscribe a square in a given circle.

Construction (Fig. 4.4)

1. With centre O, draw a circle of diameter D.
2. Through the centre O, draw two diameters, say AC and BD at right angle to each other.
3. Join A-B, B-C, C- D, and D-A. ABCD is the required square.

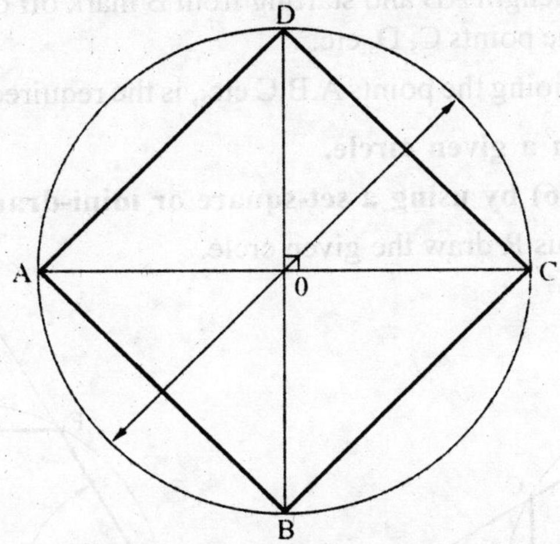

Fig. 4.4

5. To inscribe a regular polygon of any number of sides in a given circle.

Construction (Fig. 4.5)

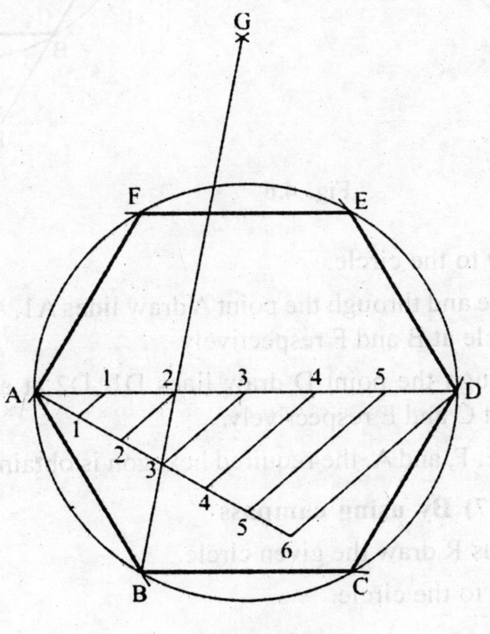

Fig. 4.5

1. Draw the given circle with AD as diameter.
2. Divide the diameter AD into N equal parts say 6.
3. With AD as radius and A and D as centres, draw arcs intersecting each other at G.
4. Join G-2 and extend to intesect the circle at B.
5. Join A-B which is the length of the side of the required polygon.
6. Set the compass to the length AB and starting from B mark off on the circumference of the circles, obtaining the points C, D, etc.

   The figure obtained by joing the points A,B,C etc., is the required polygon.

6. To inscribe a hexagon in a given circle.

(a) Construction (Fig. 4.6) by using a set-square or mini-draughter

1. With centre O and radius R draw the given crcle.

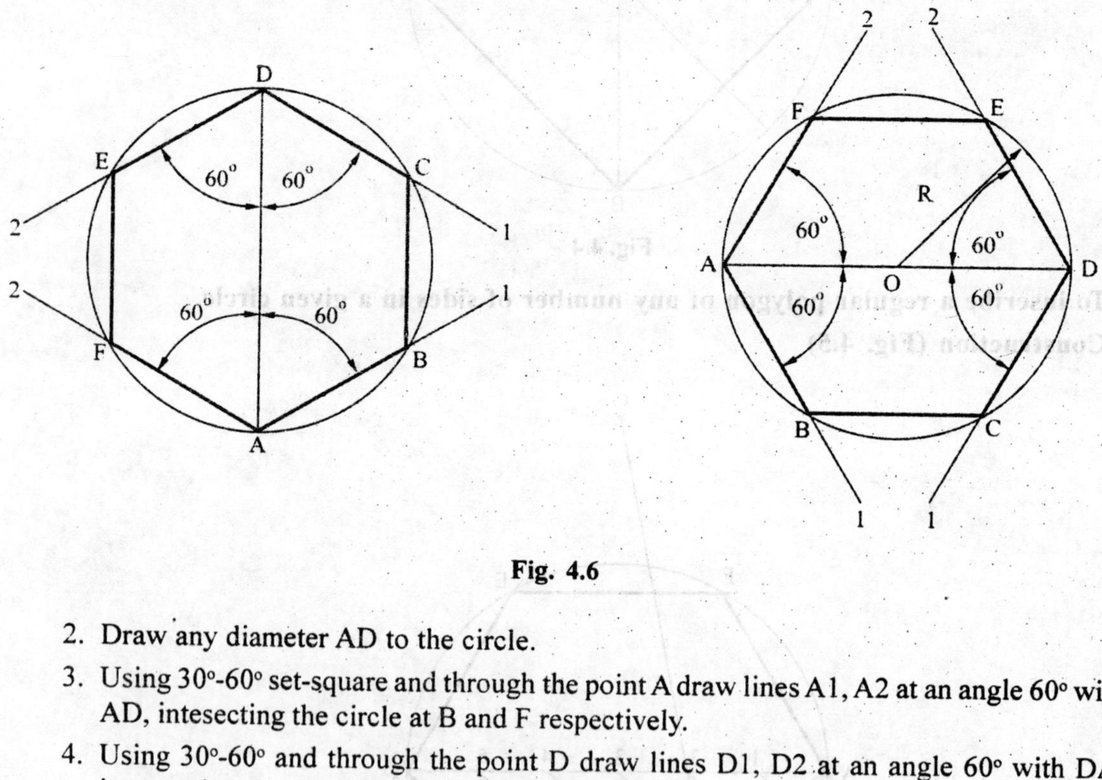

Fig. 4.6

2. Draw any diameter AD to the circle.
3. Using 30°-60° set-square and through the point A draw lines A1, A2 at an angle 60° with AD, intesecting the circle at B and F respectively.
4. Using 30°-60° and through the point D draw lines D1, D2 at an angle 60° with DA, intersecting the circle at C and E respectively.

   By joining A, B, C, D, E, F, and A, the required hexagon is obtained.

(b) Construction (Fig. 4.7) By using campass

1. With centre O and radius R draw the given circle.
2. Draw any diameter AD to the circle.

3. With centres A and D and radius equal to the radius of the circle draw arcs intesecting the circles at B, F, C and E respectively.
4. A B C D E F is the required hexagon.

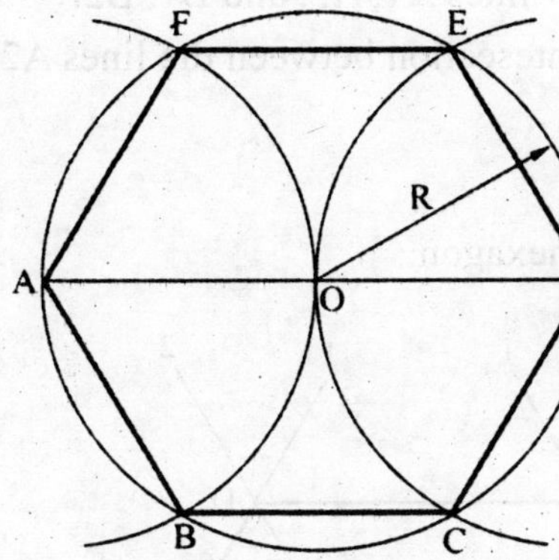

Fig. 4.7

7. To circumscribe a hexagon on a given circle of radius R

Construction (Fig. 4.8)

1. With centre O and radius R draw the given circle.
2. Using 60° position of the mini draughter or 30°-60°set square, circumscribe the hexagon as shown.

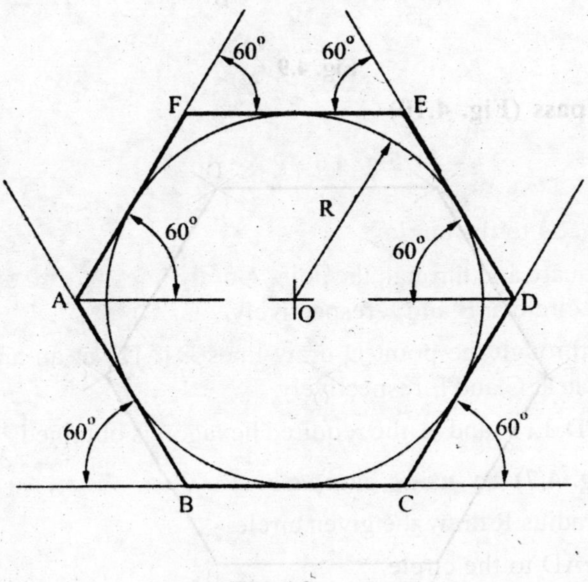

Fig. 4.8

8. To construct a hexagon, given the length of the side.

(a) contruction (Fig. 4.9) Using set square

1. Draw a line AB equal to the side of the hexagon.
2. Using 30°-60° set-square draw lines A1, A2, and B1, B2.
3. Through O get, the point of intesection between the lines A2 at D and B2 at E.
4. Join D,E
5. Obtain points and F similarly
6. A B C D E F is the required hexagon.

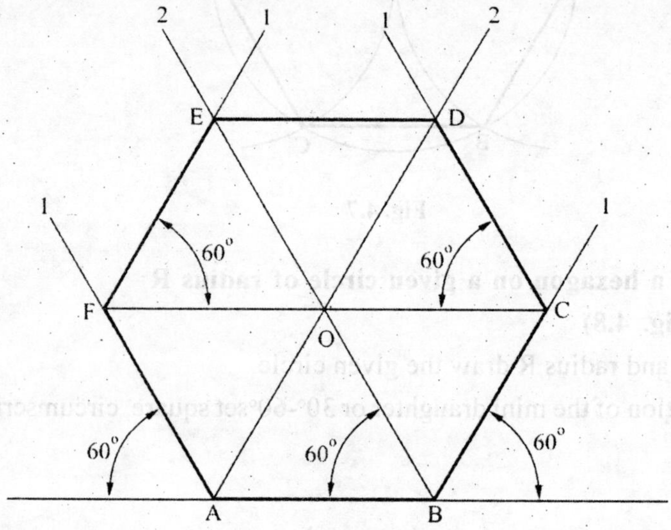

Fig. 4.9

(b) By using compass (Fig. 4.10)

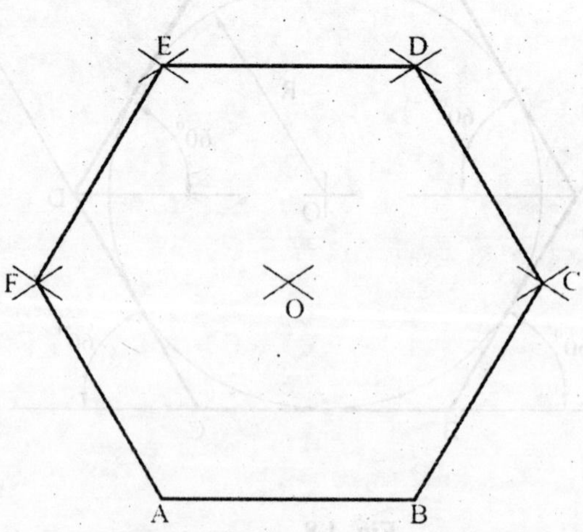

Fig. 4.10

1. Draw a line AB equal to the of side of the hexagon.
2. With centres A and B and radius AB, draw arcs intersecting at O, the centre of the hexagon.
3. With centres O and B and radius OB (= AB) draw arcs intersecting at C.
4. Obtain points D, E and F in a sinilar manner.

9. To construct a regular polygon (say a pentagon), given the length of the side.

construction (Fig. 4.11)

1. Draw a line AB equal to the side and extend to P such that AB = BP
2. Draw a semicircle on AP and divide it into 5 equal parts by trial and error.
3. Join B to second division
2. Irrespective of the number of sides of the polygon B is always joined to the second division.
4. Draw the perpendicular bisectors of AB and B2 to intersect at O.
5. Draw a circle with O as centre and OB as radius.
6. With AB as radius intersect the circle successively at D and E. Then join CD, DE and EA.

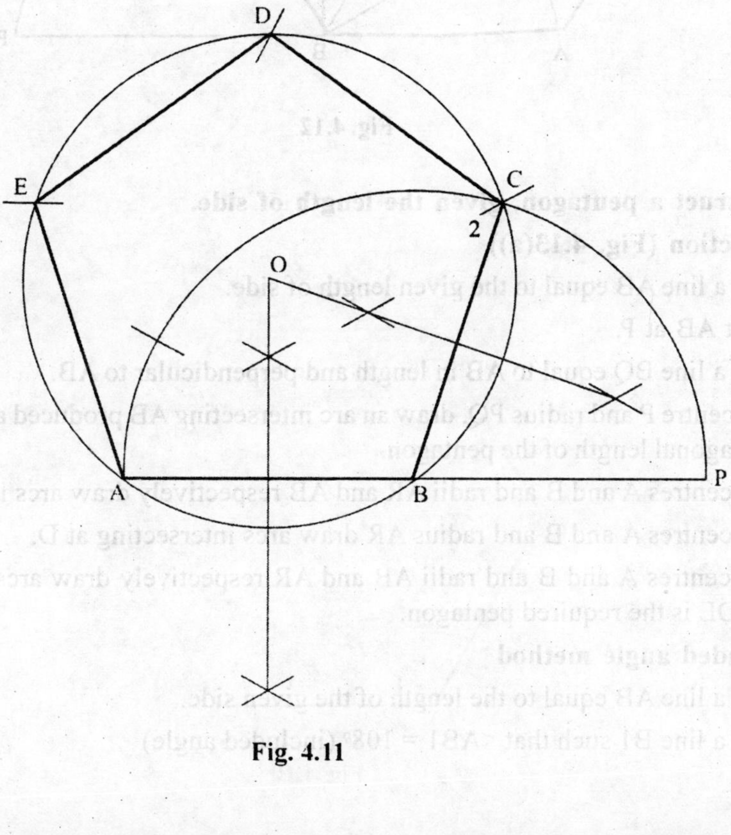

Fig. 4.11

10. To construct a regular polygon (say a hexagon), given the side AB – alternate method.

Construction (Fig. 4.12)

1. Steps 1 to 3 are same as in Example 9
2. Join B- 3, B-4, B-5 and produce them.
3. With 2 as centre and radius AB intersect the line B3 produced at D. Similarly get the point E and F.
4. Join 2- D, D-E, E-F and F-A to get the required hexagon.

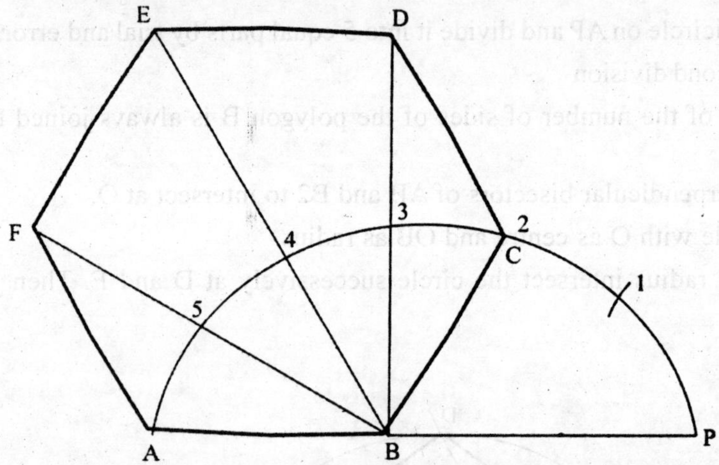

Fig. 4.12

11. To construct a pentagon, given the length of side.

(a) Construction (Fig. 4.13(a))

1. Draw a line AB equal to the given length of side.
2. Bisect AB at P.
3. Draw a line BQ equal to AB in length and perpendicular to AB.
4. With centre P and radius PQ, draw an arc intersecting AB produced at R. AR is equal to the diagonal length of the pentagon.
5. With centres A and B and radii AR and AB respectively draw arcs intersecting at C.
6. With centres A and B and radius AR draw arcs intersecting at D.
7. With centres A and B and radii AB and AR respectively draw arcs intersecting at E. ABCDE is the required pentagon.

(b) By included angle method

1. Draw a line AB equal to the length of the given side.
2. Draw a line B1 such that <AB1 = 108° (included angle)

3. Mark C on B1 such that BC = AB
4. Repeat steps 2 and 3 and complete the pentagon ABCDE

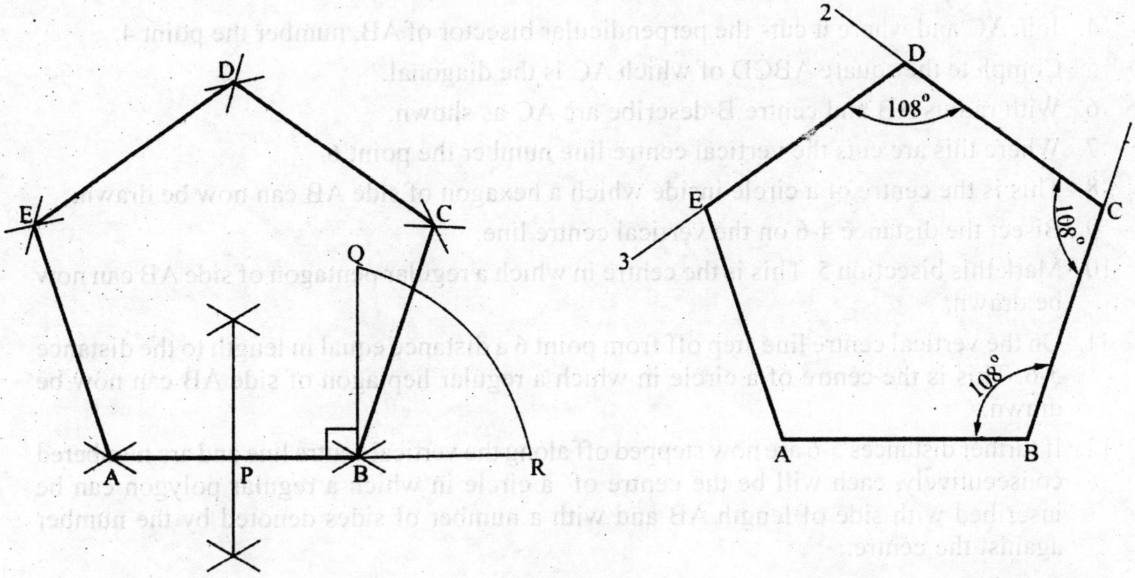

Fig. 4.13(a)&(b)

12. To construct a regular figure of given side length and of N sides on a straight line.

Construction (Fig. 4.14)

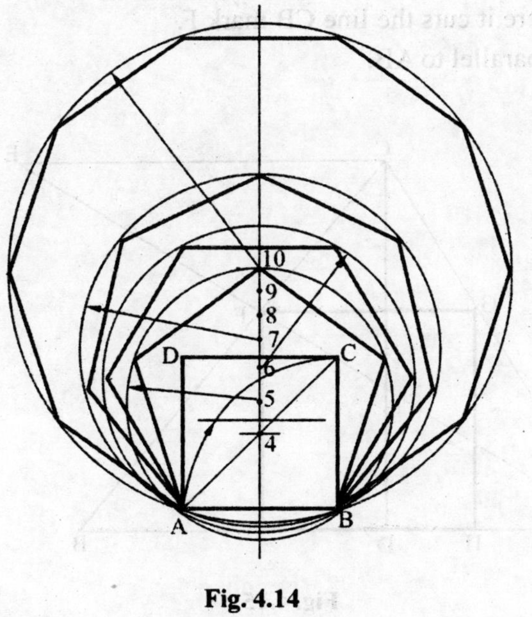

Fig. 4.14

1. Draw the given straight line AB.
2. At B erect a perpendicular BC equal in length to AB.
3. Draw perpendicular bisector of AB.
4. Join AC and where it cuts the perpendicular bisector of AB, number the point 4.
5. Complete the square ABCD of which AC is the diagonal.
6. With radius AB and centre B describe arc AC as shown.
7. Where this arc cuts the vertical centre line number the point 6.
8. This is the centre of a circle inside which a hexagon of side AB can now be drawn.
9. Bisect the distance 4-6 on the vertical centre line.
10. Mark this bisection 5. This is the centre in which a regular pentagon of side AB can now be drawn.
11. On the vertical centre line step off from point 6 a distance equal in length to the distance 5-6. This is the centre of a circle in which a regular heptagon of side AB can now be drawn.
12. If further distances 5-6 are now stepped off along the vertical centre line and are numbered consecutively, each will be the centre of a circle in which a regular polygon can be inscribed with side of length AB and with a number of sides denoted by the number against the centre.

13. To inscribe a square in a triangle.

Construction (Fig. 4.15)

1. Draw the given triangle ABC.
2. From C drop a perpendicular to cut the base AB at D.
3. From C draw CE parallel to AB and equal in length to CD.
4. Draw AE and where it cuts the line CB mark F.
5. From F draw FG parallel to AB.

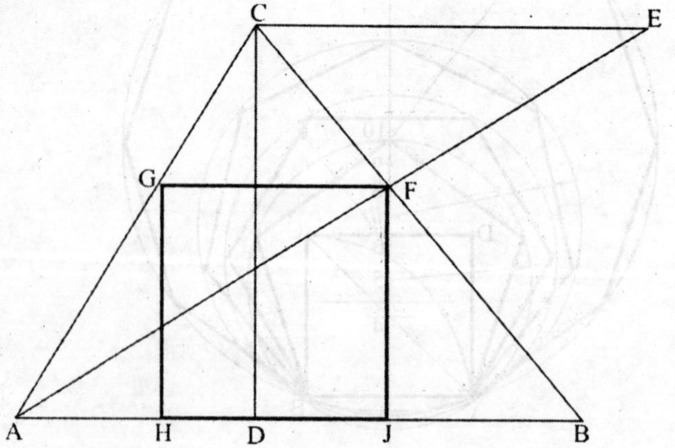

Fig. 4.15

6. From F draw FJ parallel to CD.
7. From G draw GH parallel to CD.
8. Join H to J.

   Then HJFG is the required square.

14. To inscribe within a given square ABCD, another square, one angle of the required square to touch a side of the given square at a given point

Construction (Fig. 4.16)

1. Draw the given square ABCD.
2. Draw the diagonals and where they intersect mark the point O.
3. Mark the given point E on the line AB.
4. With centre O and radius OE, draw a circle.
5. Where the circle cuts the given square mark the points G, H, and F.
6. Join the points GHFE.

Then GHFE is the required square.

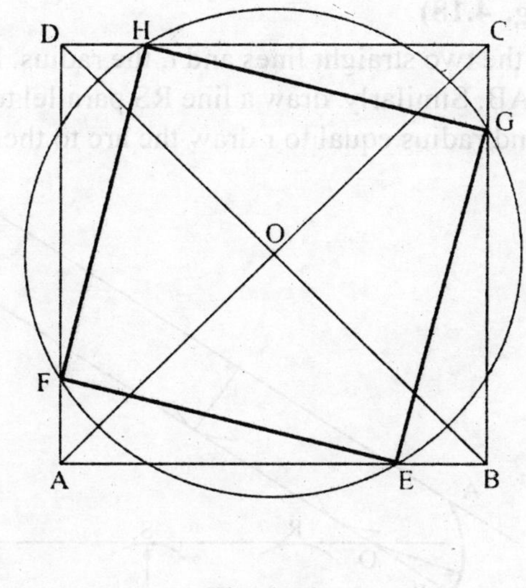

Fig. 4.16

15. To draw an arc of given radius touching two straight lines at right angles to each other.

Construction (Fig. 4.17)

Let r be the given radius and AB and AC the given straight lines. With A as centre and radius equal to r draw arcs cutting AB at P and Q. With P and Q as centres draw arcs to meet at O. With O as centre and radius equal to r draw the required arc.

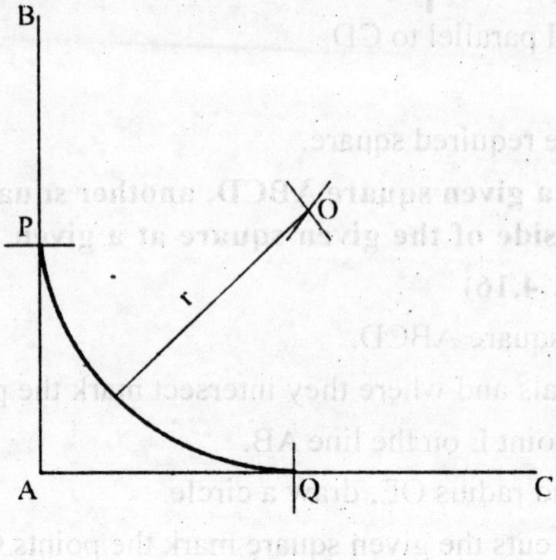

Fig. 4.17

16. To draw an arc of a given radius, touching two given straight lines making an angle between them.

Construction (Fig. 4.18)

Let AB and CD be the two straight lines and r, the radius. Draw a line PQ parallel to AB at a distance r from AB. Similarly, draw a line RS parallel to CD. Extend them to meet at O. With O as centre and radius equal to r draw the arc to the two given lines.

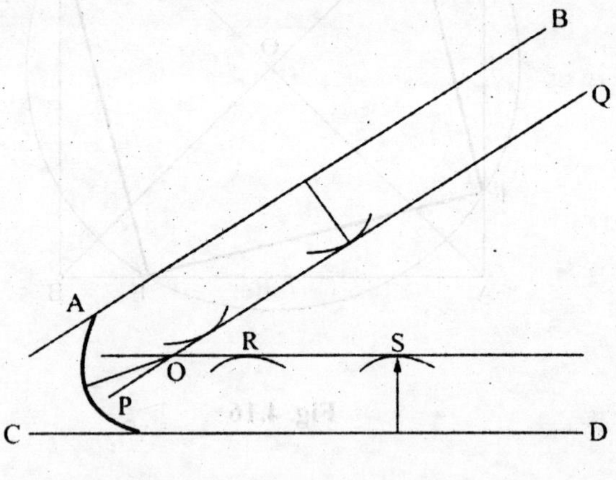

Fig. 4.18

17. To draw a tangent to a circle

Construction (Fig. 4.19(a) and (b))

(a) At any point P on the circle.

1. With O as centre, draw the given circle. P is any point on the circle at which tangent to be drawn (Fig 4.19(a))

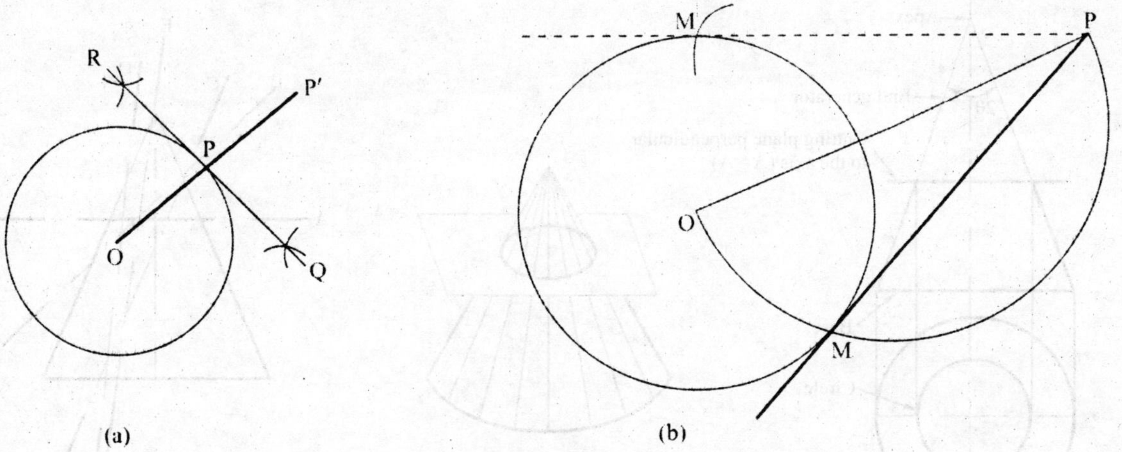

Fig. 4.19

2. Join O with P and produce it to P' so that OP = PP'
3. With O and P' as centres and a length greater than OP as radius, draw arcs intersecting each other at Q.
4. Draw a line through P and Q. This line is the required tangent that will be perpendicular to OP at P.

(b) From any point outside the circle.

1. With O as centre, draw the given circle. P is the point outside the circle from which tangent is to be drawn to the circle (Fig 4.19(b)).
2. Join O with P. With OP as diameter, draw a semi-circle intersecting the given circle at M. Then, the line drawn through P and M is the required tangent.
3. If the semi-circle is drawn on the other side, it will cut the given circle at M'. Then the line through P and M' will also be a tangent to the circle from P.

4.2 Conic Sections

Cone is formed when a right angled triangle with an apex and angle θ is rotated about its altitude as the axis. The length or height of the cone is equal to the altitude of the triangle and the radius of the base of the cone is equal to the base of the triangle. The apex angle of the cone is 2θ (Fig.4.20(a)).

When a cone is cut by a plane, the curve formed along the section is known as a conic. For this purpose, the cone may be cut by different section planes (Fig.4.20(b)) and the conic sections obtained are shown in Fig. 4.20(c), (d), and (e).

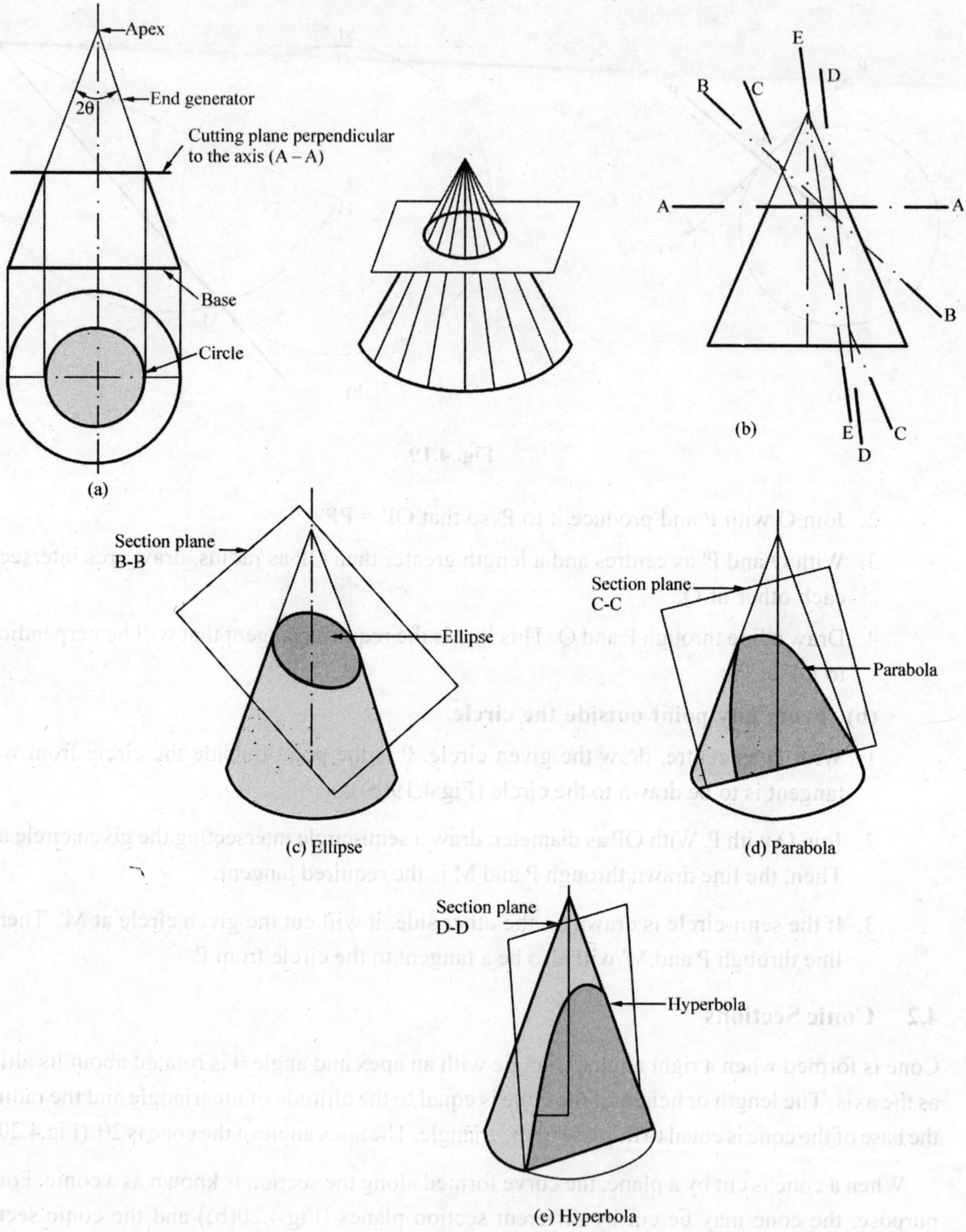
Apex
End generator
2θ
Cutting plane perpendicular to the axis (A – A)
Base
Circle
(a)
E
B
C
D
A
A
B
E
D
C
D
(b)
Section plane B-B
Ellipse
(c) Ellipse
Section plane C-C
Parabola
(d) Parabola
Section plane D-D
Hyperbola
(e) Hyperbola

Fig. 4.20

4.2.1 Circle

When a cone is cut by a section plane A-A making an angle $\alpha = 90^\circ$ with the axis, the section obtained is a circle. (Fig. 4.20(a))

4.2.2 Ellipse

When a cone is cut by a section plane B-B at an angle, α more than half of the apex angle i.e., θ less than 90°, the curve of the section is an ellipse. Its size depends on the angle α and the distance of the section plane from the apex of the cone. (Fig. 4.20(c))

4.2.3 Parabola

If the angle α is equal to θ i.e., when the section plane C-C is parallel to the slant side of the cone, the curve at the section is a parobola. This is not a closed figure like circle or ellipse. The size of the parabola depends upon the distance of the section plane from the slant side of the cone. (Fig. 4.20(d))

4.2.4 Hyperbola

If the angle α is less than θ (section plane D-D), the curve at the section is hyperbola. The curve of intersection is hyperbola, even if $\alpha = \theta$, provided the section plane is not passing through the apex of the cone. However if the section plane passes through the apex, the section produced is an isosceles triangle. (Fig. 4.20(e))

4.2.5 Conics defined by locus of a point

Conics are defined by locus of a point which moves in such a way that its distance from a fixed point (focus) is always in a constant ratio with its perpendicular distance from a fixed straight line called directrix FP:PQ. (Fig. 4.21).

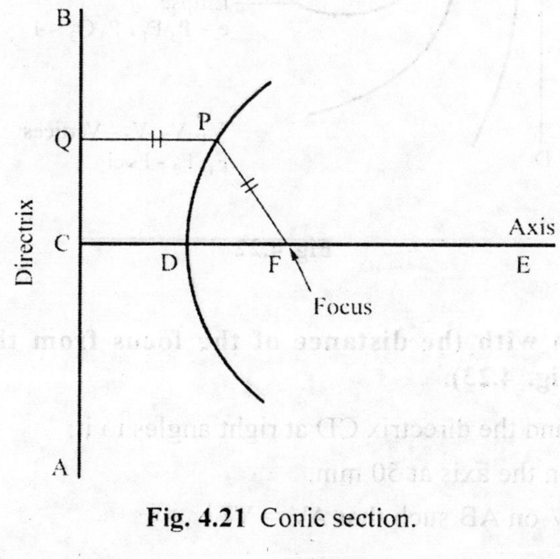

Fig. 4.21 Conic section.

Terminologies involved in the conic section (Fig. 4.21 and 4.22).

(i) ***Focus:*** The fixed point in the axis is called **Focus.**

(ii) ***Directrix:*** The fixed straight line is called as **directrix.**

(ii) ***Axis:*** The line perpendicular to the directrix and passing through the focus is called the axis.

(iv) ***Vertex:*** The point of intersection of conic section with the axis is called as vertex.

(v) ***Eccentricity (e):***

$$\text{Eccentricity (e)} = \frac{\text{Dis tan ce of the moving point from focus}}{\text{Distance of the moving point from directrix}}$$

i.e., $$e = \frac{PF}{PQ} = \text{constant}$$

If e < 1, then the curve obtained is an Ellipse.

If e = 1, then the curve obtained is a Parabola.

If e > 1, then the curve obtained is a Hyperbola

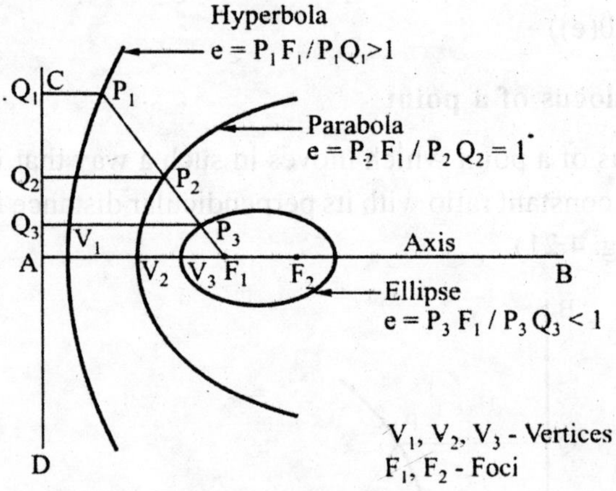

Fig. 4.22

To draw a parabola with the distance of the focus from the directrix at 50 mm (Eccentricity method Fig. 4.23).

1. Draw the axis AB and the directrix CD at right angles to it:
2. Mark the focus F on the axis at 50 mm.
3. Locate the vertex V on AB such that AV = VF

4. Draw a line VE perpendicular to AB such that VE = VF
5. Join A, E and extend. Now, VE/VA = VF/VA = 1, the **eccentricity**.
6. Locate number of points 1, 2, 3, etc., to the right of V on the axis, which need not be equi-distant.
7. Through the points 1, 2, 3, etc., draw lines perpendicular to the axis and to meet the line AE extended at 1', 2', 3' etc.
8. With centre F and radius 1-1', draw arcs intersecting the line through 1 at P_1 and P_1'.
9. Similarly, lolcate the points P_2, P_2', P_3, P_3', etc., on either side of the axis. Join the points by smooth curve, forming the required parabola.

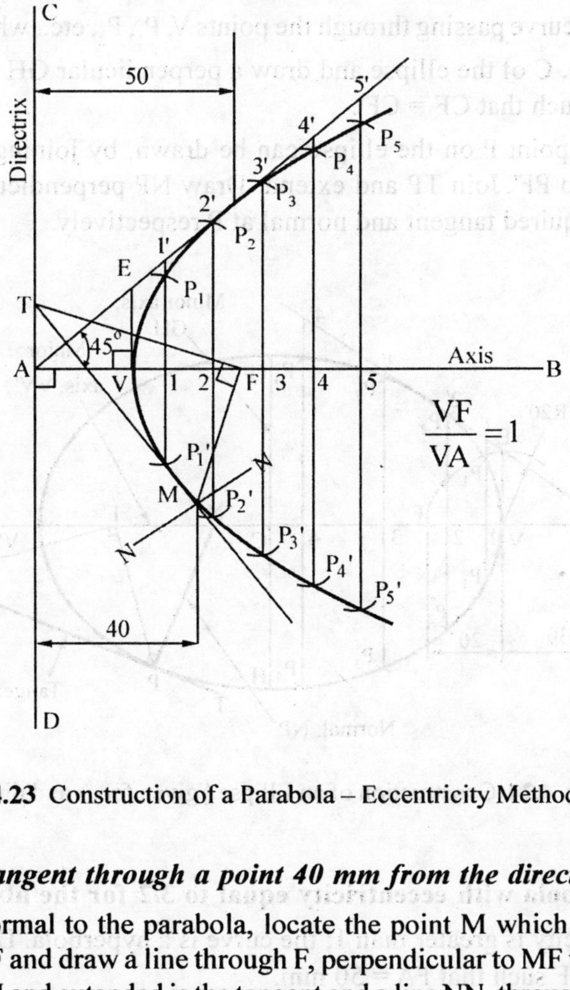

Fig. 4.23 Construction of a Parabola – Eccentricity Method.

To draw a normal and tangent through a point 40 mm from the directrix.

To draw a tangent and normal to the parabola, locate the point M which is at 40 mm from the directrix. Then join M to F and draw a line through F, perpendicular to MF to meet the directrix at T. The line joining T and M and extended is the tangent and a line NN, through M and perpendicular to TM is the normal to the curve.

To draw an ellipse with an eccentricity 2/3 for the above problem (Fig. 4.24)

1. Draw one directrix, DD and the axis, AA' perpendicular to DD and mark the focus, F such that FA = 50 mm.
2. As the eccentricity is 2/3, divide FA into 2 + 3 = 5 equal parts. By definition VF/VA = 2/3 and hence locate the vertex, V. Draw VE perpendicular to the axis such that VE = VF. Join AE and extend it as shown in Fig. 4.24. This is the eccentricity scale, which gives the distances directly in the required ratio. In triangle AVE, VE/VA=VF/VA=2/3
3. Mark any point 1 on the axis and draw a perpendicular through it to intersect AE produced at 1'. With centre F and radius equal to 1-1' draw arcs to intersect the perpendicular through 1 at P_1 both above and below the axis of the conic.
4. Similarly, mark points 2, 3, 4, etc., as described above.
5. Draw a smooth curve passing through the points V, P_1, P_2, etc., which is the required **ellipse.**
6. Mark the centre, C of the ellipse and draw a perpendicular GH to the axis. Also mark the other focus F' such that CF = CF'.
7. Tangent at any point P on the ellipse can be drawn, by joining P F' and by drawing F'T perpendicular to PF'. Join TP and extend. Draw NP perpendicular to TP. Now, TPT and NPN are the required tangent and normal at P respectively.

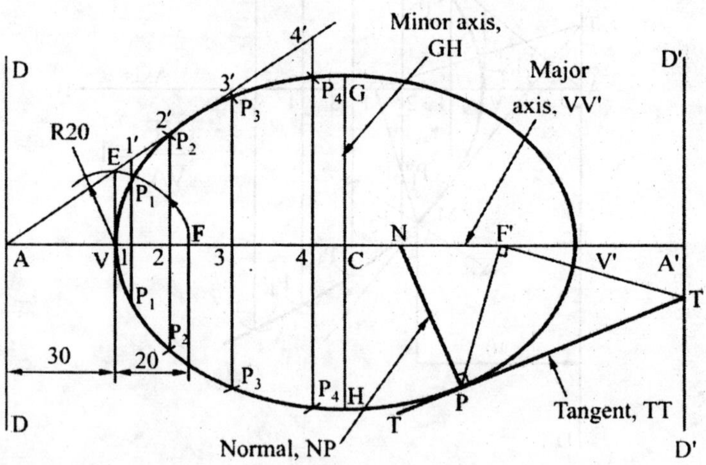

Fig. 4.24 Construction of an Ellipse (given focus and directrix).

To draw a hyperbola with eccentricity equal to 3/2 for the above problem. (Fig. 4.25)

1. As the eccentricity is greater than 1; the curve is a hyperbola. Draw one directirx DD and mark the focus F such that FA = 50 mm.
2. As the eccentricity is 3/2, divide FA into 3 + 2 = 5 equal parts. By difinition VF/VA = 3/2 and hence locate the vertexV.

3. Draw VE perpendicular to the axis such that VE = VF. Join AE and extend it as shown in Fig. 4.25.
4. This is the eccentricity scale, which gives the distances in the required ratio. In triangle AVE, EF/VA = VF/VA = 3/2
5. Mark any point 1 on the axis and proceed further as explained in earlier to get the points P_1, P_2, P_3, P_1', P_2', P_3' etc. Draw a smooth curve passing through the points V, P_1, P_2, P_3, etc. which is the required hyperbola.
6. Tangent and normal at any point P on the hyperbola can be drawn as shown.

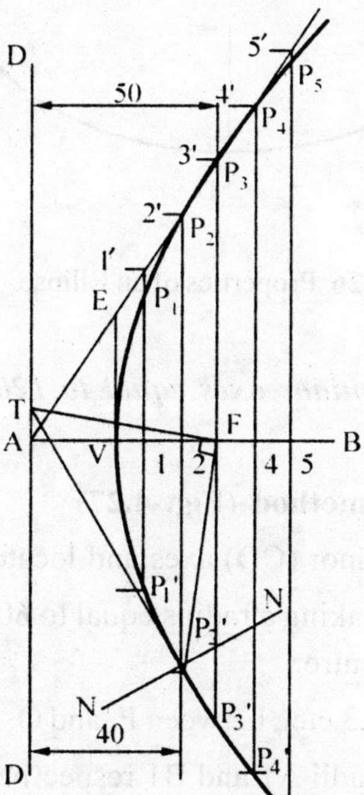

Fig. 4.25 Construction of Hyperbola (Eccentricity Method).

Other Methods of Construction of Ellipse

Given the dimensions of major and minor axes, the ellipse can be drawn by, (i) Foci method, (ii) Oblong method, (iii) Concentric circle method and (iv) Trammel method (v) Four centre method.

Definition of Ellipse

Ellipse is a curve traced by a point moving such that the sum of its distances from the two fixed points, foci, is constant and equal to the major axis.

Referring Fig. 4.26, F_1 and F_2 are the two foci, AB is the major axis and CD is the minor axis. As per the difinition, $PF_1 + PF_2 = CF_1 + CF_2 = QF_1 + QF_2 = AB$. It may also be noted that $CF_1 = CF_2 = 1/2$ AB (Major axis)

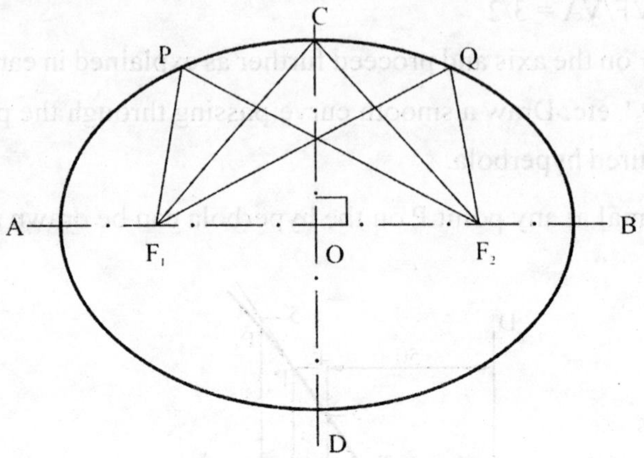

Fig. 4.26 Properties of an Ellipse.

To draw an ellipse with major and minor axes equal to 120 mm and 80 mm respectively. (5 methods)

1. Foci Method/Arcs of circles method (Fig. 4.27)

1. Draw the major (AB) and minor (CD) axes and locate the centre O.
2. Locate the foci F_1 and F_2 by taking a radius equal to 60 mm (1/2 of AB) and cutting AB at F_1 and F_2 with C as the centre.
3. Mark a number of points 1,2,3 etc., between F_1 and O, which need not be equi-distance.
4. With centres F_1 and F_2 and radii A1 and B1 respectively, draw arcs intersecting at the points P_1 and P_1'.
5. Again with centres F_1 and F_2 and radii B1 and A1 respectively, draw arcs intersecting at the points Q_1 and Q_1'.
6. Repeat the steps 4 and 5 with the remaining points 2, 3, 4 etc., and obtain additional points on the curve.

   Join the points by a smooth curve, forming the required ellipse.

To mark a Tangent and Normal to the ellipse at any point, say M on it, join the foci F_1 and F_2 with M and extend F_2M to E and bisect the angle <EMF$_1$. The bisector TT represents the required tangent and a line NN drawn through M and perpendicular to TT is the normal to the ellipse.

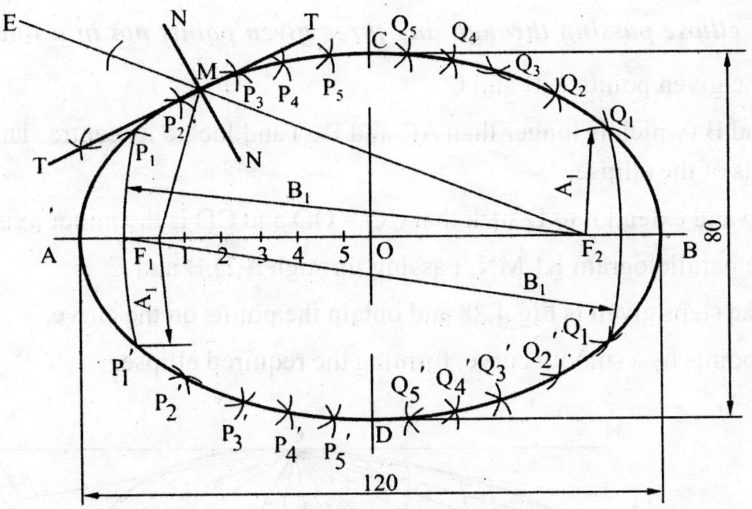

Fig. 4.27 Construction of Ellipse - Foci or Arc's of circles Method.

2. Oblong Method (Fig. 4.28)

1. Draw the major and minor axes AB and CD and locate the centre O.
2. Draw the rectangle KLMN passing through A,D,B,C.
3. Divide AO and AN into same mumber of equal parts, say 4.
4. Join C with the points 1', 2', 3'.
5. Join D with the points 1, 2, 3 and extend till they meet the lines C1', C2', C3', respectively at P_1, P_2 and P_3.
6. Repeat steps 3 to 5 to obtain the points in the remaining three quadrants.
7. Join the points by a smooth curve forming the required ellipse.

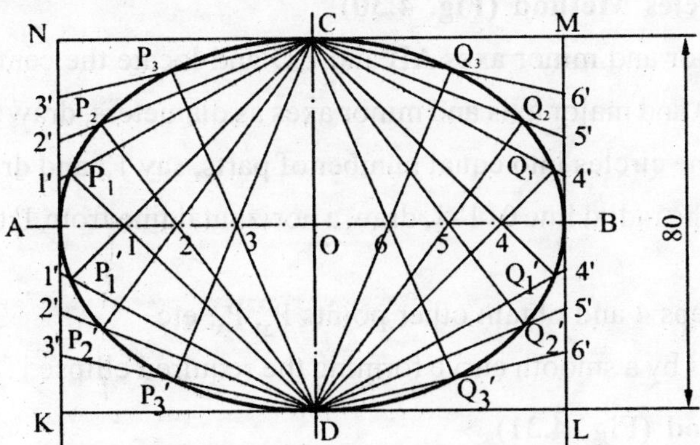

Fig. 4.28 Oblong Method. (Rectangle)

To draw an ellipse passing through any three given points not in a line. (Fig. 4.29)

1. Locate the given points A,B and C
2. Join A and B (which is longer than AC and BC) and locate its centre. This becomes the major axis of the ellipse.
3. Draw CO and extend it to D such that CO = OD and CD is the minor axis of the ellipse.
4. Draw the parallelogram KLMN, Passing through A,D,B and C.
5. Follow the steps given is Fig.4.28 and obtain the points on the curve.
6. Join the points by a smooth curve, forming the required ellipse.

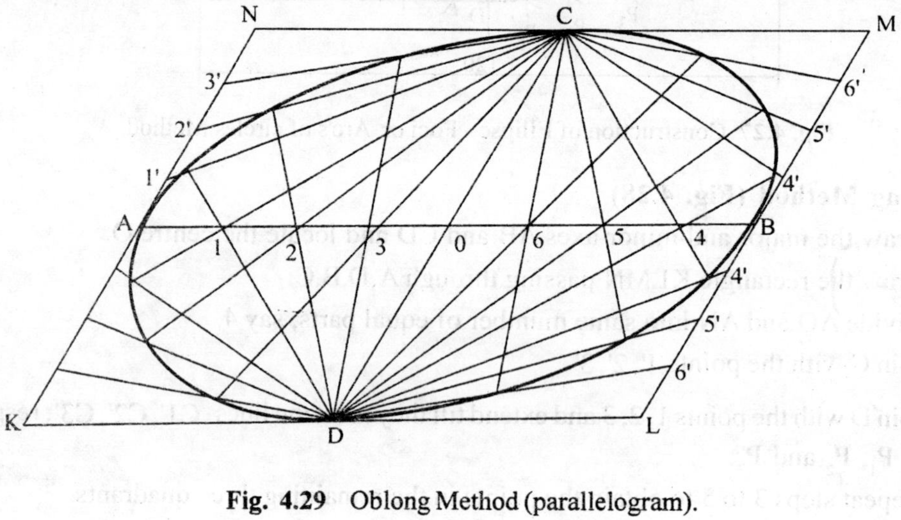

Fig. 4.29 Oblong Method (parallelogram).

3. Concentric Circles Method (Fig. 4.30)

1. Draw the major and minor axes AB and CD and locate the centre O.
2. With centre O and major axis and minor axes as diameters, draw two concentric circles.
3. Divide both the circles into equal number of parts, say 12 and draw the radial lines.
4. Considering the radial line 0- 1'-1, draw a horizontal line from 1' to meet the vertical line from 1 at P_1.
5. Repeat the steps 4 and obtain other points P_2, P_3, etc.
6. Join the points by a smooth curve forming the required ellipse.

4. Trammel Method (Fig. 4.31)

1. Draw the major and minor axes AB and CD and then locate the centre O.
2. Take a strip of stiff paper and prepare a trammel as shown. Mark the semi-major and semi-minor axes PR and PQ on the trammel.

3. Position the trammel so that the points R and Q lie respectively on the minor and major axes. As a rule, the third point P will always lie on the ellipse required.
4. Keeping R on the minor axis and Q on the mojor axis, move the trammel to other position and locate other points on the curve.
5. Join the points by a smooth curve forming the required ellipse.

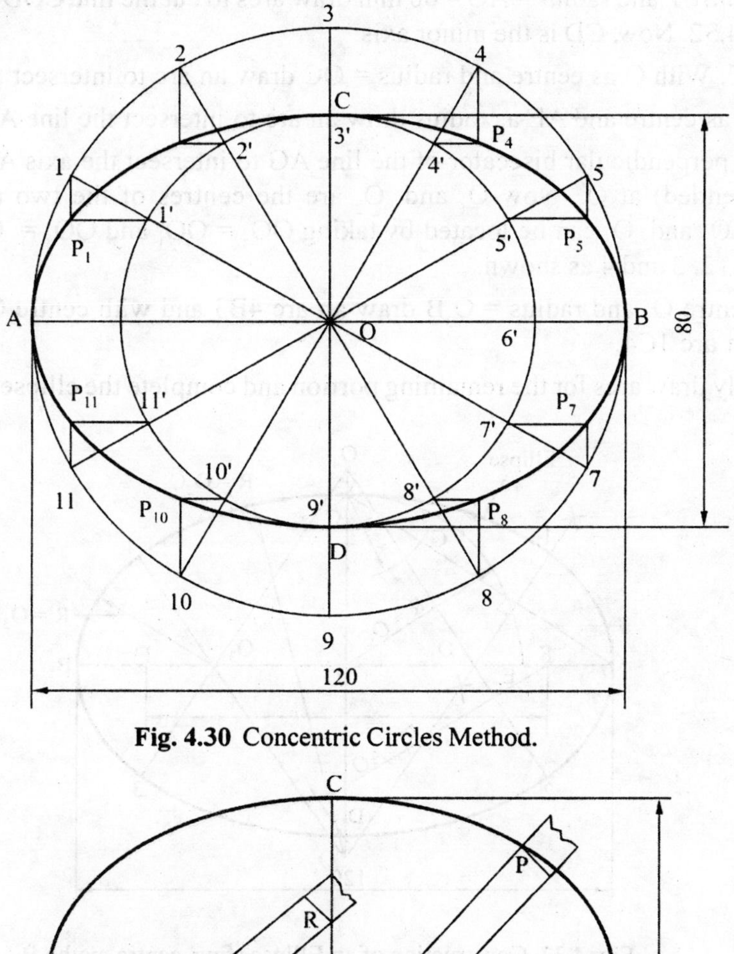

Fig. 4.30 Concentric Circles Method.

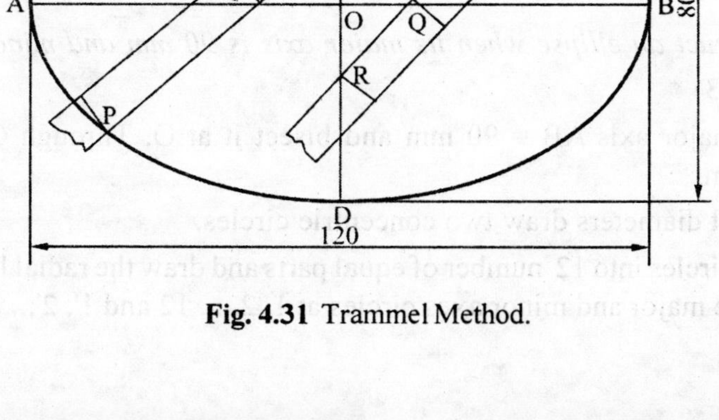

Fig. 4.31 Trammel Method.

5. Four Centre Method (Fig. 4.32)

The foci of an ellipse are 90 mm apart and the major axis is 120 mm long. Draw the ellipse by four centre method.

1. Draw the major axis AB = 120 mm. Draw a perpendicular bisector COD. Mark the foci F and F' such that FO = F'O = 45 mm.
2. With centre F and radius = AO = 60 mm draw arcs to cut the line COD at C and D as shown in Fig. 4.32. Now, CD is the minor axis.
3. Join AC. With O as centre and radius = OC draw an arc to intersect the line AB at E.
4. With C as centre and AE as radius draw an arc to intersect the line AC at G.
5. Draw a perpendicular bisecator of the line AG to intersect the axis AB at O_1 and the axis CD(extended) at O_2. Now O_1 and O_2 are the centres of the two arcs. The other two centres O_3 and O_4 can be located by taking $OO_3 = OO_1$ and $OO_4 = OO_2$. Also locate the points 1, 2, 3 and 4 as shown.
6. With centre O_3 and radius = O_3B draw an arc 4B3 and with centre O_2 and radius = O_2C draw an arc 1C4.
7. Similarly draw arcs for the remaining portion and complete the ellipse.

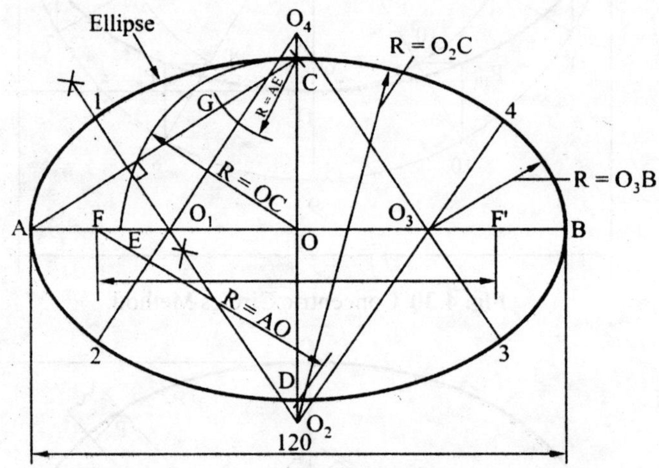

Fig. 4.32 Construction of an Ellipse (four-centre method).

Problem : *Construct an ellipse when its major axis is 90 mm and minor axis is 55 mm.*

Solution: (Fig. 4.33)

1. Draw the major axis AB = 90 mm and bisect it at O. Through O draw a vertical line CD = 55 mm.
2. To represent diameters draw two concentric circles.
3. Divide the circles into 12 number of equal parts and draw the radial lines. These radial lines intersect the major and minor axes circles at 1, 2, ... 12 and 1', 2',....12' respectively.

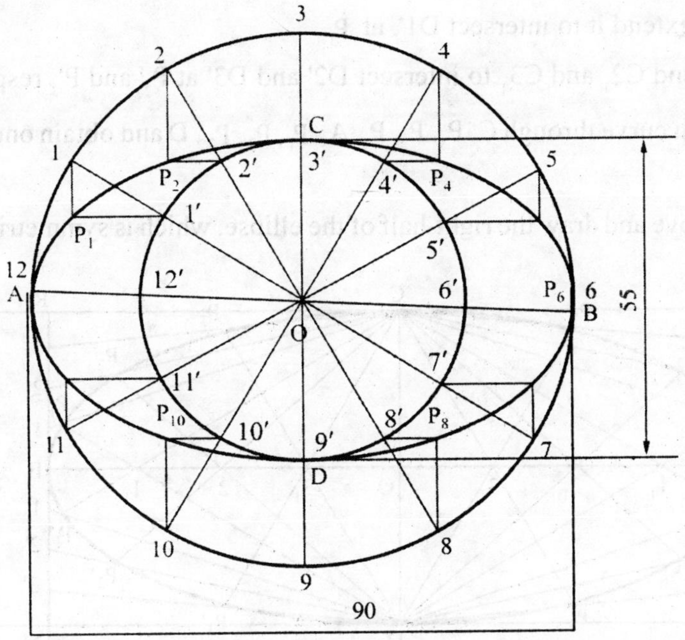

Fig. 4.33 Concentric Circles Method.

4. From 1 draw a vertical line (parallel to CD) and from 1' draw a horizontal line (parallel to AB). Both intersect at P_1.
5. Repeat the abve and obtain the points P_2,....P_{12} corresponding to 2 and 2',...12, and 12' respectively.
6. Draw a smooth ellipse through P_1,P_2,.....P_{12}, P_1.

Problem : A ground is in the shape of a rectangle 120 m × 60 m. Inscribe an elliptical lawn in it to a suitable scale.

***Solution :* (Fig. 4.34)**

1. Draw the major axis AB = 120 m and minor axis CD = 60 m. Both axes bisect each other at O.
2. Through A and B draw lines parallel to CD.
3. Through C and D draw lines parallel to AB and construct the rectangle PQRS. Now PS=AB and SR = CD.
4. Divide AQ and AP into any number of equal parts (4 say) and name the points as 1, 2, 3 and 1' 2' 3' respectively starting from A on AQ and AP.
5. Divide AO into same number of equal parts, and name the points as 1_1, 2_1, 3_1 starting from A on AO.
6. Join 1, 2, 3 with C. Join D1_1 and extend it to intersect at P_1.
7. Similarly extend D2_1 and D3_1 to intersect C2 and C3 at P_2 and P_3 respectively. Join 1' ,2' ,3' with D.

8. Join $C1_1$ and extend it to intersect D1', at P_1'.
9. Similarly extend $C2_1$ and $C3_1$ to intersect D2' and D3' at P_2' and P'_3 respectively.
10. Draw a smooth curve through C, P_3, P_2, P_1, A, P_1', P_2', P_3', D and obtain one half (left-half) of the ellipse.
11. Repeat the above and draw the right-half of the ellipse, which is symmetrical to the left-half.

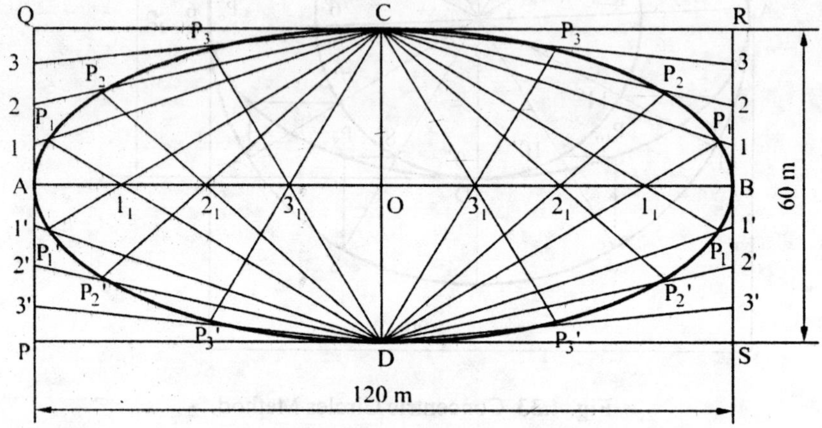

Fig. 4.34 Rectangle (or) Oblong Method.

Problem : Construct an ellipse when a pair of conjugate diameters AB and CD are equal to 120 mm and 50 mm respecitively. The angle between the conjugate diameters is 60°.

Solution : **(Fig. 4.35)**

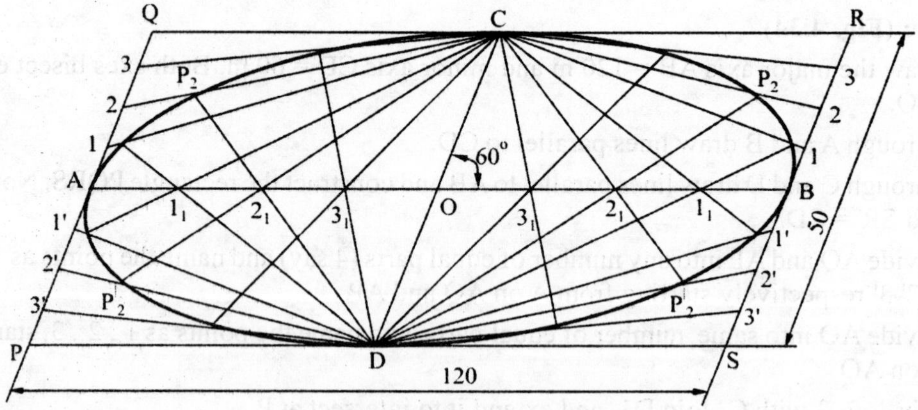

Fig. 4.35 Parallelogram Method.

To construct the ellipse using conjugate diameters :

1. Draw a conjugate diameter AB = 120 mm and bisect it at O.
2. Angle between the conjugte diameters is 60°. Therefore draw another conjugate diameter CD through O such that the angle COB = 60°.
3. Through A and B draw lines parallel to CD. Through C and D draw lines parallel to AB and construct a parallelogram PQRS as shown in Fig. 4.35.
4. Repeat the procedure given in steps 4 to 10 in above problem and complete the construction of the ellipse inside the parallelogram PQRS.

Problem : *Draw an ellipse with mojor axis 120 mm and minor axis 80 mm. Determine the eccentricity and the distance between the directrices.*

Solution: **(Fig. 4.36)**

Eccentricity $e = V_1F_1 / V_1A = V_1F_2 / V_1B$

therefore $V_1F_2 - V_1F_1 / V_1B - V_1A = F_1F_2 / V_1V_2$

From the triangle F_1CO

OC = 40 mm (half of minor axis)

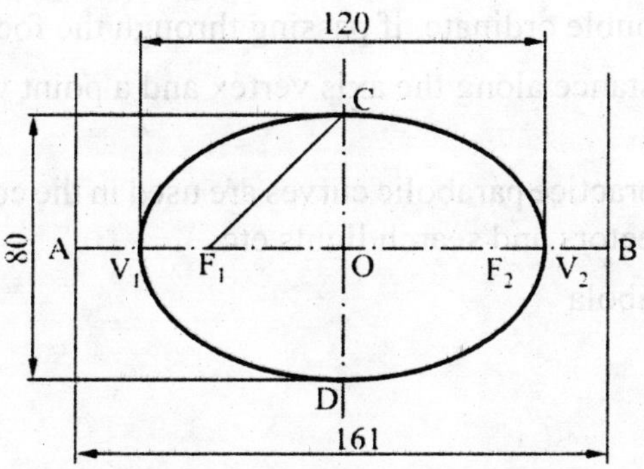

Fig. 4.36

F_1C = 60 mm (half of major axis)

Thus $F_10 = \sqrt{60^2 - 40^2} = 44.7$ mm

Hence $F_1F_2 = 2F_1O = 89.4$ mm

on substitution $e = \frac{89.4}{120} = 0.745$. Also, eccentricity $e = V_1V_2/AB$, Hence, the distance between the directrices $AB = V_1V_2/e = 161$ mm.

Parabola

The parabola is a plane curve generated by a point moving so that at any position, its distance from a fixed point (focus) is always exactly equal to its distance from a fixed straight line (directrix).

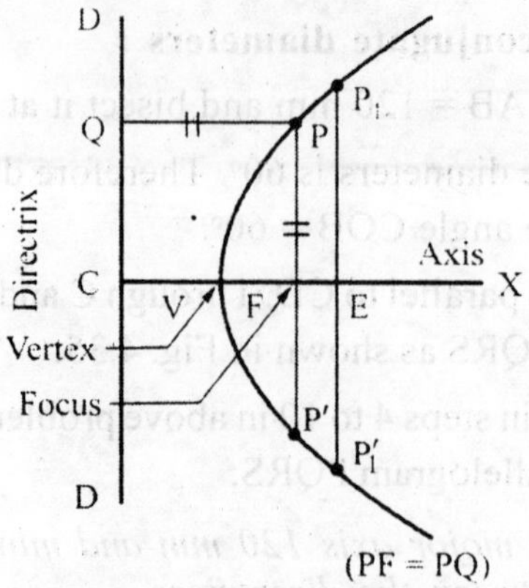

Fig. 4.37 Parabola (PF = PQ).

The following terms are used while solving problems of construction of parabola.

Ordinate (PF or P_1E): Any line from the curve which is perpendicular to the axis is called ordinate.

Double ordinate (PP' or P_1 P_1'): Twice the length of the ordinates is called double ordinate.

Latus rectum (PP'): The double ordinate, if passing through the focus is called latus rectum.

Absciss (VF or VE): The distance along the axis vertex and a point where the ordinate passes is called abscissa.

Application: In engineering practice parabolic curves are used in the construction of bridge arches, road sections, parabolic reflectors and search lights etc.

Methods of constructing parabola

1. Eccentricity method
2. Rectangle method
3. Tangent method

1. Eccentricity method

Problem : Construct a parabola when the distance between its focus and its directrix is equal to 60 mm. Draw a tangent at any point on the curve.

Solution : **(Fig. 4.38)**

1. As the eccentricity of the conic is one, the curve is a **parabola.**
2. Draw the directrixDD and the axis AB perpendicular to DD. Mark the focus F such that AF = 60 mm. By definition, VF/VA = 1 and hence mark the point V, the vertex at the midpoint of AF as shown in Fig. 4.38.
3. Mark any number of points (say 6) on VB and draw verticals through these points.
4. With F as centre and A1 as radius draw an arc to cut the vertical through point 1 at P_1. Similarly obtain points P_2, P_3, P_4, etc.

5. Draw a smooth curve passing through these points to obtain the required parabola.
6. To draw tangent at any point see Fig. 4.23.

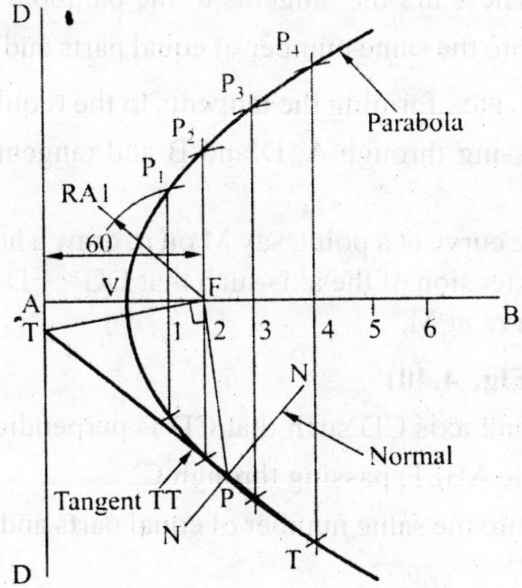

Fig. 4.38 Construction of a Parabola.

To draw a parabola with 70 mm as base and 30 mm as the length of the axis.

1. Tangent Method (Fig. 4.39)

1. Draw the base AB and locate its mid-point C.

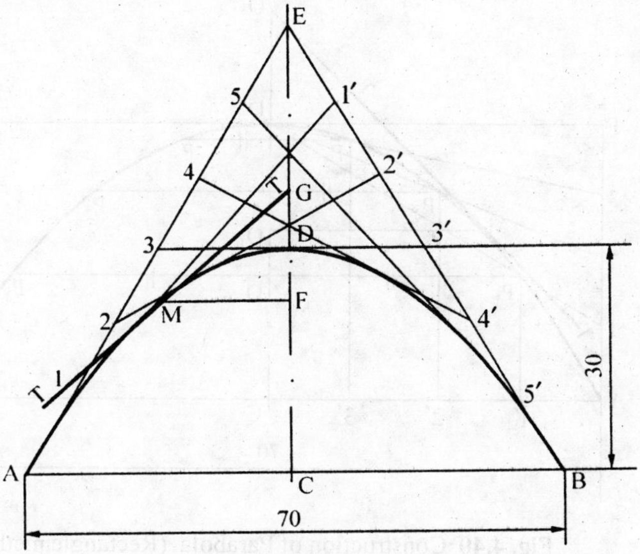

Fig. 4.39 Tangent Method.

2. Through C, draw CD perpendicular to AB forming the axis
3. Produce CD to E such that DE = CD
4. Join E-A and E-B. These are the tangents to the parabola at A and B.
5. Divide AE and BE into the same number of equal parts and number the points as shown.
6. Join 1-1', 2-2', 3-3', etc., forming the tangents to the required parabola.
7. A smooth curve passing through A, D and B and tangential to the above lines is the required parabola.

Note : To draw a tangent to the curve at a point, say M on it, draw a horizontal through M, meeting the axis at F. mark G on the extension of the axis such that DG = FD. Join G, M and extend, GMT forming the tangent to the curve at M.

2. Rectangle Method (Fig. 4.40)

1. Draw the base AB and axis CD such that CD is perpendicular bisector to AB.
2. Construct a rectangle ABEF, passing through C.
3. Divide AC and AF into the same number of equal parts and number the points as shown.
4. Join 1, 2 and 3 to D.
5. Through 1', 2' and 3', draw lines parallel to the axis, intersecting the lines 1D, 2D and 3D at P_1, P_2 and P_3 respectively.
6. Obtain the points P_1, P_2 and P_3, which are symmetrically placed to P_1, P_2 and P_3 with respect to the axis CD.
7. Join the points by a smooth curve forming the required parabola.

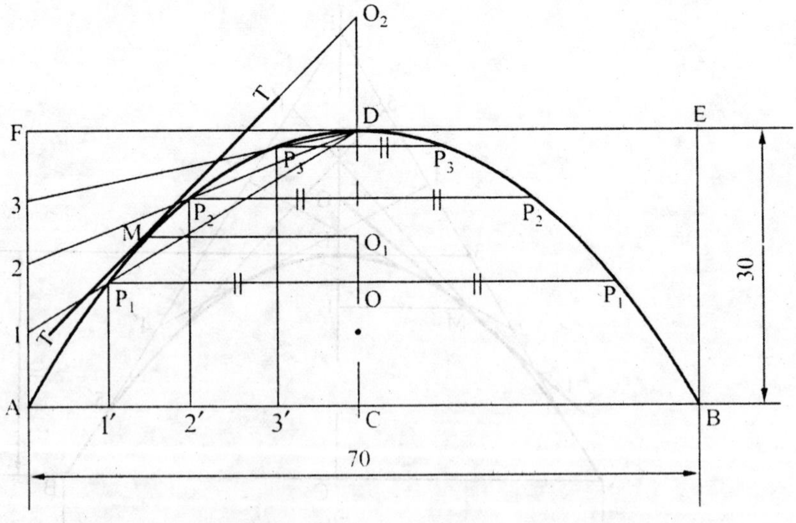

Fig. 4.40 Construction of Parabola. (Rectangle method)

Note : Draw a tangent T-T at M following the method indicated in Fig.4.31. (DO_1 = DO_2)

Problem: *A fountain jet is dicharged from the ground level at an inclination of 45°. The jet travels a horizontal distance of 10 m from the point of discharge and falls on the ground. Trace the path of the jet.*

***Solution:* (Fig. 4.41)**

1. Draw the base AB of 10 m long and locate its mid-point C.
2. Through C draw a line perpendicular to AB forming the axis.
3. Through A and B, draw lines at 45°, to the base intersecting the axis at D.
4. Divide AD and BD into the same number of equal parts and number the points as shown.
5. Join 1-1', 2-2', 3-3' etc., forming the tangents to the required path of jet.
6. A smooth curve passing through A and B and tangential to the above lines is the required path of the jet which is parabolic in shape.

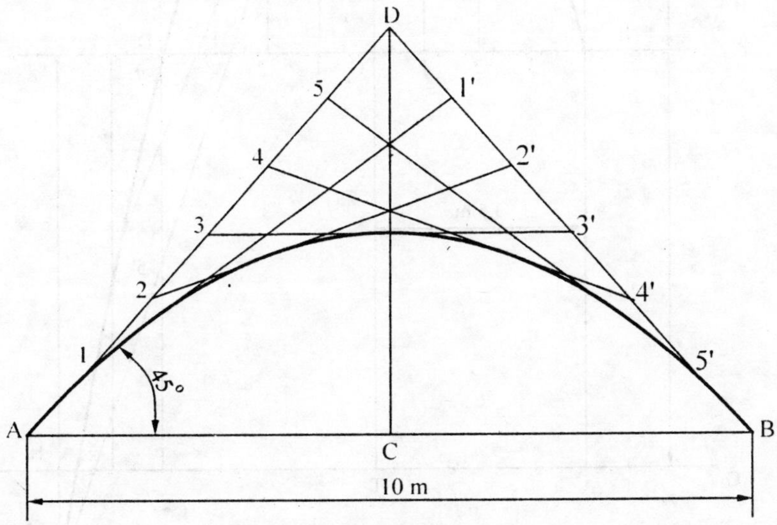

Fig. 4.41

Problem: *A stone is thrown from a building of 7 m high and at its highest flight it just crosses a plam tree 14 m high. Trace the path of the stone, if the distance between the building and the tree measured along the ground is 3.5 m.*

***Solution:* (Fig. 4.42)**

1. Draw lines AB and OT, representing the building and plam tree respectively, 3.5 m apart and above the ground level.
2. Locate C and D on the horizontal line through B such that CD=BC=3.5 and complete the rectangle BDEF.
3. Inscribe the parabola in the rectangle BDEF, by rectangular method.

4. Draw the path of the stone till it reaches the ground (H) extending the principle of rectangle method.

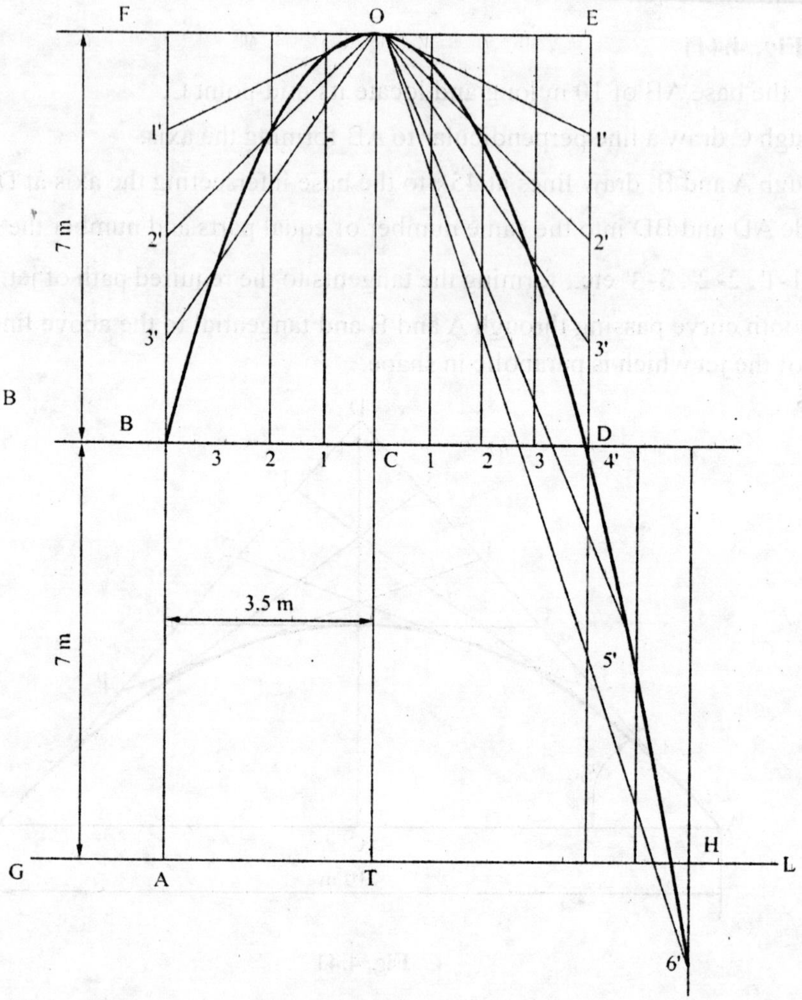

Fig. 4.42

Problem: A ball thrown from the ground level reaches a maximum height of 5 m and travels a horizontal distance of 12 m from the point of projection. Trace the path of the ball.

Solution : **(Fig. 4.43)**

1. The ball travels a horizontal distance of 12 m. By taking a scale of 1:100, draw PS = 12 cm to represent the double ordinate. Bisect PS at O.
2. The ball reaches a maximum height of 5 m. So from O erect vertical and mark the vertex V such that OV = 5cm. Now OV is the abscissa.

3. Construct the rectangle PQRS such that PS is the double ordinate and PQ = RS = VO(abscissa).
4. Divide PQ and RS into any number of (say 8) equal parts as 1, 2,... 8 and 1' 2' ....8' respectively, starting from P on PQ and S on SR. Join 1, 2, ...8 and 1' ,2' ....8' with V.
5. Divide PO and OS into 8 equal parts as 1_1 2_1 ......8_1 and $1'_1$ $2'_1$......$8'_1$ respectively, starting from P on PO and from S on SO.
6. From 1_1 erect vertical to meet the line V1 at P_1.
7. Similarly from 2_1, ...8_1 erect verticals to meet the lines V2, ...V8 at P_2....P_8 respectively.
8. Also erect verticals from $1'_1$ $2'_1$ .....$8'_1$ to meet the lines V1' ....V2' .....V8' at P_1' ....P_2' .....P8' respectively.
9. Join P, P_1, P_2, ... P'_7.V_1.....P'_1 and S to represent the path of the ball which is a **parabola.**

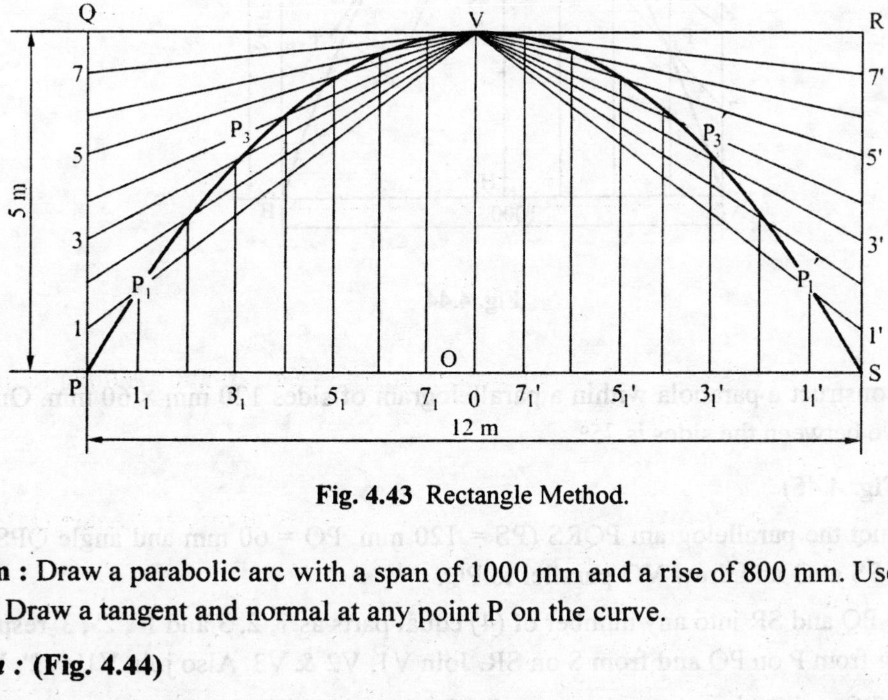

Fig. 4.43 Rectangle Method.

Problem : Draw a parabolic arc with a span of 1000 mm and a rise of 800 mm. Use rectangular method. Draw a tangent and normal at any point P on the curve.

***Solution :* (Fig. 4.44)**

1. Draw an enclosing rectangle ABCD with base AB = 1000 mm and height BC = 800 mm using a suitable scale.
2. Mark the axis VH of the parabola, where V is the vertex and mid point of line CD. Divide DV and AD into the same number of equal parts (say 4).
3. Draw a vertical line through the point 1' lying on the line DV. Join V with 1 lying on the line AD. These two lines intersect at point P_1 as shown in Fig. 4.44.

4. Similarly obtain other points P_2, P_3, etc.
5. Draw a smooth curve passing through these points to obtain the required parabola.
6. Tangent at any point P on the parabola can be drawn as follows.

From point P draw the ordinate PE. With V as centre and VE as radius draw a semicircle to cut the axis produced at G. Join GP and extend it to T. Draw NP perpendicular to TP. Now, TPT and NPN are the required tangent and normal at P.

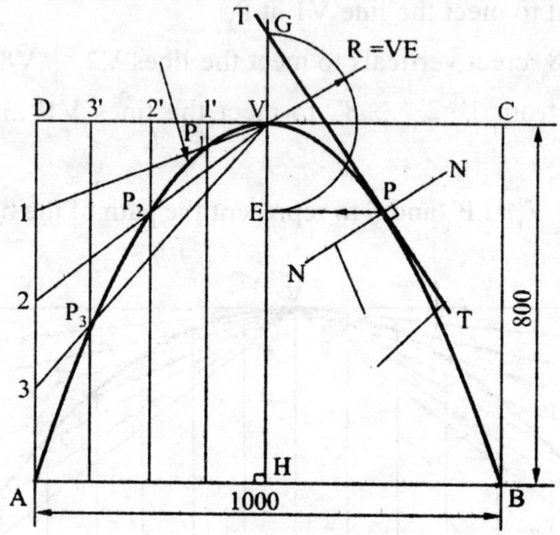

Fig. 4.44

Problem : Construct a parabola within a parallelogram of sides 120 mm × 60 mm. One of the included angle between the sides is 75°.

***Solution :* (Fig. 4.45)**

1. Construct the parallelogram PQRS (PS = 120 mm, PQ = 60 mm and angle QPS = 75°). Bisect PS at O and draw VO parallel to PQ.
2. Divide PQ and SR into any number of (4) equal parts as 1, 2, 3 and 1', 2', 3' respectively starting from P on PQ and from S on SR. Join V1, V2 & V3. Also join V1', V2', V3'.
3. Divide PO and OS into 4 equal parts as 1_1, 2_1, 3_1 and $1'_1$, $2'_1$, $3'_1$ respectively starting from P on PO and from S on SO.
4. From 1_1 draw a line parallel to PQ to meet the line V1 at P_1. Similarly obtain the points P_2 and P_3.
5. Also from $1'_1$, $2'_1$, $3'_1$ draw lines parallel to RS to meet the lines V1', V2', and V3' at P_1', P_2', and P_3' respectively and draw a smooth parabola

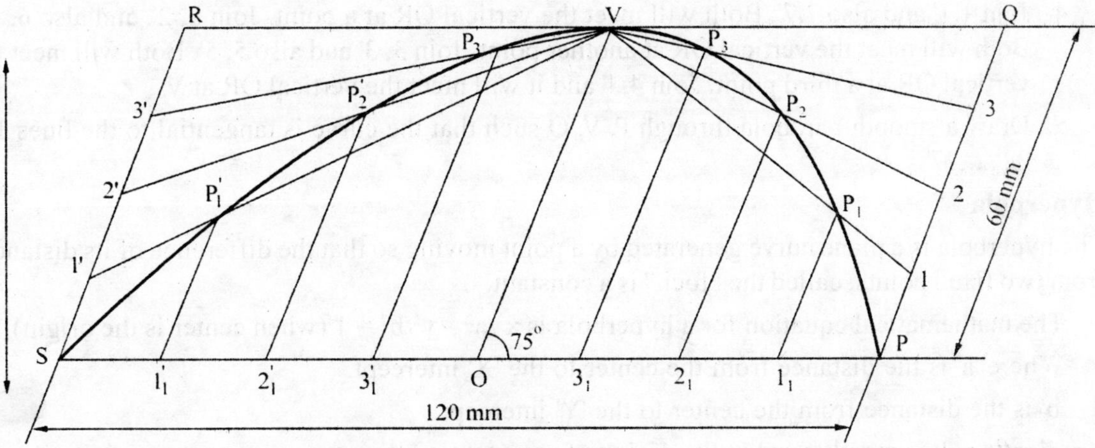

Fig. 4.45

Problem : A fountain jet discharges water from ground level at an inclination of 55° to the ground. The jet travels a horizontal distance of 10 m from the point of discharge and falls on the ground. Trace the path of the jet.

Solution : **(Fig. 4.46)**

1. Taking the scale as 1:100 draw PQ = 10 cm. Jet discharges water at 55° to the ground. So, at P and Q draw 55° lines to intersect at R. PQR is an isosceless triangle.
2. Bisect PQ at O. At O, erect vertical to pass through R. Bisect OR at V, the vertex.
3. Divide PR into any number of (say 8) equal parts as 1, 2, ...7 starting from P on PR. Divide RQ into same number of (8) equal parts as 1', 2' .... 7' starting from R on RQ.

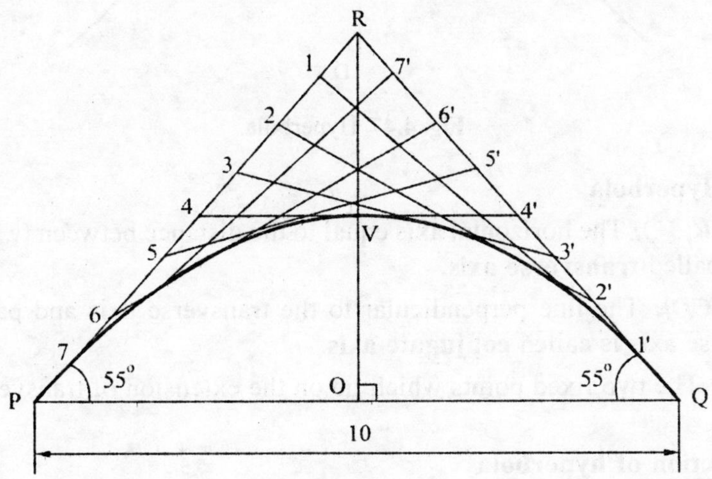

Fig. 4.46

4. Join 1,1' and also 7,7'. Both will meet the vertical OR at a point. Join 2, 2', and also 6, 6',. Both will meet the vertical OR at another point. Join 3, 3' and also 5, 5'. Both will meet the vertical OR at a third point. Join 4,4' and it will meet the vertical OR at V.
5. Draw a smooth parabola through P, V, Q such that the curve is tangential to the lines 11', 22',....77'.

Hyperbola

The hyperbola is a plane curve generated by a point moving so that the difference of its distances from two fixed points, called the "foci," is a constant.

The mathematical equation for a hyperbola is $x^2/a^2 - y^2/b^2 = 1$ (when center is the origin)

Where 'a' is the distance from the center to the 'X' intercept.

'b' is the distance from the center to the 'Y' intercept.

Application: It is mostly used in the design of channels, cooling towers.

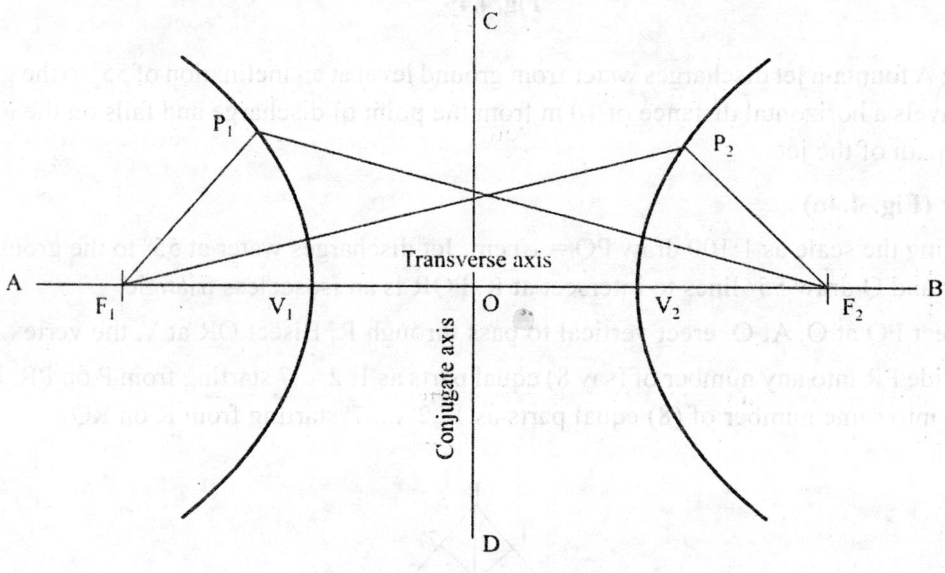

Fig. 4.47 Hyperbola.

Terms used in Hyperbola

Transverse axis ($V_1 V_2$): The horizontal axis equal to the distance between two vertices of a pair of hyperbolas is called **transverse axis.**

Conjugate axis (CD): The line perpendicular to the transverse axis and passing through the centre of transverse axis is called **conjugate axis.**

Foci (F_1 and F_2): The two fixed points which lie on the extension of transverse axis are called **foci.**

General construction of hyperbola

General construction of hyperbola by eccentricity method is given in Fig. 4.25.

A hyperbola is a curve generated by a point moving such that the difference of its distances from two fixed points called foci is always constant and equal to the distance between the vertices of the two branches of hyperbola. This distance is also known as the major axis of the hyperbola.

Refering Fig. 4.48, the difference between $P_1F_1 \sim P_1F_2 = P_2F_2 \sim P_2F_1 = V_1V_2$ (major axis)

The axes AB and CD are known as transverse and conjugate axes of the hyperbola. The curve has two branches which are symmetric about the conjugate axis.

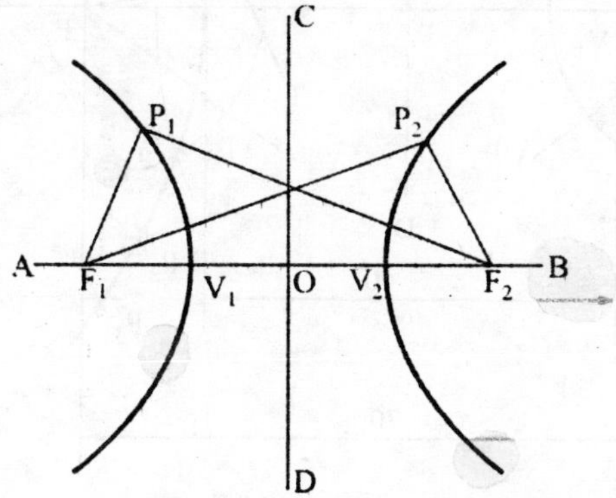

Fig. 4.48 Properties of Hyperbola.

Other methods

Method of constructing a hyperbola, given the foci and the distance between the vertices. (Fig. 4.48)

Problem : *Construct a hyperbola with its foci 70 mm apart and the mojor axis (distance between the vertices)as 40 mm. Draw a tangent to the curve at a point 20 mm from the focus.*

Construction (Fig. 4.49)

1. Draw the transverse and conjugate axes AB and CD of the hyperbola and locate F_1 and F_2, the foci and V_1 and V_2, the vertices.
2. Mark number of points 1,2,3 etc., on the transverse axis, which need not be equi-distant.
3. With centre F_1 and radius V_11, draw arcs on either side of the transverse axis.
4. With centre F_2 and radius V_21, draw arcs intersecting the above arcs at P_1, and P_1'.
5. With centre F_2 and radius V_11, draw arcs on either side of the transverse axis.
6. With centre F_1 and radius V_21, draw arcs intersecting the above arcs at Q_1, Q'_1.
7. Repeat the steps 3 to 6 and obtain other points P_2, P^1_2, etc. and Q_2, Q'_2, etc.
8. Join the points $P_1, P_2, P_3, P_1', P_2', P_3'$ and $Q_1, Q_2, Q_3, Q_1', Q_2', Q_3'$ forming the two branches of hyperbola.

Note : To draw a tangent to the hyperbola, locate the point M which is at 20 mm from the focus say F_2. Then, join M to the foci F_1 and F_2. Draw a line TT, bisecting the $<F_1MF_2$ forming the required tangent at M.

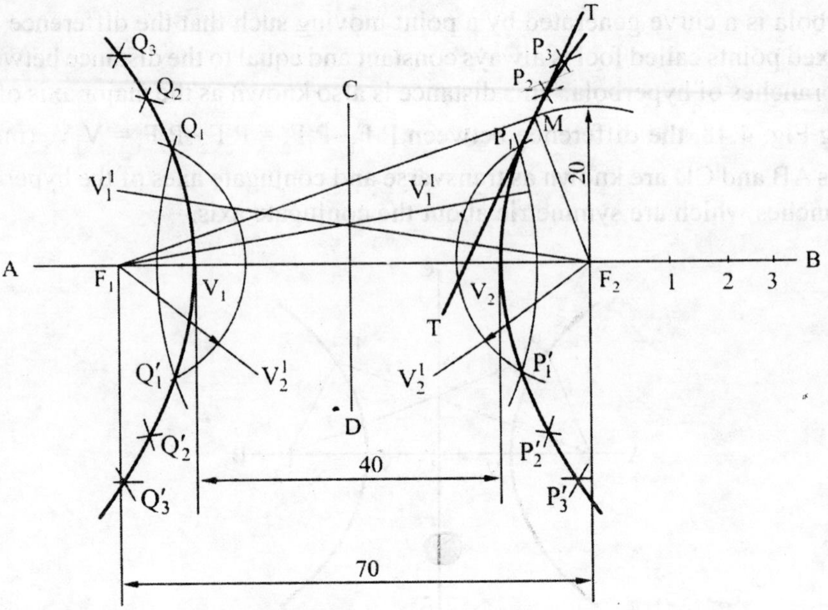

Fig. 4.49 Construction of a Hyperbola.

To draw the asymptotes to the given hyperbola

Lines passing through the centre and tangential to the curve at infinity are known as asymptotes.

Construction (Fig. 4.50)

1. Through the vertices V_1 and V_2 draw perpendiculars to the transverse axis.

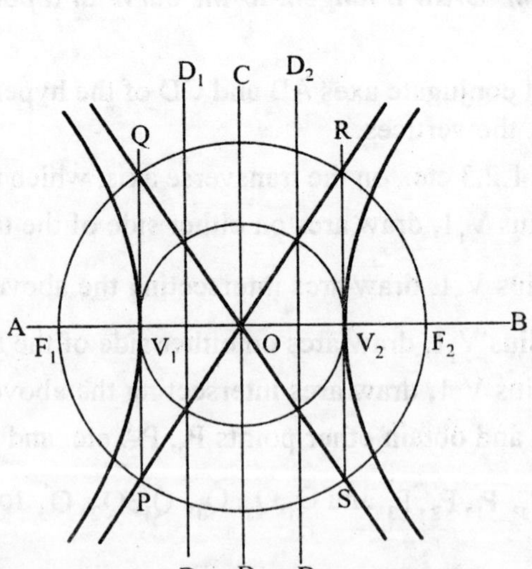

Fig. 4.50 Drawing Asymptotes to a Hyperbola.

2. With centre O and radius $OF_1 = (OF_2)$, draw a circle meeting the above lines at P, Q and R,S.
3. Join the points P,O,R and S,O,Q and extend, forming the asymptotes to the hyperbola.

Note : The circle drawn with O as centre and V_1 V_2 as diameters is known as auxiliary circle. Asymptotes intersect the auxiliary circle on the directrix. Thus, D_1 D_1 and D_2 D_2 are the two directrices for the two branches of hyperbola.

Rectangular Hyperbola

When the asymptotes to the hyperbola intersect each other at right angles, the curve is known as a rectangular hyperbola.

Problem : Two points F_1 and F_2 are located on a plane sheet 100 mm apart. A point P on the curve moves such that the difference of its distances from F_1 and F_2 always remains 50 mm. Find the locus of the point and name the curve. Mark asymptotes and directrices.

Solution : **(Fig. 4.51)**

1. A curve traced out by a point moving in the same plane in such a way that the difference of the distances from two fixed points is constant, is called a hyperbola.
2. Draw a horizontal line and mark the fixed points F_2 and F_1 in such a way that F_2F_1 = 100 mm. Draw a perpendicular bisector C_1OC_2 to F_2F_1 as shown in Fig. 4.51.
3. Mark the points V_2 and V_1 on the horizontal line such that V_2V_1 = 50 mm and $V_2O = V_1O$.
4. With centre O and radius equal to F_2O draw a circle. Draw tangents at V_2 and V_1 to intersect the above circle at J, M, K and L as shown. Draw a line joining JOL and produce it and this line is one asymptote.
5. The other asymptote is the line passingt through KOM.

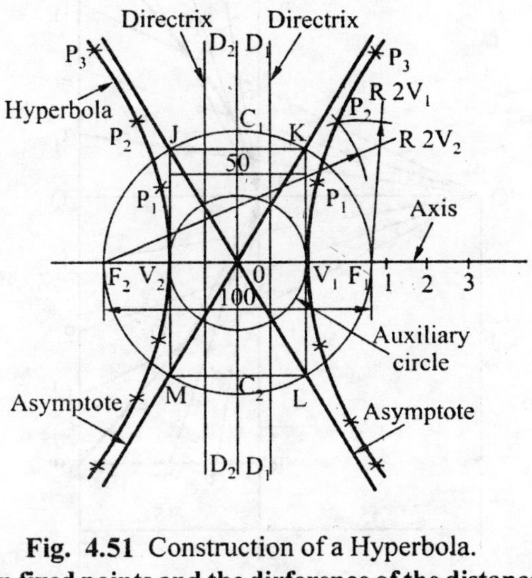

Fig. 4.51 Construction of a Hyperbola.
(given fixed points and the dirference of the distances)

6. Mark any number of points 1, 2, 3, etc., on the axis of the hyperbola. With F_2 as centre and radius equal to $2V_2$ draw an arc to cut the arc drawn with F_1 as centre and radius equal to $2V_1$. The point of intersection is marked as P_2. Similarly obtain other points of intersection $P_1 P_3 P_4$, etc. It may be noted that $P_2F_2 - P_2F_1 = P_3F_2 - P_3F_1 = 50$ mm. Draw a smooth curve passing through the points V, $P_1 P_2 P_3$, etc., which is the required hyperbola. Also draw another hyperbola on the other side of the axis as shown.

Problem : Draw a hyperbola when its double ordinate is 90 mm, abscissa is 35 mm and half the transverse axis is 45 mm.

Solution: **(Fig. 4.52)**

1. Draw OBQ such that OB = half the transverse axis = 45 mm and BQ = abscissa = 35 mm. Through Q erect vertical such that PP' = double ordinate = 90mm = 2PQ.
2. Construct the rectangle PP' R'R. Divide PR and P'R' into any number of equal parts (say 4) as 1, 2, 3, and 1' 2' 3' starting from P on PR and P' on P' R' respectively. Join B1, B2, B3, B1', B2' and B3'.
3. Divide the ordinates PQ and QP' into the same number of equal parts as 1_1 2_1 3_1 and $1_1'$ $2_1'$ $3_1'$ starting from P on PQ and P' on P'Q respectively.
4. Join 01_1 to meet B1 at P_1. Join 02_1 and 03_1 to meet B2 and B3 at P_2 and P_3 respectively, Similarly join $01'_1$, $02'_1$ and $03'_1$ to meet B1' B2' B3' at P'_1 P'_2 P'_3 respectively.
5. Join P, P_1, P_2, P_3, B, P'_3, P'_2, P'_1 and P' by a smooth hyperbola.

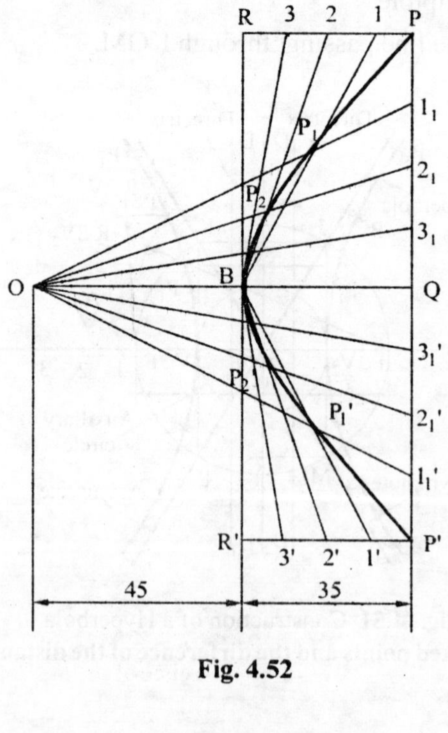

Fig. 4.52

Problem : Construct a rectangular hyperbola when a point P on it is at a distance of 30 mm and 40 mm resepctively from the two asymptotes.

Solution: **(Fig. 4.53)**

1. For a rectangular hyperbola, angle between the asymptotes is 90°. So, draw OR_1 and OR_2 such that the angle R_1OR_2 is 90°.
2. Mark A and B along OR_2 and OR_1 respectively such that OA = 40 mm and OB = 30 mm. From A draw AX parallel to OR_1 and from B draw BY parallel to OR_2. Both intersect at P.
3. Along BP mark 1, 2, and 3 at approximately equal intervals. Join O1, O2, and O3, and extend them to meet AX at 1_1, 2_1 and 3_1 respectively.
4. From 1_1 draw a line parallel to OR_2 and from 1 draw a line parallel to OR_1. From 2 and 3 draw lines parallel to OR_1. They intersect at P_2 and P_3 respectively.
5. Then along PA mark points 4_1 and 5_1 at approximately equal inervals. Join $O4_1$ and $O5_1$ and extend them to meet BY at 4 and 5 respectively.
6. From 4_1 and 5_1 draw lines parallel to OR_2 and from 4 and 5 draw lines parallel to OR_1 to intersect at P_4 and P_5 respectively
7. Join P_1, P_2, P_3, P, P_4, P_5 by smooth rectangular hyperbola.

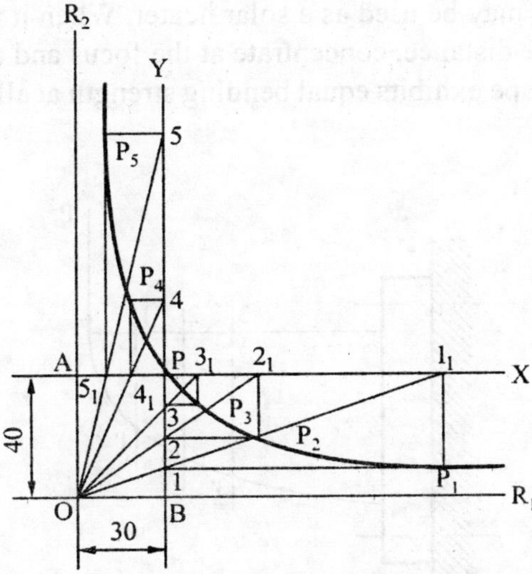

Fig. 4.53 Rectangular Hyperbola.

Application of Conic Curves

An ellipsoid is generated by rotating an ellipse about its major axis. An ellipsoidal surface is used as a head-lamp reflector. The light source (bulb) is placed at the first focus F_1 (Fig. 4.54). This works effectively, if the second focus F_2 is at a sufficient distance from the first focus. Thus, the light rays reflecting from the surface are almost parallel to each other.

Parabolic Curve

The parabolic curve finds its application for reflecting surfaces of light, Arch forms, cable forms in suspension bridges, wall brackets of uniform strength, etc.

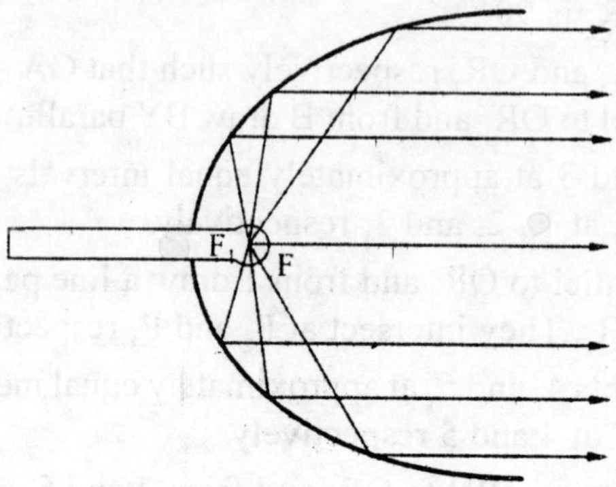

Fig. 4.54 Ellipsoidal Reflector.

The paraboloid reflector may be used as a solar heater. When it is properly adjusted, the sun rays emanating from infinite distance, concentrate at the focus and thus produce more heat. The wall bracket of parabolic shape exhibits equal bending strength at all sections (Fig. 4.55).

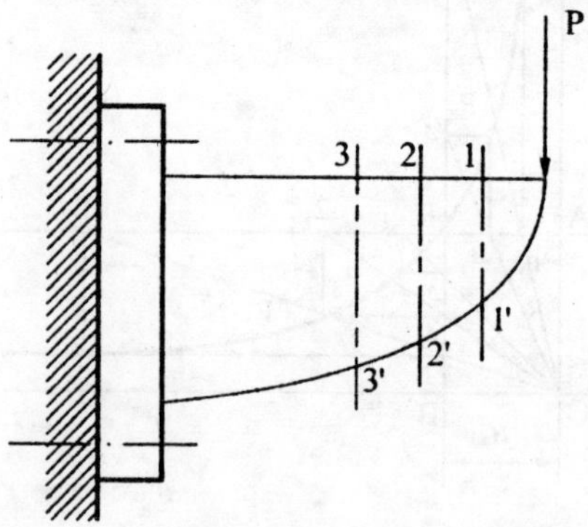

Fig. 4.55 Wall Bracket of Uniform Strength.

Hyperbola

A rectangular hyperbola is a graphical representation of Boyles law, PV=Constant. This curve also finds its application in the design of water channels.

4.3 Special Curves

Cycloidal Curves

Cycloidal curves are generated by a fixed point in the circumference of a circle when it rolls without slipping along a fixed straight line or circular path. The rolling circle is called the generating circle, the fixed straight line, the directing line and the fixed circle, the directing circle.

4.3.1 Cycloid

A cycloid is a curve generated by a fixed point on the circumference of a circle, when it rolls without slipping along a straight line.

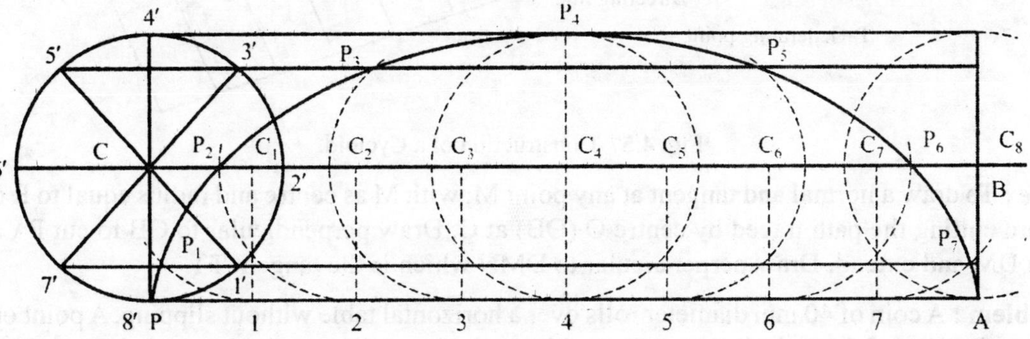

Fig. 4.56 Cycloid Formation.

To draw a cycloid, given the radius R of the generating circle. (Fig. 4.57).

1. With centre O and radius R, draw the given generating circle.
2. Assuming point P to be the initial position of the generating point, draw a line PA, tangential and equal to the circumferance of the circle.
3. Divide the line PA and the circle into the same number of equal parts and number the points.
4. Draw the line OB, parallel and equal to PA. OB is the locus of the centre of the generating circle.
5. Errect perpendiculars at 1',2',3', etc., meeting OB at O_1, O_2, O_3, etc.
6. Through the points 1,2,3 etc., draw lines parallel to PA.
7. With centre O_1 and radius R, draw an arc intersecting the line through 1 at P_1, P_1 is the position of the generating point, when the centre of the generating circle moves to O_1.
8. Similarly locate the points P_2, P_3 etc.
9. A smooth curve passing through the points P,P_1, P_2,P_3 etc., is the required cycloid.

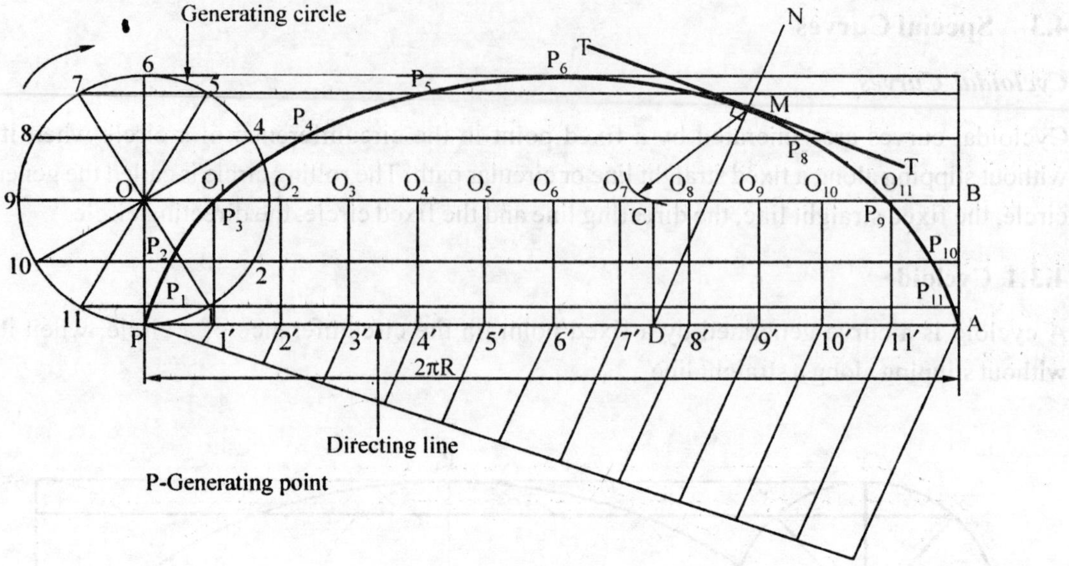

Fig. 4.57 Construction of a Cycloid.

Note : To draw a normal and tangent at any point M; with M as centre and radius equal to R draw an arc cutting the path traced by centre O (OB) at C. Draw perpendicular to OB to cut PA at D. Join DM and extend. Draw perpendicular to DMN which is the tangent TT.

Problem : A coin of 40 mm diameter rolls over a horizontal table without slipping. A point on the circumference of the coin is in contact with the table surface in the beginning and after one complete revolution. Draw the cycloidal path traced by the point. Draw a tangent and normal at any point on the curve.

Solution: (Fig. 4.58)

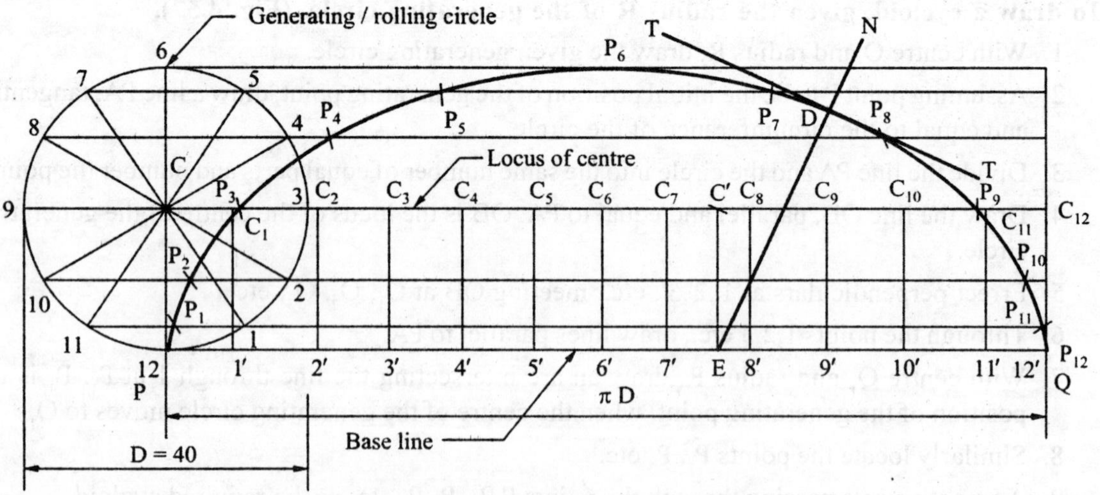

Fig. 4.58 Costruction of a cycloid.

1. Draw a generating circle of diameter 40 mm. Draw the base line **PQ** equal to the circumference of the generating circle from **P** as shown.
2. Divide the generation circle into **12** equal parts as **1, 2, 3**... etc., as shown and draw the horizontal lines from the points **1, 2, 3** ... etc.
3. Divide the base line **PQ** into the same number of equal parts and name the points as **1', 2', 3'** ... etc.
4. Draw the perpendicular lines from the points **1', 2', 3'** ... etc to which will intersect the horizontal lines drawn from the point **C** (locus of the centre) at $\mathbf{C_1, C_2, C_3}$... etc respectively.
5. Taking $\mathbf{C_1, C_2, C_3}$... etc as centres and radius equal to radius of generating circle (20 mm), draw arcs to cut the horizontal lines through **1, 2, 3...** etc. at $\mathbf{P_1, P_2, P_3}$... etc.
6. Now draw a smooth curve through $\mathbf{P, P_1, P_2, P_3}$... etc. The curve obtained is the cycliod.

Construction of normal and tangent at a given point D

Taking **D** as centre and radius equal to radius of the generaing circle (20 mm), cut the line of locus of centre at **C'**.

From **C'**, draw a perpendicular line to PQ to get the point E on the base line. Join the points **D** and **E**, which is the **normal** for the given point.

To get the required tangent **(TT)**, draw a perpendicular line from the point **D** to the line **DE**.

4.3.2 Epi-cycloid

If the generating point P is on the circumference of a circle that rolls along the **convex side** of a larger circle, the curve generated is an epi-cycloid.

Application of epi-cycloid

(i) It is used in the design of profiles of gear tooth system.

(ii) Epicycloid curves are also used in some mechanisms.

Practical application of epi-cycloid.

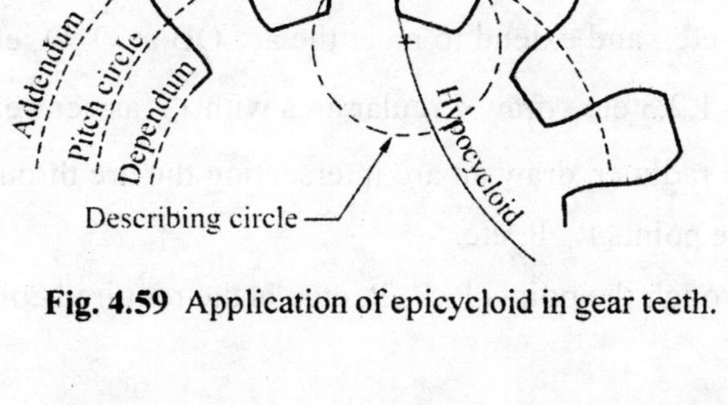

Fig. 4.59 Application of epicycloid in gear teeth.

Construction of Epi-cycloid (Fig. 4.60)

These curves are drawn in a manner similar to the cycloids.

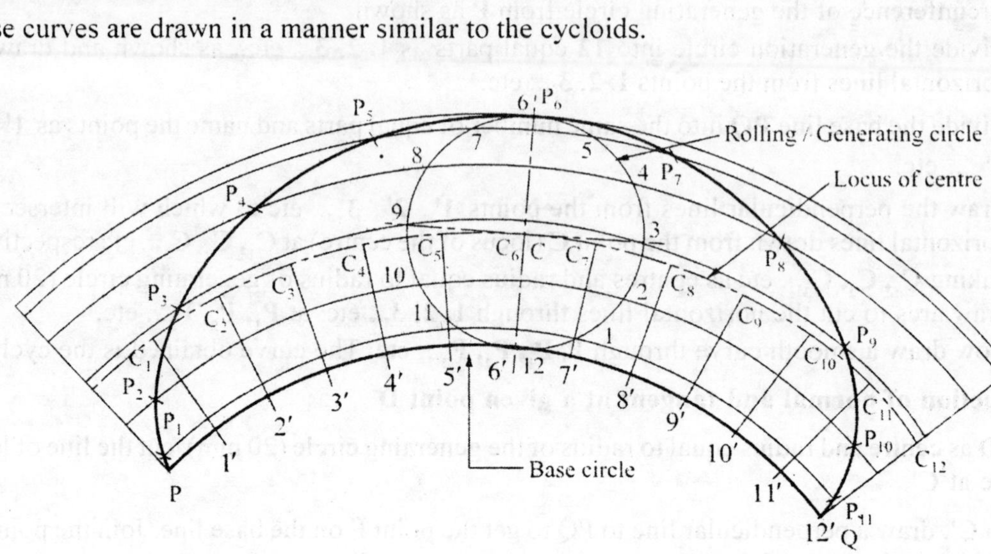

Fig. 4.60

To draw an epi-cyloid, given the radius 'r' of the generating circle and the radious 'R' of the directing circle. (Fig. 4.61)

1. With centre O' and radius R, draw a part of the directing circle.
2. Draw the generating circle, by locating the centre O of it, on any radial line O' P extended such that OP = r.
3. Assuming P to be the generating point, locate the point, A on the directing circle such that the arc length PA is equal to the circumference of the generating circle. The angle subtended by the arc PA at O' is given by $\theta = <PO'A = 3600 \times r/R$.
4. With centre O' and radius O'O, draw an arc intersecting the line O'A produced at B. The arc OB is the locus of the centre of the generating circle.
5. Divide the arc PA and the generating circle into the same number of equal parts and number the points.
6. Join O'-1', O'-2', etc., and extend to meet the arc OB at O_1,O_2 etc.
7. Through the points 1,2,3 etc., draw circular arcs with O' as centre.
8. With centre O_1 and radius r, draw an arc intersecting the arc through 1 at P_1.
9. Similarly, locate the points P_2, P_3 etc.
10. A smooth curve through the points P_1,P_2,P_3 etc., is the required epi-cycloid.

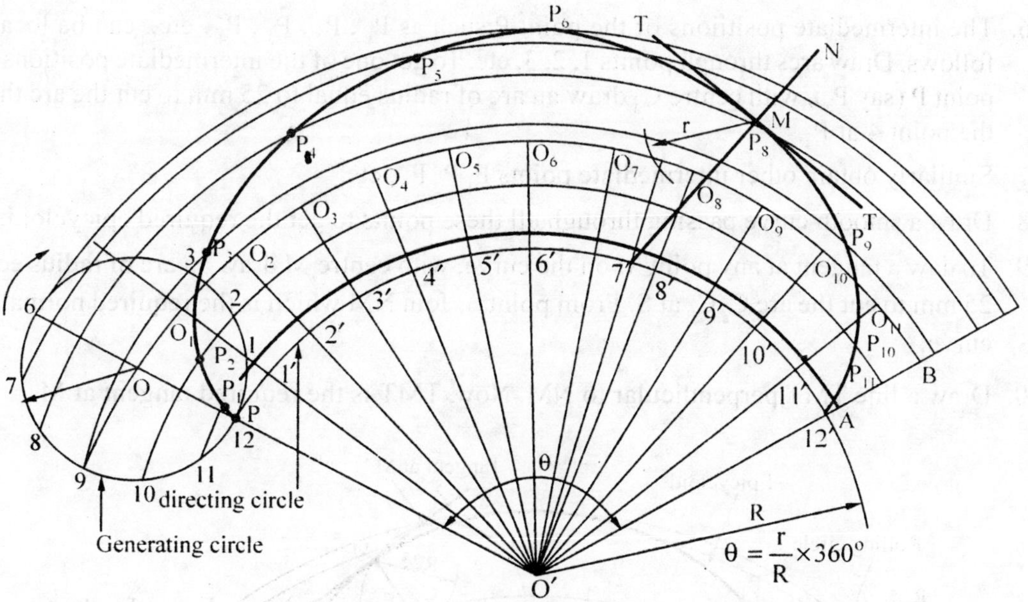

Fig. 4.61 Construction of an Epi-Cycloid.

Problem : Draw an epicycloid having a generating circle of diameter 50 mm and a directing curve of radius 100 mm. Also draw a normal and a tangent at any point M on the curve.

Solution : **(Fig. 4.62)**

1. Let, AB be the circumference of the generating circle of radius, r = 25 mm. Let, θ be the angle subtended at the centre of the directing (base) circle of radius = 100 mm by the arc AB. Then,

   (Angle AOB)/360° = (Arc, AB/(Circumference of directing circle)

   i.e. $\theta/360 = (2\pi r)/(2\pi R)$

   $= (2\pi \times 25)/2\pi \times 100)$

   ∴ $\theta = (25 \times 360°)/100$

   $= 90°$

2. Draw the arc AB with centre O and radius = 100 mm in such a way that the angle AOB = 90°. Join OA and extend it to C such that AC is equal to the radius of the rolling circle.
3. With centre C_2 and radius = 25 mm draw the rolling circle. Draw an arc C_aC_b with centre O and radius = OC_a. Here, C_aC_b represents the locus of the centre of the rolling circle.
4. Divide the rolling circle into any number of equal parts (say 12). Also divide the arc C_aC_b into the same number of equal parts and mark the points as C_1, C_2, C_3, etc., as shown in Fig. 4.62.
5. The required curve (epicycloid) is the path of the point P on the circumference of the circle which rolls over C_a C_b. Let P_o be the initial position of the point P and it coincides with the point A. When the rolling circle rolls once on arc AB, the point P will coincide with B and it is marked by P_o.

6. The intermediate positiions of the point P such as P_1, P_2, P_3, P_4, etc., can be located as follows. Draw arcs through points 1, 2, 3, etc. To get one of the intermediate positions of the point P (say P_4), with centre C_4 draw an arc of radius equal to 25 mm to cut the arc through the point 4 at P_4.
7. Similarly obtain other intermediate points P_1 P_2 P_3, etc.
8. Draw a smooth curve passing through all these points to get the required epicycloid.
9. To daw a tangent at any point M on the curve, with centre M draw an arc of radius equal to 25 mm to cut the arc C_a C_b at S. From point S, Join NM which is the required normal to the curve.
10. Draw a line TMT perpendicular to NM. Now, TMT is the required tangent at M.

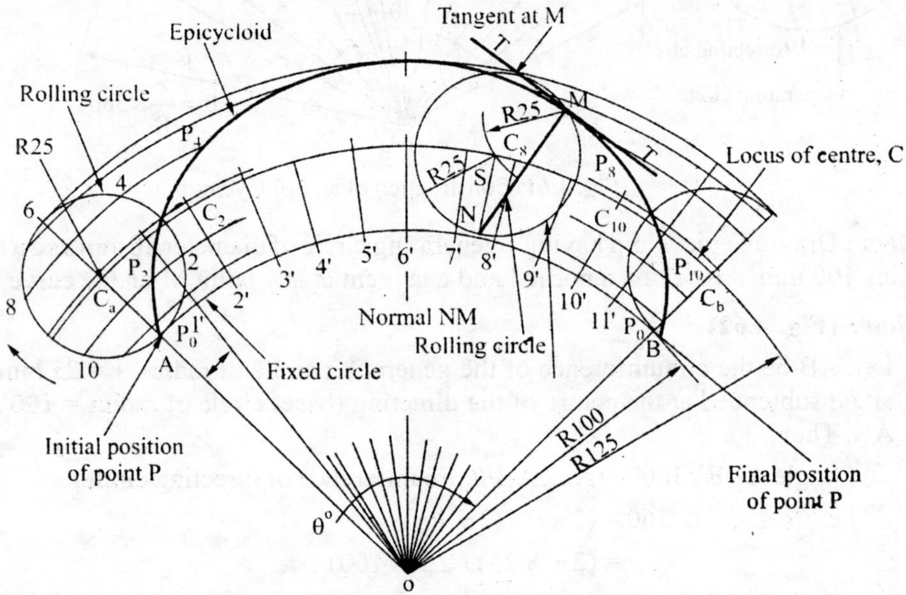

Fig. 4.62 Epicycloid.

Problem : Draw an epicycloid of rolling circle of diameter 40 mm which rolls outside another circle (base circle) of 150 mm diameter for one revolution. Draw a tangent and normal at any point an the curve.

Solution : **(Fig. 4.63)**

1. In one revolution of the generating circle, the generatin point P will move to a point Q, so that the arc PQ is equal to the circumference of the generating circle. θ is the angle subtended by the arc PQ at the centre O.

$$\text{To calculate } \theta: \frac{P\hat{O}Q}{360^\circ} = \frac{\text{Arc PQ}}{\text{circumference of directing circle}} = \frac{2\pi\ r}{2\pi\ R} = \frac{r}{R}$$

$$\therefore \hat{POQ} = \theta \times 360^{o} = \frac{r}{R} \times 360^{o} = \frac{20}{75} \times 360 = 96^{o}$$

2. Taking any pont O as centre and radius (R) 75 mm, draw an arc PQ which subtends an angle $\theta = 96^{0}$ at O.
3. Let P be the generating point. On OP produced, mark PC = r = 20 mm = radius of the rolling circle. Taking centre C and radius r (20 mm) draw the rolling circle.
4. Divide the rolling circle into 12 equal prats and name them as 1, 2, 3, etc., in the counter clock wise direciton, since the rolling circle is assumed to roll clockwise.
5. With O as centre, draw concentric arcs passing through 1, 2, 3, ... etc.
6. With O as centre and OC as radius draw an arc to represent the locus of centre.
7. Divide the arc PQ into same number of equal parts (12) and name them as 1' 2' .. etc.
8. Join O1' , O2'.... etc., and produce them to cut the locus of centre at C_1, C_2 ... etc.
9. Taking C_1 as centre and radius equal to r, draw an arc cutting the arc through 1 at P_1. Similarly obtain the other points and draw a smooth curve through them.

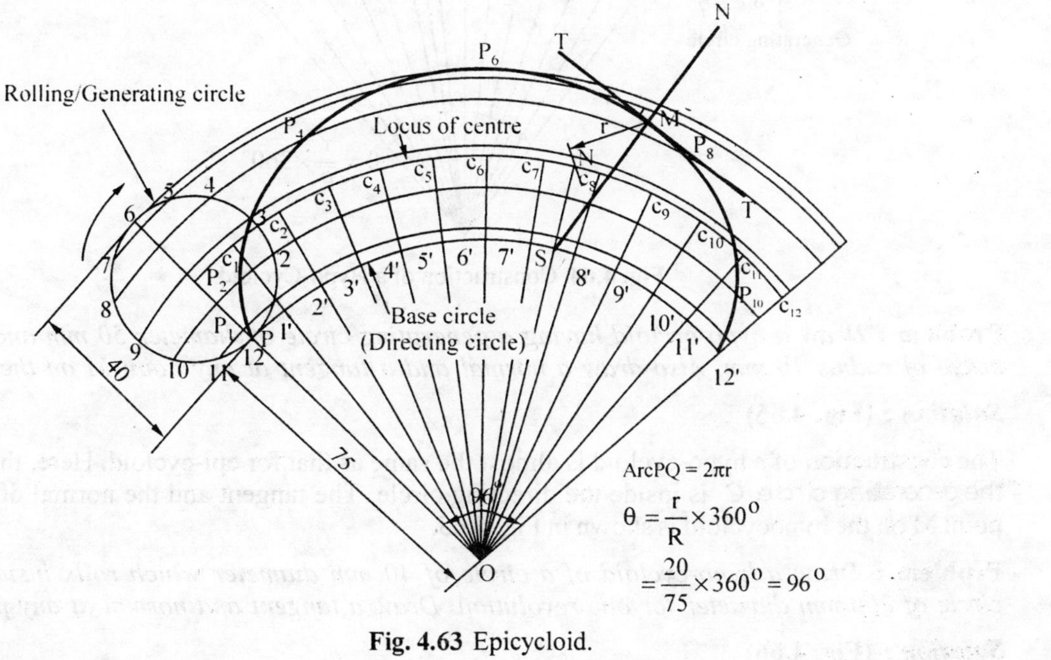

Fig. 4.63 Epicycloid.

4.3.3 Hypo-cycloid

If the generating point P is on the circumference of a circle that rolls along the **concave side** of a larger circle, the curve generated is a **hypo-cycloid**.

Application of hypo-cycloid

(i) It is used in the design of profiles of gear tooth system

(ii) It is also used in the design of flat disk cam profile for metal cutting machine tools.

Construction of hypo-cycloid

Note 1 : The above procedure is to be followed to construct a hypo-cycloid with the generating circle rolling inside the directing circle (Fig. 4.64).

Note 2 : T-T is the tangent and NM is the normal to the curve at the point M.

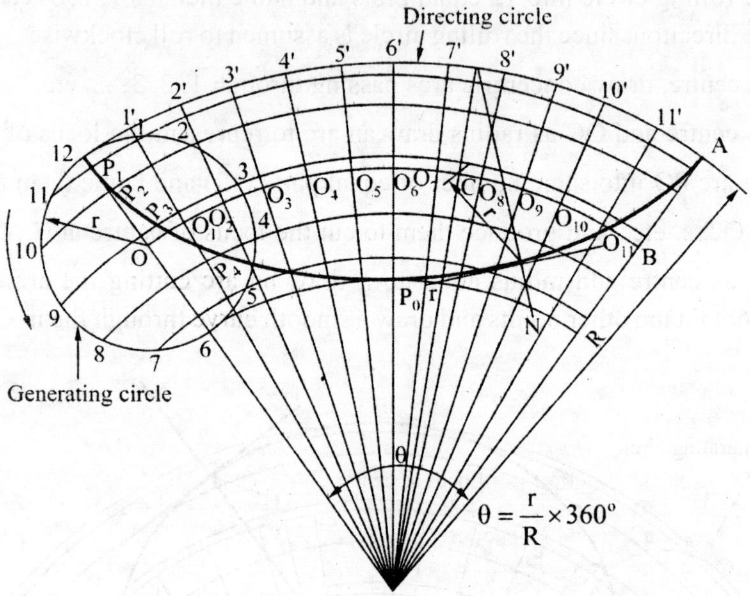

Fig. 4.64 Construction of a Hypo-Cycloid.

Problem : *Draw a hypo-cycloid having a generating circle of diameter 50 mm and directing circle of radius 10 mm. Also draw a normal and a tangent at any point M on the curve.*

***Solution :* (Fig. 4.65)**

The construction of a hypo-cycloid is almost the same as that for epi-cycloid. Here, the centre of the generating circle, C_a is inside the directing circle. The tangent and the normal drawn at the point M on the hypocycloid is shown in Fig. 4.65.

Problem : *Draw a hypo-cycloid of a circle of 40 mm diameter which rolls inside another circle of 200 mm diameter for one revolution. Draw a tangent and normal at any point on it.*

***Solution :* (Fig. 4.66)**

1. Taking any point O as centre and radius (R) 100 mm draw an arc PQ which subtends an angle θ = 72° at O.
2. Let P be the generating point. On OP mark PC = r = 20 mm, the radius of the rolling circle.

3. With C as centre and radius r (20 mm) draw the rolling circle. Divide the rolling circle into 12 equal parts as 1, 2, 3 etc., in clock wise direction, since the rolling circle is assumed to roll counter clock wise.
4. With O as centre, draw concentric arcs passing through 1, 2, 3 etc.
5. With O as centre and OC as radius draw an arc to represent the locus of centre.
6. Divide the arc PQ into same number of equal parts (12) as 1' 2' 3' etc.
7. Join O1' O2' etc., which intersect the locus of centre at $C_1C_2C_3$ etc.
8. Taking centre C_1 and radius r, draw an arc cutting the arc through 1 at P_1. Similarly obtain the other points and draw a smooth curve through them.

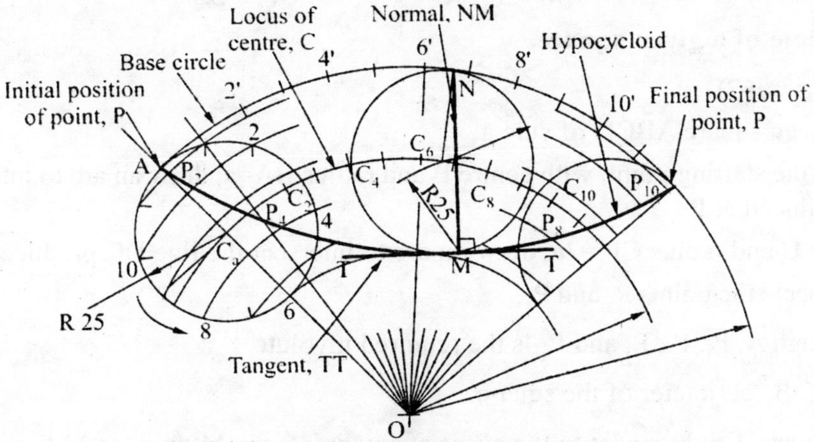

Fig. 4.65 Hypo-cycloid.

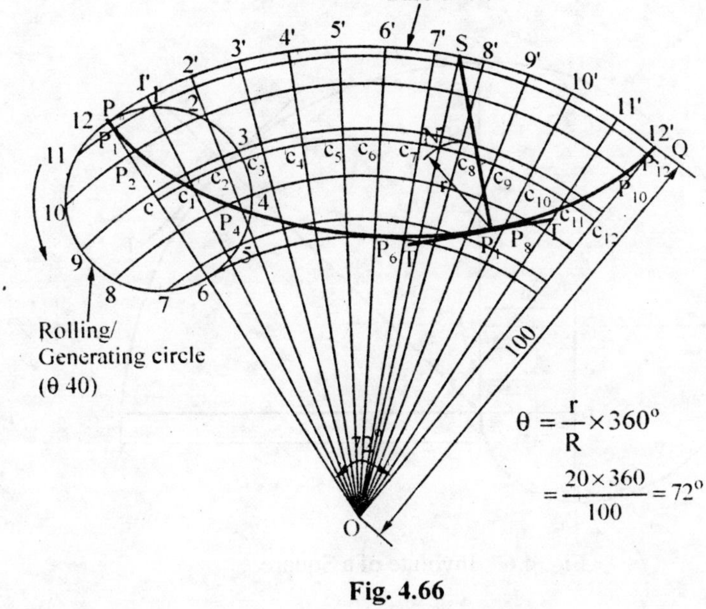

$$\theta = \frac{r}{R} \times 360^\circ$$

$$= \frac{20 \times 360}{100} = 72^\circ$$

Fig. 4.66

To draw a tangent and normal at a given point M:

1. With M as centre and radius r = CP cut the locus of centre at the point N.
2. Join ON and extend it to intersect the base circle at S.
3. Join MS, the normal.
4. At M, draw a line perpendicular to MS to get the required tangent.

4.4 Involutes

An involute is a curve traced by a point on a perfectly flexible string, while unwinding from around a circle or polygon, the string being kept taut (tight). It is also a curve traced by a point on a straight line while the line is rolling around a circle or polygon without slipping.

To draw an involute of a given square.

Construction (Fig. 4.67)

1. Draw the given square ABCD of side a.
2. Taking A as the starting point, with centre B and radius BA=a, draw an arc to intersect the line CB produced at P_1.
3. With Centre C and radius CP_1=2a, draw on arc to intersect the line DC produced at P_2.
4. Similarly, locate the points P_3 and P_4.

The curve through A, P_1, P_2, P_3 and P_4 is the required involute.

A P_4 is equal to the perimeter of the square.

Note : To draw a normal and tangent to the curve at any point, say M on it, as M lies on the arc P_3P_4 with its centre at A, the line AMN is the normal and the line TT drawn through M and perpendicular to MA is the tangent to the curve.

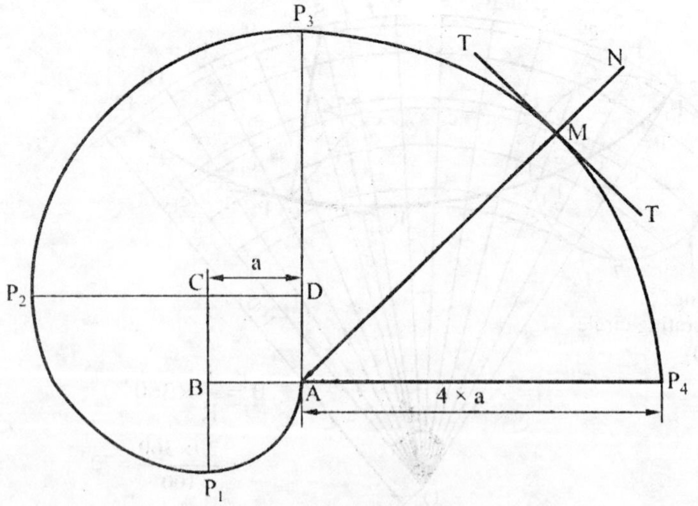

Fig. 4.67 Involute of a Square.

Involutes of a Triangle, Pentagon and Hexagon are shown in Figs. 4.68 to 4.70.

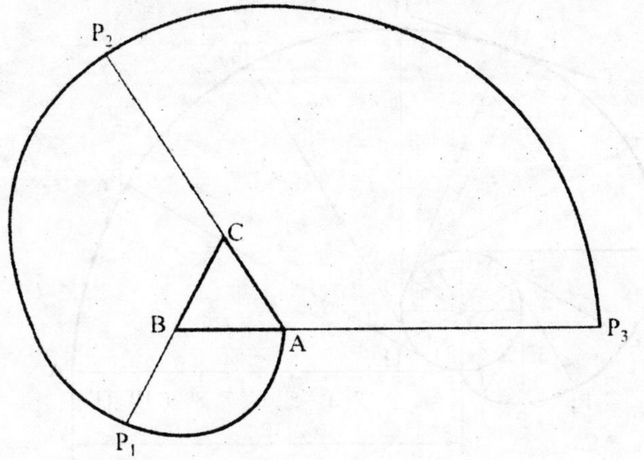

Fig. 4.68 Involute of a Triangle.

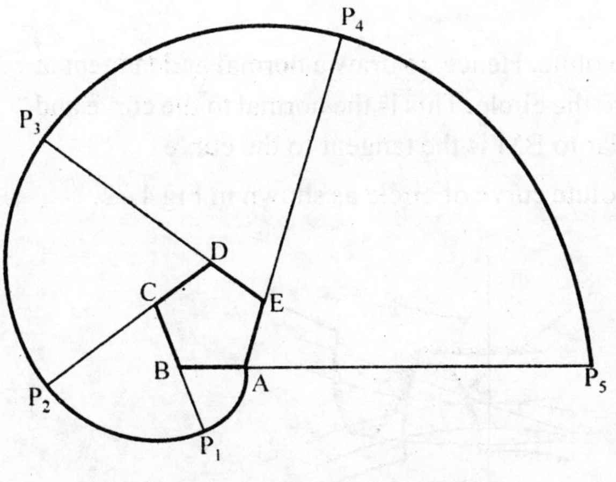

Fig. 4.69 Involute of a Pentagon.

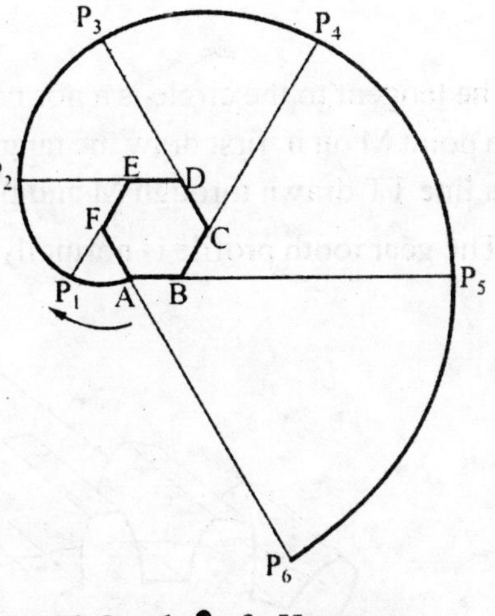

Fig. 4.70 Involute of a Hexagon.

To draw an involute of a given circle of radus R.

Construction (Fig. 4.71)

1. With O as centre and radius R, draw the given circle.
2. Taking P as the starting point, draw the tangent PA equal in length to the circumference of the circle.
3. Divide the line PA and the circle into the same number of equal parts and number the points.
4. Draw tangents to the circle at the points 1,2,3 etc., and locate the points P_1, P_2, P_3 etc., such that $1\ P_1 = P_1'$, $2\ P_2 = P_2'$ etc.

   A smooth curve through the points P, P_1, P_2 etc., is the required involute.

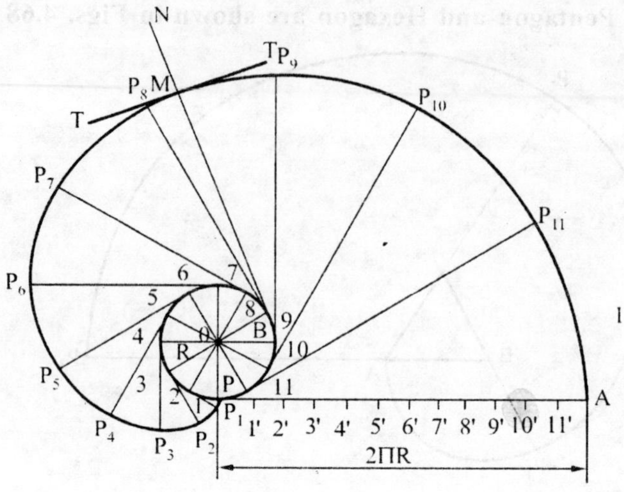

Fig. 4.71

Note :

1. The tangent to the circle is a normal to the involute. Hence, to draw a normal and tangent at a point M on it, first draw the tangent BMN to the circle. This is the normal to the curve and a line TT drawn through M and perpendicular to BM is the tangent to the curve.
2. The gear tooth profile is normally of the involute curve of circle as shown in Fig 4.72.

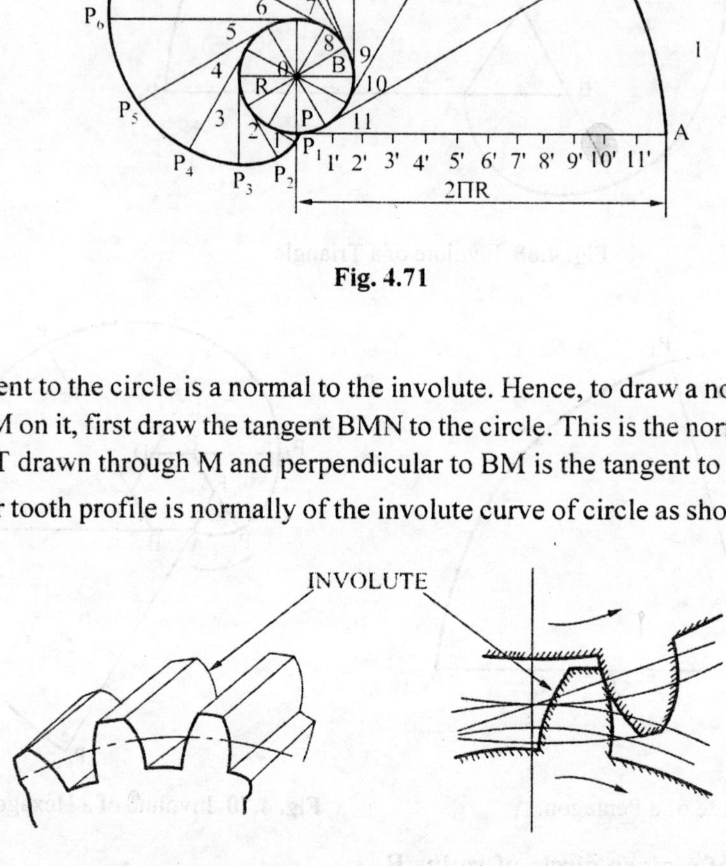

Fig. 4.72 Involute of a Circle as Gear Tooth Profile.

4.5 Helix

The helix is the curve generated by a point that revolves uniformly around and up or down the surface of a cylinder.

Types of helixes

(i) Cylindrical helix (Helix) (ii) Conical helix

(i) ***Cylindrical helix :*** It is defined as a curve generated by a point moving around and along the surface of a right circular cylinder.

(ii) ***Conical helix :*** A cone with a uniform angular velocity about the axis and with a uniform linear velocity in the direction of the axis.

Lead or pitch : It is the axial distance moved by the generating point in one revolution.

Applications of Helix

(i) It is used in the design of screw threads.

(ii) It is also used in the design of springs, spiral stair cases, conveyors, etc.

4.5.1 Construction of Cylinderical Helix

1. Given the diameter of the cylinder and the lead draw the top and front views.
2. Divide the circumference (top view) into a convenient, number of parts (normally divided into 12) and lable them.
3. Project lines down to the front view
4. Divide the lead into the same number of equal parts and label them as shown in Fig. 4.73.
5. The points of intersection of lines with corresponding numbers lie on the helix. Since points 8 to 12 lie on the back portion of the cylinder, the helix curve starting at point 7 and passing through points 8, 9, 10, 11, 12, to the point 1 which appear as a hidden line.

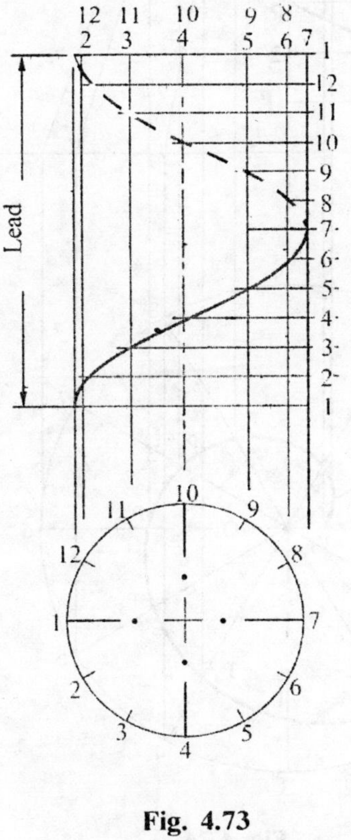

Fig. 4.73

6. If the development of the cylinder is drawn, the helix will appear as a straight line on the development.

4.5.2 Construction of Conical Helix (pitch, diameter, length of the cone are given) (Fig. 4.74)

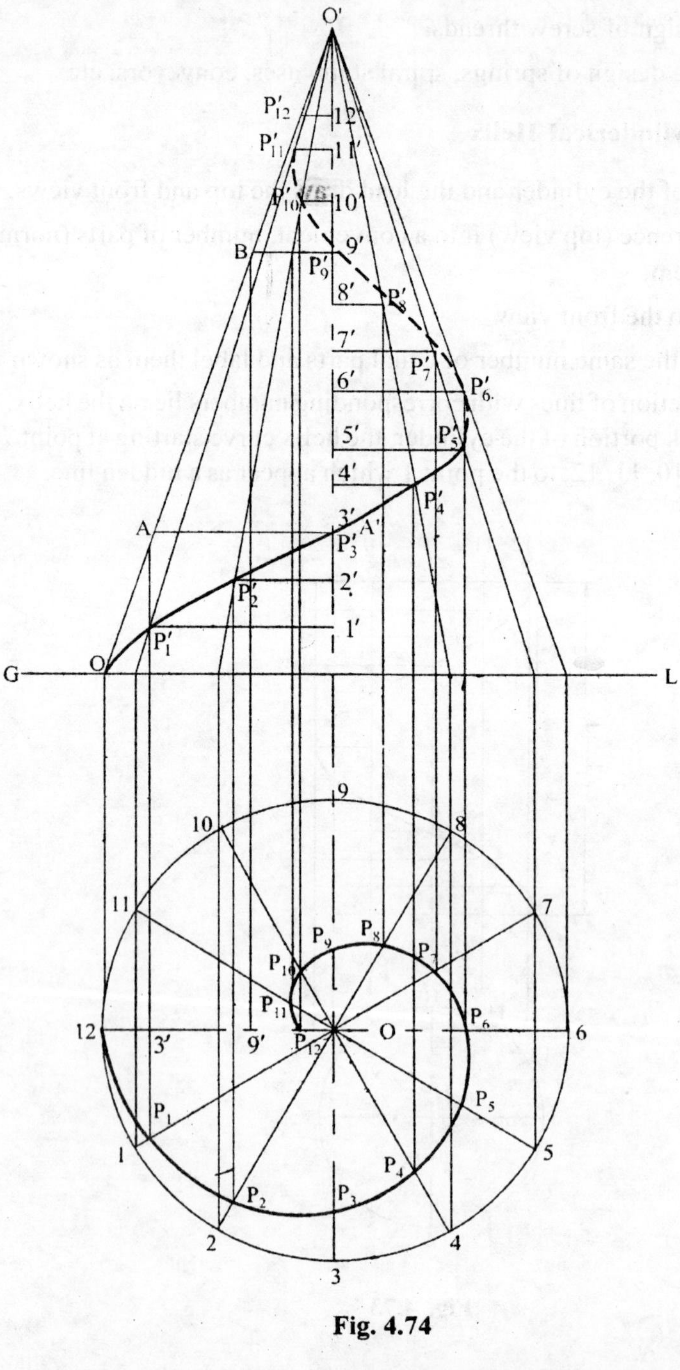

Fig. 4.74

1. First draw the top view of the cone as a circle for the given diameter and divide the circle into 12 equal parts. Join **1, 2.** etc.with centre **O**. Draw projectors from these points to the base line and draw a triangle of given base and height. Mark **O'** as apex. Mark a point **A** on the axis for the given pitch of the helix and divide the pitch distance into 12 equal parts.

2. Draw a horizontal line through the point **A** to cut the generator **(O'12')** at **A'**.

3. Let P be the starting point. When the point P moves around 30°, it should have moved up through one division to a point P_1' on the generator **O'1'** obtained by drawing a horizontal line from the point **1'**.

4. Project **P_1'** downwards to cut the top view of the generator O1 at P_1. Similarly get the points P_2, P_3.... P_{12}. Draw a smooth curve through these points in both top view and front view.

Problem : Draw the projection of two complete turns of a wire spring of sqaure cross section of side 20 mm with a pitch of 50 mm and the outside and inside diameter of the spring as 100 mm and 60 mm respectively.

***Solution :* (Fig. 4.75)**

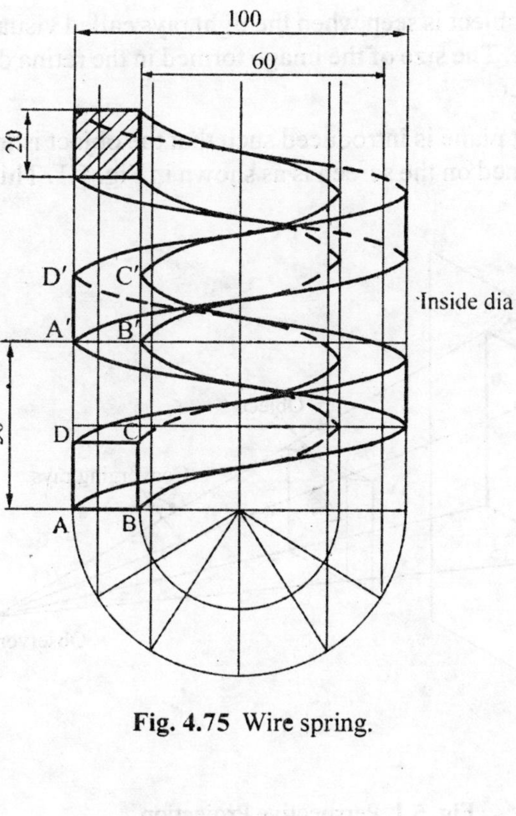

Fig. 4.75 Wire spring.

CHAPTER 5

Orthographic Projections

5.1 Introduction

In the preceding chapters 1 to 4 **plane geometry,** where the constructions of the geometrical figures having only two dimensions are discussed, **solid geometry** is delt within the following chapters.

Engineering drawing, particularly solid geometry is the graphic language used in the industry to record the ideas and informations necessary in the form of blue prints to make machines, buildings, structures etc., by engineers and technicians who design, develop, manufacture and market the products.

5.1.1 Projection

As per the optical physics, an object is seen when the light rays called visual rays coming from the object strike the observer's eye. The size of the image formed in the retina depends on the distance of the observer from the object.

If an imaginary transparent plane is introduced such that the object is in between the observer and the plane, the image obtained on the screen is as shown in Fig. 5.1. This is called **perspective**

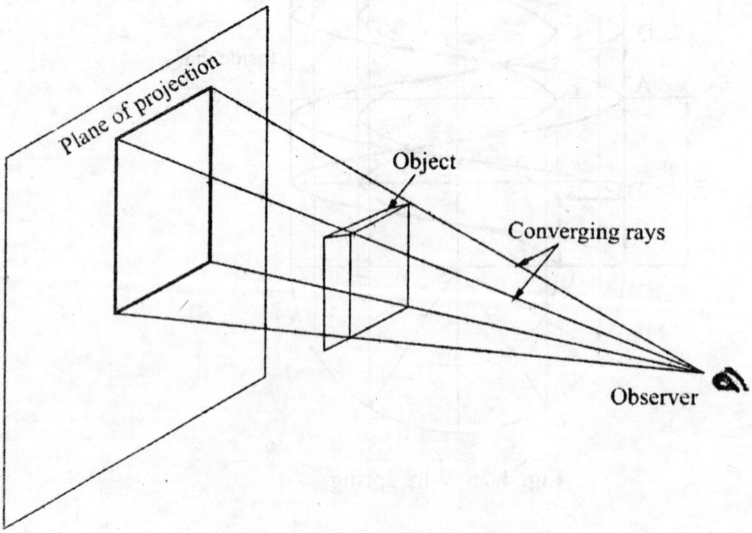

Fig. 5.1 Perspective Projection.

view of the object. Here, straight lines (rays) are drawn from various points on the contour of the object to meet the transparent plane, thus the object is said to be **projected** on that plane.

The figure or view formed by joining, in correct sequence, the points at which these lines meet the plane is called the projection of the object. The lines or rays drawn from the object to the plane are called projectors. The transparent plane on which the projections are drawn is known as plane of projection.

5.2 Types of Projections

1. Pictorial projections

(i) Perspective projection

(ii) Isometric projection

(iii) Oblique projection

The projections in which the description of the object is completely understood in one view is known as **pictorial projection**. They have the advantage of conveying an immediate impression of the general shape and details of the object, but not its true dimensions or sizes.

2. Orthographic Projection

'ORTHO' means right angle and orthographic means right angled drawing. When the projectors are perpendicular to the plane on which the projection is obtained, it is known as orthographic projection.

5.2.1 Method of Obtaining Front View

Imagine an observer looking at the object from an infinite distance (Fig. 5.2). The rays are parallel to each other and perpendicular to both the front surface of the object and the plane. When the observer is at a finite distance from the object, the rays converge to the eye as in the case of perspective projection. When the observer looks from the front surface F or the block, its true shape and size is seen. When the rays or porjectors are extended further they meet the **vertical plane**(VP) located behind the object. By joining the projectors meeting the plane in correct sequence the **Front view** is obtained.

Front view shows only two dimensions of the object, viz. length L and height H. It does not show the breadth B. Thus one view or projection is insufficient for the complete description of the object.

As Front view alone is insufficient for the complete description of the object, another plane called **Horizontal plane** (HP) is assumed such that it is hinged and perpendicular to VP and the object is in front of the VP and above the HP as shown in Fig. 5.3(a).

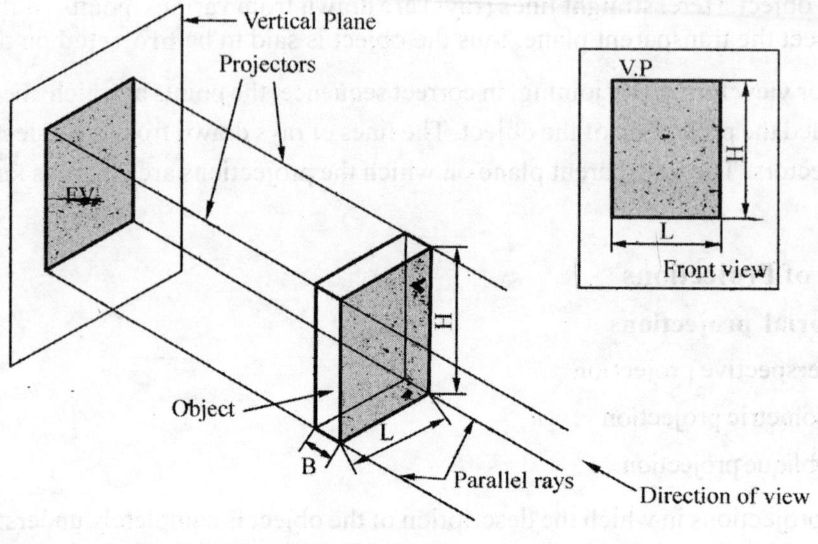

Fig. 5.2 Method of Obtaining Orthographic Front View.

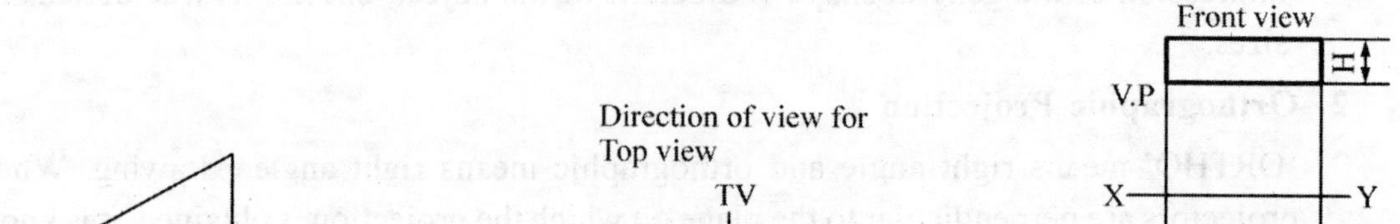

Fig. 5.3 Method of Obtaining Orthographic Top View.

5.2.2 Method of Obtaining Top View

Looking from the top, the projection of the top surface is the **Top view** (TV). Both top surface and Top view are of exactly the same shape and size. Thus, Top view gives the **True length L** and breadth B of the block but not the height H.

Note 1. Each projection shows that surface of the object which is nearer to the observer and far away from the plane.

2. Orghographic projection is the standard drawing form of the industrial world.

X Y Line : The line of intersection of VP and HP is called the reference line and is denoted as XY.

Obtaining the Projectin on the Drawing Sheet

It is convention to rotate the HP through 90° in the clockwise direction about XY line so that it lies in the extension of VP as shown in Fig. 5.3(a). The two projections – Front view and Top view may be drawn on the two dimensional drawing sheet as shown in Fig.5.3(b).

Thus, all details regarding the shape and size, viz. Length (L), Height (H) and Breadth (B) of any object may be represented by means of orthographic projections i.e., Front view and Top view.

Terms Used

VP and HP are called as **Principal planes of projection or reference planes**. They are always transparent and at right angles to each other. The projection on VP is designated as **Front view** and the projection on HP as **Top view.**

Four Quadrants

When the planes of projections are extended beyond their line of intersection, they form **Four Quadrants**. These quadrants are numbered as I, II, III and IV in clockwise direction when rotated about reference line XY as shown in Fig. 5.4.

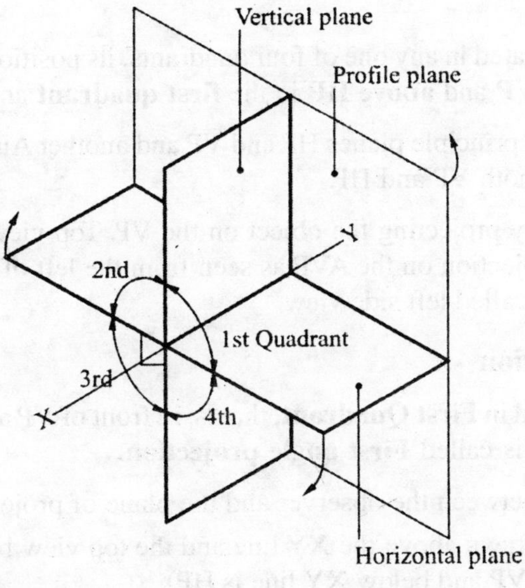

Fig. 5.4 Four Quadrants.

In the Fig. 5.5 the object is in the **first quadrant** and the projections obtained are **"First angle projections"** i.e,, the object lies in between the observer and the planes of projection. Front view shows the length(L) and height(H) of the object, and Top view shows the length(L) and the breadth(B) of it.

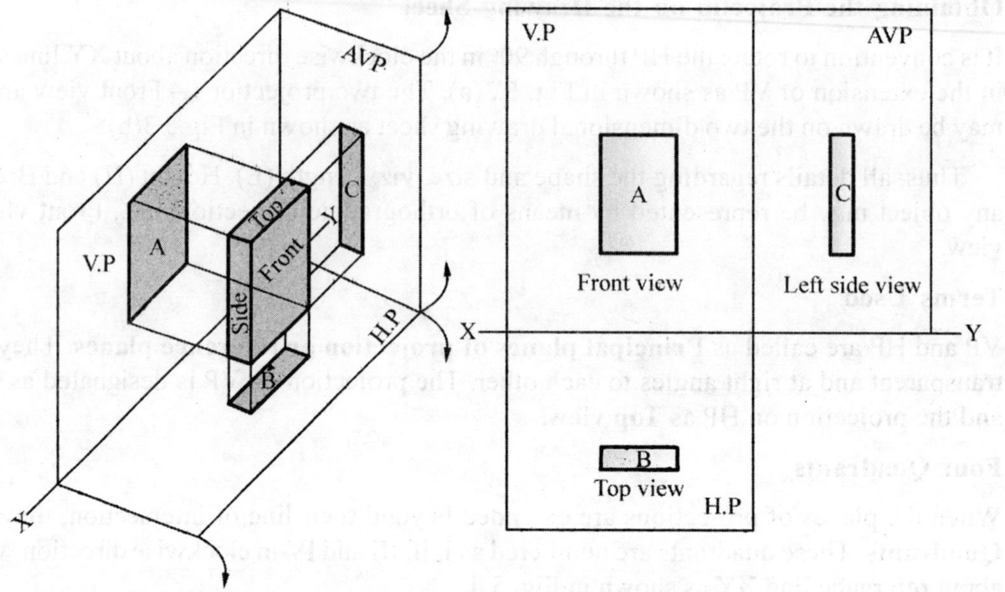

Fig. 5.5 Orthographic Projection of Front, Top and Side Views.

The object may be situated in any one of four quadrants, its position relative to the planes being described as **in front of VP** and **above HP** in the **first quadrant** and so on.

Fig. 5.5 shows the two principle planes HP and VP and another Auxiliary vertical plane (AVP). AVP is perpendicular to both VP and HP.

Front view is drawn by projecting the object on the VP. Top view is drawn by projecting the object on the HP. The projection on the AVP as seen from the left of the object and drawn on the right of the front view, is called left side view.

5.3 First Angle Projection

When the object is situated in **First Quadrant**, that is, in front of VP and above HP, the projections obtained on these planes is called **First angle projection.**

(i) The object lies in between the observer and the plane of projection.

(ii) The front view is drawn above the XY line and the top view below XY. (Above XY line is VP and below XY line is HP).

(iii) In the front view, HP coincides with XY line and in top view VP coincides with XY line.

(iv) Front view shows the length (L) and height (H) of the object and Top view shows the length (L) and breadth (B) or width(W) or thicknes(T) of it.

5.4 Third Angle Projection

In this, the object is situated in **Third Quadrant**. The planes of porjection lie between the object and the observer. The front view comes below the XY line and the top view above it.

BIS Specification (SP46 : 2003)

BIS has recommended the use of First angle projection in line with the specifications of ISO adapted by all countries in the world.

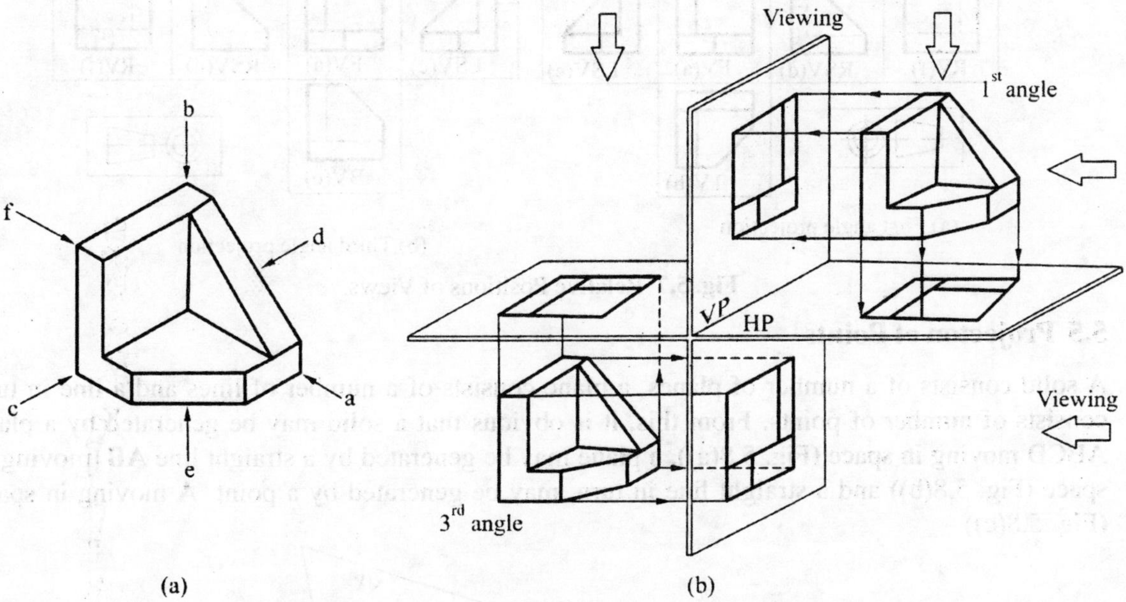

Fig. 5.6 Principles of Orthographic Projection.

Designation and Relative Position of Views

An object in space may be imagined as surrounded by six mutually perpendicular planes. So, it is possible to obtain six different views by viewing the object along the six directions, normal to the six planes. Fig. 5.7 shows an object with the six possible directions to obtain the six different views which are designated as follows.

1. View in the direction a = front view
2. View in the direction b = top view
3. View in the direction c = left side view
4. View in the direction d = right side view
5. View in the direction e = bottom view
6. View in the direction f = rear view

The relative position of the views in **First angle and third angle projection** are shown in Fig. 5.7.

Note: A study of the Fig. 5.7 reveals that in both the methods of projection, the views are identical in shape and size but their location with respect to the front view only is different.

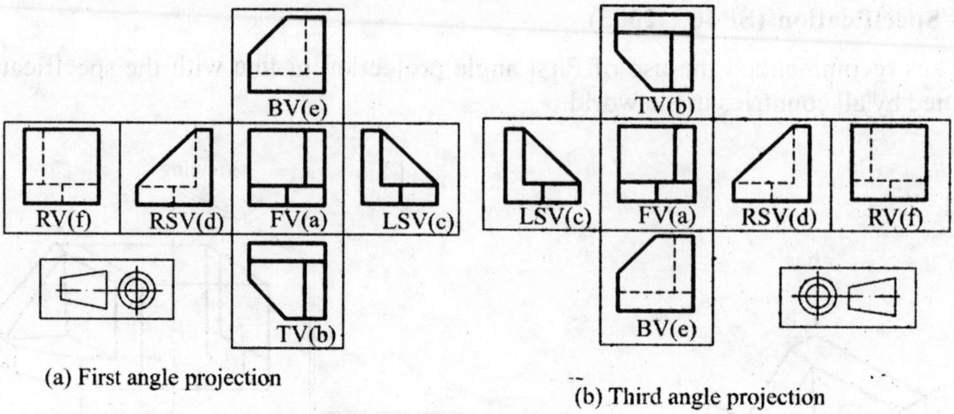

Fig. 5.7 Relative Positions of Views.

5.5 Projecton of Points

A solid consists of a number of planes, a plane consists of a number of lines and a line in turn consists of number of points. From this, it is obvious that a solid may be generated by a plane ABCD moving in space (Fig. 5.8(a)), a plane may be generated by a straight line **AB** moving in space (Fig. 5.8(b)) and a straight line in turn, may be generated by a point **A** moving in space (Fig. 5.8(c))

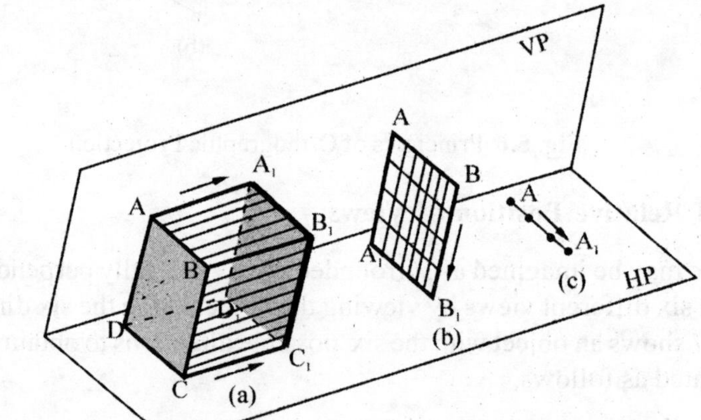

Fig. 5.8

Points in Space

A point may lie in space in any one of the four quadrants. The positions of a point are:

1. First quadrant, when it lies **above HP** and **in front of VP.**
2. Second quadrant, when it lies **above HP and behind VP.**
3. Third quadrant, when it lies **below HP and behind VP.**
4. Fourth quadrant, when it lies **below HP and in front of VP.**

Knowing the distances of a point from HP and VP, projections on HP and VP are found by extending the projections perpendicular to both the planes. Projection on HP is called **Top view** and projection on VP is called **Front view**

Notation followed

1. Actual points in space are denoted by capital letters A, B, C.
2. Their front views are denoted by their corresponding lower case letters with dashes a', b', c', etc., and their top views by the lower case letters a, b, c etc.
3. Projectors are always drawn as continuous thin lines.

Note:

1. **Students are advised to make their own paper/card board/perplex model of HP and VP as shown in Fig. 5.4. The model will facilitate developing a good concept of the relative position of the points lying in any of the four quadrants.**
2. **Since the projections of points, lines and planes are the basic chapters for the subsequent topics on solids viz, projection of solids, development, pictorial drawings and conversion of pictorial to orthographic and vice versa, the students should follow these basic chapters carefully to draw the projections.**

Problem : Point A is 40 mm above HP and 60 mm in front of VP. Draw its front and top views.

***Solution* : (Fig. 5.9)**

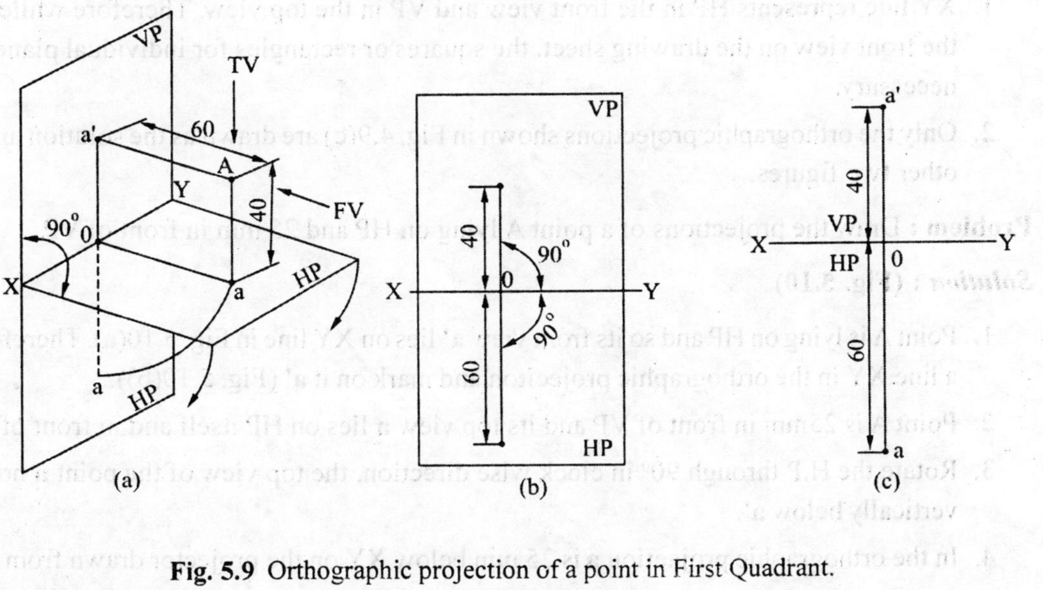

Fig. 5.9 Orthographic projection of a point in First Quadrant.

1. The point **A** lies in the **I Quadrant**
2. Looking from the front, the point lies 40 mm above HP. A-a' is the projector perpendicular to V.P. Hence **a'** is the front view of the point A and it is 40 mm above the XY line.
3. To obtain the top view of **A,** look from the top. Point A is 60 mm in front of VP. Aa is the projector perpendicular to HP. Hence, **a** is the top view of the point A and it is 60 mm in front of XY.
4. To convert the projections **a'** and **a** obtained in the pictorial view into orthographic projections the following steps are needed.
   (a) Rotate the H.P about the XY line through 90° in the clock wise direction as shown.
   (b) After rotation, the first quadrant is opened out and the HP occupies the position verically below the VP line. Also, the point **a** on HP will trace a quadrant of a circle with **o** as centre and **o-a** as radius. Now **a** occupies the position just below **o**. The line joining **a'** and **a**, called the projector, is perpendicular to XY (Fig. 5.9(b)).
5. To draw the orthographic projections
   (a) *Front view :* Draw the XY line and draw a projectior at any point on it. Mark **a'** 40 mm above XY on the projector.
   (b) *Top view :* on the same projector, mark **a** 60 mm below XY. (Fig. 5.9(c))

Note:

1. XY line represents HP in the front view and VP in the top view. Therefore while drawing the front view on the drawing sheet, the squares or rectangles for individual planes are not necessary.
2. Only the orthographic projections shown in Fig. 4.9(c) are drawn as the solution and not the other two figures.

Problem : Draw the projections of a point A lying on HP and 25 mm in front of VP.

***Solution* : (Fig. 5.10)**

1. Point A is lying on HP and so its front view **a'** lies on XY line in Fig. 5.10(a). Therefore, mark a line XY in the orthographic projeciton and mark on it **a'** (Fig. 5.10(b)).
2. Point A is 25mm in front of VP and its top view **a** lies on HP itself and in front of XY.
3. Rotate the H.P through 90° in clock wise direction, the top view of the point **a** now comes vertically below **a'**.
4. In the orthographic projection **a** is 25 mm below XY on the projector drawn from **a'**.

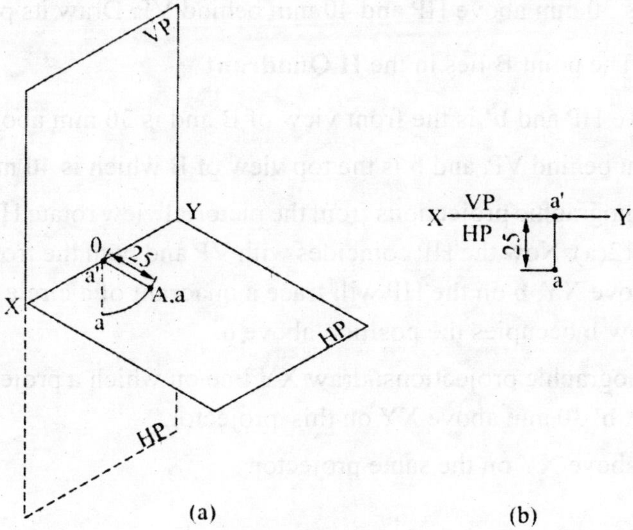

Fig. 5.10

Problem : Draw the projections of a point A lying on VP and 70 mm above HP.

***Solution* : (Fig. 5.11)**

1. Looking at the pictorial view from the front (Fig. 5.11(a)) the point A is 70 mm above H.P and so **a'** is 70 mm above XY. Hence, mark **a'** the orthographic projection 70 mm above XY (Fig. 5.11(b)).
2. Looking at the pictorial view from the top, point **a** is on VP and its view lies on XY itself. The top view **a** does not lie on the HP. So in this case the HP need not be rotated. Therefore mark **a** on XY on the projector drawn from **a'**

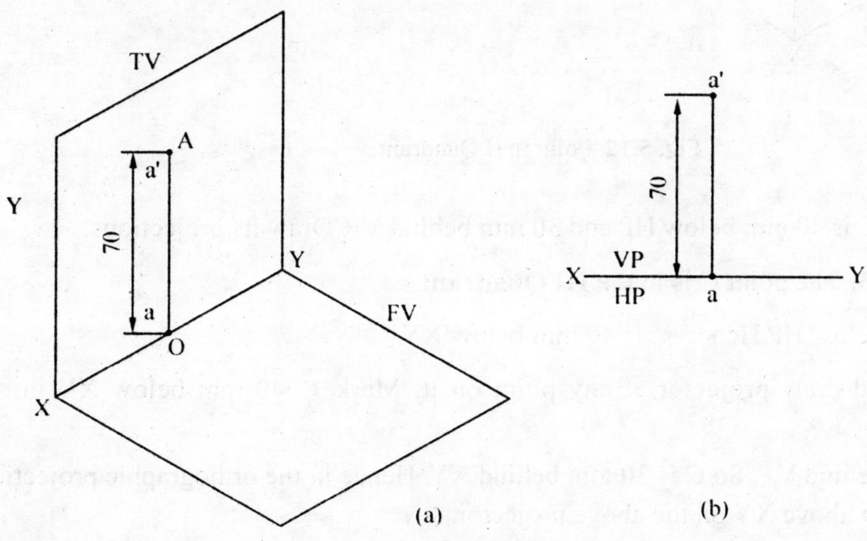

Fig. 5.11

Problem : A Point B is 30 mm above HP and 40 mm behind VP. Draw its projection.

***Solution* : (Fig. 5.12)** The point B lies in the **II Quadrant**

1. It is 30 mm above HP and **b'** is the front view of B and is 30 mm above XY.
2. Point B is 40 mm behind VP. and **b** is the top view of B which is 40 mm behind XY.
3. To obtain the orthographic projections from the pictorial view rotate HP by 90° about XY as shown in Fig. 5.12(a). Now the HP coincides with VP and both the front view and top view are now seen above XY. **b** on the HP will trace a quadrant of a circle with **o** as centre and **ob** as radius. Now **b** occupies the position above **o**.
4. To draw the orthographic projections; draw XY line on which a projectior is drawn at any point. Mark on it **b'** 30 mm above XY on this projector.
5. Mark **b** 40 mm above XY on the same projector.

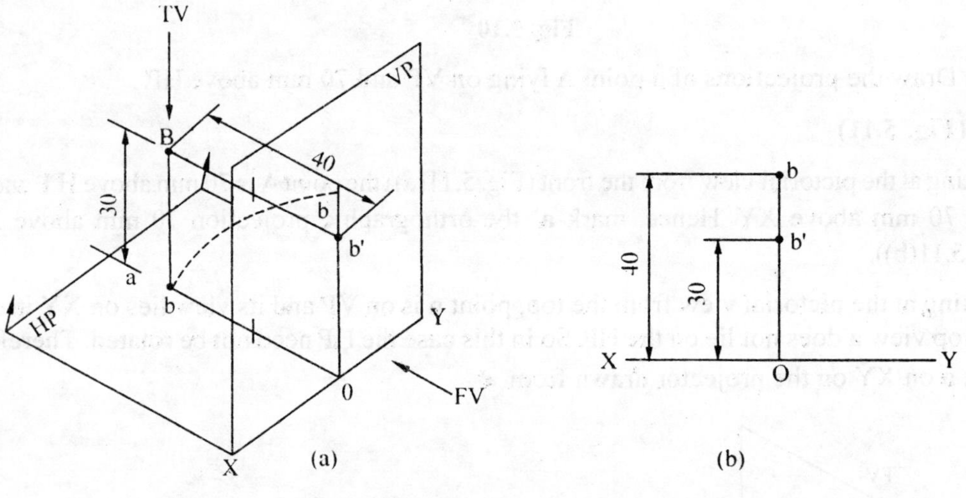

Fig. 5.12 Point in II Quadrant.

Problem : A point C is 40 mm below HP and 30 mm behind VP. Draw its projections.

***Solution* : (Fig. 5.13)** The point C is in the **III Quadrant**

1. **C** is 40 mm below HP Hence c' is 40 mm below XY.
2. Draw XY and draw projector at any point on it. Mark c' 40 mm below XY on the projector.
3. **C** is 30 mm behind V.P. So **C** is 30 mm behind XY. Hence in the orthographic projections mark **c** 30 mm above XY on the above projector.

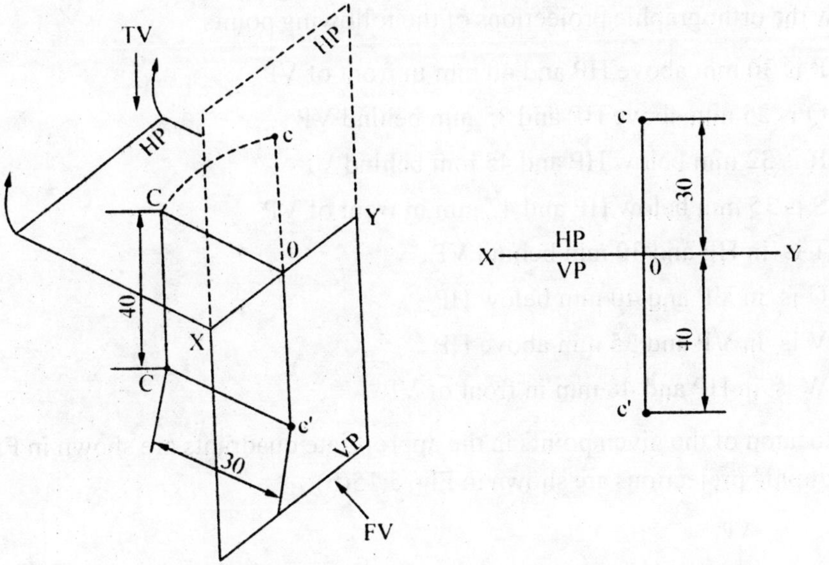

Fig. 5.13 Point in III Quadrant.

Problem : A point D is 30 mm below HP and 40 mm in front of VP. Draw its projeciton.

Solution : **(Fig. 5.14)** The point D is in the **IV Quadrant.**

1. D is 30 mm below HP. Hence, **d'**, is 30 mm below XY. Draw XY line and draw a projector perpendicular to it. Mark **d'** 30 mm below XY on the projector.
2. D is 40 mm in front of VP; so **d** is 40 mm in front of XY. Therefore, mark **d** 40 mm below XY.

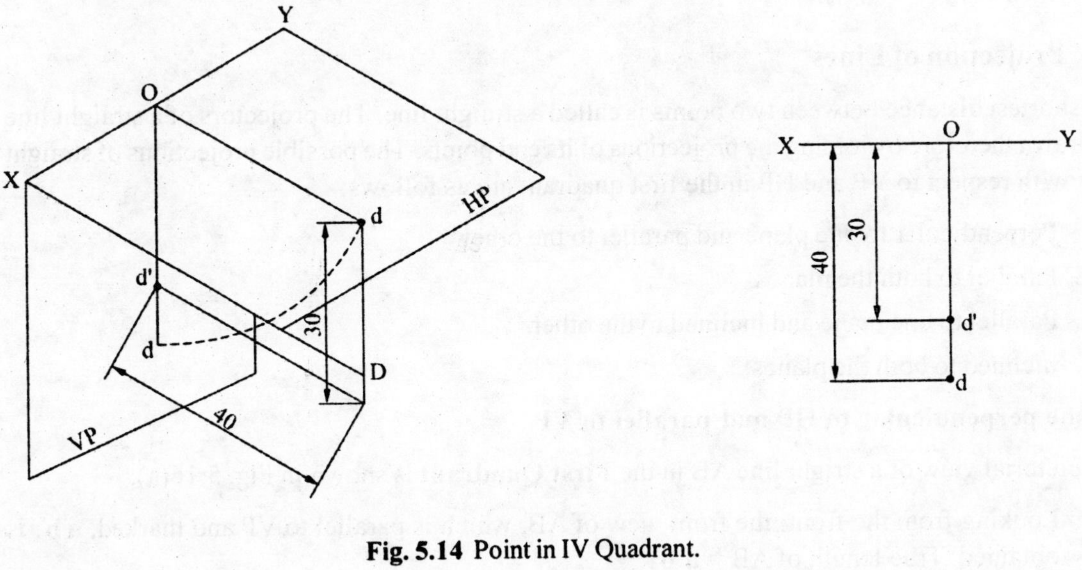

Fig. 5.14 Point in IV Quadrant.

Problem : Draw the orthographic projections of the following points.

(a.) Point P is 30 mm above HP and 40 mm in front of VP

(b.) Point Q is 25 mm above HP and 35 mm behind VP

(c.) Point R is 32 mm below HP and 45 mm behind VP

(d.) Point S is 35 mm below HP and 42 mm in front of VP

(e.) Point T is in HP and 30 mm behind VP

(f.) Point U is in VP and 40 mm below HP

(g.) Point V is in VP and 35 mm above HP

(h.) Point W is in HP and 48 mm in front of VP

Solution : The locaton of the given points in the appropriate quadrants are shown in Fig. 5.15(a) and their orthographic prejections are shown in Fig. 5.15(b).

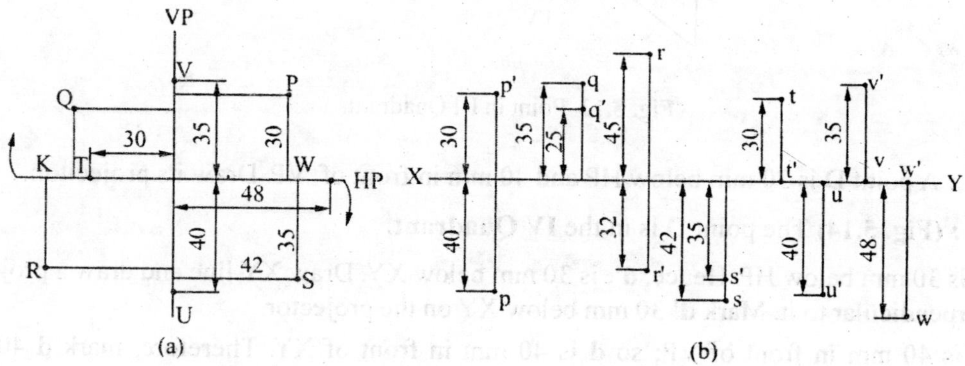

Fig. 5.15 Points in Various Quadrants.

5.6 Projection of Lines

The shortest distance between two points is called a straight line. The projectors of a straight line are drawn therefore by joining the projections of its end points. The possible projections of straight lines with respect to VP and HP in the first quadrant are as follows:

1. Perpendicular to one plane and parallel to the other.
2. Parallel to both the planes.
3. Parallel to one plane and inclined to the other.
4. Inclined to both the planes.

1. Line perpendicular to HP and parallel to VP

The pictorial view of a stright line AB in the **First Quadrant** is shown in Fig. 5.16(a).

1. Looking from the front; the front view of AB, which is parallel to VP and marked, **a'b'**, is obtained. True length of AB = **a'b'**.

2. Looking from the top; the top view of AB, which is perpendicular to HP is obtained **a** and **b** coincide.
3. The Position of the line AB and its projections on HP and VP are shown in Fig. 5.16(b).
4. The HP is rotated through 90° in clock wise direction as shown in Fig.5.16(b).
5. The projection of the line on VP which is the front view and the projection on HP, the top view are shown in Fig. 5.16(c).

Note : Only Fig. 5.16(c) is drawn on the drawing sheet as a solution.

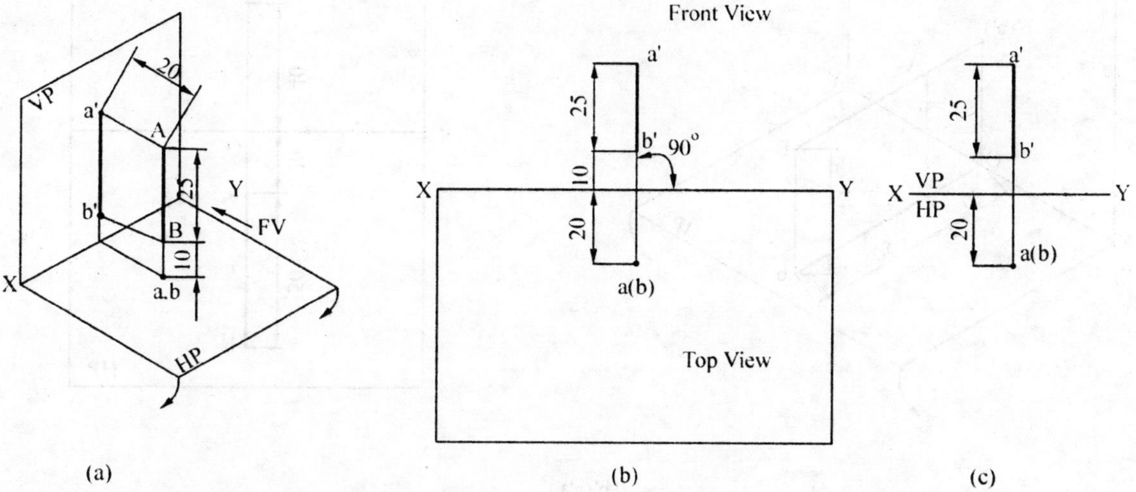

Fig. 5.16 Line Perpendicular to HP and Parallel to VP.

1. Line perpendicular to VP and parallel to HP.

Problem : A line AB 50 mm long is perpendicular to VP and parallel to HP. Its end A is 20 mm in front of VP and the line is 40 mm above HP. Draw the projectons of the line.

Solution : **(Fig. 5.17)** The line is parallel to HP. Therefore the true length of the line is seen in the top view. So, top view is drawn first.

1. Draw XY line and draw a projector at any point on it.
2. Point A is 20 mm in front of VP. Mark **a** which is the top view of A at a distance of 20 mm below XY on the projector.
3. Mark the point **b** on the same projector at a distance of 50 mm below **a. ab** is the top view which is true length of AB.
4. To obtain the front view; mark **b'** at a distance 40 mm above XY line on the same projector.

5. The line AB is perpendicular to VP. So, the front view of the line will be a point. Point A is hidden by B. Hence the front view is marked as **b' (a')**. **b'** coincides with **a'**.
6. The final projections are shown in Fig. 5.17(c).

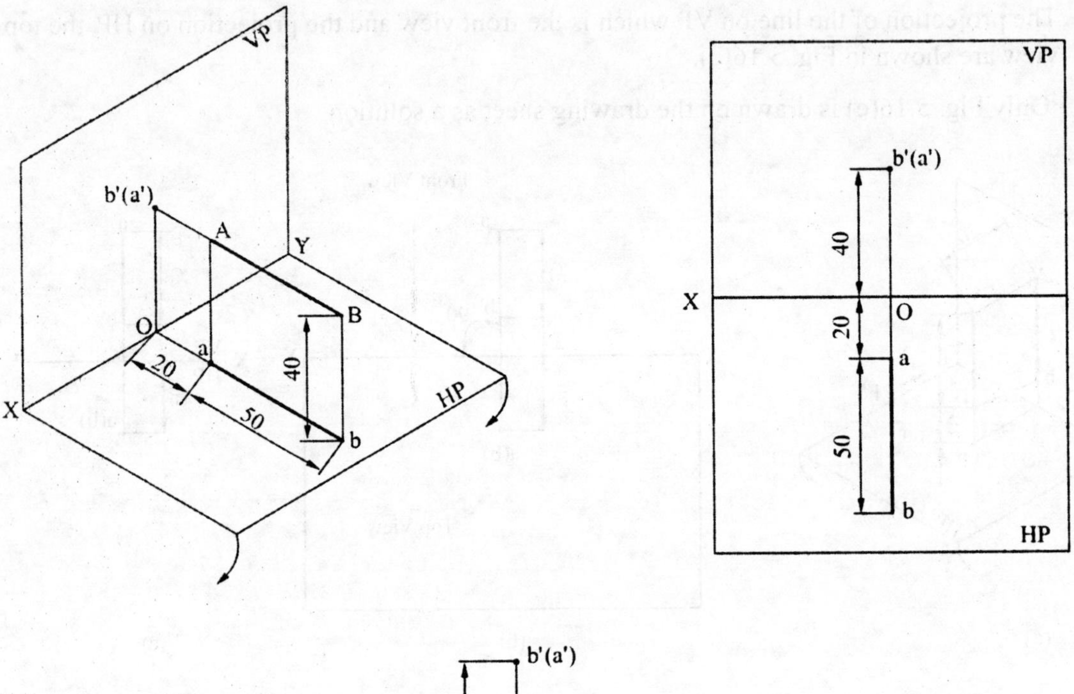

Fig. 5.17 Line Perpendicular VP and Parallel to HP.

2. Line parallel to both the planes

Problem : A line CD 30 mm long is parallel to both the planes. The line is 40 mm above HP and 20 mm in front of VP. Draw its projection.

Solution **: (Fig. 5.18)**

1. Draw the XY line and draw a projector at any point on it.
2. To obtain the front view mark **c'** at a distance of 40 mm abvoe XY (HP). The line CD is parallel to both the planes. Front view is true length and is parallel to XY. Draw **c' d'** parallel to XY such that **c' d'** = CD = 30 mm, which is the true length.
3. To obtain the top view; the line is also parallel to VP and 20 mm in front of V.P. Therefore on the projector from **c'**, mark **c** at distance 20 mm below XY line.
4. Top view is also true length and parallel to XY. Hence, **cd** parallel to XY such that **cd** = CD = 30 mm is the true length.(Fig. 5.18).

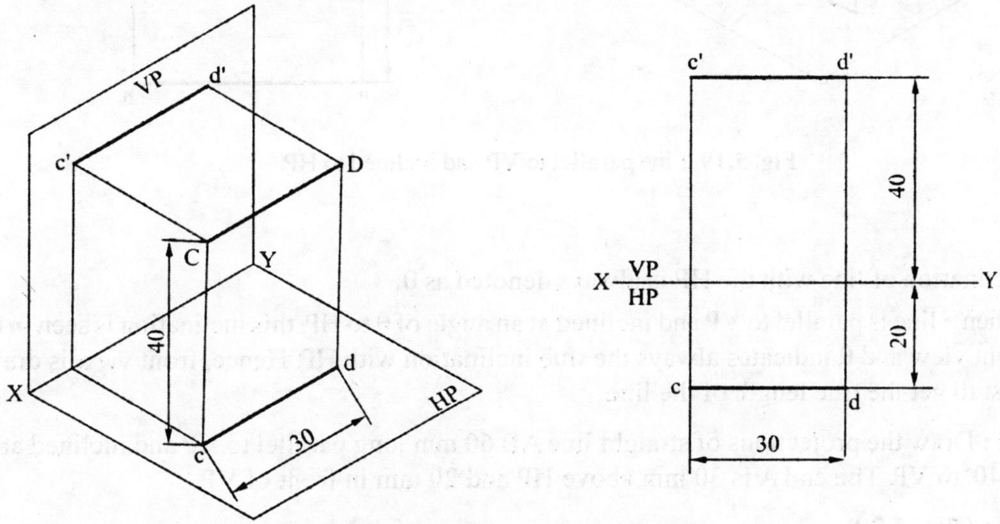

Fig. 5.18 Line Parallel to both the Planes.

3. Line parallel to VP and inclined to HP.

Problem : A line AB 40 mm long is parallel to VP and inclined at an angle of 30° to HP. The end A is 15 mm above HP and 220 mm in front of VP. Draw the projections of the line.

Solution **: (Fig. 5.19)**

1. A is 15 mm above HP, mark **a'**, 15 mm above XY.
2. A is 20 mm in front of VP. Hence mark **a** 20 mm below XY.
3. To obtain the front view **a' b'**; as AB is parallel to VP and inclined at an angle θ to HP, **a'b'** will be equal to its **true length** and inclined at an angle of 30° to HP. Therefore draw a line from **a'** at an angle 30° to XY and mark **b'** such taht **a' b'** = 40 mm = true length.
4. To obtain the top view ab; since the line is inclined to HP its projection on HP (its top veiw) is reduced in length. From **b'** draw a projector to intersect the horizontal line drawn from **a** at **b**. **ab** is the top view of AB.

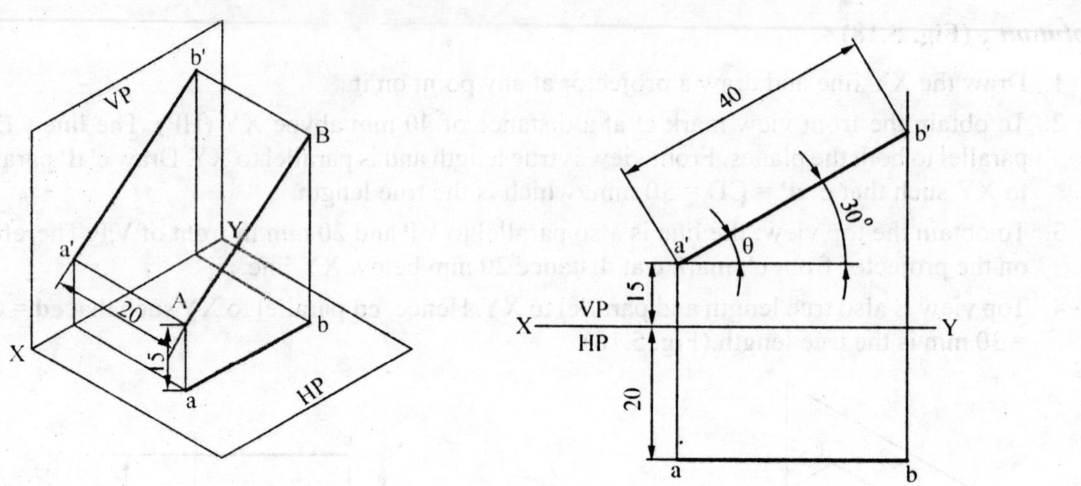

Fig. 5.19 Line parallel to VP and inclined to HP.

Note :

1. Inclination of line with the HP is always denoted as θ.
2. When a line is parallel to VP and inclined at an angle of θ to HP, this inclination is seen in the front view and θ indicates always the true inclination with HP. Hence, front view is drawn first to get the true length of the line.

Problem : Draw the projections of straight line AB 60 mm long parallel to HP and inclined at an angle of 40° to VP. The end A is 30 mm above HP and 20 mm in front of VP.

***Solution* : (Fig. 5.20)**

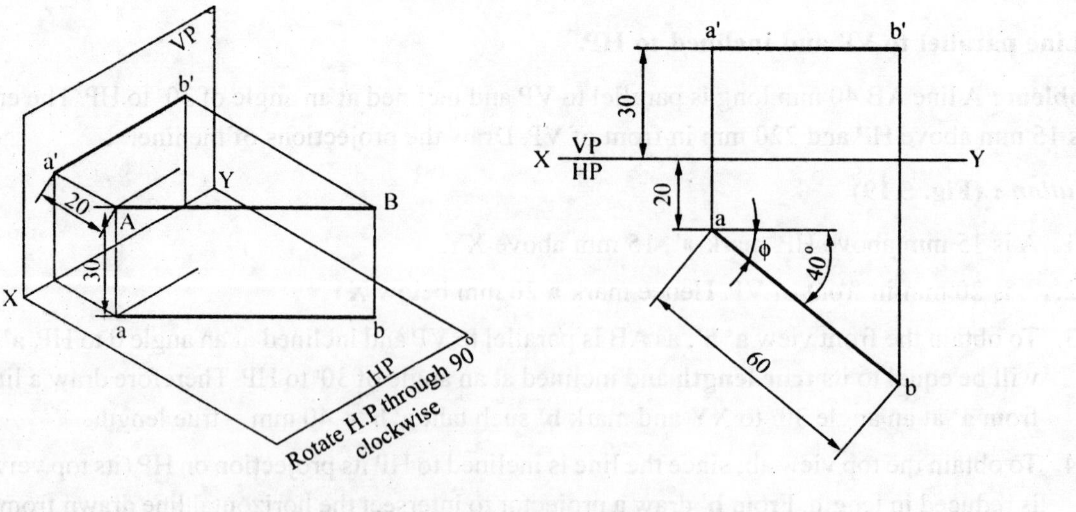

Fig. 5.20 Line Parallel to HP and Inclined to VP.

1. A is 30 mm above HP, mark **a'** , 30 mm above XY.
2. A is 20 mm in front of VP, mark **a** 20 mm below XY.
3. To obtain the top view; as AB is parallel to HP and inclined at an angle ϕ to VP, **ab** will be equal to the true length of AB, and inclined at angle ϕ to XY. Therefore, draw a line from **a** at 40° to XY and mark **b** such that **ab**=60 mm true length.
4. To obtain the front view **a'b'**, since the line is inclined to VP its projection on VP i.e., the front view will be reduced in length. Draw from **b** a projector to intersect the horizontal line drawn from **a** at **b'**. **a'b'** is the front view of AB.

Note :

1. Inclination of a line with VP is always denoted by ϕ.
2. When a line is paralel to HP and inclined at an angle of ϕ to VP, this inclination ϕ is seen in the top view and hence top view is drawn first to get the true length of the line.

Problem : A line AB of 50 mm long is parallel to both HP and VP. The line is 40 mm above HP and 30 mm in front of VP. Draw the projections of the line.

Solution **: (Fig. 5.21)**

Fig. 5.21(a) shows the position of the line AB in the first quadrant. The points a', b' on VP and a, b on HP are the front and top views of the ends A and B of the line AB. The lines a'b' and ab are the front and top views of the line AB respectively.

Fig. 5.21(b) shows the relative positions of the views along with the planes, after rotating HP, till it is in-line with VP. Fig. 5.21(c) shows the relative positions of the view only.

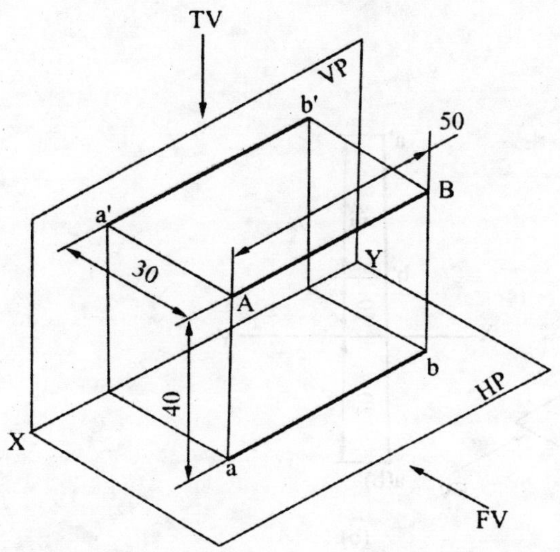

(a)

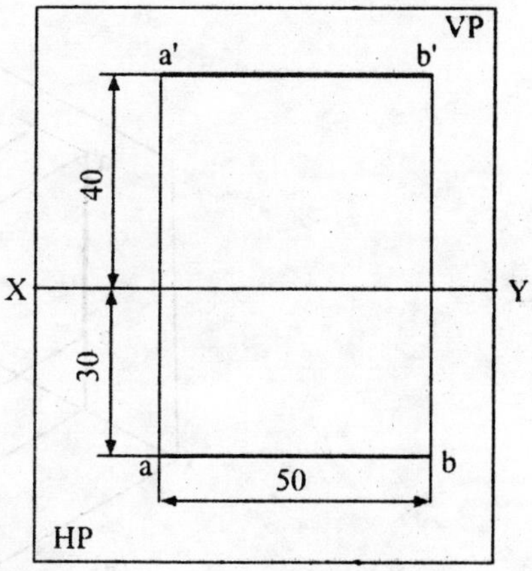

(b)

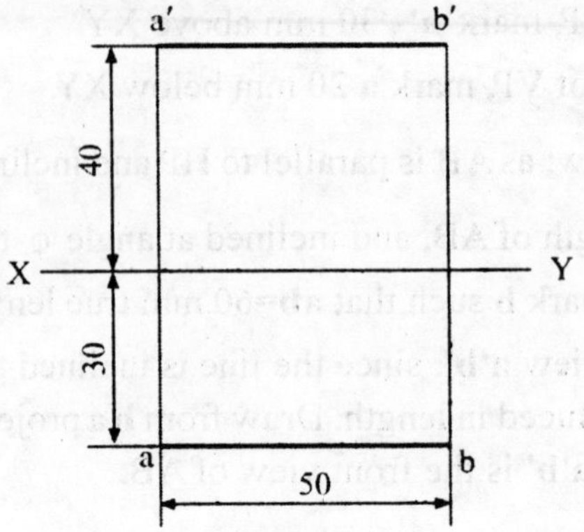

(c)

Fig. 5.21

Problem : A line AB of 25 mm long is perpendicular to HP and parallel to VP. The end points A and B of the line are 35 mm and 10 mm above HP respectively. The line is 20 mm in front of VP. Draw the projections of the line.

***Solution* : (Fig. 5.22)**

Fig. 5.22(a) shows the position of the line AB in the first quadrant. As the line is parallel to VP, the length of the front view is equal to the true length of the line and the top view appears as a point. Fig. 4.2(b) shows the projection.

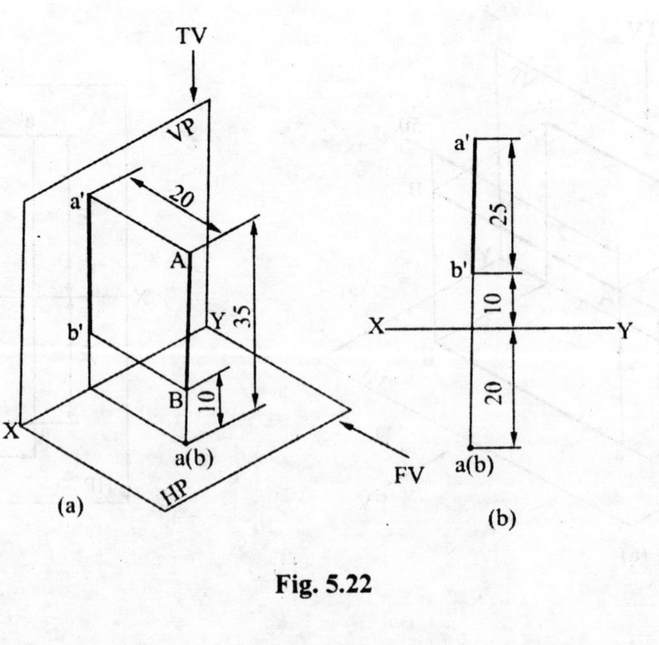

Fig. 5.22

Problem : A line AB of 25 mm long is perpendicular to VP and parallel to HP. The end points A and B of the line are 10 mm and 35 mm in front of VP respectively. The line is 20 mm above HP. Draw its projections.

Solution **: (Fig. 5.23)**

Fig. 5.23(a) shows the position of the line AB in the first quadrant. As the line is parallel to HP, the length of the top view is equal to the true length of the line and the front view appears as a point. Figure 5.23(b) shows the projection of the line.

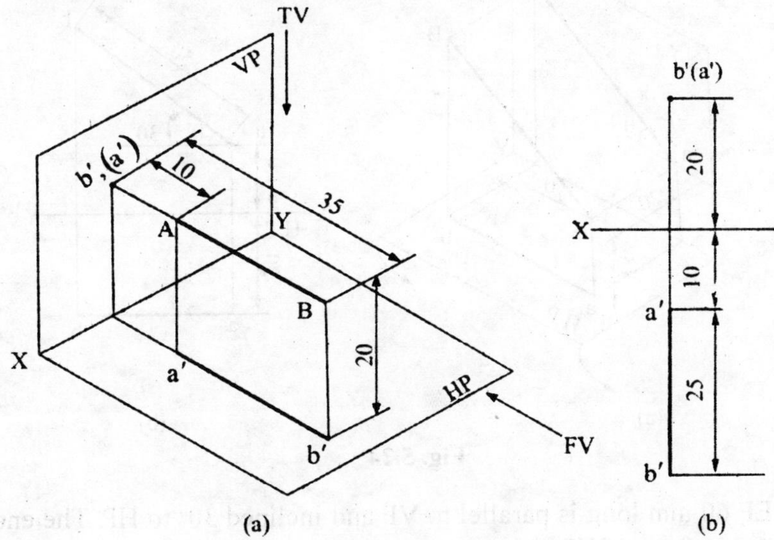

Fig. 5.23

Problem : A line AB 50 mm long is parallel to VP and inclined at an angle of 30° to HP. The end A is 15 mm above HP and 20 mm in front of VP. Draw the projections of the line.

Solution **: (Fig. 5.24)**

1. A is 15 mm above HP. Hence mark **a'** 15 mm above XY.
2. A is 20 mm in front of VP. Hence mark **a** 20 mm below XY.

To obtain the front view a'b', look from the front (FV):

3. As AB is parallel to VP and inclined at an angle of 30° to HP, **a'b'** will be equal to its true length and inclined at an angle of 30° to XY.

Note : When a line is parallel to VP and inclined at an angle of θ to HP, this inclination θ will be seen in the front view. θ **denotes always the true inclination with HP.**

3. Therefore from **a'** draw a line at an angle of 30° to XY and mark **b'** such that **a'b'** 50 mm = true length.

To obtain the top view ab look from the top TV:

Since the line is inclined to HP, its projection on HP i.e., the top view will be in reduced length.

4. From **b'** draw a projector to intersect the horizontal line drawn from **a** at **b**, **ab** is the top view of AB.

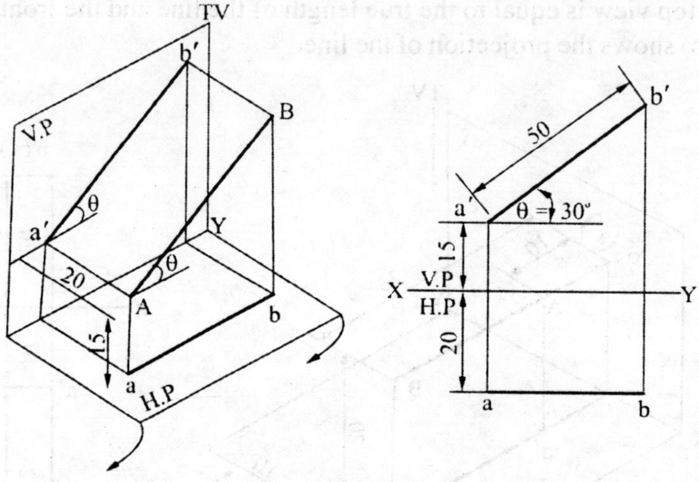

Fig. 5.24

Problem : A line EF 60 mm long is parallel to VP and inclined 30° to HP. The end E is 10 mm above HP and 20 mm in front of VP. Draw the projections of the line.

***Solution* : (Fig. 5.25).**

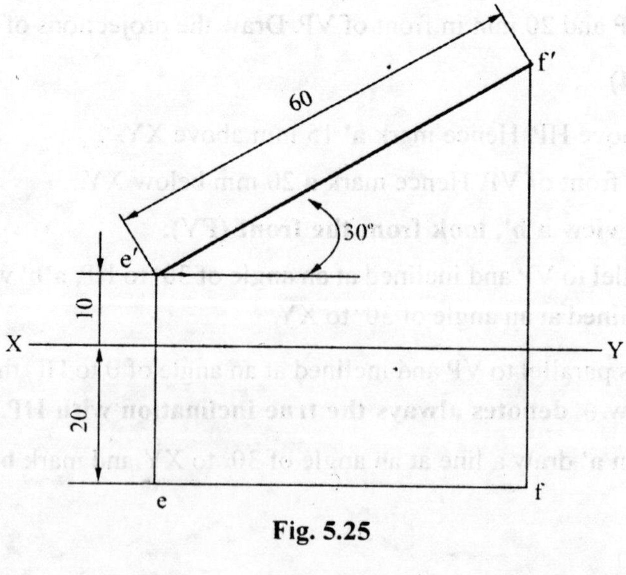

Fig. 5.25

Problem : The length of the front view of a line CD which is parallel to HP and inclined 30° to VP, is 50 mm. The end C of the line is 15 mm in front of VP and 25 mm above HP. Draw the projections of the line and find its ture length.

Solution **: (Fig. 5.26).**

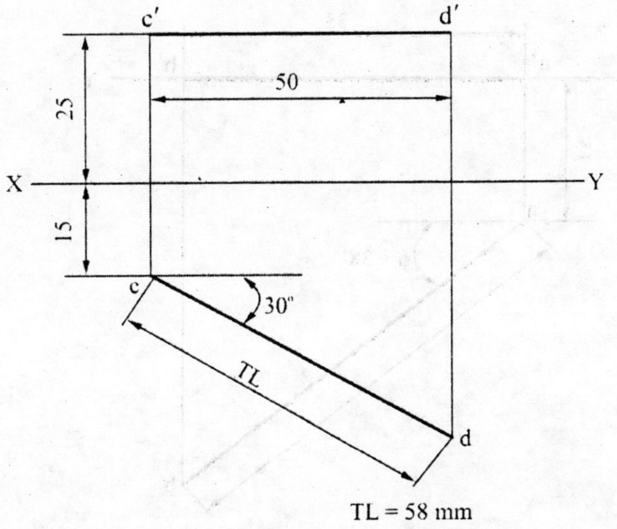

Fig. 5.26

Problem : A line CD 40 mm long is in VP and inclined to HP. The top view measures 30 mm. The end C is 10 mm above HP. Draw the projections of the line. Determine its inclination with HP.

Solution **: (Fig. 5.27).**

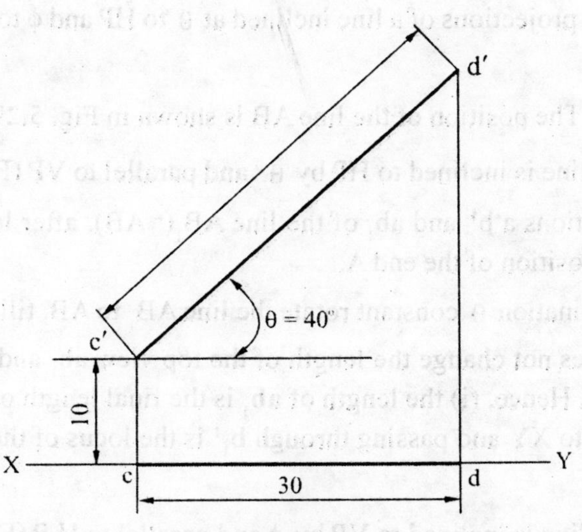

Fig. 5.27

Problem : A line AB 45 mm long is in HP and inclined to VP. The end A is 15 mm in front of VP. The length of the front view is 35 mm. Draw the projections of the line. Determine its inclination with VP.

Solution : **(Fig. 5.28).**

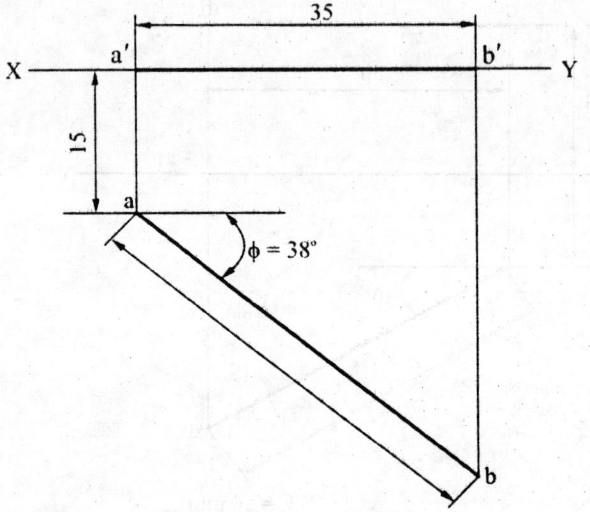

Fig. 5.28

4. Line inclined to both the planes

When a line is inclined to both HP and VP, it is called an oblique line. The solution to this kind of problem is obtained in three stages, as described below.

Problem : To draw the projections of a line inclined at θ to HP and ϕ to VP, given the position of one of its ends.

Solution : **(Fig. 5.29).** The position of the line AB is shown in Fig. 5.29a.

Step 1 Assume the line is inclined to HP by θ^0 and parallel to VP (Fig.5.29b).

1. Draw the projections a'b', and ab_1 of the line AB_1(=AB), after locating projections and **a** from the given position of the end A.

   Keeping the inclination θ constant rotate the line AB_1 to AB, till it is inclined at ϕ^0 to VP. This rotation does not change the length of the top view ab_1 and the distance of the point B_1=(B) from HP. Hence, (i) the length of $\mathbf{ab_1}$ is the final length of the top view and (ii) the line f-f, parallel to XY and passing through $\mathbf{b_1}^1$ is the locus of the front view of the end of point B.

Step 2 Assume the line is inclined to VP by ϕ and parallel to H.P (Fig.5.29c)

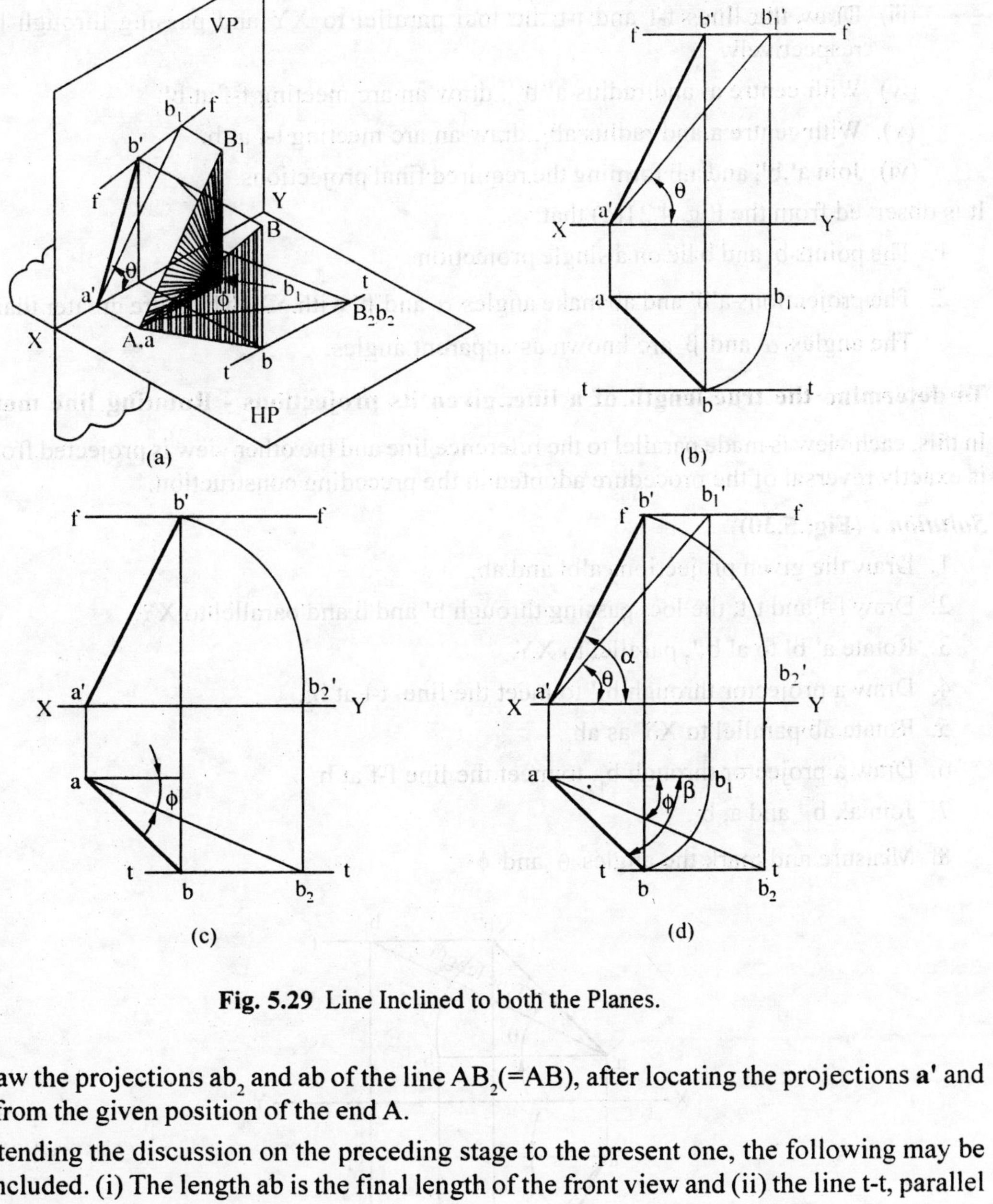

Fig. 5.29 Line Inclined to both the Planes.

2. Draw the projections ab_2 and ab of the line AB_2(=AB), after locating the projections **a'** and **a**, from the given position of the end A.

   Extending the discussion on the preceding stage to the present one, the following may be concluded. (i) The length ab is the final length of the front view and (ii) the line t-t, parallel to XY and pasing through b_2 is the locus of the top view of the end point B.

Step 3 Combine Stage I and Stage II (Fig. 5.29(d)),

3. Obtain the final projections by combining the results from stage 1 and II as indicated below:

   (i) Draw the projections a' b', and ab_2 making an angle θ and φ respectively with XY, after location of the projections a' and a, from the given position of the end point A.

(ii) Obtain the projections a' b_2' and ab_1, parallel to XY, by rotation.

(iii) Draw the lines f-f and t-t the loci parallel to XY and passing through b_2' and b_2 respectively.

(iv) With centre a' and radius a' b_2', draw an arc meeting f-f at b'.

(v) With centre a and radius ab_1, draw an arc meeting t-t at b.

(vi) Join a',b', and a,b forming the required final projections.

It is observed from the Fig. 4.21(c) that:

1. The points b' and b lie on a single projection
2. The projections a'b' and ab make angles α and β with XY, which are greater than θ and ϕ. The angles α and β are known as apparent angles.

To determine the true length of a line, given its projections - Rotating line method

In this, each view is made parallel to the reference line and the other view is projected from it. This is exactly reversal of the procedure adopted in the preceding construction.

***Solution :* (Fig. 5.30)**

1. Draw the given projections a'b' and ab.
2. Draw f-f and t-t, the loci passing through b' and b and parallel to XY.
3. Rotate a' b' to a' b_2', parallel to XY.
4. Draw a projector through b_2' to meet the line t-t at b_2.
5. Rotate ab parallel to XY as ab_1.
6. Draw a projector through b_1, to meet the line f-f at b_1'.
7. Join a', b_1' and a, b_2.
8. Measure and mark the angles θ and ϕ

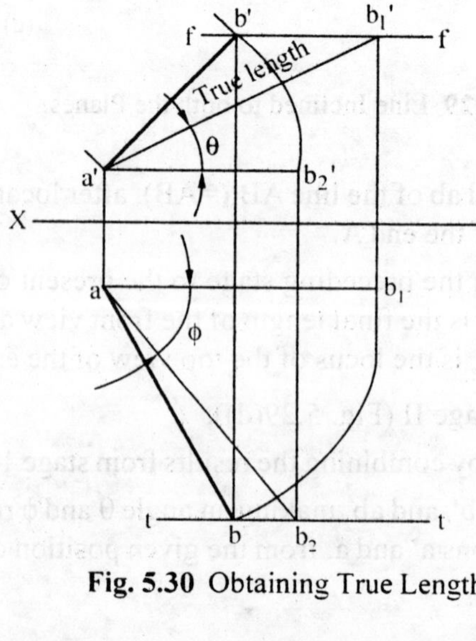

Fig. 5.30 Obtaining True Length.

Problem : Line inclined to both the planes

Solution : **(Fig. 5.31)**

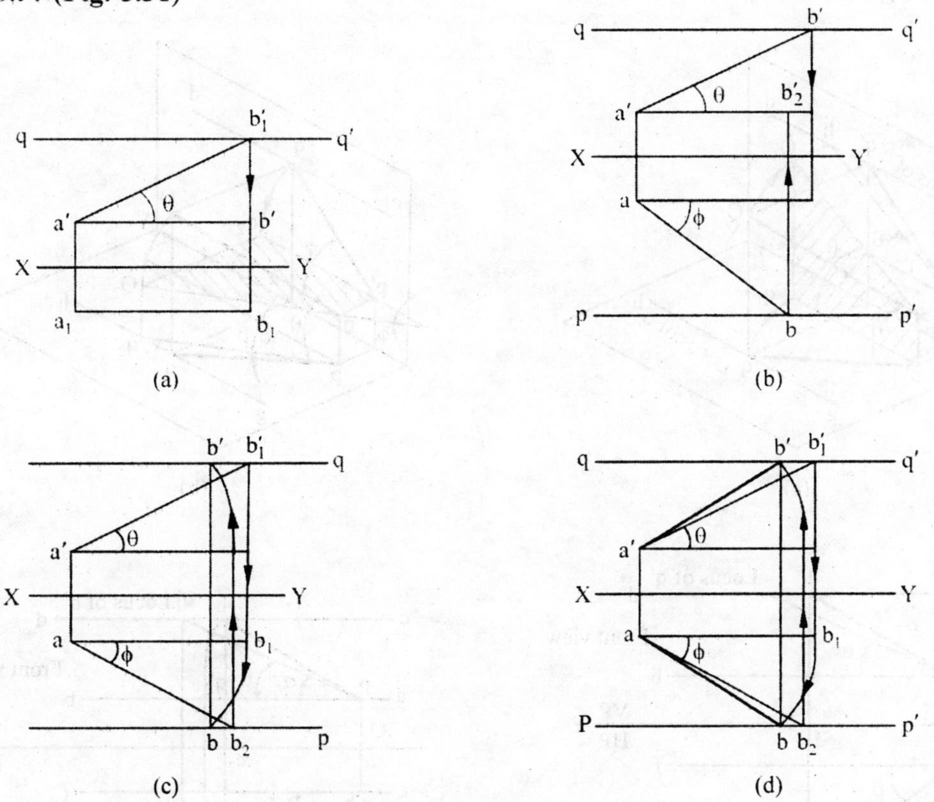

Fig. 5.31

Step 1: When a line is inclined to both the planes, (inclined θ° to H.P and ϕ° to V.P) its projections are drawn by first assuming it to be inclined at θ° to H.P and parallel to V.P. The projectins of line **AB** in this position are **a'b'** (front view) and **ab** (top view) as shown in Fig. 5.31(b).

Step II: Next projections of the line AB are drawn by assuming the line inclined at an angle ϕ° to V.P and parallel to H.P. The front view and top view of line in this portion are **a'b'** and ab as shown in Fig. 5.31(b).

Step III: Draw the locus of **b** as **pp'** and draw another locus of **b_1** as **qq'**.

Step IV: Draw an arc with **a** as centre **ab_1** as radius to cut the locus of **pp'** at **b**. Similarly draw an arc **a'** as centre, **$a'b_2'$** as radius to cut the locus of **qq'** at **b'**. Join **ab** and **a'b'** which gives the required top view and front view of the line **AB** respectively.

Problem : To determine the true length and true inclination from projection.

Solution: **(Fig. 5.32a&b).**

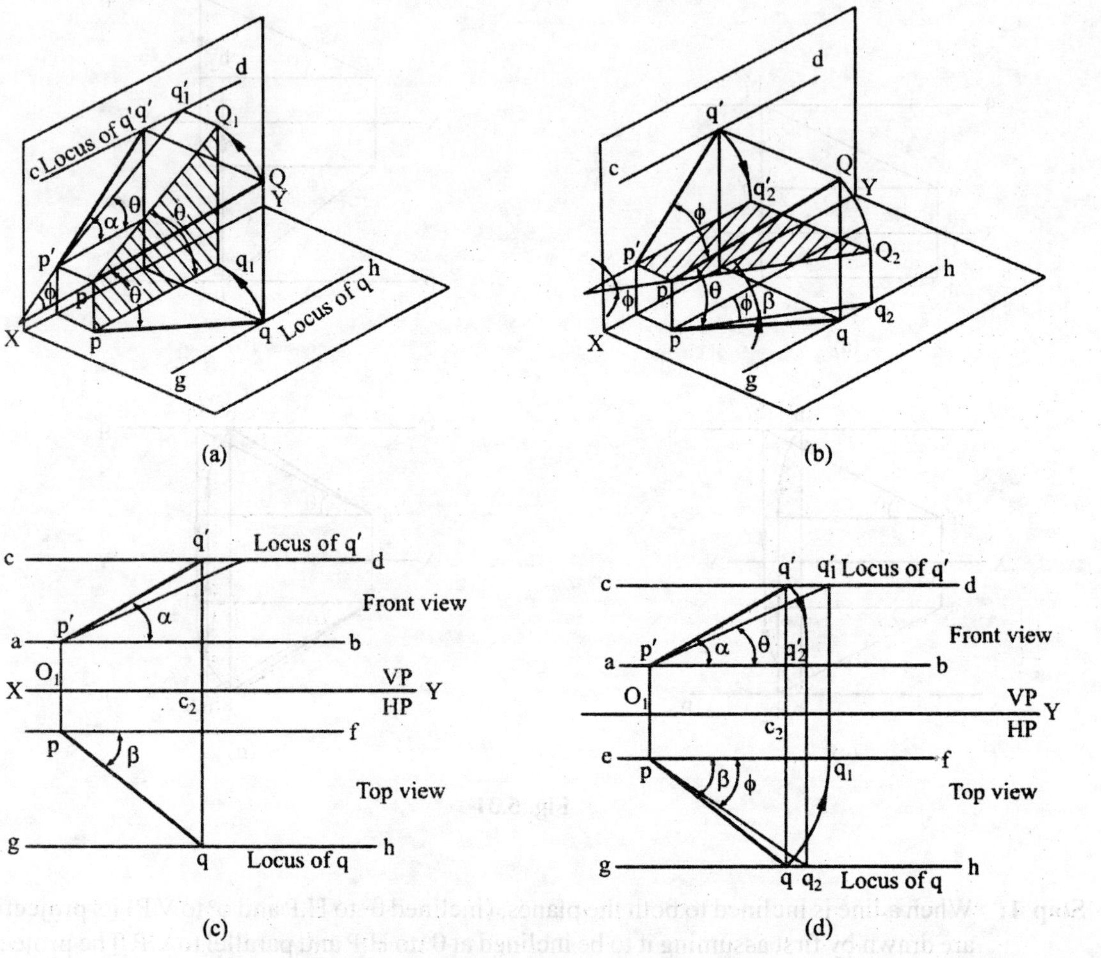

Fig. 5.32 Rotating line method.

Step I: To get the true length and true inclination from final proejction first draw the loci of two ends in both the front and top views. (Locus is the line parallel to XY and passing through that point) (Fig. 5.32a&b).

Step II: Make the top view parallel to XY by keeping any one of the end fixed and rotate the line parallel to XY line and project it on the other view to get true length and true inclination (θ) with H.P.

Step III: Make the front view parallel to XY by keeping any one of the end fixed and rotate the line parallel to XY line and project the titled end to the other view to get true inclination (ϕ) with V.P.

Problem : One end **P** of a line PQ 70 mm long is 35 mm infront of V.P and 25 mm above H.P. the line is inclined to 40° to the H.P. and 30° to the V.P. Draw the projections of PQ.

***Solution:* (Fig. 5.33)**

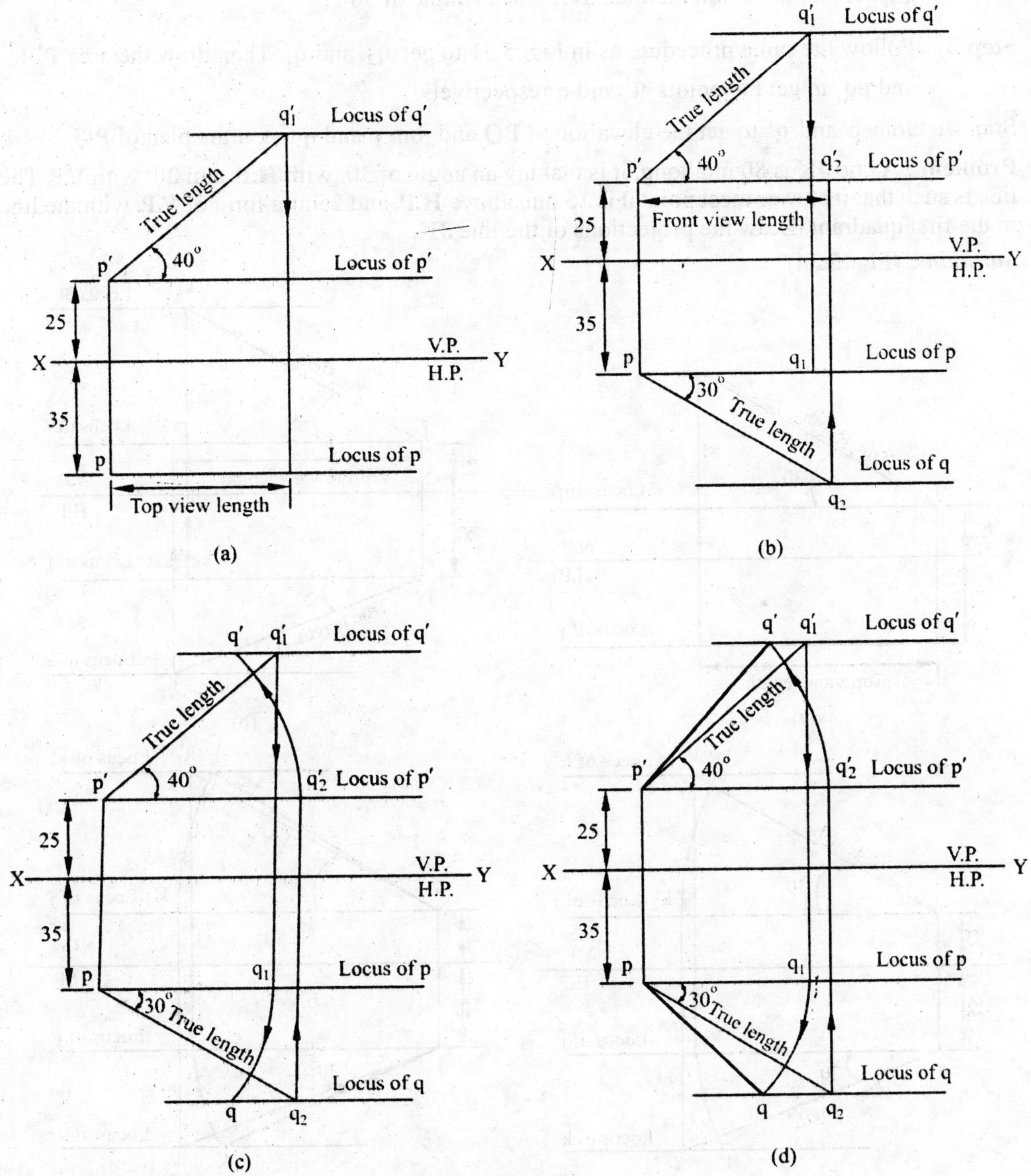

Fig. 5.33

Step 1: Draw the reference line XY and locate **p'** and p at a distance of 25 m above and 35 mm below XY line respectively.

Step 2: Draw the line **p'q₁'** of length 70 mm inclined to XY at an angle of 40° and draw the **pq₂** of length 70 mm inclined to XY at an angle of 30°.

Step 3: Follow the same procedure as in Fig. 5.31 to get **q₂'** and **q₁**. Then draw the arcs **p'q₂'** and **pq₁** to get the points **q'** and **q** respectively.

Step 4: Join **p'** and **q'** to get the elevation of **PQ** and join **p** and **q** to get the plan of **PQ**.

Problem : A line **JK** is 80 mm long, it is making an angle of 30° with **H.P.** and 20° with V.P. The line is such that its lower most point **J** is 15 mm above **H.P.** and 25 mm fornt of **V.P.** with the line in the first quadrant. Draw the projections of the line **JK**.

Solution : **(Fig. 5.34)**

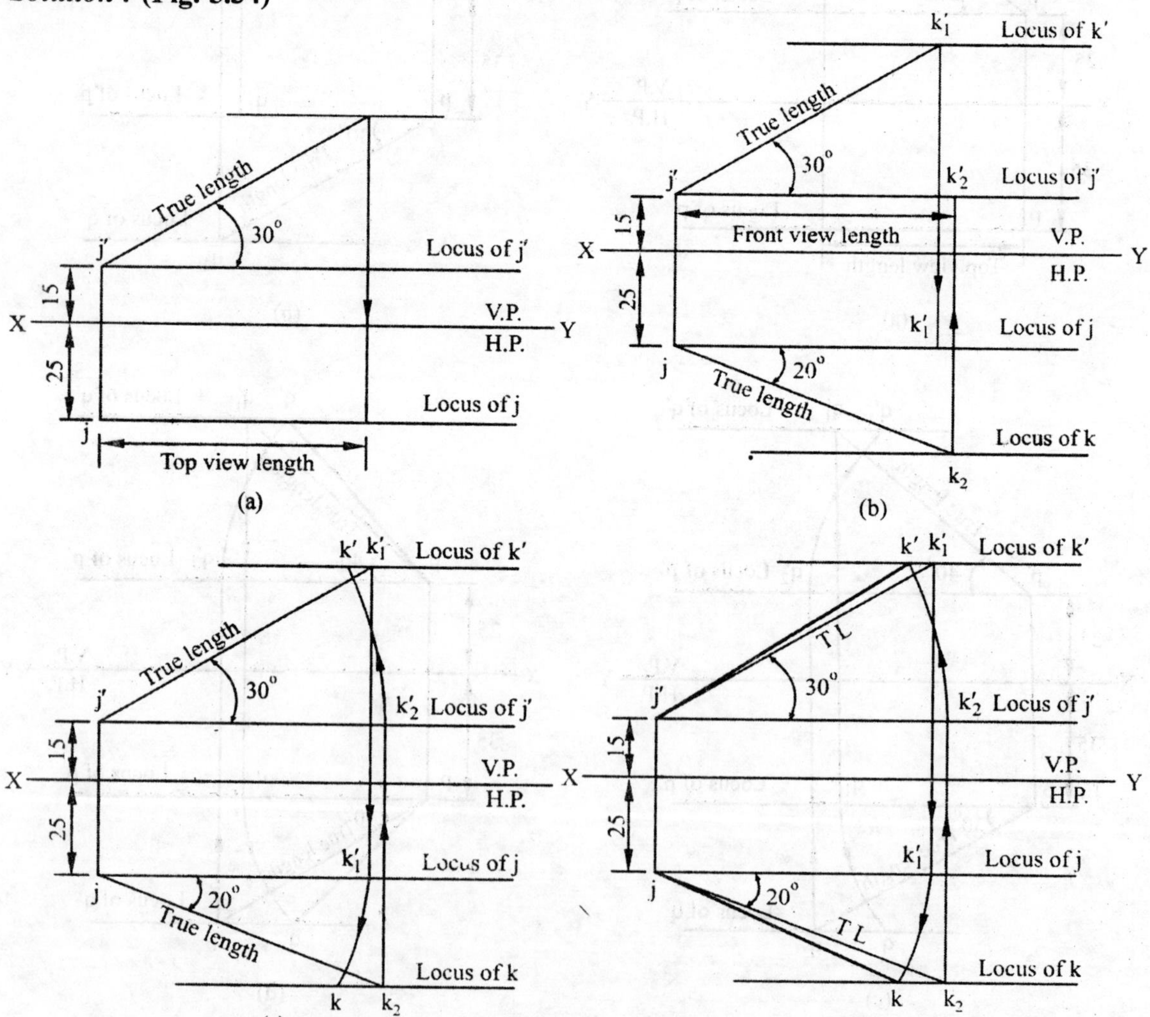

Fig. 5.34

Step 1: Draw the reference line XY and locate **j'** and **j** at a distance of 15 mm above and 25 mm below XY line respectively.

Step 2: Draw the line **$j'k_1'$** of length 80 mm inclined 30° to XY and draw the line **jk_2** of length 80 mm inclined 20° to XY.

Step 3: Follow the same procedure as in Fig. 5.3 to get **k_2'** and **k_1**. Then draw the arcs **jk** and **jk_1** to get the points **k'** and **k** repsectively.

Step 4: Join **j'** and **k'** to get the elevation of **JK** and join **j** and **k** to get the plan of **JK**.

Problem : End A of the line AB is 15 mm above H.P. and 20 mm infront of V.P. The other end is 50 mm above H.P. and 65 mm infront of V.P. The distance between the end projector is 50 mm. Draw the projection and find true inclinations and true length.

Solution: **(Fig. 5.35).**

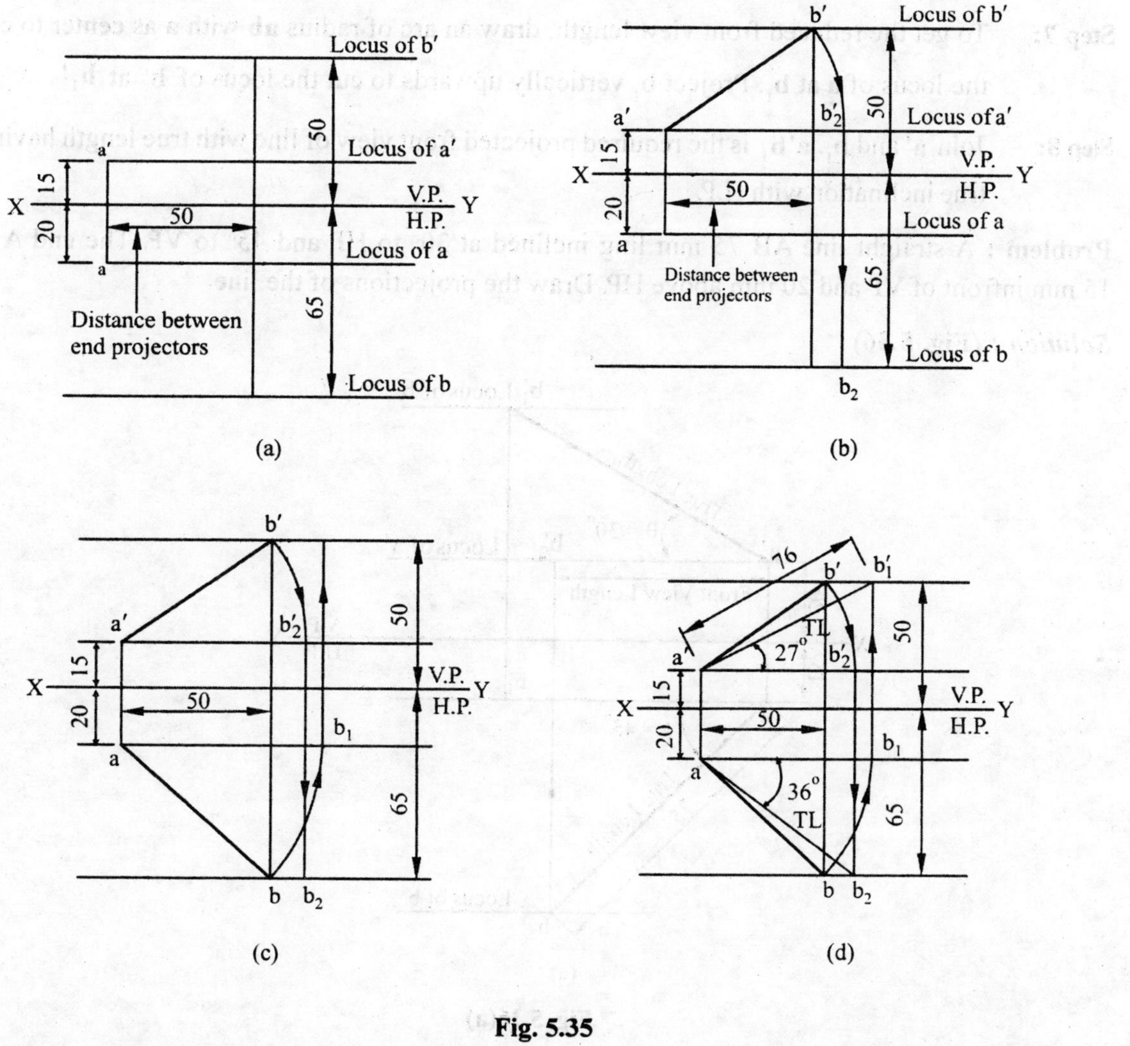

Fig. 5.35

Step 1: Draw the reference line XY and locate **a'** and **a** at a distance of 15 mm above and 20 mm blow respectively. Draw the locus of **a** and **a'**.

Step 2: To locate the other end, draw the locus of other end at a distance of 50 mm above and 65 mm below XY.

Step 3: Draw the vertical projector of other end at a distance of 50 mm.

Step 4: Join **a'** and **b'**. **a' b'** is the front view of **AC**. Join **a** and **b ab** is the top view of **AB**.

Step 5: To gets reduced top view length, draw an arc of radius **a' b'** with **a'** as centre to cut the locus of **a'** at b_2'. Project b_2' vertically downwards to cut the locus of **b** at b_2.

Step 6: Join **a** and b_2. ab_2 is the required projected top view line with true length and the inclination with V.P.

Step 7: To get the reduced front view length, draw an arc of radius **ab** with **a** as center to cut the locus of **a** at b_1. Project b_1 vertically upwards to cut the locus of **b'** at b_1'.

Step 8: Join **a'** and b_1. $a'b_1$ is the required projected front view of line with true length having true inclination with H.P.

Problem : A straight line AB 75 mm ling inclined at 30° to HP and 45° to VP. The end A is 15 mm infront of VP and 20 mm above HP. Draw the projections of the line.

Solution : **(Fig. 5.36)**

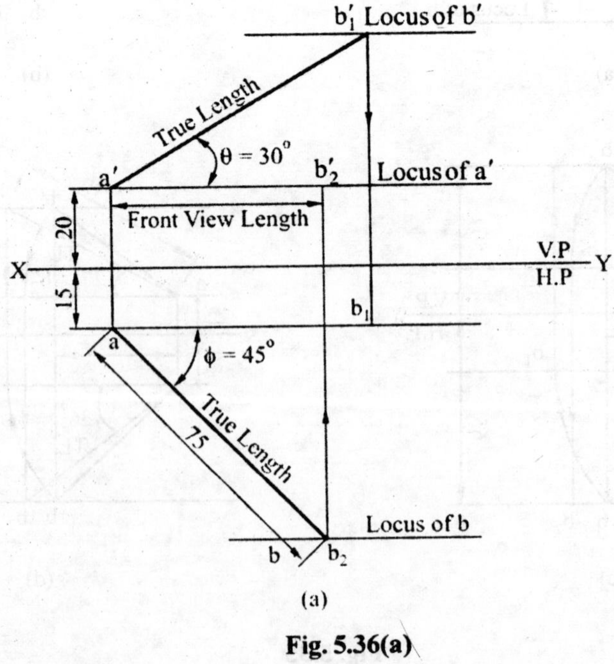

(a)

Fig. 5.36(a)

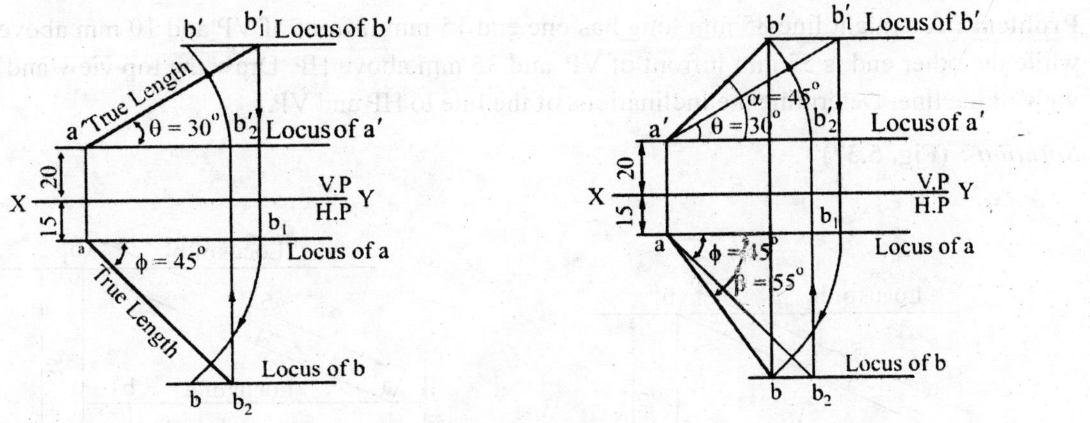

Fig. 5.36(b)&(c)

Step 1:

1. Draw reference line XY and locate **a'** at a distance 20 mm above XY and locate **a** at a distance 15 mm below XY.
2. Through **a'** draw a line of length 75 mm inclined at an angle of 30° to XY. Mark the end of the straight line as **b_1'** draw the locus through **a, a'** and **b'** parallel to XY.
3. Draw the projections from **b_1'** vertically downwards to the locus of a. Now **ab_1** is the reduced top view of 30° inclined line to HP.
4. Through **a** draw a line of length 75 mm inclined at an angle of 45° to XY. Mark the end of line b_2 and draw the locus of b_2.
5. Draw the projector from **b_2** vertically upwards to the locus of **a', a'b_2'** is the reduced front view length of 45° inclination with VP.

Step 2:

6. Draw an arc **a'** as centre and **ab'** as radius to meet the locus of **b_2** at **b**.
7. Draw an arc **a'** as centre and **a'b_2'** as radius to meet the locus of **b_1'** at **b'**.
8. Join **a'** and **b'**. It gives the front view of the line **AB**. To get the apparent inclination with H.P (α) measure the angle between a'b' and XY.
9. Join **a** and **b**. It gives the top view of line **AB**. To get the apparent inclination with VP. (β) measure the angle between ab and XY.

Problem : A straight line 85 mm long has one end 15 mm infront of VP and 10 mm above HP., while the other end is 55 mm infront of VP. and 35 mm above HP. Draw the top view and front view of the line. Determine the inclinations of the line to HP and VP.

Solution : **(Fig. 5.37)**

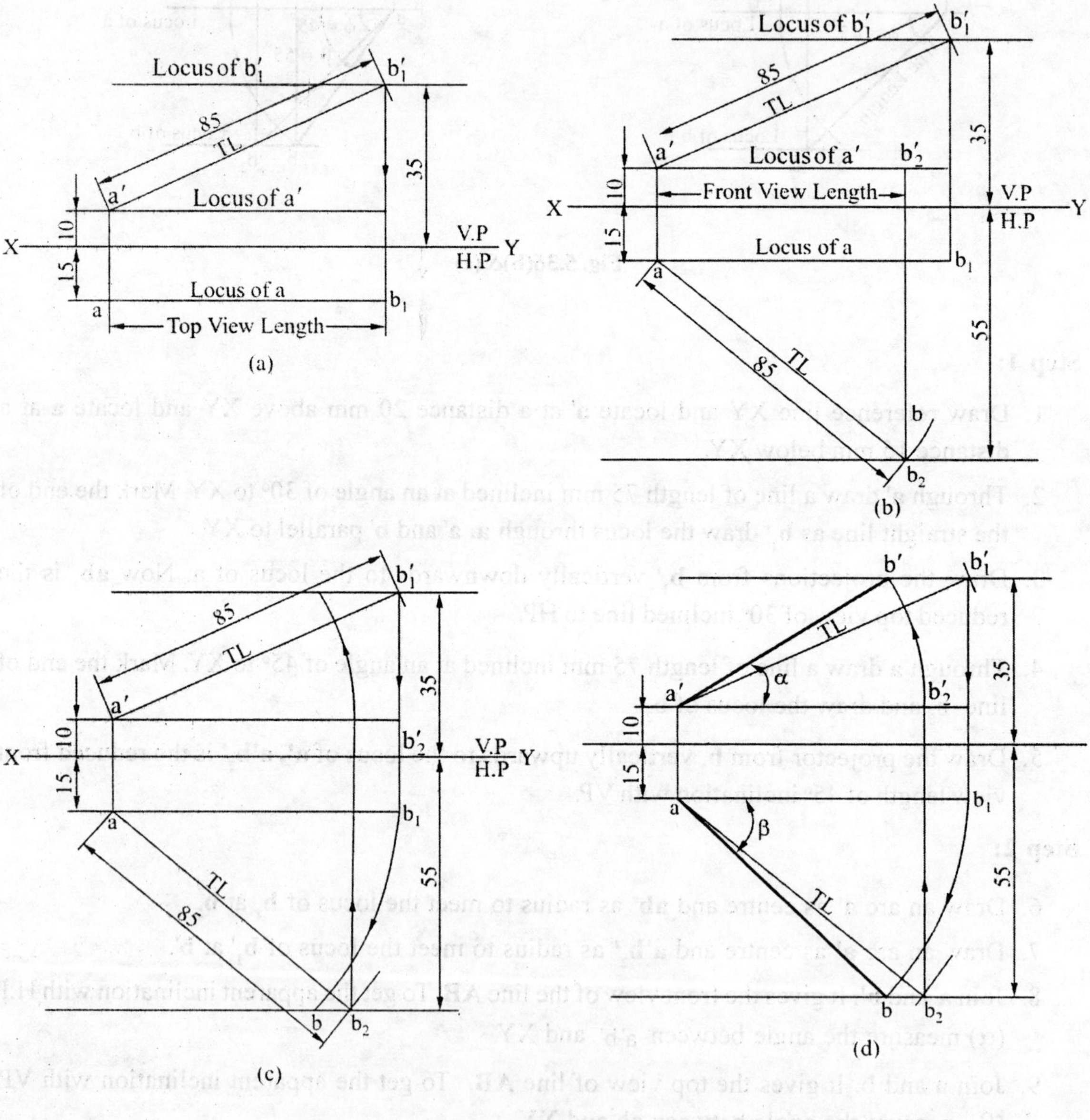

Fig. 5.37

Step 1:

1. Draw the reference line XY and locate point **a'** and a at a distance of 10 mm above and 15 mm below XY line. Draw the locus of **a'** and **a**.
2. Take the true length of the line as radius and a' as centre, cut an arc on the locus of $\mathbf{b_1}$**'** at $\mathbf{b_1}$'. Project $\mathbf{b_1}$**'** vertically downwards to cut the locus of **a** at $\mathbf{b_1}$.

Step 2:

3. Draw the locus of $\mathbf{b_1}$**'** at a distanceof 35 mm above XY line and draw another locus of $\mathbf{b_2}$ at a distance of 55 mm below XY line.
4. Draw an arc of true length of line as radius and **a** as centre to cut the locus of $\mathbf{b_2}$ at $\mathbf{b_2}$. Project $\mathbf{b_2}$ vertically upwards to cut the locus of a at $\mathbf{b_2}$**'**.

Step 3:

5. Draw an arc of radius **a'**$\mathbf{b_2}$**'** with centre **a'** to cut the locus of $\mathbf{b_1}$**'** at **b'**.
6. Draw an arc of radius **a**$\mathbf{b_1}$ with centre **a** to cut the locus of $\mathbf{b_2}$ at **b**.

***Step* 4:**

7. Join **a** and **b**, **ab** is the plan of line **AB**.

Problem : A line PQ 60 mm long, has its end P 15 mm above HP. and 20 mm infront of VP. Its top view and front view measures 50 mm and 40 mm respectively. Draw its projections and determine its true inclinations with HP and VP.

***Solution :* (Fig. 5.38)**

1. Draw the reference line XY and locate **p'** and **p** at a distance of 15 mm above and 20 mm below XY line respectively. Draw locus of **p** and **p'**.
2. Mark $\mathbf{q_1}$ of distance of 50 mm (top view length) from **p** on the locus of **p**.
3. Project $\mathbf{q_1}$ vertically upwards and cut it with an arc having true length 60 mm. Mark it as $\mathbf{q_1}$**'** and join **p** and $\mathbf{q_1}$**'** . Draw the locus of $\mathbf{q_2}$.
4. Mark $\mathbf{q_2}$**'** on the locus of **p'** at a distance of 40 mm (front view length).
5. Project $\mathbf{q_2}$**'** vertically downwards and cut it with an arc having true length 60 mm. Mark it as $\mathbf{q_2}$ and join **p** and $\mathbf{q_2}$.
6. From **p'**, cut an arc of radius **p'**$\mathbf{q_1}$**'** with **p'** as centre on the locus of $\mathbf{q_1}$**'** to mark **q'**. Join **p'** and **q'**. **p'q'** is the required elevation of line **PQ**.
7. From **p**, cut an arc of radius **p**$\mathbf{q_1}$ with **p** as centre on the locus of $\mathbf{q_2}$ to mark **q**. Join **p** and **q**. **pq** is the required plan of line **PQ**.
8. Angle between **p'q'** and XY lines is the true inclination of **PQ** with HP. Angle between **p**$\mathbf{q_2}$ and XY line is the true inclination of **PQ** with VP.

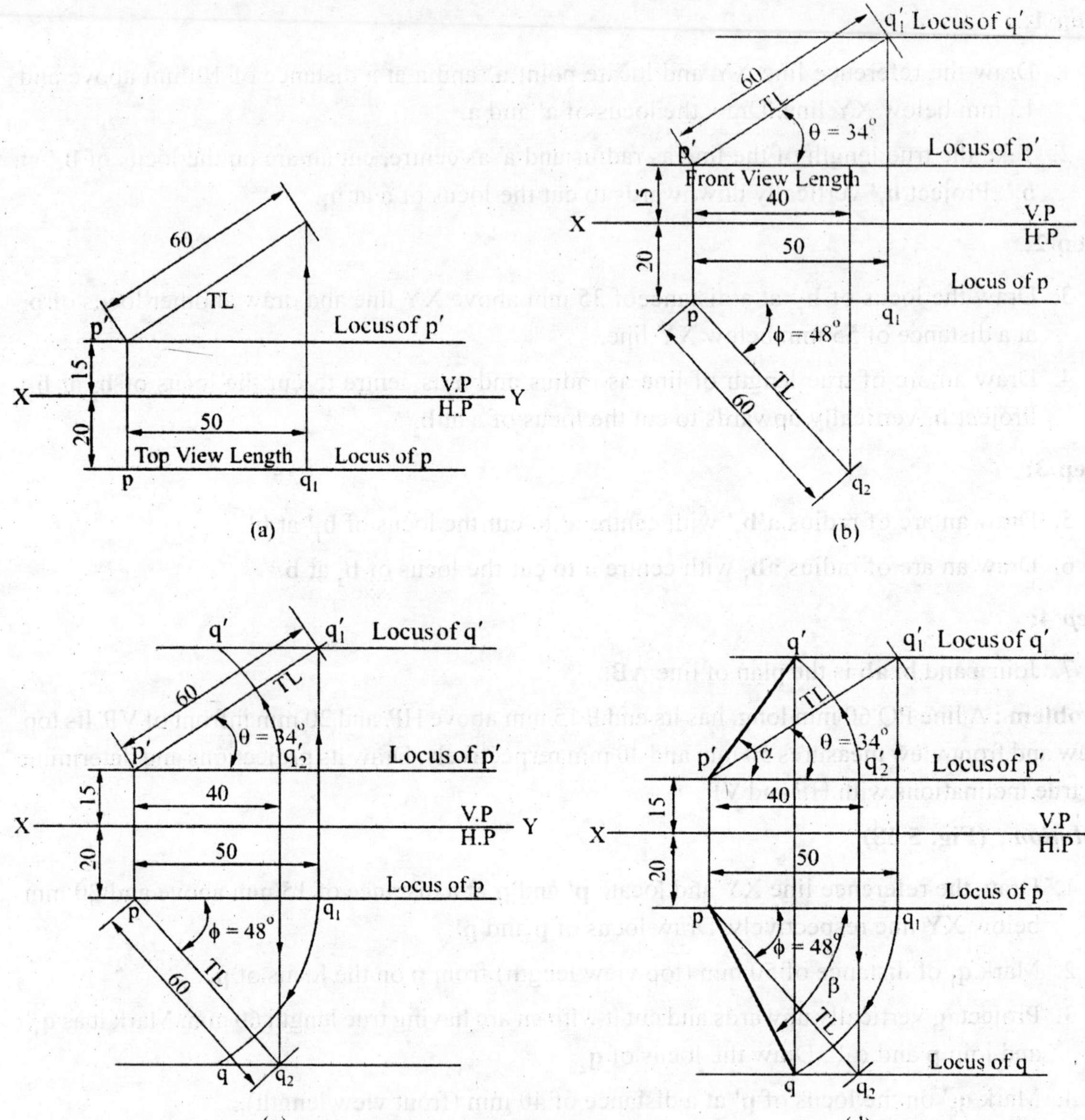

Fig. 5.38

Problem : A line AB 60 mm long, has its end A 30 mm above HP. and 25 mm infront of VP. The top view and front view has a length of 40 mm and 55 mm respectively. Draw its projections.

Solution : **(Fig. 5.39)**

1. Draw the reference line XY and locate **a'** and a at a distance of 30 mm and 25 mm below XY respectively. Draw locus of a and **a'**.

2. Mark **b_1** at a distance of 40 mm (top view length) from **a** on the locus of **a**. Mark **b_2'** at a distance of (front view length) 55 mm from **a'** on the locus of **a'**.
3. Follow the same procedure as in previous problems to find the ends **b_1'** and **b_2**, draw locus of **b_1'** and **b_2** and mark **b'** and **b** on them.
4. **a'b'** gives the required elevation of line **AB** and ab gives the required plan of line **AB**.

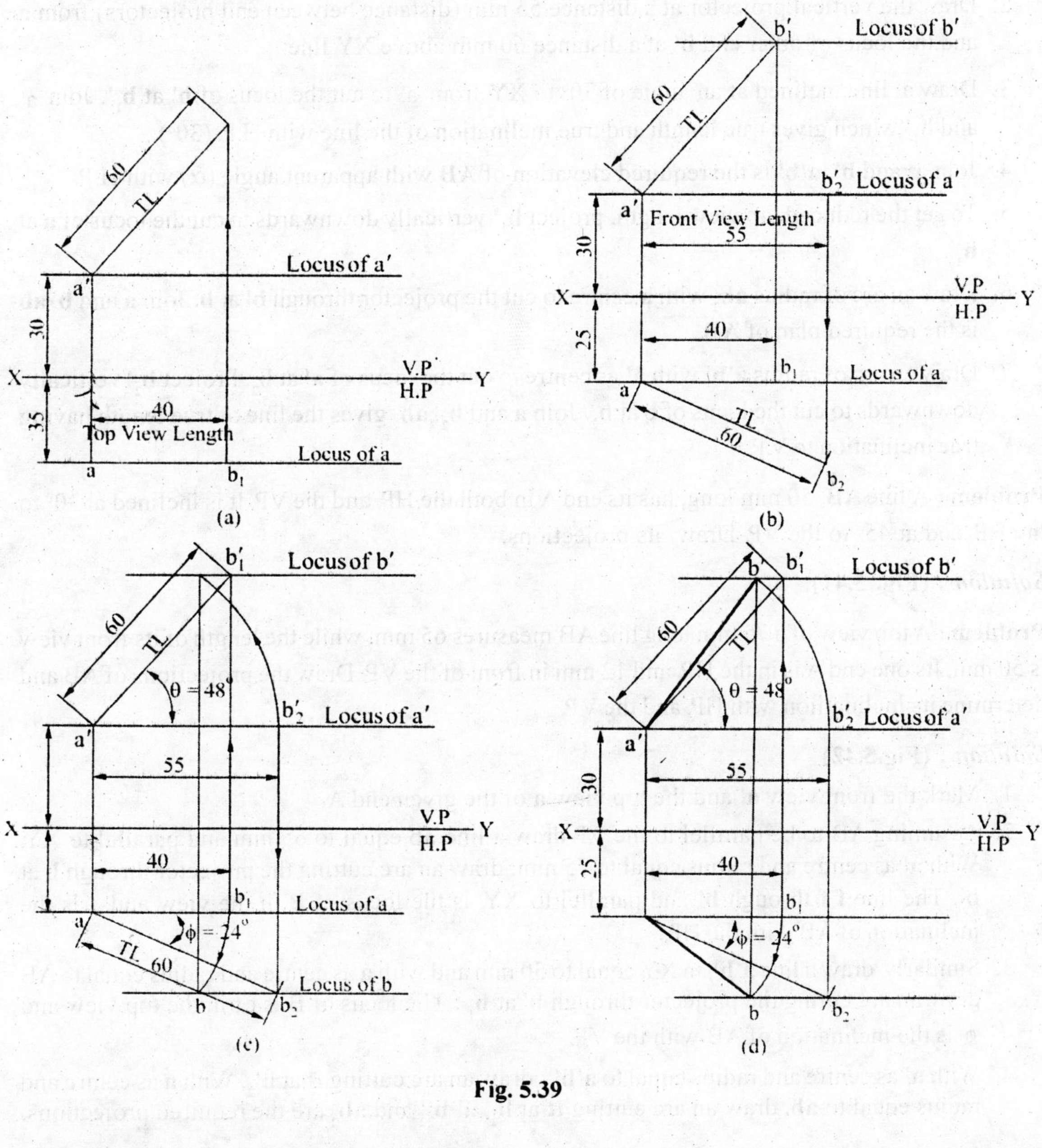

Fig. 5.39

Problem : A line AB has its end A 15 mm above HP. and 20 mm infront of VP. The end B is 60 mm above HP. and the line is inclined 30° to HP. The distance between and end projectors of the line is 55 mm. Draw the projections and find its inclination with VP. and HP.

Solution : **(Fig. 5.40)**

1. Draw the reference line XY and locate **a'** and a at a distance of 15 mm above and 20 mm below XY respectively. Draw the locus of **a'** and **a**.
2. Draw the vertical projector at a distance 55 mm (distance between end projectors) from **a** and the locus of other end **b'** at a distance 60 mm above XY line.
3. Draw a line inclined at an angle of 30° to XY from a' to cut the locus of **b'** at **b_1'**. Join a' and **b_1'** which gives true length and true inclination of the line with H.P. (30°).
4. Join **a'** and **b'**. **a'b'** is the required elevation of **AB** with apparent angle (α) with H.P.
5. To get the reduced top view length, project **b_1'** vertically downwards to cut the locus of **a** at **b_1**.
6. Draw an arc of radius **ab_2** with **a** centre to cut the projector through **b'** at **b**. Join **a** and **b**. **ab** is the required plan of **AB**.
7. Draw an arc of radius **a'b'** with **a'** as centre to cut the locus of **a'** at **b_2**. Project **b_2'** vertically downwards to cut the locus of **b** at **b_2**. Join **a** and **b_2**. **ab_2** gives the line of true length having true inclination to V.P.

Problem : A line AB, 50 mm long, has its end A in both the HP. and the VP. It is inclined at 30° to the HP. and at 45° to the VP. Draw its projections.

Solution : **(Fig. 5.41).**

Problem : A top view of a 75 mm long line AB measures 65 mm, while the length of its front view is 50 mm. Its one end A is in the HP. and 12 mm in front of the VP. Draw the projections of AB and determine its inclinatiion with HP. and the VP.

Solution : **(Fig.5.42)**

1. Mark the front view **a'** and the top view **a** of the given end A.
2. Assuming AB to be parallel to the VP draw a line **ab** equal to 65 mm and parallel to XY. With **a'** as centre and radius equal to 75 mm, draw an arc cutting the projector through **b** at **b'**. The line **f f** through **b'** and parallel to XY. is the locus of B in the view and θ is the inclination of AB with the HP.
3. Similarly, draw a line **a'b'**, in XY equal to 50 mm and with **a** as centre and radius equal to AB draw an arc cuting the projector through **b'**,at **b_1**. The locus of B is **t t** in the top view and ϕ is the inclination of AB with the VP.
4. With **a'** as centre and radius equal to **a'b_1'**, draw an arc cutting **ff** at **b_2'**. With **a** as centre and radius equal to **ab**, draw an arc cutting **tt** at **b_2**. **a' b_2'** and **ab_2** are the required projections.

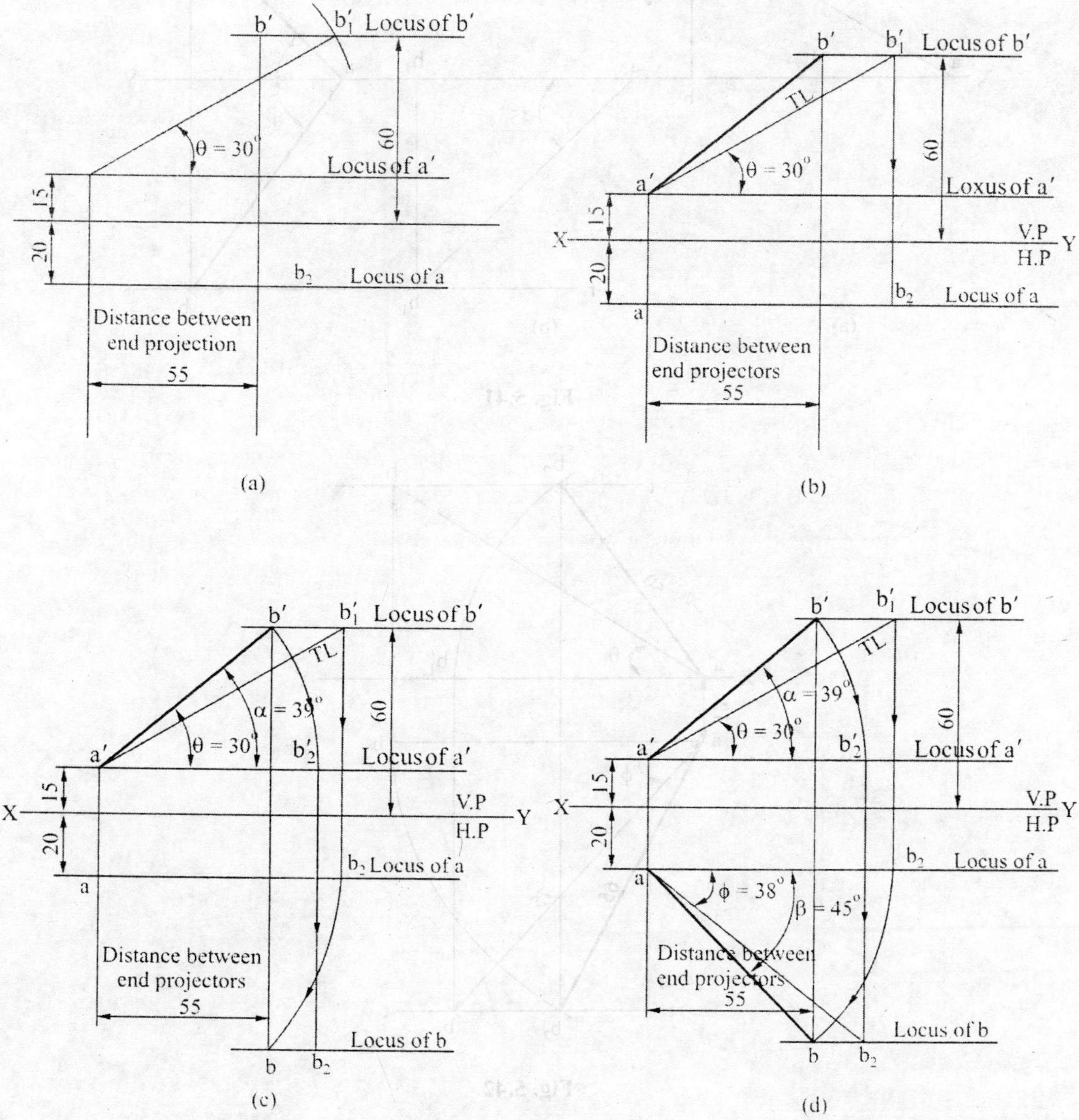
b′
b′1 Locus of b′
θ = 30°
60
Locus of a′
15
20
b2 Locus of a
Distance between
end projection
55
(a)
b′
b′1 Locus of b′
TL
θ = 30°
60
a′
Loxus of a′
15
X
V.P
H.P
Y
20
b2 Locus of a
a
Distance between
end projectors
55
(b)
b′
b′1 Locus of b′
TL
α = 39°
θ = 30°
60
a′
b′2
Locus of a′
15
X
V.P
H.P
Y
20
a
b2 Locus of a
Distance between
end projectors
55
Locus of b
b
b2
(c)
b′
b′1 Locus of b′
TL
α = 39°
θ = 30°
60
a′
b′2
Locus of a′
15
X
V.P
H.P
Y
20
a
b2 Locus of a
ϕ = 38°
β = 45°
Distance between
end projectors
55
Locus of b
b
b2
(d)

Fig. 5.40

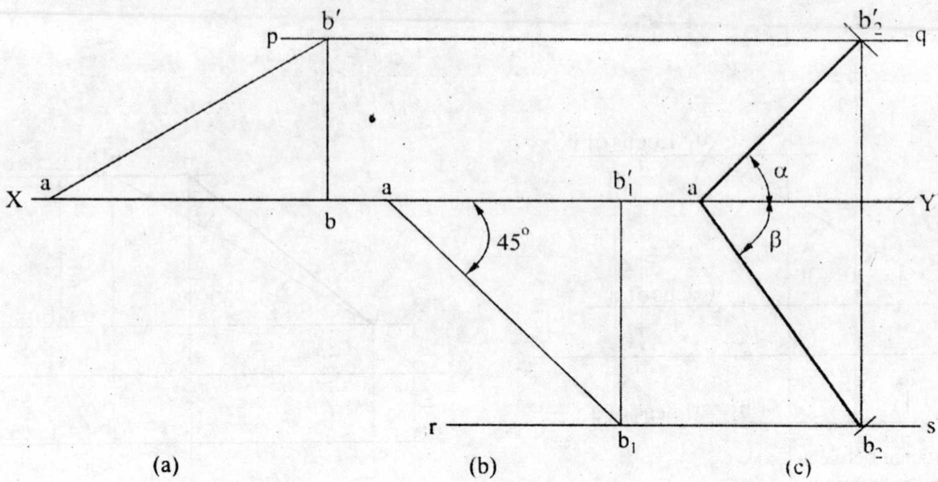

Fig. 5.41

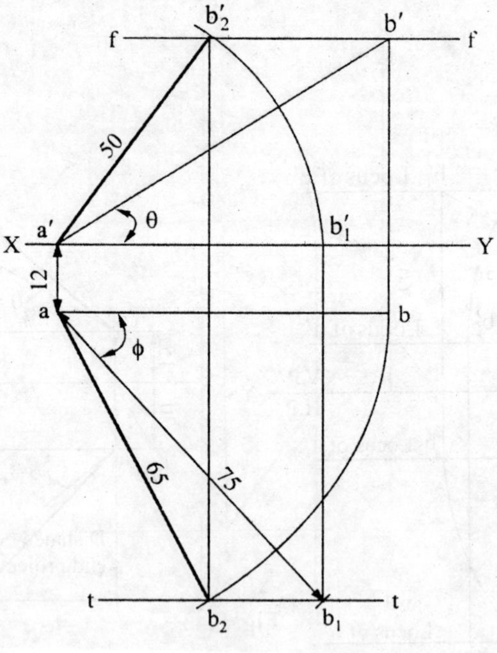

Fig. 5.42

Problem : A line AB, 90 mm long, is inclined at 30° to the HP. Its end A is 12 mm above the HP. and 20 mm in front of the VP. Its front view measures 65 mm. Draw the top view of AB and determine its inclination with the VP.

***Solution* : (Fig. 5.43)**

1. Mark **a** and **a'** the projections of the end A. Through **a'**, draw a line **a'b'** 90 mm long and making an angle of 30° with XY.
2. With **a'** as centre and radius equal to 65 mm, draw an arc cutting the path of **b'** at b_1'. $a'b_1'$ is the front view of AB.
3. Project b' to b_1 so that ab is parallel to XY. ab is the length of AB in the top view.
4. With a as centre and radius equal to ab_1 draw an arc cutting the projector through **b'**, at b_1. Join **a** with b_1. ab_1 is the required top view.
5. ϕ is the inclination of AB with VP.

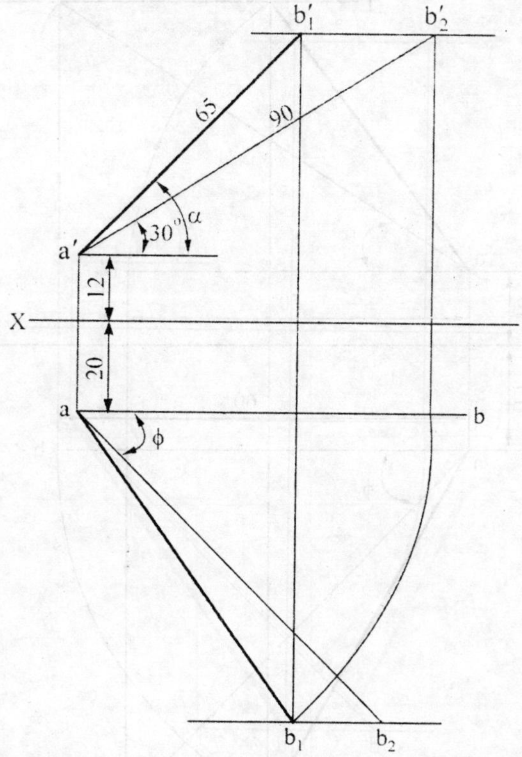

Fig. 5.43

Problem : A line AB of 70 mm long, has its end A at 10 mm above HP and 15 mm in front of VP. Its front view and top view measure 50 mm and 60 mm respectively. Draw the projections of the line and dermine its inclinations with HP. and VP.

***Solution* : (Fig. 5.44)**

1. Draw the reference line XY and locate the projections **a, a'** of the end A.
2. Draw $\mathbf{a'b_2'}$**=50 mm**, parallel to XY, representing the length of the front view.

3. With centre **a** and radius 70 mm (true length), draw an arc intersecting the projector through **b'$_2$ at b$_2$**.
4. Join **a', b$_2$.**
5. Draw ab$_1$ (=60 mm), parallel to XY, representing the length of the top view.
6. With centre **a'** and radius 70 mm (true length), draw an arc intersecting the projector through **b$_1$ at b$_1$'**.
7. Through **b$_1$'**, draw the line f-f, representing the locus of front view of B.
8. Through **b$_2$**, draw the line t-t, representing the locus of top view of B.

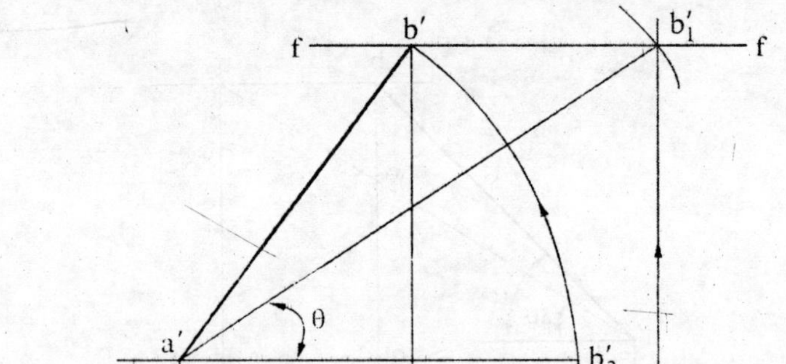

Fig. 5.44

9. With centre **a** and radius **a b$_2$'** , draw an arc intersecting f-f at **b'**.
10. Join **a', b'**, representing the front view of the line.
11. With centre **a** and radius **ab$_1$,** draw an arc intersecting t-t at **b**.
12. Join **a,b** repesenting the top view of the line.

5.7 TRACES

Traces of a line

When a line is inclined to a plane, it will meet that plane when produced if necessary. The point at which the line or line produced meets the plane is called its trace.

The point of intersection of the line or line produced with HP is called Horizontal Trace (HT) and that with VP is called Vertical Trace (VT).

To find HT and VT of a line for its various positions with respect to HP and VP.

1. Line parallel to HP and perpendicular to VP.

Problem : A line AB 25 mm long is parallel to HP. and perpendicular to VP. The end is 10 mm in front of VP and the line is 20 mm above HP. Draw the projections of the line and find its traces.

***Solution :* (Fig. 5.45)**

1. Draw the front view **a' (b')** and top view **ab**.
2. AB is perpendicular to VP.

   Therefore mark VT, in the front view to coincide with **a'(b')**.
3. AB is parallel to HP. Therefore it has no HT.

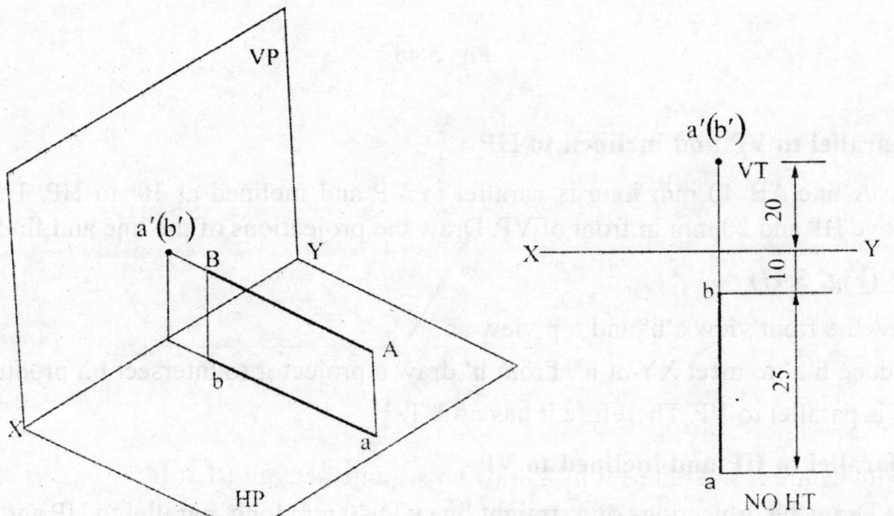

Fig. 5.45

2. Line parallel to VP. and perpendicular to HP.

Problem : A line CD 25 mm long is parallel to VP and perpendicular to HP. End C is 35 mm above HP. and 20 mm in front of VP. End D is 10 mm above HP. Draw the projections of the line CD and find its traces.

***Solution* : (Fig. 5.46)**

1. Draw the front view **c'd'** and top view **cd.**
2. Produce **c'd'** to meet XY at **h'**. From **h'** draw a projector to intersect **CD** produced at HT.
3. CD is parallel to VP. Therefore it has no VT.

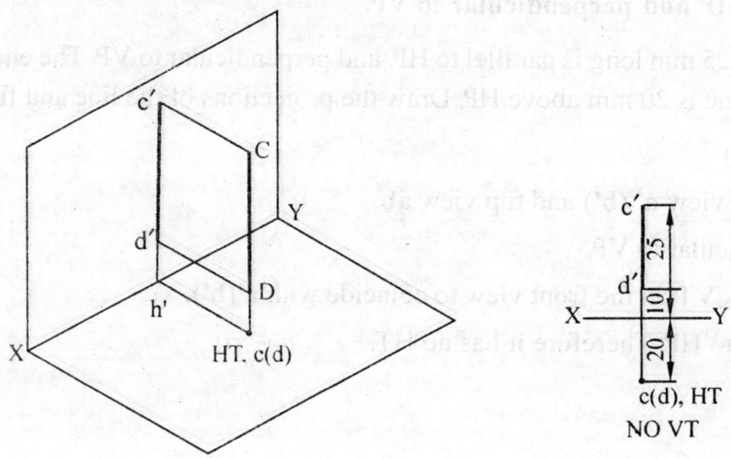

Fig. 5.46

3. Line parallel to VP and inclined to HP

Problem : A line AB 40 mm long is parallel to VP and inclined at 30° to HP. The end A is 15 mm above HP and 20 mm in front of VP. Draw the projections of the line and find its traces.

***Solution* : (Fig. 5.47)**

1. Draw the front view a'b' and top view ab.
2. Produce **b'a'** to meet XY at **h'**. From **h'** draw a projector to intersect **ba** produced at HT.
3. AB is parallel to VP. Therefore it has no VT.

4. Line parallel to HP and inclined to VP

Problem : Draw the projections of a straight line CD 40 mm long, parallel to HP and inclined at 35° to VP. The end C is 20 mm above HP. and 15 mm in front of VP. Find its traces.

***Solution* : (Fig. 5.48)**

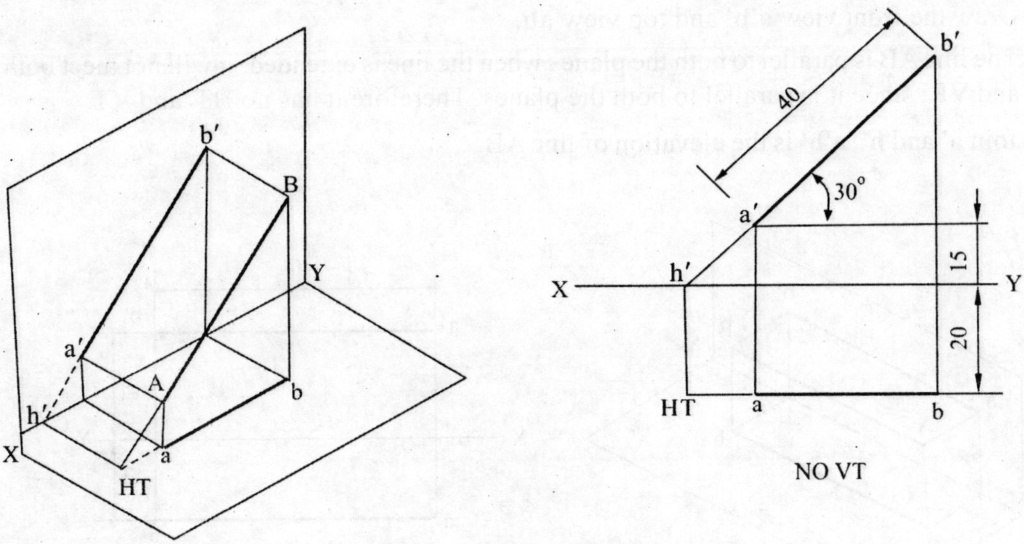

Fig. 5.47

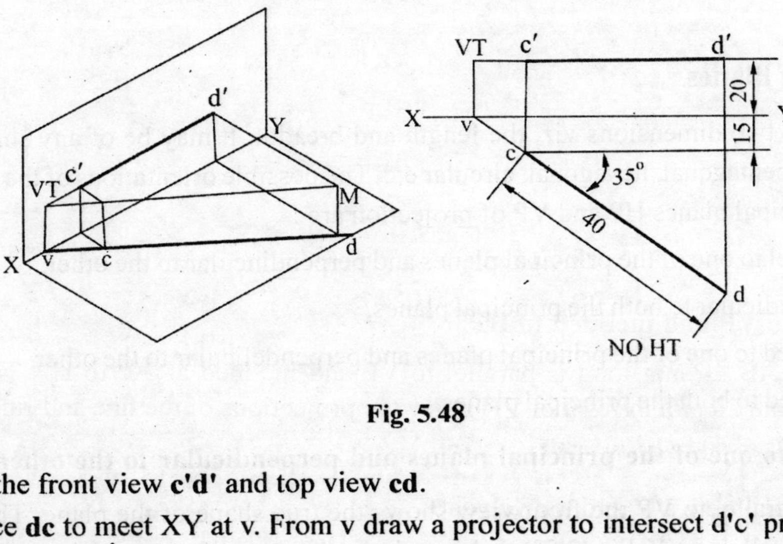

Fig. 5.48

1. Draw the front view **c'd'** and top view **cd**.
2. Produce **dc** to meet XY at v. From v draw a projector to intersect d'c' produced at VT.
3. CD is parallel to HP.
4. Therefore it has no HT.

5. Line parallel to both HP and VP

Problem : *A line AB 40 mm long is parallel to both the planes. The line is 20 mm above HP and 15 mm in front of VP. Draw the projections and find its traces.*

Solution **: (Fig. 5.49)**

1. Draw the front view **a'b'** and top view **ab.**
2. The line AB is parallel to both the planes when the line is extended, it will not meet both HP. and VP., since it is parallel to both the planes. Therefore it has no HT. and VT.
8. Join **a'** and **b'**. **a'b'** is the elevation of line AB.

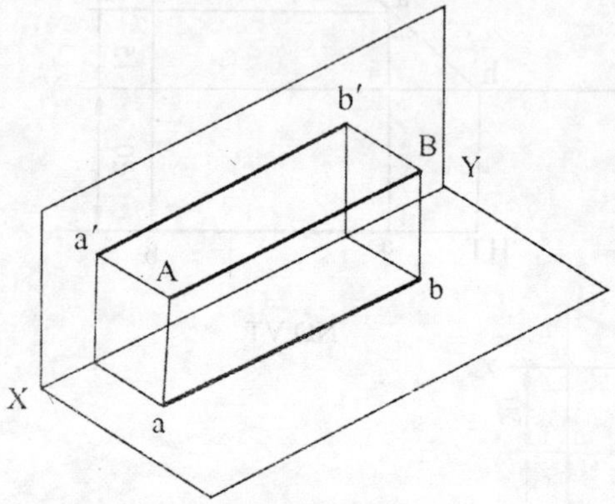

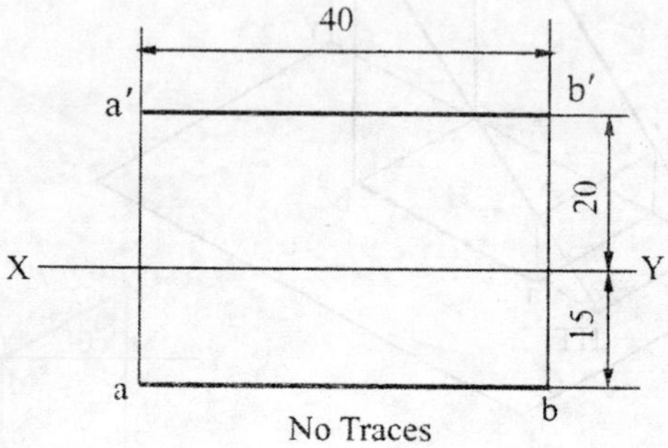

Fig. 5.49

5.8 Projection of Planes

A plane figure has two dimensions viz. the length and breadth. It may be of any shape such as triangular, square, pentagonal, hexagonal, circular etc. The possible orientations of the planes with respect to the principal planes HP and VP of projection are :

1. Plane parallel to one of the principal planes and perpendincular to the other,
2. Plane perpendicular to both the principal planes,
3. Plane inclined to one of the principal planes and perpendclicular to the other,
4. Plane inclined to both the principal planes.

1. Plane parallel to one of the principal planes and perpendicular to the other

When a plane is parallel to VP the front view shows the true shape of the plane. The top view appears as a line parallel to XY. Fig. 5.50(a) shows the projections of a square plane ABCD, when it is parallel to VP and perpendicular to HP. The distances of one of the edges above HP and from the VP are denoted by d_1 and d_2 respecively.

Fig. 5.50(b) shows the projections of the plane. Fig. 5.50(c) shows the projections of the plane, when its edges are equally inclined to HP.

Fig. 5.51 shows the projections of a circular plane, parallel to HP and perpendicular to VP

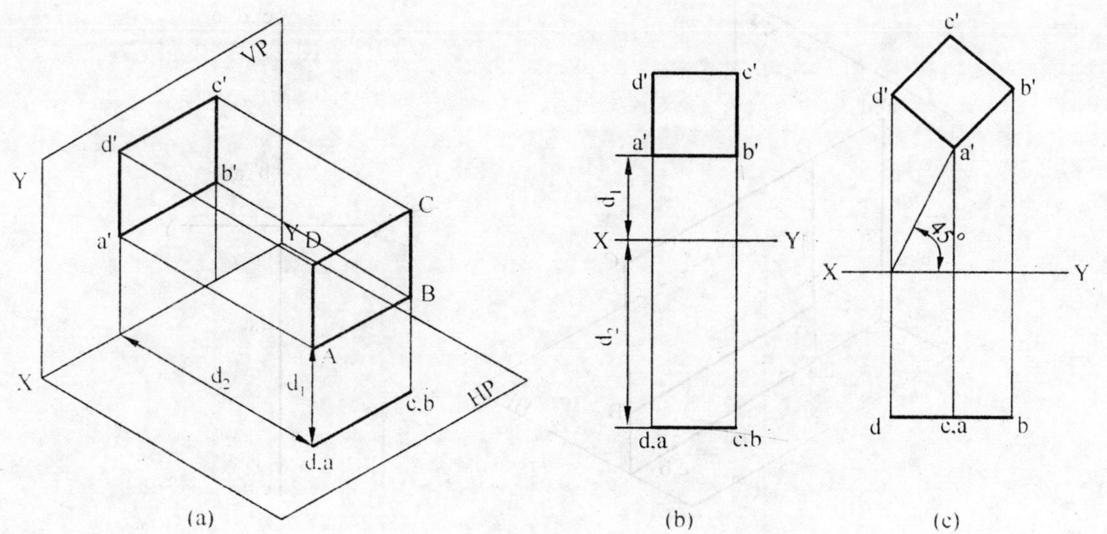

Fig. 5.50

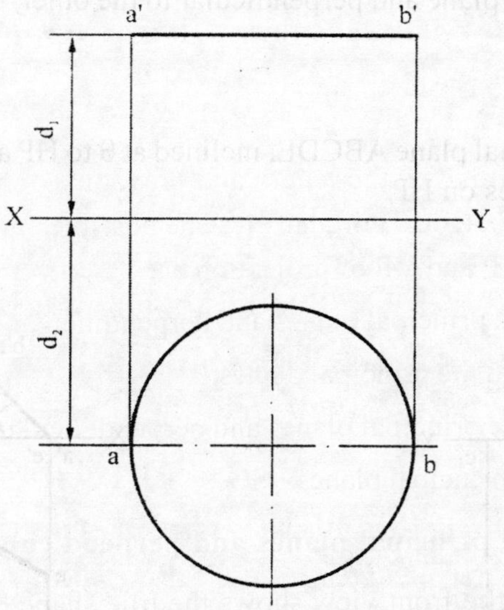

Fig. 5.51

2. Plane perpendicular to both HP and VP.

When a plane is perpendicular to both HP. and VP, the projections of the plane appear as straight lines. Fig. 5.52 shows the projections of a rectangular plane ABCD, when one of its longer edges is parallel to HP. Here, the lengths of the front and top views are equal to the true lengths of the edges.

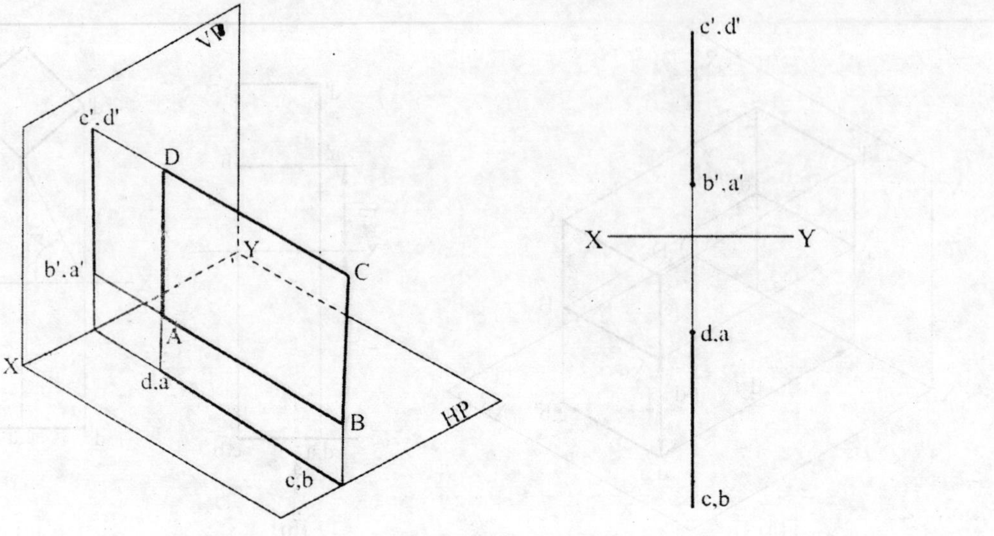

Fig. 5.52

3. Plane inclined to one of the principal planes and perpendicular to the other

When a plane is inclined to one plane and perpendicular to the other, the projections are obtained in two stages.

Problem :

(i) Projections of a pentagonal plane ABCDE, inclined at θ to HP and perpendicular to VP and resting on one of its edges on HP.

***Solution :* (Fig. 5.53)**

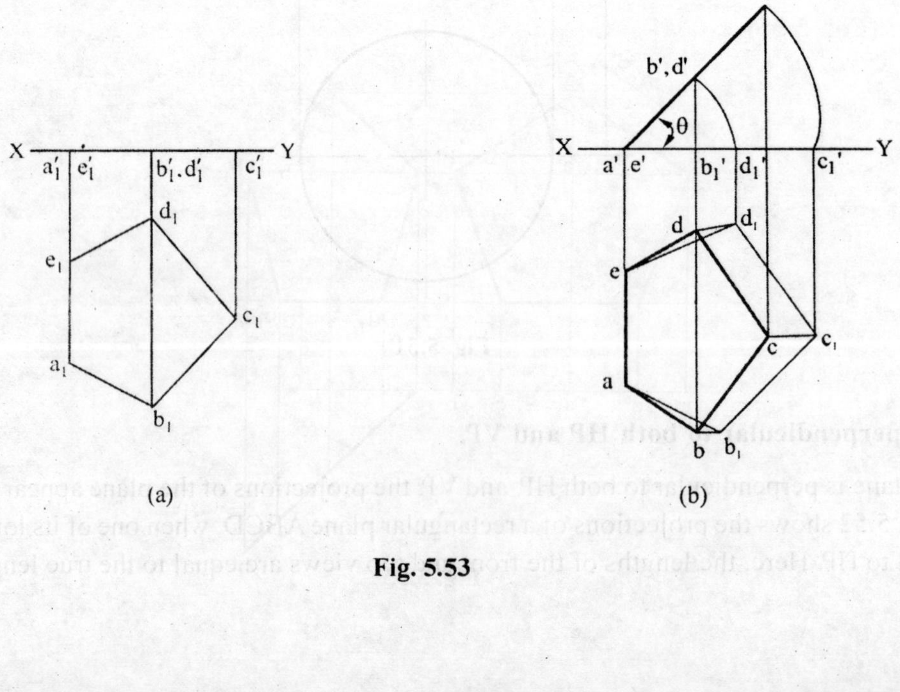

Fig. 5.53

Step II Rotate the plane (front view) till it makes the given angle with HP.

2. Rotate the front view till it makes the given angle θ with XY which is the final front view.
3. obtain the final top view abcde by projection.

Problem : Following the method similar to the above, the projections are obtained in Fig.5.54 for hexagonal plane, inclined at ϕ to VP and perpendicular to HP, with the edge parallel to HP.

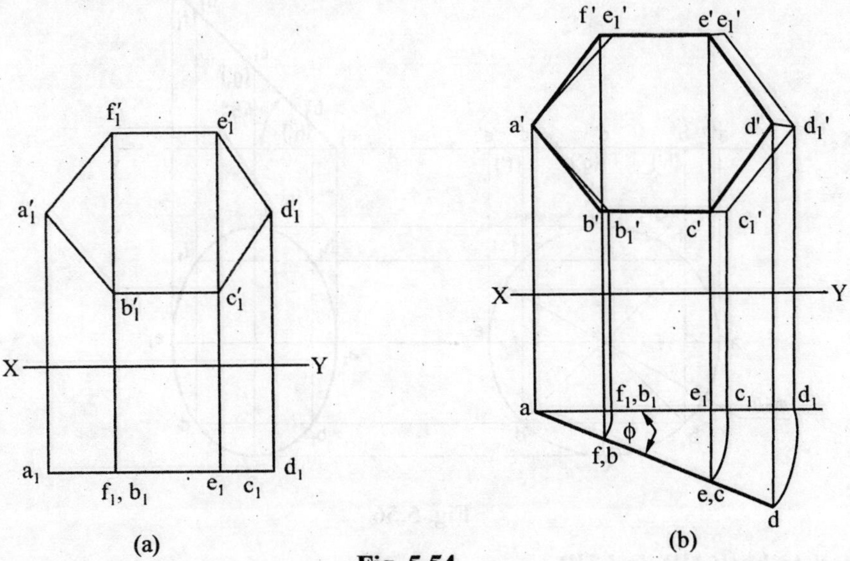

Fig. 5.54

Problem : A regular pentagon ABCDE, of side 25 mm side has its side CD on ground. Its plane is perpendicular to H.P and inclined at 45° to the V.P. Draw the projections of the pentagon when its corner nearest to V.P. is 15 mm from it.

Solution **: (Fig. 5.55)**

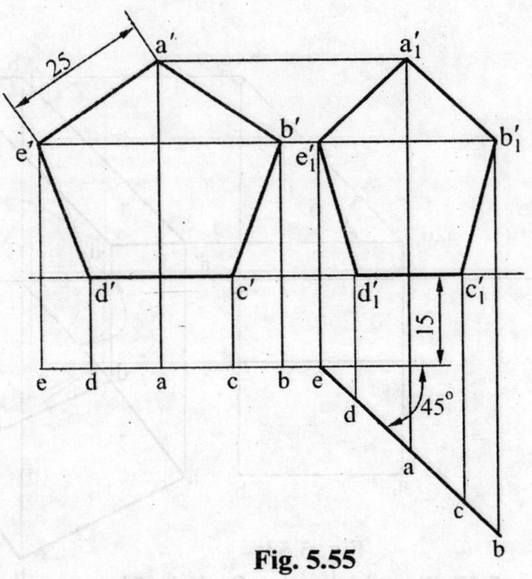

Fig. 5.55

Problem: A circular plate of diameter 50 mm is resting on *HP* on a point on the circumference with its surface inclined at 45° to *HP* and perpendicular to *VP*. Draw its projections.

Solution (Fig. 5.56)

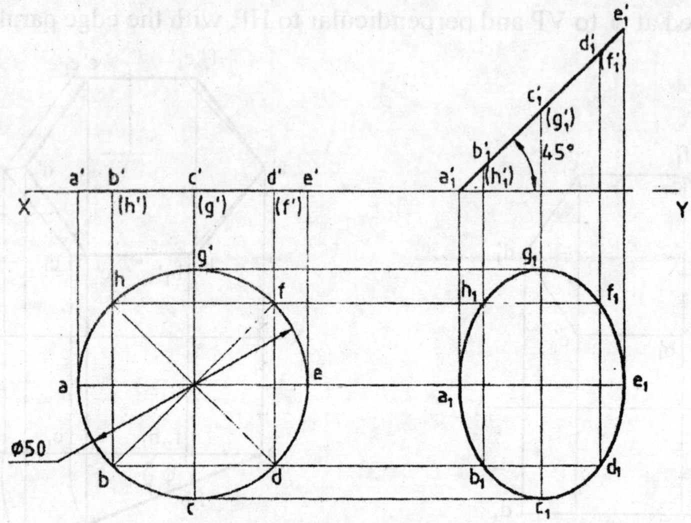

Fig. 5.56

Plane inclined to both HP and VP

If a plane is inclined to both HP and VP, it is said to be an oblique plane. Projections of oblique planes are obtained in three stages.

Problem : A rectangular plane ABCD is inclined to HP by an angle θ, its shorter edge being parallel to HP and inclined to VP by an angle ϕ. Draw its projections.

Solution : **(Fig. 5.57)**

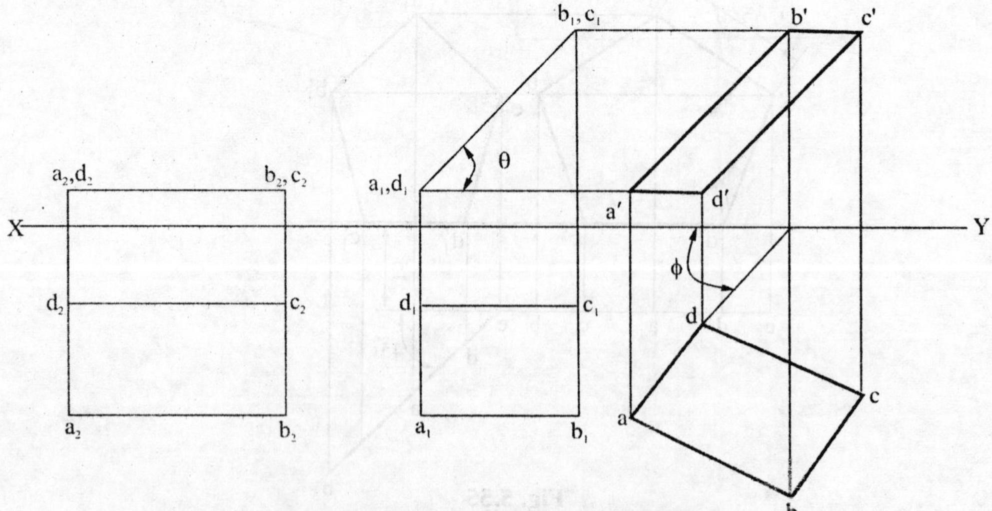

Fig. 5.57 Plane inclined to Both the Planes.

Step 1: Assume that the plane is parallel to HP and a shorter edge of it is perpendicular to VP.

1. Draw the projections of the plane.

Step 2 : Rotate the plane till it makes the given angle with HP.

2. Redraw the front view, making given angle θ with XY and then project the top view.

Step 3 : Rotate the plane till its shorter edge makes the given angle ϕ with VP.

3. Redraw the top view abcd such that the shorter edge **ad**, is inclined to XY by ϕ.
4. Obtain the final front view a'b'c'd', by projection.

Problem : A pentagonal plane ABCDE of 35 mm side has its plane inclined 50° to H.P. Its diameter joining the vertex B to the mid point F of the base DE is inclined at 25° to the XY-line. Draw its projections keeping the corner B nearer to V.P.

Solution **: (Fig. 5.58).**

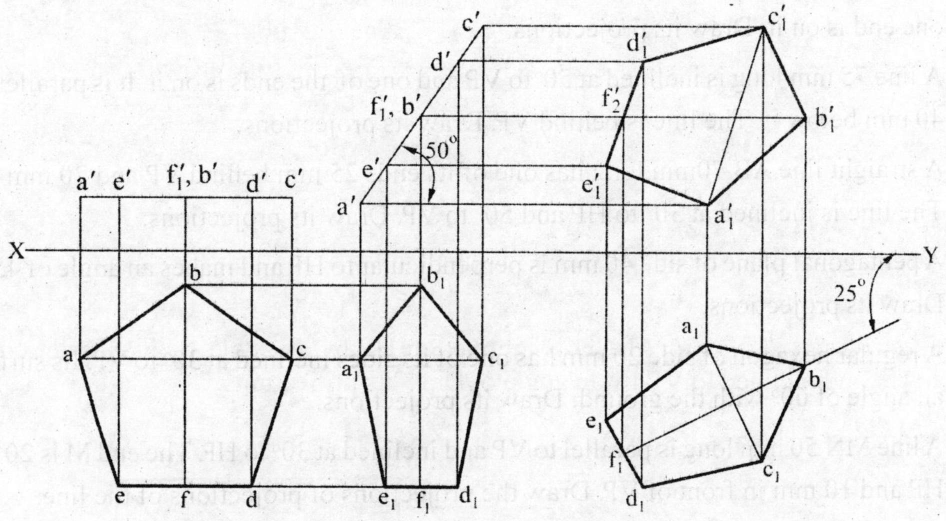

Fig. 5.58

EXERCISES

1. Draw the projections of A point a which is at 40 mm above HP and 25 mm in front of VP.
2. A point A is at 55 mm above HP and 40 mm behind VP. Draw its projection.
3. A point A is lying at 30 mm behind VP and 60 mm below HP. Draw its projections.
4. Draw the projections of a point A which lies at 40 mm below the HP and 70 mm in front of VP.
5. Draw the projections of a straight line 70 mm long when it is parallel to both HP and VP. It is 15 mm in front of VP and 40 mm above HP.
6. A straight line of length 70 mm is parallel to VP and perpendicular to HP. It's one end is 20 mm below the HP and 50 mm behind VP. Draw its orthographic projections.
7. A line of length 70 mm is parallel and 20 mm in front of VP. It is also inclined at 45° to HP and one end is on it. Draw its projections.
8. A line 75 mm long is inclined at 50° to VP and one of the ends is on it. It is parallel to HP and 40 mm below it. The line is behind VP. Draw its projections.
9. A straight line AB 70mm long has one of its ends 25 mm behind VP and 20 mm below HP. The line is inclined at 30° to HP and 50° to VP. Draw its projections.
10. A pentagonal plane of side 40 mm is perpendicular to HP and makes an angle of 45° with VP. Draw its projections.
11. A regular hexagon of side 20 mm has one of its sides inclined at 30° to VP. Its surface makes an angle of 60° with the ground. Draw its projections.
12. A line MN 50 mm long is parallel to VP and inclined at 30° to HP. The end M is 20 mm above HP and 10 mm in front of VP. Draw the projections of projections of the line.
13. A line PQ 40 mm long is parallel to VP and inclined at an angle of 30° to HP. The end P is 15 mm above HP and 20 mm in front of VP. Draw the projections of the line.

CHAPTER 6

Projection of Solids

6.1 Introduction

A solid has three dimensions, the length, breadth and thickness or height. A solid may be represented by orthographic views, the nuber of which depends on the type of solid and its orientation with respect to the planes of projection. solids are classified into two major groups. (i) Polyhedra, and (ii) Solids of revolution

6.1.1 Polyhedra

A polyhedra is defined as a solid bounded by plane surfaces called faces. They are :

(i) Regular polyhedra (ii) Prisms and (iii) Pyramids.

6.1.2 Regular Polyhedra

A polyhedron is said to be regular if its surfaces are regular polygons. The following are some of the regular plolyhedra.

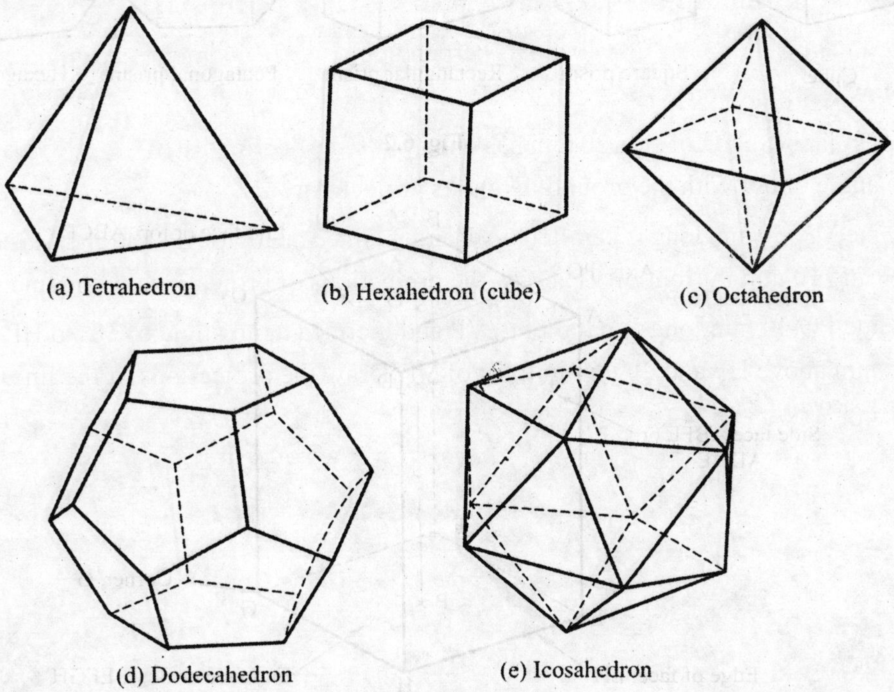

(a) Tetrahedron (b) Hexahedron (cube) (c) Octahedron

(d) Dodecahedron (e) Icosahedron

Fig. 6.1

(a) Tetrahedron: It consists of four equal faces, each one being an equilateral triangle.

(b) 'Hexa hedron(cube): It consists of six equal faces, each a square.

(c) Octahedron : It thas eight equal faces, each an equilateral triangle.

(d) Dodecahedron : It has twelve regular and equal pentagonal faces.

(e) Icosahedron : It has twenty equal, equilateral triangular faces.

6.2 Prisms

A prism is a polyhedron having two equal ends called the bases prallel to each other. The two bases are joined by faces, which are rectangular in shape. The imaginary line passing through the centres of the bases is called the axis of the prism.

A prism is named after the shape of its base. For example, a prism with square base is called a square prism, the one with a pentagonal base is called a pentagonal prism, and so on (Fig. 6.2). The nomenclature of the prism is given in Fig. 6.3.

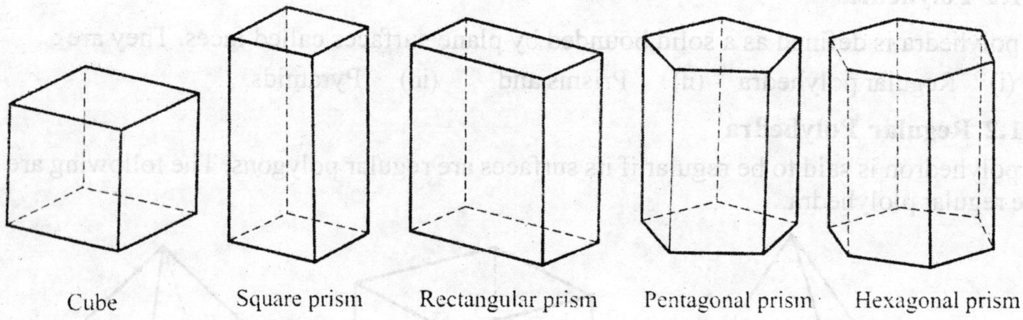

Fig. 6.2

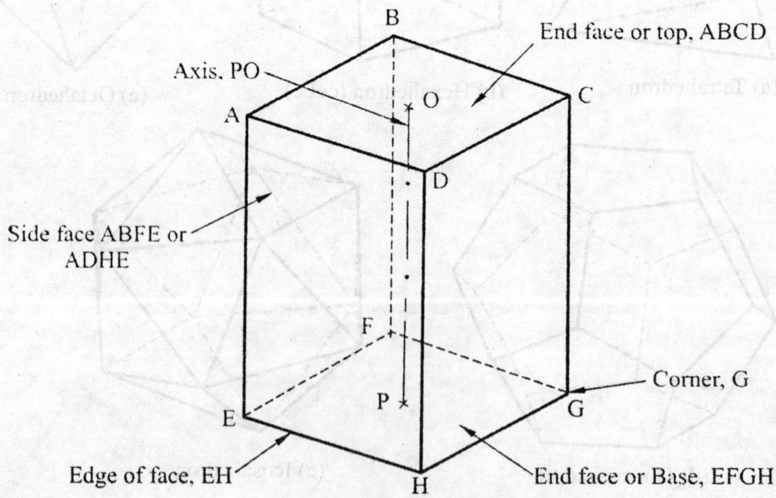

Fig. 6.3 Nomenclature of a Square Prism.

6.3 Pyramids

A pyramid is a polyhedron having one base, with a number of isosceles triangular faces, meeting at a point called the apex. The imaginary line passing through the centre of the base and the apex is called the axis of the pyramid.

The pyramid is named after the shape of the base. Thus, a square pyramid has a square base and pentagonal pyramid has pentagonal base and so on (Fig. 6.4(a)). The nomenclature of a pyramind is shown in Fig. 6.4(b).

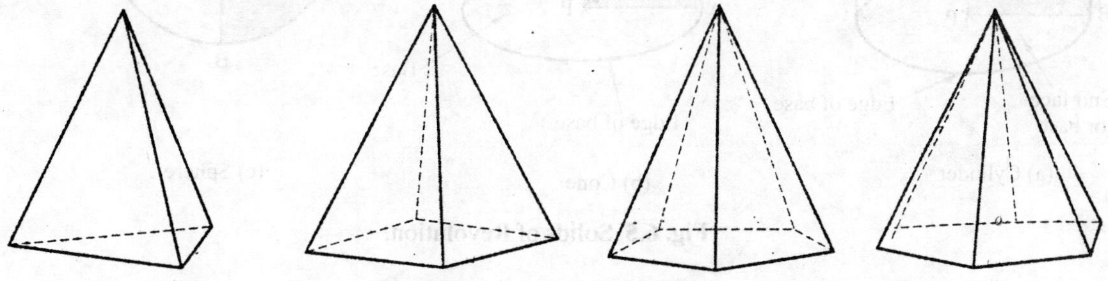

Fig. 6.4(a) Pyramids.

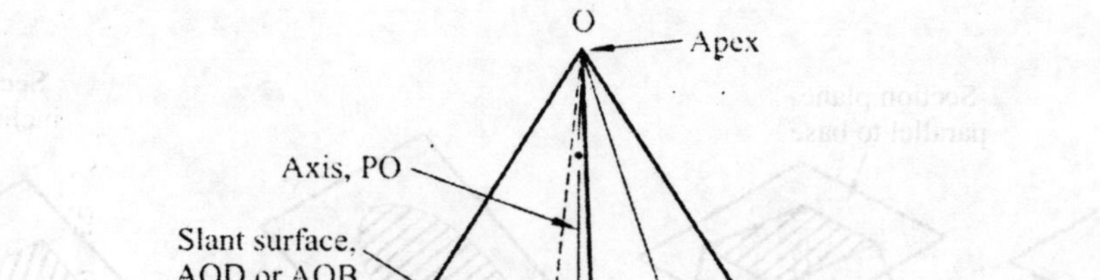

Fig. 6.4(b) Nomenclature of a Square Pyramid.

6.4 Solids of Revolution

If a plane surface is revolved about one of its edges, the solid generated is called a solid of revolution. The examples are (i) Cylinder, (ii) Cone, (iii) Sphere (Fig. 6.5).

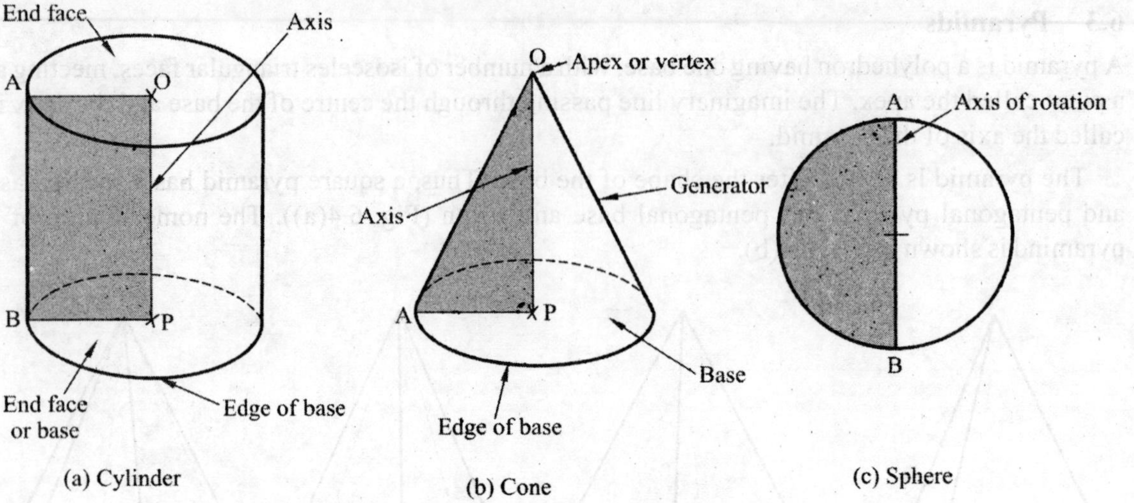

Fig. 6.5 Solids of Revolution.

6.5 Frustums and Truncated Solids

If a cone or pyramid is cut by a section plane parallel to its base and the portion containing the apex or vertex is removed, the remaining portion is called frustum of a cone or pyramid. When the cutting plane is inclined to the axis they are called truncated solids (Fig. 6.6).

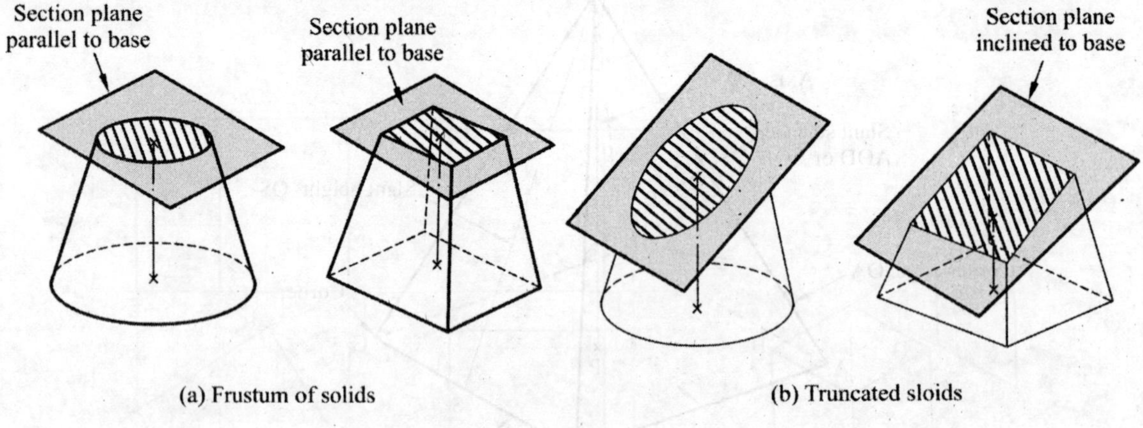

Fig. 6.6 Frustum of a Solids and Truncated Solids.

6.6 Examples in Prisms
Position of a Solid with Respect to the Reference Planes

The position of solid in space may be specified by the location of either the axis, base, edge, diagonal or face with the principal planes of projection. The following are the positions of a solid considered.

1. Axis perpendicular to one of the principal planes.
2. Axis parallel to both the principal planes.
3. Axis inclined to one of the principal planes and parallel to the other.
4. Axis inclined to both the principal planes.

The position of solid with reference to the principal planes may also be grouped as follows:

1. Solid resting on its base.
2. Solid resting on any one of its faces, edges of faces, edges of base, generators, slant edges, etc.
3. Solid suspended freely from one of its corners, etc.

1. Axis perpendicular to one of the principal planes

When the axis of a solid is perpendicular to one of the planes, it is parallel to the other. Also, the projection of the solid on that plane will show the true shape of the base.

When the axis of a solid is perpendicular to HP, the top view must be drawn first and then the front view is projected from it. Similarly when the axis of the solid is perpendicular to VP, the front view must be drawn first and then the top view is projected from it.

Problem : *Draw the projections of a cube of 35 mm side, resting on one of its faces (bases) on HP., such that one of its vertical faces is parallel to and 10 mm in front of VP.*

Solution : **(Fig. 6.7(b))**

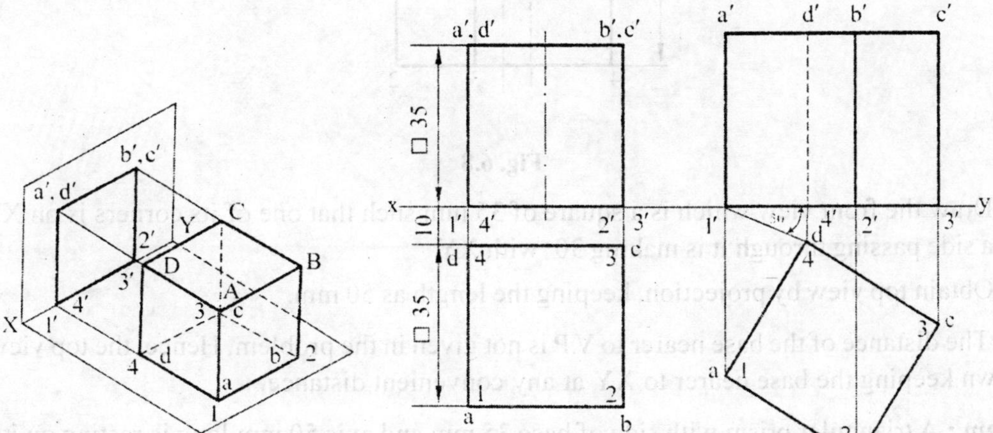

Fig. 6.7

Fig 6.7(a) shows the cube positioned in the first quadrant.

1. Draw the top view such that one of its edges is 10 mm below XY.
2. Obtain the front view by projecrtion, keeping one of its bases on XY.

Note : (i) For the cube considered ABCD is the top base and 1234 the bottom base, (ii) Fig. 6.7(c) shows the projections of a cube, resting on one of its bases on HP. such that an edge of its base is inclined at 30° to VP.

Problem : A square prism with side of base 35 mm and axis 50 mm long, lies with one of its longest edges on HP such that its axis is perpendicular to VP. Draw the projections of the prism when one of its rectangular faces containing the above longer edge is inclined at 30° to HP.

Solution : **(Fig. 6.8)**

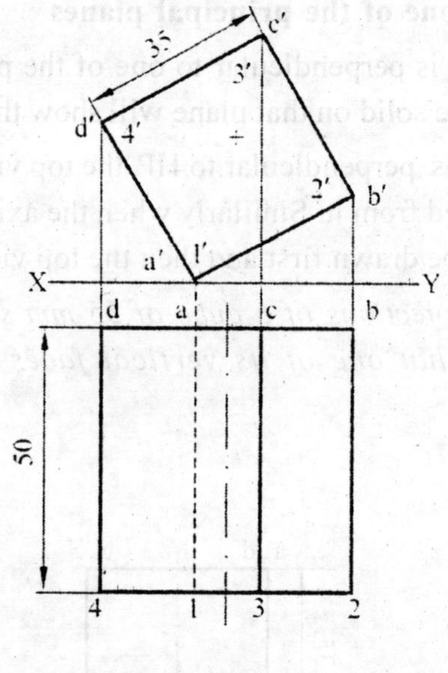

Fig. 6.8

1. Draw the front view which is a square of 35 mm such that one of its corners is on XY and a side passing through it is making 30° with XY.
2. Obtain top view by projection, keeping the length as 50 mm.

Note : The distance of the base nearer to V.P is not given in the problem. Hence, the top view may be drawn keeping the base nearer to XY at any convenient distance.

Problem : A triangular prism with side of base 35 mm and axis 50 mm long is resting on its base on HP. Draw the projections of the prism when one of its rectangular faces is perpendicular to VP and the nearest edge parallel to VP is 10 mm from it.

Solution : **(Fig. 6.9)**

1. Draw the top view keeping one edge perpendicular to XY and one corner at 10mm from XY.

2. Obtain the front view by projection, keeping the height equal to 50 mm.

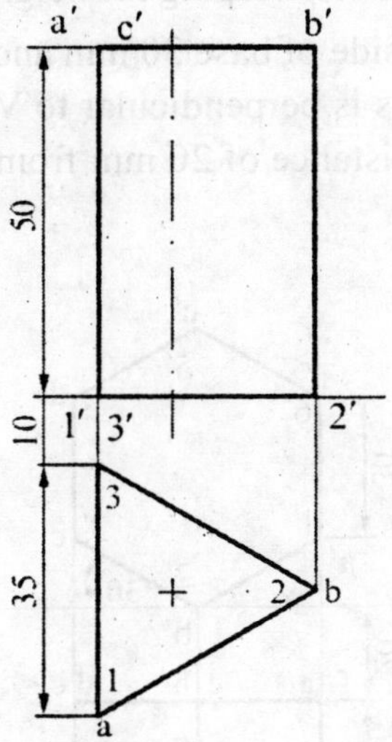

Fig. 6.9

Problem : *A pentagonal prism with side of base 30 mm and axis 60 mm long is resting on its base on HP such that one of its rectangular faces is parallel to VP and 15 mm away from it. Draw the projections of the prism.*

Solution : **(Fig. 6.10)**

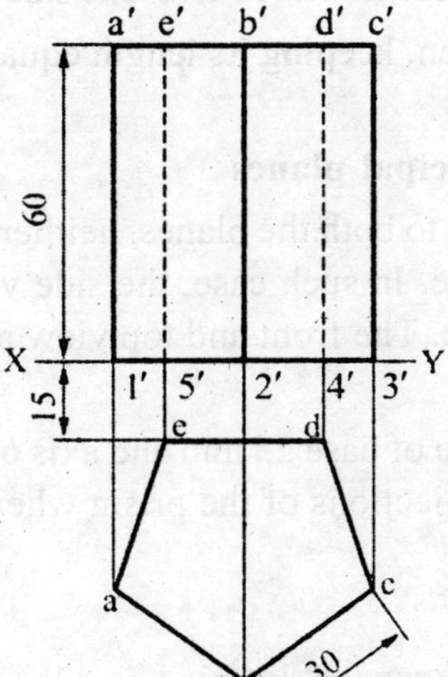

Fig. 6.10

1. Draw the top view keeping one edge of the base parallel to XY and 15mm away from it.
2. Obtain the front view by projection keeping the height equal to 60mm.

Problem : A hexagonal prism with side of base 30 mm and axis 60 mm long lies with one of its longer edges on HP such that its axis is perpendicular to VP. Draw the projections of the prism when the base nearer to VP is at a distance of 20 mm from it.

Solution : **(Fig. 6.11)**

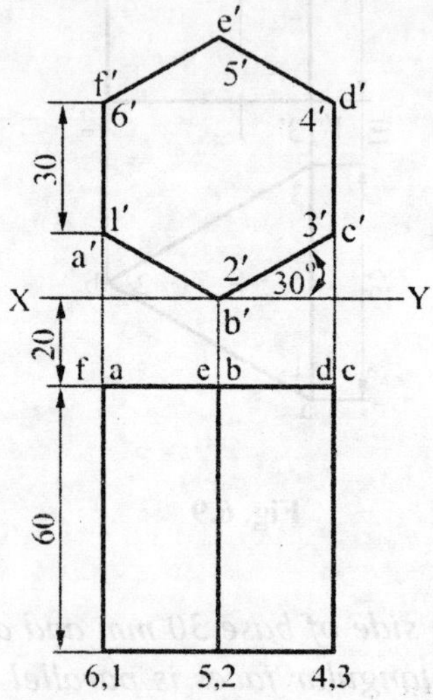

Fig. 6.11

1. Draw the front view keeping one corner on XY and one side making an angle of 30° with XY.
2. Obtain the top view by projection, keeping its length equal to 60 mm and one of its bases 20 mm from XY.

2. **Axis parallel to both the principal planes**

   when the axis of solid is parallel to both the planes, neither the front view nor the top view reveal the true shape of the base. In such case, the side view must be drawn first which shows the true shape of the base. The front and top view are then projected from the side view.

Problem : A hexagonal prism with side of base 25 mm and axis 60 mm long is lying on one of its rectangular faces on HP. Draw the projections of the prism when its axis is parallel to both HP and VP.

Solution : **(Fig. 6.12)**

1. Draw the right side view of the hexagon, keeping an edge on XY.
2. Draw the second reference line X_1Y_1 perpendicular to XY and to the rigtht of the above view at any convenient location.

3. **Obtain the front view by projection, keeping its length equal to 60mm**
4. **Obtain the top view by projecting the above views.**

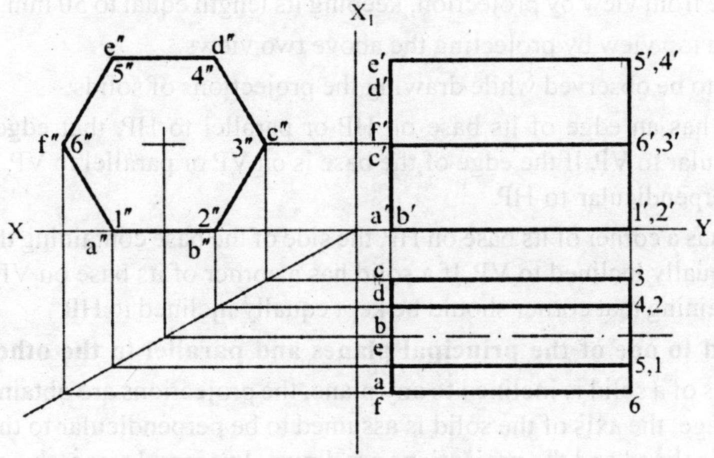

Fig. 6.12

Problem : Draw the three views of a triangular prism of side 25 mm and length 50 mm when its axis is parallel to HP.

***Solution* : (Fig. 6.13)**

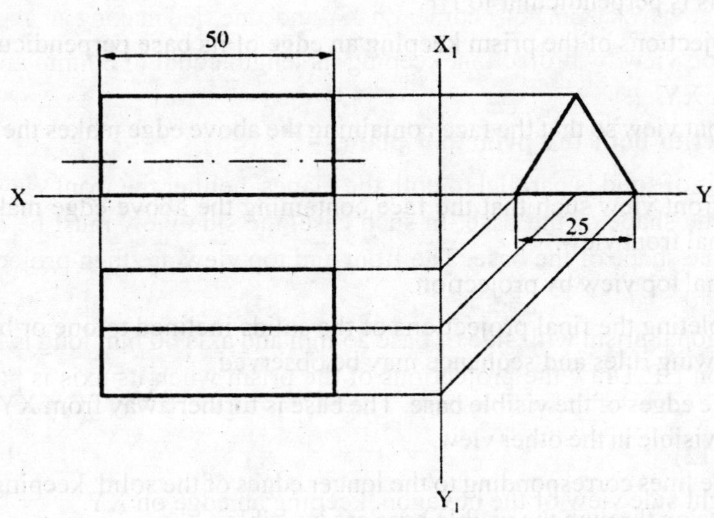

Fig. 6.13

1. Draw the left side view, an equilateral triangle of side 25 mm, keeping one edge on XY.
2. Draw the reference line X_1Y_1 perpendicular to XY and to the left of the above view at any convenient location.
3. Obtain the front view by projection, keeping its length equal to 50 mm.
4. Obtain the top view by projecting the above two views.

Note : Rules to be observed while drawing the projections of solids.

(i) If a solid has an edge of its base on HP or parallel to HP, that edge should be kept perpendicular to VP. If the edge of the base is on VP or parallel to VP, that edge should be kept perpendicular to HP.

(ii) If a solid has a corner of its base on HP, the side of the base containing that corner should be kept equally inclined to VP. If a solid has a corner of its base on VP, the sides of the base containing that corner should be kept equally inclined to HP.

3. Axis inclined to one of the principal planes and parallel to the other

When the axis of a solid is inclined to any plane, the projections are obtained in two stages. In the first stage, the axis of the solid is assumed to be perpendicular to the plane to which it is actually inclined and the projections are drawn. In second stage, the position of one of the projections is altered to statisfy the given condition and the other view is projected from it. This method of obtaining the projections is known as the change of position method.

Problem : A pentagonal prism with side of base 30 mm and axis 60 mm long is resting with an edge of its base on HP, such that the rectangular face containing that edge is inclined at 60° to HP. Draw the projections of the prism when its axis is parallel to VP.

Solution : **(Fig. 6.14)**

Step 1

Assume that the axis is perpendicular to HP.

1. Draw the projections of the prism keeping an edge of its base perpendicular to VP.

Step 2

1. Rotate the front view so that the face containing the above edge makes the given angle with the HP.
2. Redraw the front view such that the face containing the above edge makes 60° with XY. This is the final front view.
3. Obtain the final top view by projection.

Note : For completing the final projections of the solids inclined to one or both the principal planes, the following rules and sequence may be observed.

(i) Draw the edges of the visible base. The base is further away from XY in one view will be fully visible in the other view.

(ii) Draw the lines corresponding to the longer edges of the solid, keeping in mind that the lines passing through the visible base are invisible.

(iii) Draw the edges of the other base.

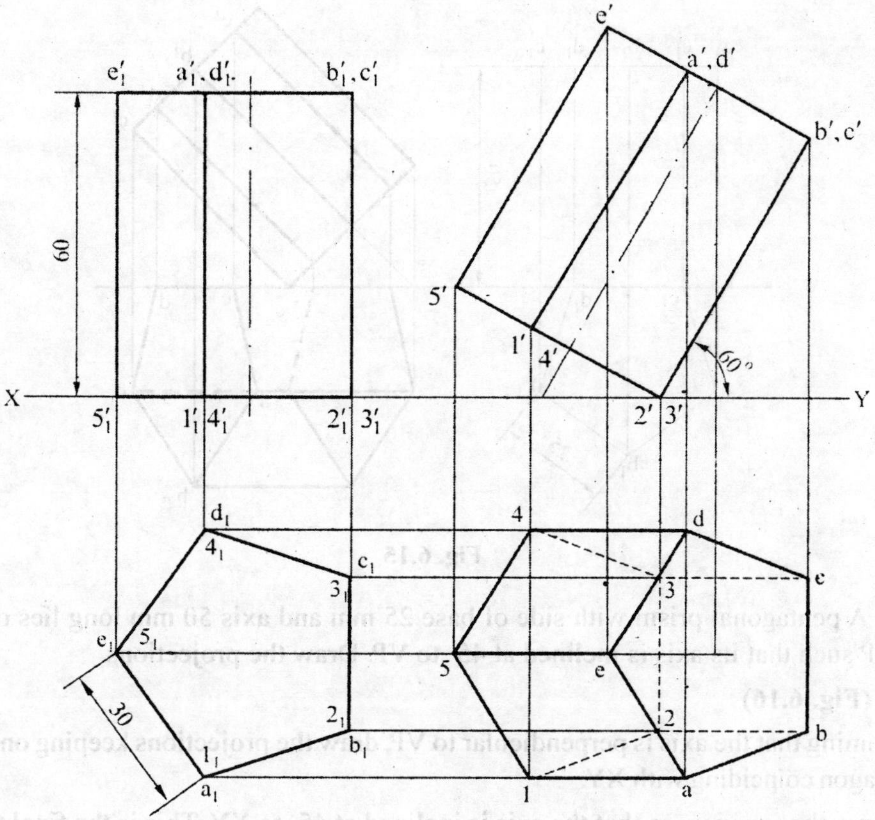

Fig. 6.14

Problem : *Draw the projections of a pentagonal prism of base 25 mm side and 50 mm long. The prism is resting on one of its rectangular faces in VP with its axis inclined at 45⁰ to HP.*

Solution : **(Fig. 6.15)**

Step 1

Assume that the axis is perpendicular to HP.

1. Draw the projections of the prism keeping one of its bases on HP and a rectangular face in VP.

Step 2

1. Rotate the front view so that the axis makes the given angle with HP.
2. Redraw the front view such that the axis makes 45° wth XY. This is the final front view.
3. Obtain the final top view by projection.

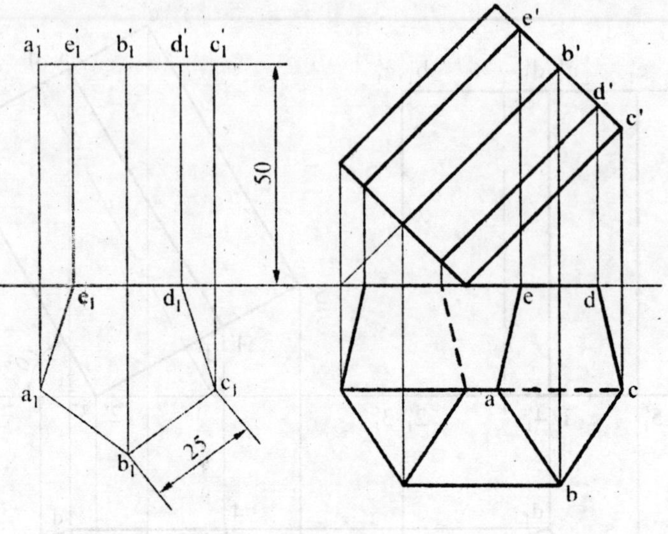

Fig. 6.15

Problem : **A pentagonal prism with side of base 25 mm and axis 50 mm long lies on one of its faces on HP such that its axis is inclined at 45° to VP. Draw the projections.**

Solution : **(Fig. 6.16)**

1. **Assuming that the axis is perpendicular to VP, draw the projections keeping one side of the pentagon coinciding with XY.**
2. **Redraw the top view so that the axis is inclined at 45° to XY. This is the final top view.**
3. **Obtain the final front view by projection.**

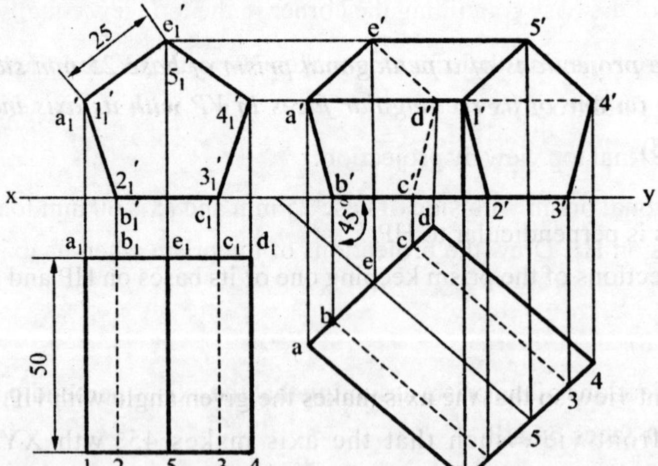

Fig. 6.16

Problem : A hexagonal prism with side of base 25 mm and 50 mm long is resting on a corner of its base on HP. Draw the projections of the prism when its axis is making 30° with HP and parallel to VP.

Solution : **(Fig. 6.17)**

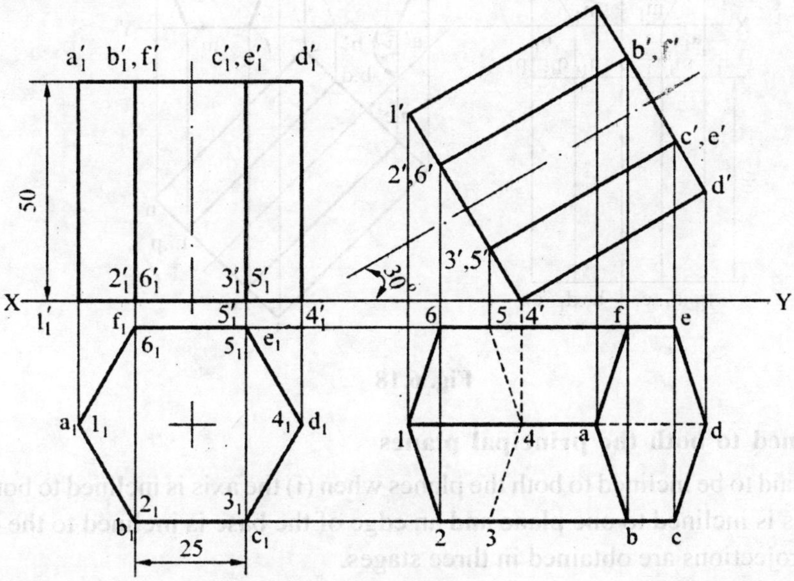

Fig. 6.17

1. **Assuming that the axis is perpendicular to HP, draw the projections of the prism, keeping two sides of the base containing the corner in the top view equally inclined to XY.**
2. **Redraw front view so that the axis makes 30° with XY and the corner 4' lies on XY. This is the final front view.**
3. **Obtain the final top view by projection.**

Problem : A Hexagonal prism with side of base 25 mm and axis 60 mm long is resting on one of its rectangular faces on HP. Draw the projections of the prism when its axis is inclined at 45° to VP.

Solution : **(Fig. 6.18)**

1. **Draw the projections of the prism assuming that the axis is perpendicular to VP, with one of its rectangular faces on HP.**
2. **Redraw the top view such that the axis makes 45° to XY. This is final top view.**
3. **Obtain the final front view by projection.**

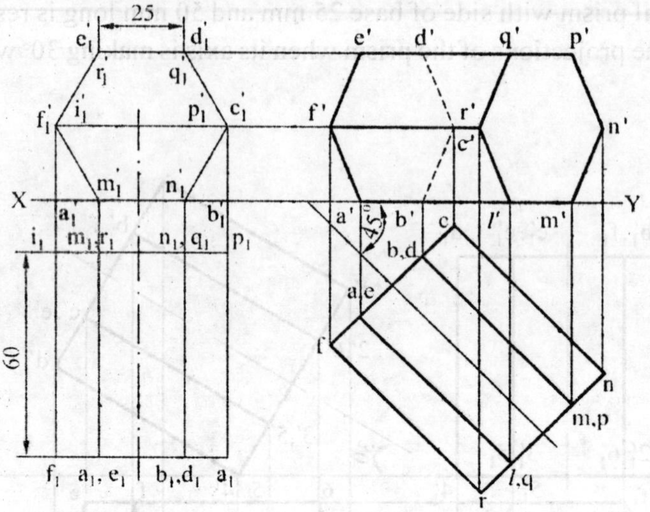

Fig. 6.18

4. **Axis inclined to both the principal planes**

   A solid is said to be inclined to both the planes when (i) the axis is inclined to both the planes, (ii) the axis is inclined to one plane and an edge of the base is inclined to the other. In this case the projections are obtained in three stages.

   Step I

   Assume that the axis is perpendicular to one of the planes and draw the projections.

   Step II

   Rotate one of the projections till the axis is inclined at the given angle and project the other view from it.

   Step III

   Rotate one of the projections obtained in Stage II, satisfiying the remaining condition and project the other view from it.

Problem : A square prism with side of base 30 mm and axis 50 mm long has its axis inclined at 60° to HP., on one of the edges of the base which is inclined at 45° to VP.

***Solution :* (Fig. 6.19)**

1. Draw the projections of the prism assuming it to be resting on one of its bases on HP with an edge of it perpendicular to VP.
2. Redraw the front view such that the axis makes 60° with XY and project the top view from it.
3. Redraw the top view such that the edge on which the prism is resting on HP is inclined at 45° to XY. This is the final top view.
4. Obtain the final front view by projection.

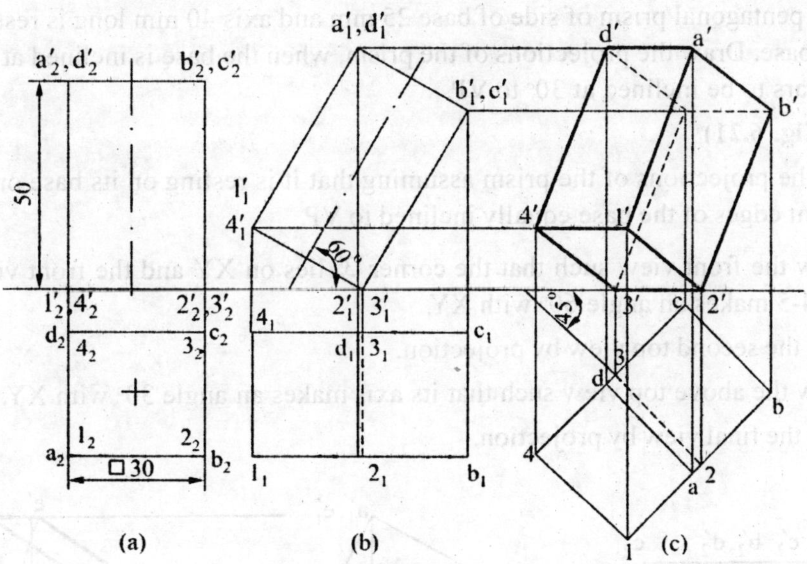

Fig. 6.19

Problem : A tetrahedron of 40 mm side rests with one of its edges on HP and inclined at 45° to VP. The triangular face containing that edge is inclined at 30° to HP. Draw the top and front views of the solid.

Solution : **(Fig. 6.20)**

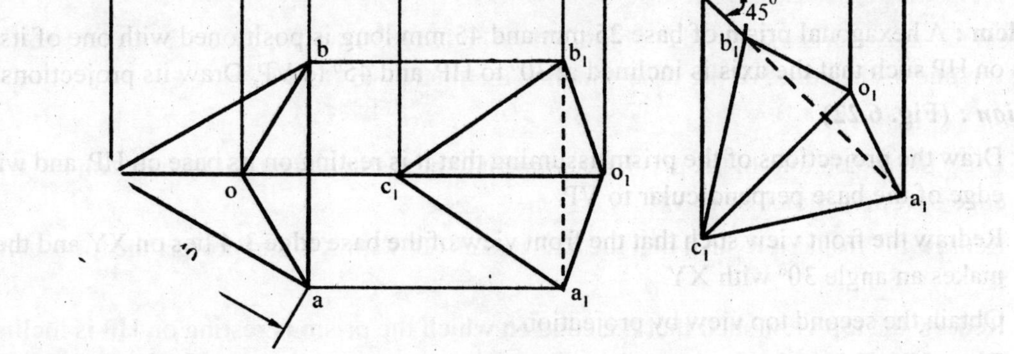

Fig. 6.20

Problem : A pentagonal prism of side of base 25 mm and axis 40 mm long is resting on HP on a corner of its base. Draw the projections of the prism, when the base is inclined at 60° to HP., and the axis appears to be inclined at 30° to VP.

***Solution :* (Fig. 6.21)**

1. Draw the projections of the prism assuming that it is resting on its base on HP., with two adjacent edges of the base equally inclined to VP.
2. Redraw the front view such that the corner 3' lies on XY and the front view of the base 1-2-3-4-5 makes an angle 60° with XY.
3. Obtain the second top view by projection.
4. Redraw the above top view such that its axis makes an angle 30° with XY.
5. Obtain the final view by projection.

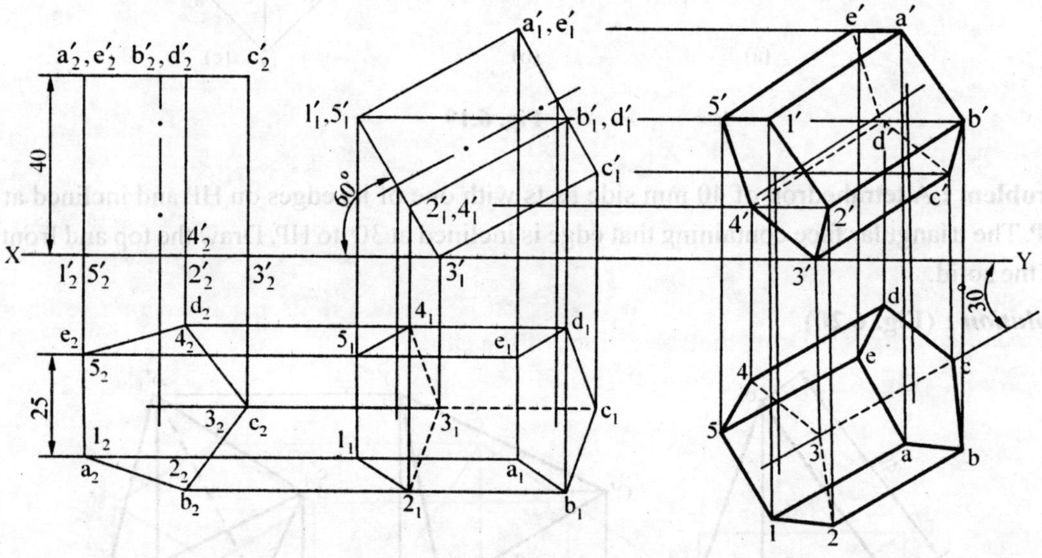

Fig. 6.21

Problem : A hexagonal prism of base 25 mm and 45 mm long is positioned with one of its base edges on HP such that the axis is inclined at 30° to HP. and 45° to VP. Draw its projections.

***Solution :* (Fig. 6.22)**

1. Draw the projections of the prism assuming that it is resting on its base on HP. and with an edge of the base perpendicular to VP.
2. Redraw the front view such that the front view of the base edge 3-4 lies on XY and the axis makes an angle 30° with XY.
3. Obtain the second top view by projection.
4. Determine the apparent angle β, the inclination the axis makes with XY in the final top view.

5. Redraw the top view such that its axis makes angle 45° with XY.
6. Obtain the final front view by projection.

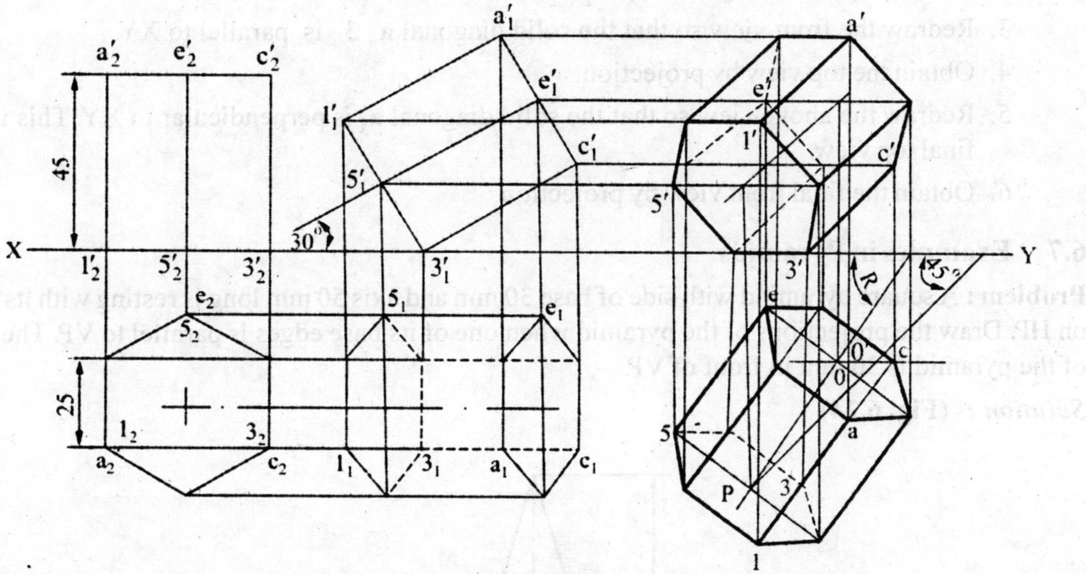

Fig. 6.22

Problem : A cube of edge 35 mm is resting on HP on one of its corners with a solid diagonal perpendicular to VP. Draw the porjections of the cube.

***Solution :* (Fig. 6.23)**

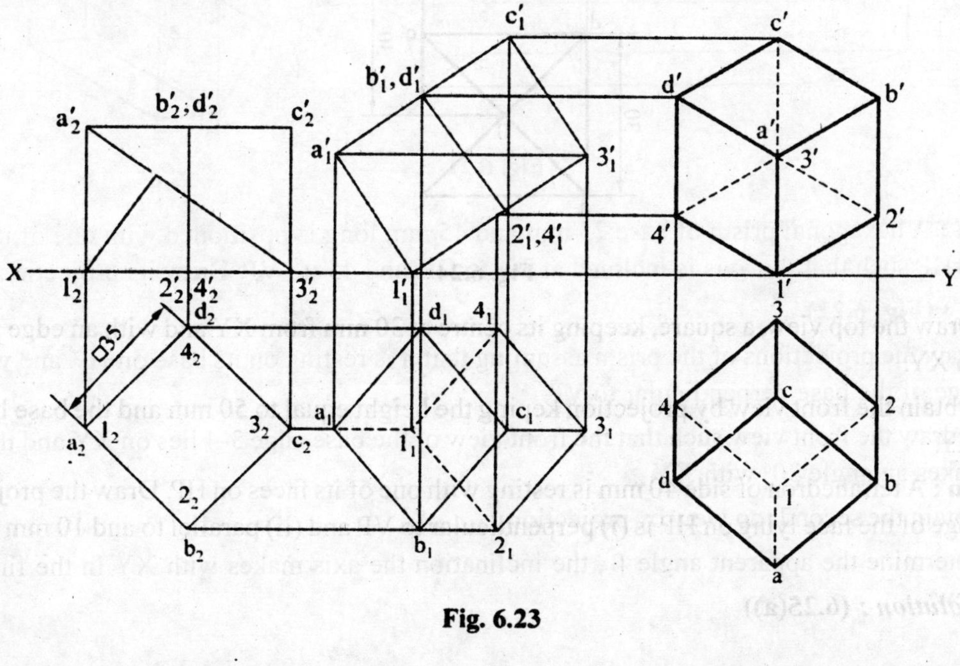

Fig. 6.23

1. Draw the projections of the cube assuming that it is lying on HP on one of its bases and vertical faces and vertical faces are equally inclined to VP.
2. Locate any solid diagonal say $a'_2\ 3'_2$
3. Redraw the front view so that the solid diagonal $a'_1\ 3'_1$ is parallel to XY.
4. Obtain the top view by projection.
5. Redraw the above view so that the solid diagonal a_3 is perpendicular to XY. This is the final top view.
6. Obtain the final front view by projection.

6.7 Examples in Pyramids

Problem : A square pyramind with side of base 30 mm and axis 50 mm long is resting with its base on HP. Draw the projections of the pyramid when one of its base edges is parallel to VP. The axis of the pyramid is 30 mm in front of VP.

Solution : **(Fig. 6.24)**

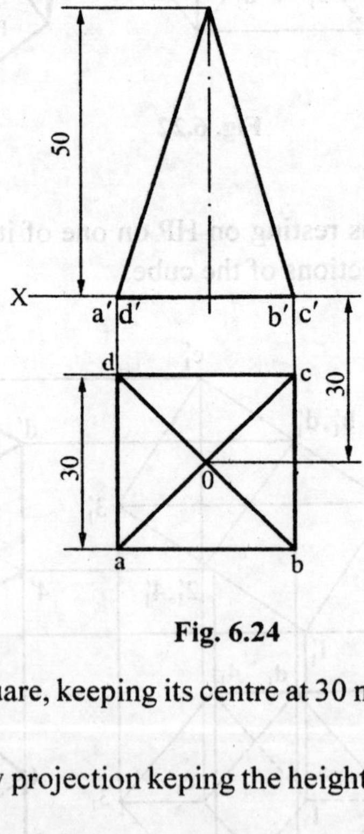

Fig. 6.24

1. Draw the top view, a square, keeping its centre at 30 mm from XY and with an edge parallel to XY.
2. Obtain the front view by projection keping the height equal to 50 mm and the base lying on XY.

Problem : A tetrahedron of side 40 mm is resting with one of its faces on HP. Draw the projections when edge of the face lying on HP is (i) perpendicular to VP and (ii) parallel to and 10 mm in front of HP.

(i) ***Solution :*** **(6.25(a))**

1. Draw the top view keeping one side perpendicular to XY. The lines oa, ob, oc represent the slant edges of the tetrahedron. The line ob is the top view of the stant edge OB. As it is parallel to XY, the length of its front view represents the true length of the edge.
2. Project the front view of the base a' b' c' on to XY.
3. Draw a projector through o.
4. With centre b'_1 and radius equal to the length of side, draw an arc intersecting the above projector at o^1.
5. Join o^1, a', (c') and o',b' forming the front view.

(ii) *Solution* : (Fig. 6.25(b))

1. Draw the top view of the tetrahedron, keeping one side of the base parallel to and 10 mm below XY.
2. Obtain the front view of the base a'b'c' on to XY by projection.
3. Draw a projector through o.
4. Rotate ob about o to ob_1 prallel to XY.
5. Through b_1 draw a projector to meet XY at b_1'.
6. With centre b'_1 and radius equal to the length of side, draw an arc meeting the projector through **o** at o^1.
7. Join **o'-a', o'-b', o'-c'**, forming the front view.

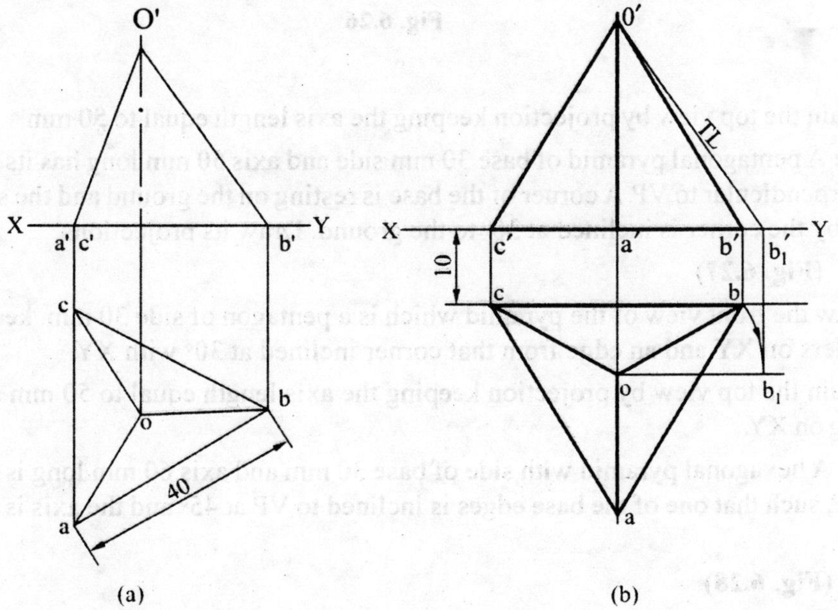

Fig. 6.25

Problem : Draw the projections of a pentagonal pyramid of side of base 30 mm and axis 50 mm long when its axis is perpendicular to VP and an edge of its base is perpendicular to HP.

Solution : **(Fig. 6.26)**

1. Draw the front view of the pyramind which is a pentagon, keeping one of its sides perpendicular to XY.

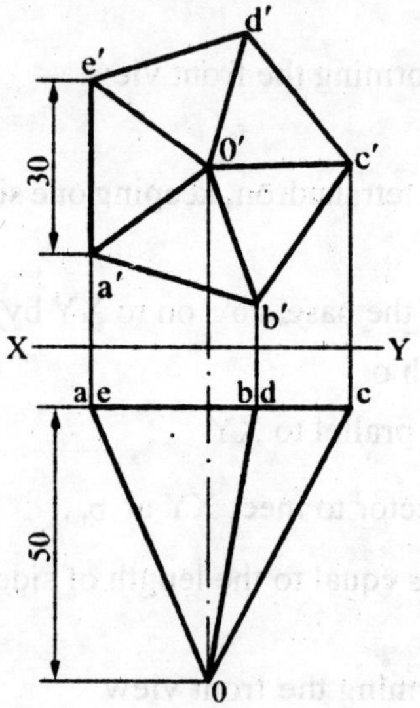

Fig. 6.26

2. Obtain the top view by projection keeping the axis length equal to 50 mm

Problem : A pentagonal pyramid of base 30 mm side and axis 50 mm long has its apex in VP and the axis perpendicular to VP. A corner of the base is resting on the ground and the side of the base contained by the corner is inclined at 30° to the ground. Draw its projections.

Solution : **(Fig. 6.27)**

1. Draw the front view of the pyramid which is a pentagon of side 30 mm, keeping one of its corners on XY and an edge from that corner inclined at 30° with XY.
2. Obtain the top view by projection keeping the axis length equal to 50 mm and the apex O lying on XY.

Problem : A hexagonal pyramid with side of base 30 mm and axis 60 mm long is resting with its base on HP, such that one of the base edges is inclined to VP at 45° and the axis is 50 mm in front of VP.

Solution : **(Fig. 6.28)**

1. Draw the top view, keeping one side of the base inclined at 45° to XY and the centre of the hexagon at 50 mm below XY.

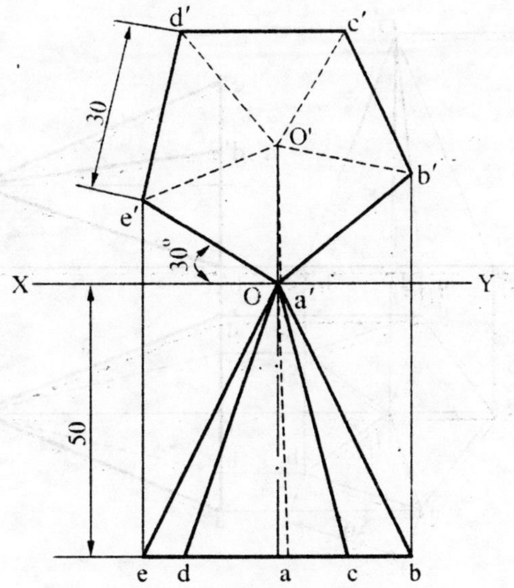

Fig. 6.27

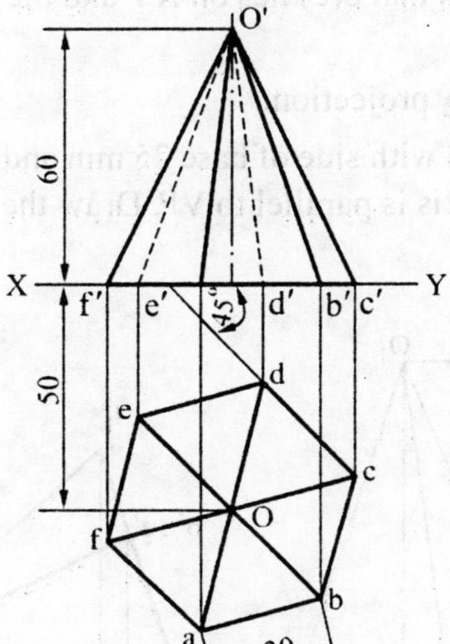

Fig. 6.28

2. Obtain the front view by projection, keeping the axis length equal to 60 mm.

Problem : A pentagonal pyramid with side of base 30 mm and axis 60 mm long rests with an edge of its base on HP such that its axis is parallel to both HP and VP. Draw the projection of the solid.

Solution : **(Fig. 6.29)**

1. Draw the projections of the pyramid with its base on HP and an edge of the base (BC) perpendicular to VP.

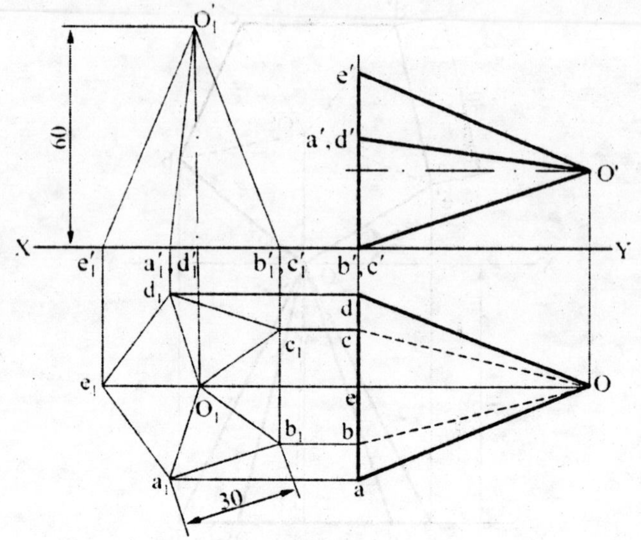

Fig. 6.29

2. Redraw the front view such that b(c) lies on XY and the axis is parallel to XY which is the final front view.
3. Obtain the final top view by projection.

Problem : A pentagonal pyramid with side of base 25 mm and axis 50 mm long is resting on one of its faces on HP such that its axis is parallel to VP. Draw the projections.

Solution : **(Fig. 6.30)**

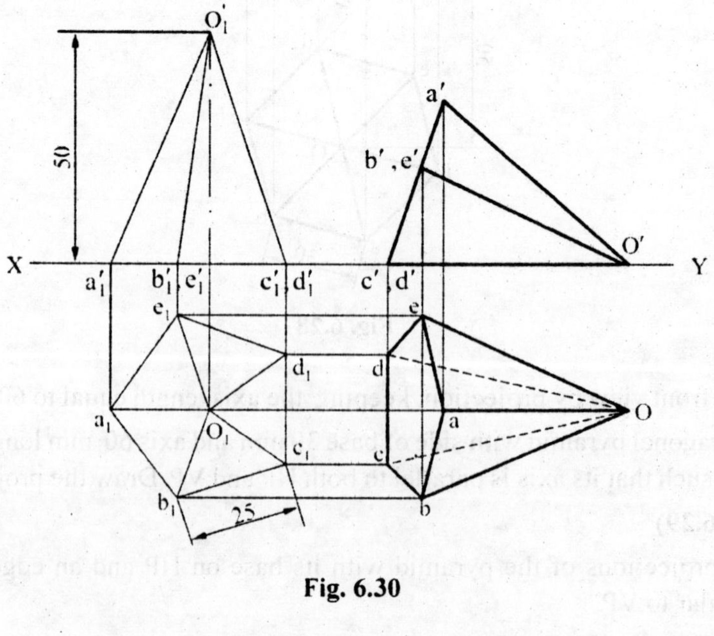

Fig. 6.30

1. Assuming the axis is perpendicular to HP, draw the projections keeping one edge of the base perpendicular to VP.
2. Redraw the front view so that the line o'cd representing the slant face, coincides with XY. This is the final front view.
3. Obtain the final top view by projection.

Problem : A pentagonal pyramid with side of base 35 mm and axis 70 mm long is lying on one of its base edges on HP so that the highest point of the base is 25 mm above HP, and an edge of the base is perpendicular to VP.

***Solution :* (Fig. 6.31)**

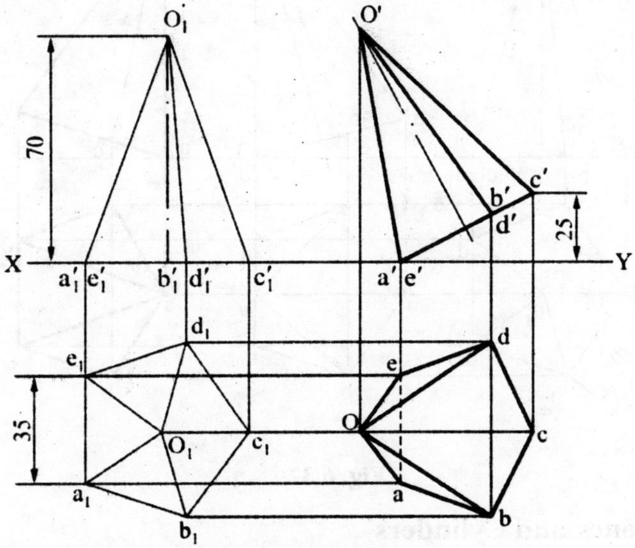

Fig. 6.31

1. Draw the projections of the pyramid, assuming that it is resting on its base on HP and one edge of the base is perpendicular to VP.
2. Redraw the front view so that the corner c' of the base is 25 mm above XY forming final front view.
3. Obtain the final top view by projection.

Problem : Draw the projections of a pentagonal pyramid with a side of base 30mm and axis 70 mm long when (i) one of its triangular faces is perpendicular to HP and (ii) one of its slant edge is vertical.

***Solution :* (Fig. 6.32)**

1. Draw the projections of the pyramid assuming that it is resting on its base on HP with an edge of the base perpendicular to VP.

   Case (i)

2. Redraw the front view such that the front view of the face OCD is perpendicular to XY (the line o'c'(d').

3. Obtain the top view by projection.

Case (ii)

4. Redraw the front view such that front view of the edge OA is perpendicular to XY (the line o'a')
5. Obtain the top view by projection.

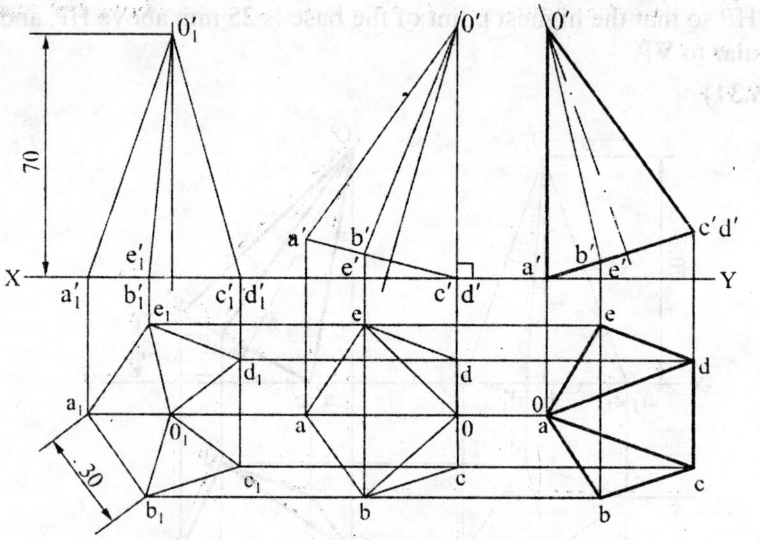

Fig. 6.32

6.8 Examples in Cones and Cylinders

Problem : Draw the projection of a cone of base 40 mm diameter, axis 60 mm long when it is resting with its base on HP.

Solution : **(Fig. 6.33)**

1. Draw the reference line XY and locate o at a convenient distance below it.
2. With centre **o** and radius 20 mm draw a circle forming the top view.
3. Obtain the front view by projection, keeping the height equal to 60 mm and the base coinciding with XY.

Problem : Draw the projections of a cone with diameter of the base as 40 mm and axis 70 mm long with its apex on HP and 35 mm from VP. The axis is perpendicular to HP.

Solution : **(Fig. 6.34)**

1. Draw the reference line XY and locate o at 35 mm below it.
2. With o as centre draw a circle of diameter 40 mm which is the top view of the cone.
3. Obtain the front view by projection, keeping the height equal to 70 mm and the apex **o'** on XY.

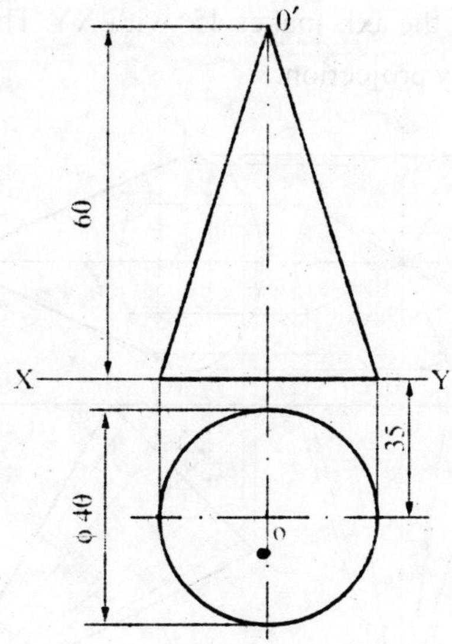

Fig. 6.33

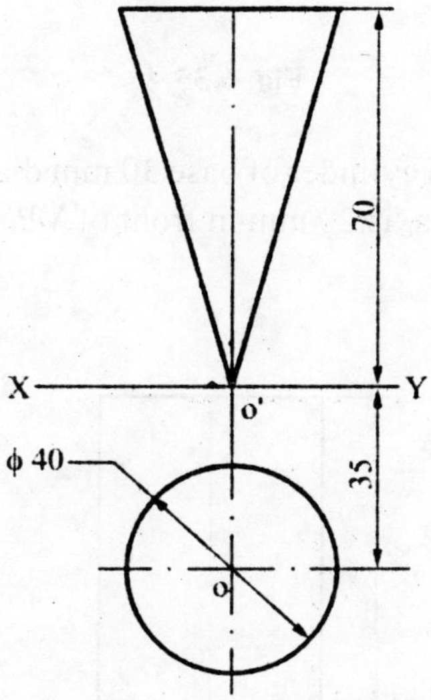

Fig. 6.34

Problem : A cone with base 30 mm diameter and axis 45 mm long lies on a point of its base on VP such that the axis makes an angle 45° with VP. Draw the projections of the cone.

***Solution :* (Fig. 6.35)**

1. Draw the projections of the cone assuming that the cone is resting with its base on VP.
2. Divide the circle into a number of equal parts and draw the corresponding generators in the top view.

3. Redraw the top view so that the axis makes 45° with XY. This is the final top view.
4. Obtain the final front view by projection.

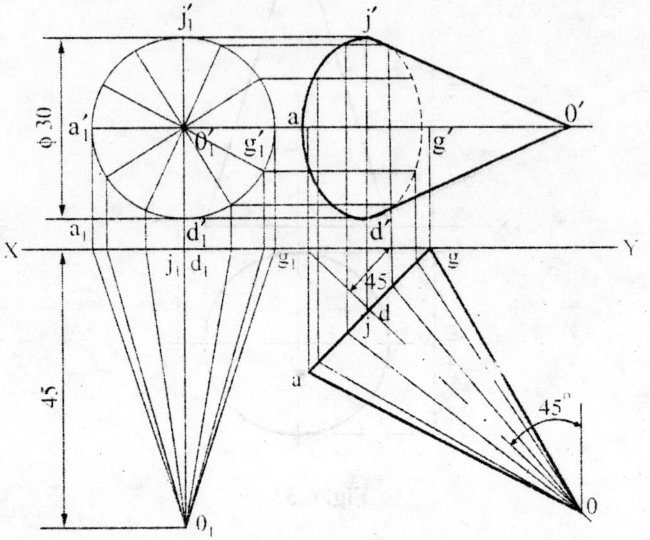

Fig. 6.35

Problem : Draw the projections of a cylinder of base 30 mm diameter and axis 45 mm long when it is resting with its base on HP and axis 20 mm in front of VP.

Solution : **(Fig. 6.36)**

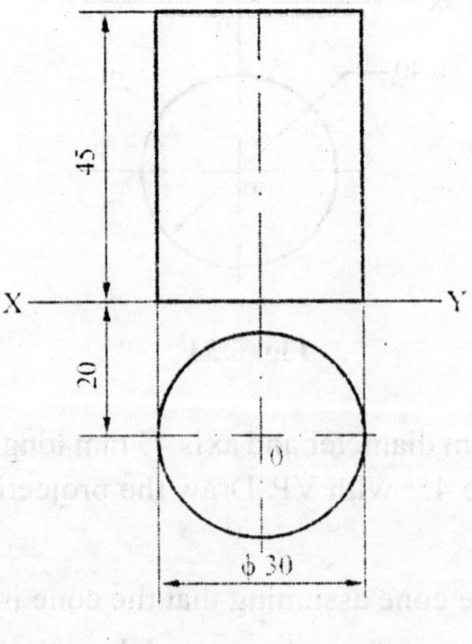

Fig. 6.36

1. Draw the reference line XY and locate o at 20 mm below it.
2. With centre **o** and radius 15 mm draw a circle forming the top view.
3. Obtain the front view by projection, keeping the height equal to 45 mm and the base coinciding with XY.

Problem : A cylinder with base 40 mm diameter and 50 mm long rests on a point of its base on HP such that the axis makes an angle of 30° with HP. Draw the projections of the cylinder.

***Solution :* (Fig. 6.37)**

1. Draw the projection of the cylinder assuming that the cylinder is resting with its base on HP.
2. Divide the circle into a number of equal parts and obtain the corresponding generators in the front view.
3. Redraw the front view such that its axis makes 30° with XY. This is the final front view.
4. Obtain the final top view by projection.

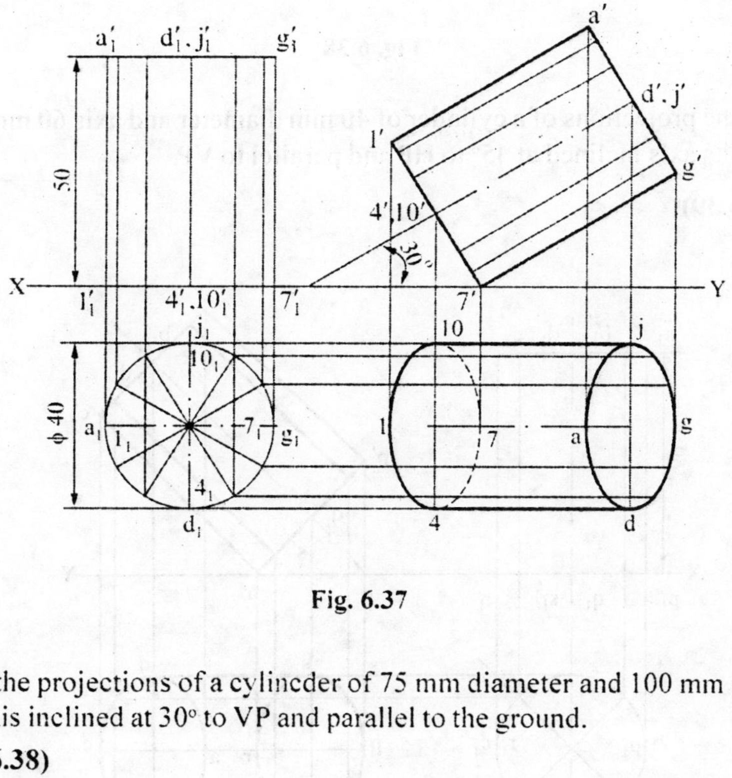

Fig. 6.37

Problem : Draw the projections of a cylincder of 75 mm diameter and 100 mm long lying on the ground with its axis inclined at 30° to VP and parallel to the ground.

***Solution :* (Fig. 6.38)**

1. Draw the projection of the cylinder assuming that the cylinder is resting on HP. with its axis perpendicular to VP.
2. Redraw the top view such that its axis makes 30° with XY. This is the final top view.
3. Obtain the final front view by projection.

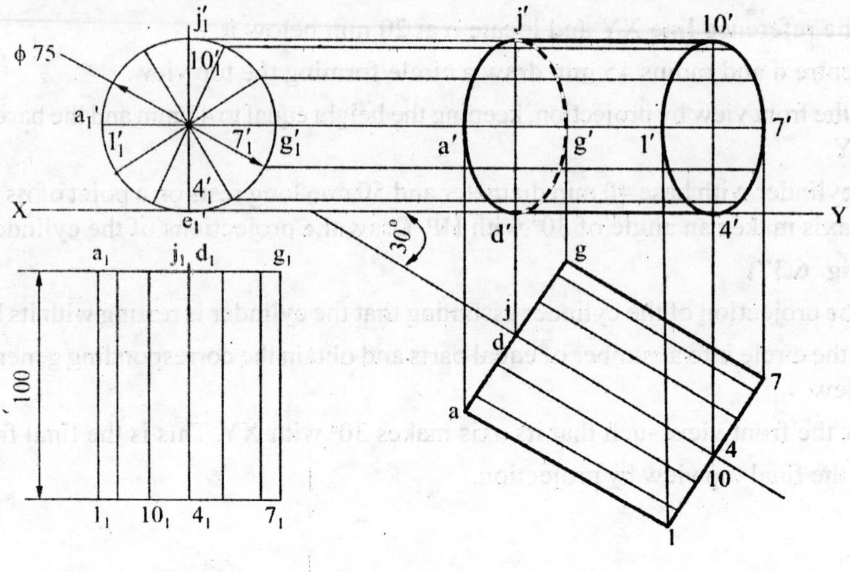

Fig. 6.38

Problem : Draw the projections of a cylinder of 40 mm diameter and axis 60 mm long, when it is lying on HP, with its axis inclined at 45° to HP and parallel to VP.

***Solution :* (Fig. 6.39)**

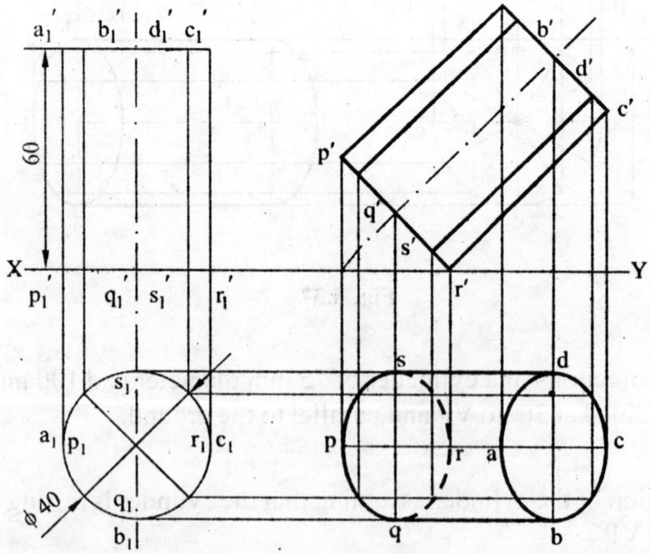

Fig. 6.39

Problem : A cylinder of base 30 mm diameter and axis 45 mm long is resting on a point of its base on HP so that the axis is inclined at 30° with HP. Draw the projections of the cylinder when the top view of the axis is inclined at 45° with XY.

***Solution* : (Fig. 6.40)**

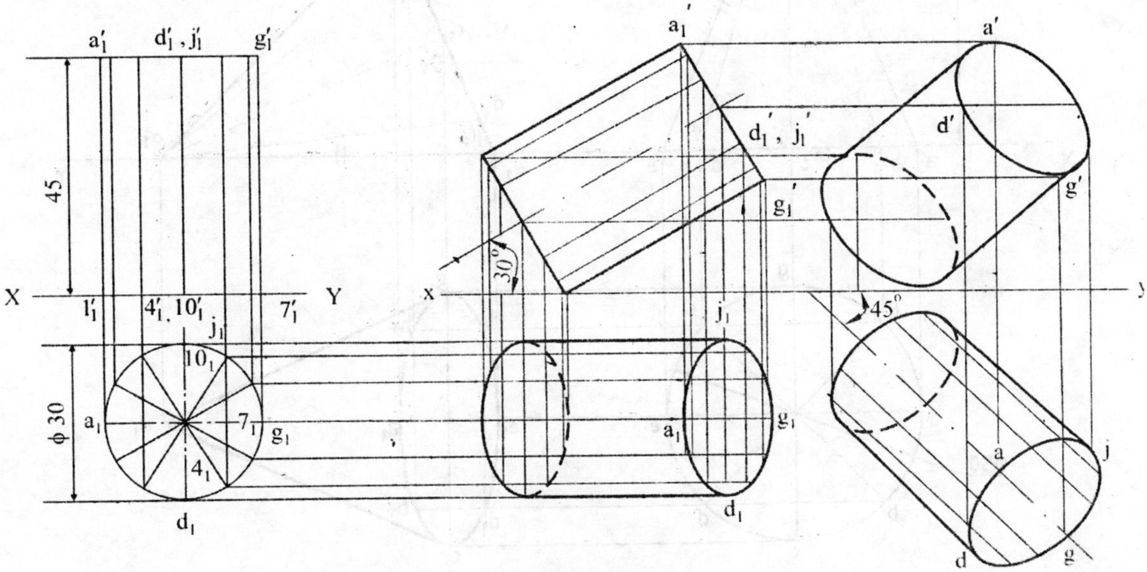

Fig. 6.40

1. Draw the projection of the cylinder assuming it to be resting on its base on HP.
2. Redraw the front view so that the axis is inclined at 30° with XY.
3. Obtain the top view by projection.
4. Redraw the above view so that the top view of the axis is inclined at 45° with XY. This final top view.
5. Obtain the final front view by projection.

Problem: A cone of base diameter 50 mm and axis length 60 mm is resting on *HP* on one of its generators with its axis parallel to *VP*. Draw its projections.

***Solution* : (Fig. 6.41)**

When a cone is resting on *HP* on one of its generators, its axis is inclined to *HP* and is given parallel to *VP*.

To draw the top and front views in step I

1. Assume the axis is perpendicular to *HP* and parallel to *VP*, the top view is drawn as a circle with the apex *o* at the centre and the front view is projected and obtained as a triangle.

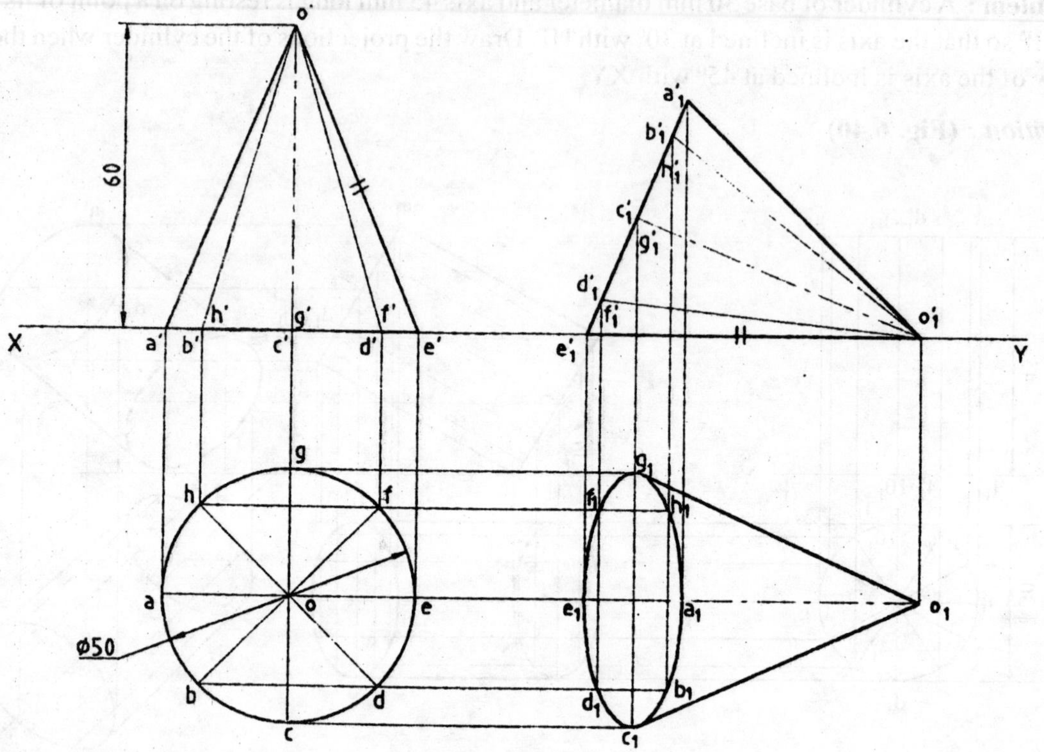

Fig. 6.41

2. Consider 8 generators and project them to the front view.

To draw the front and top views in step 2

1. Tilt and reproduce the front view in such a way that one of the generator *e'o'* is resting on *HP* and the triangular shape is reproduced and the generators are marked appropriately.
2. Draw vertical lines from a_1', b_1', c_1' etc. and horizontal lines from *a, b, c* etc. to get a_1, b_1, c_1 etc. and the apex o_1 are obtained in the top view.
3. Join a_1, b_1, c_1 etc. by drawing a smooth curve and draw the extreme generators to complete the top view.

Problem: A cone of base diameter 50 mm and axis length 60 mm is resting on *VP* on one of its generators with its axis parallel to *HP*. Draw its projections.

***Solution* : (Fig. 6.42)**

Problem: A cone of base diameter 50 mm and axis length 60 mm is resting on *HP* on a point on the circumference of the base. Its base is inclined at 50° to *HP* and perpendicular to *VP*. Draw its projections.

***Solution* : (Fig. 6.43)**

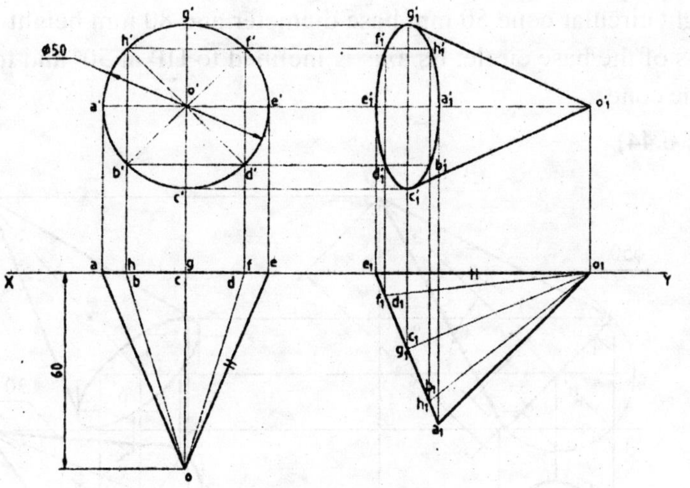

Fig. 6.42

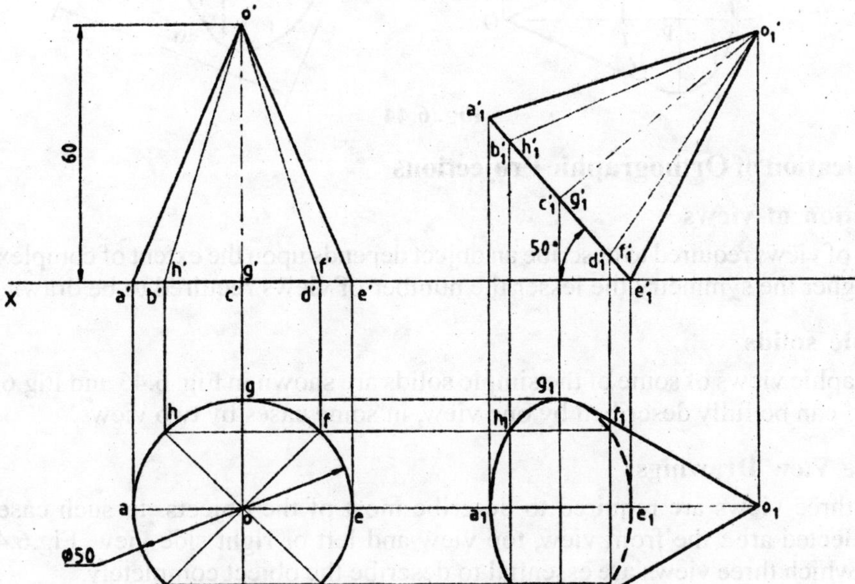

Fig. 6.43

When the base of the solid is kept inclined to *HP* and perpendicular to *VP*, its axis is inclined to *HP* and parallel to *VP*.

To draw the top and front views in step 1

1. **Assume the axis is perpendicular to *HP* and parallel to *VP*, the top view is a circle with the apex *o* at the centre and the front view is projected and obtained as a triangle.**
2. **Consider generators by dividing the circle into equal number of parts and project them to front view.**

Problem : A right circular cone 50 mm base diameter and 80 mm height rests on the ground on one of the points of the base circle. Its axis is inclined to HP at 50° and to VP at 30°. Draw the projections of the cone.

Solution : **(Fig. 6.44)**

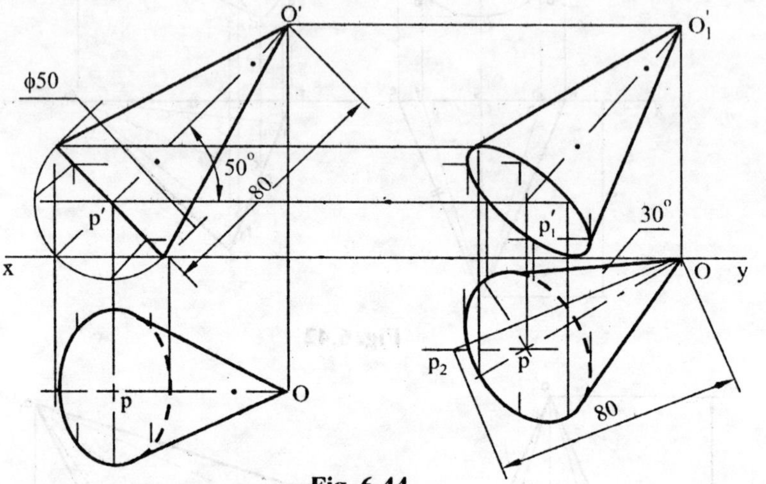

Fig. 6.44

6.9 Application of Orthographic Projections

6.9.1 Selection of views

The number of views required to describe an object depends upon the extent of complexity involved in it. The higher the symmetry the lesser the number of views required to be drawn.

6.9.2 Simple solids

The orthographic views of some of the simple solids are shown in Fig. 6.45 and Fig.6.46. In some cases a solid can be fully described by one view, in some cases by two views.

6.9.3 Three View Drawings

In general, three views are required to describe most of the objects. In such cases the views normally selected are : the front view, top view and left or right side view. Fig.6.47 shows an example in which three views are essential to describe the object completely.

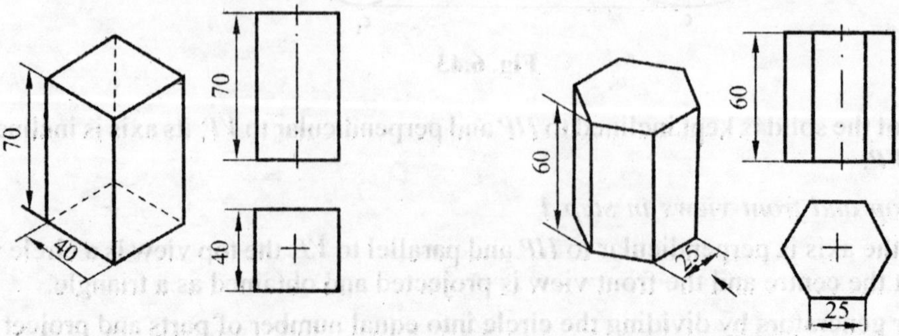

(a) (b)

Fig. 6.45

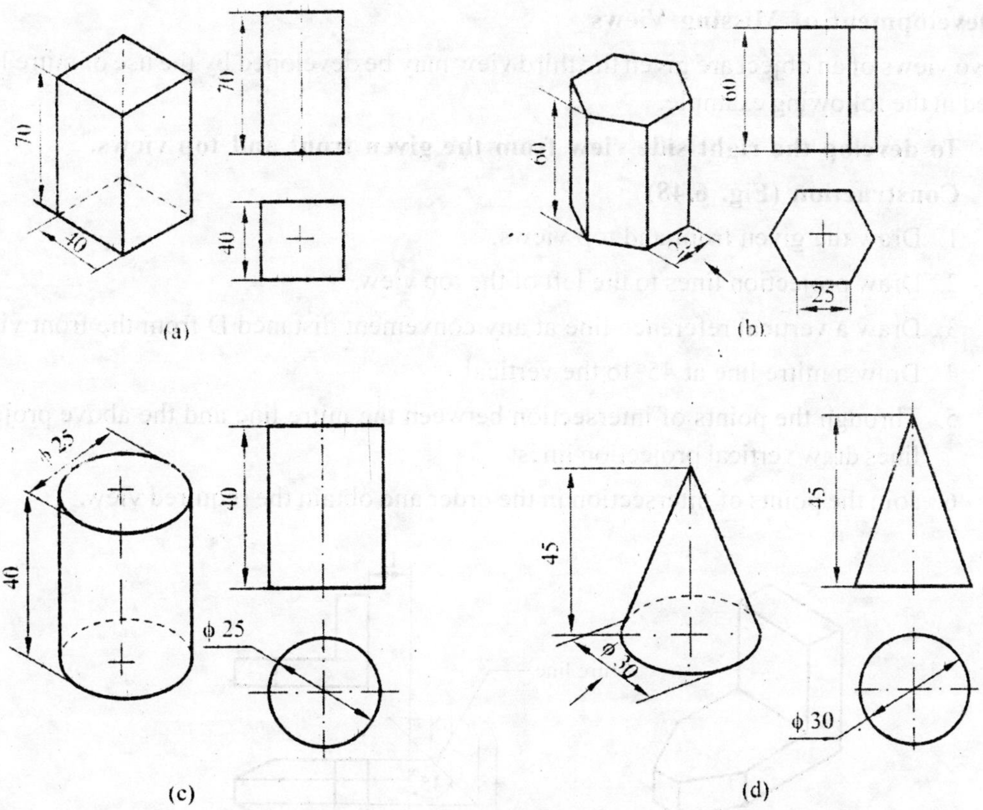

Fig. 6.46

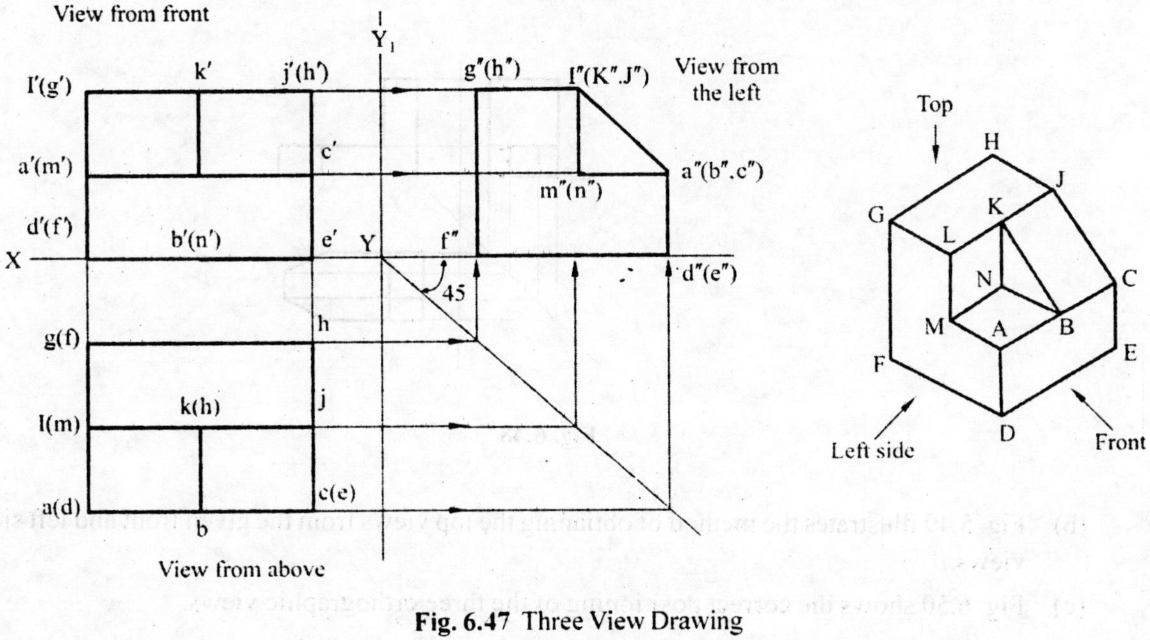

Fig. 6.47 Three View Drawing

6.9.4 Development of Missing Views

When two views of an object are given the third view may be developed by the use of mitre line as described in the following example.

(a) **To develop the right side view from the given front and top views.**

Construction (Fig. 6.48)

1. Draw the given front and top views.
2. Draw projection lines to the left of the top view.
3. Draw a vertical reference line at any convement distancd D from the front view.
4. Draw a mitre line at 45° to the vertical
5. Through the points of intersection between the mitre line and the above projection lines draw vertical projection lines.
6. Join the points of intersection in the order and obtain the required view.

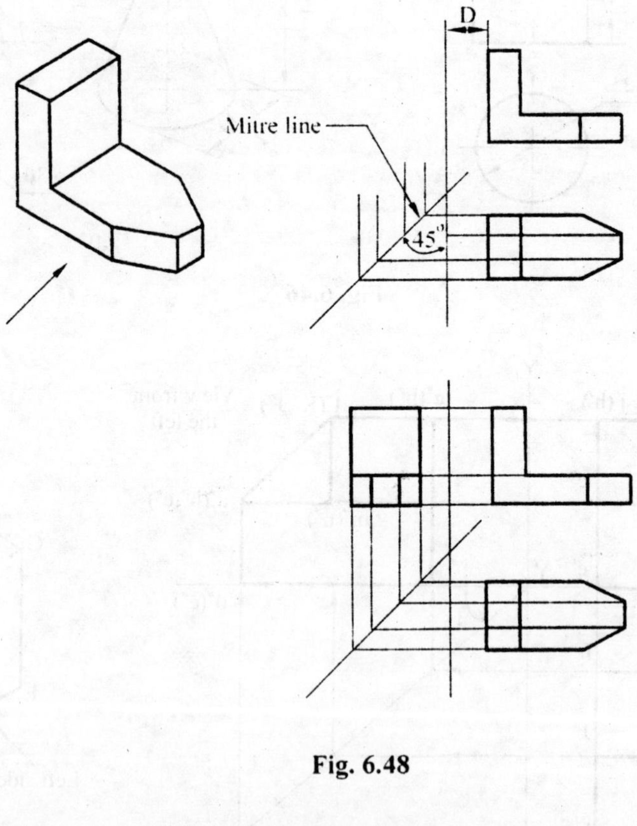

Fig. 6.48

(b) Fig. 5.49 illustrates the method of obtianing the top views from the given front and left side views.

(c) Fig. 6.50 shows the correct positioning of the three orthographic views.

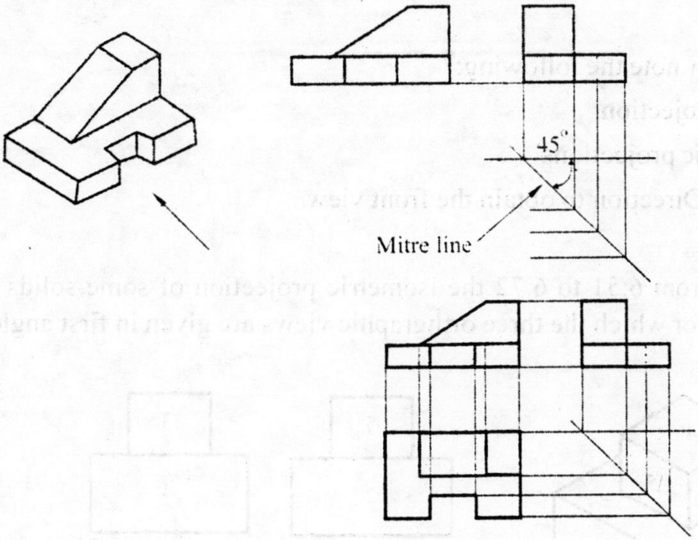

Fig. 6.49

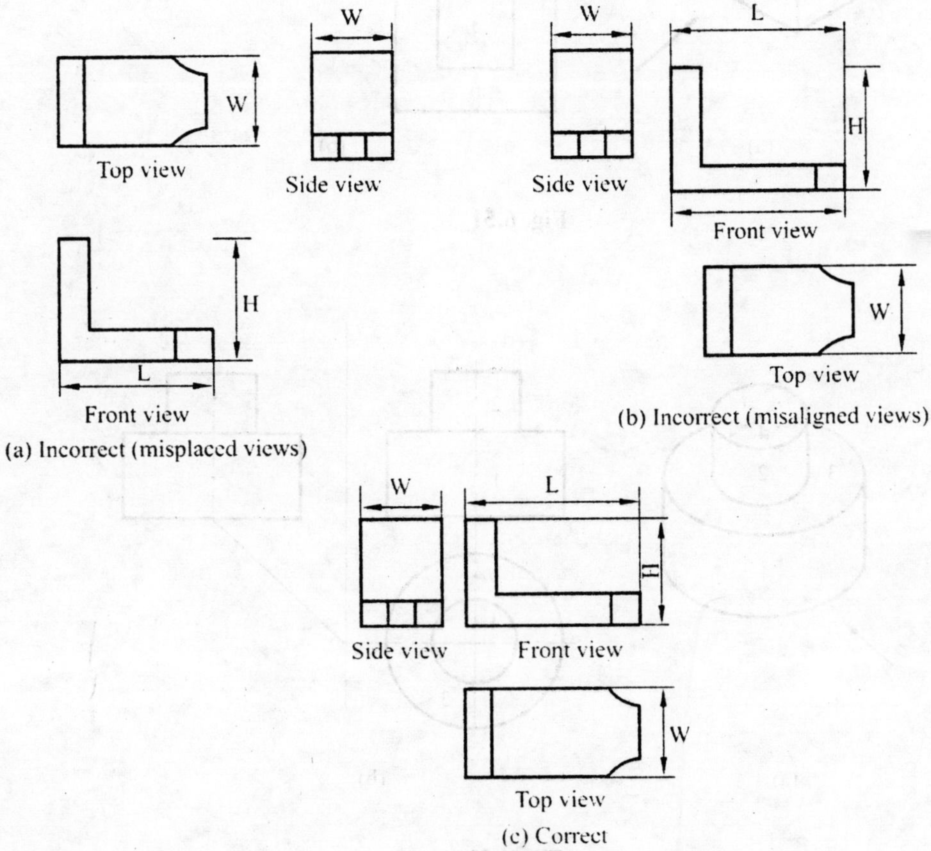

Fig. 6.50

Examples

In the examples below note the following:

Fig. (a) - Isometric projection

Fig. (b) - Orthographic projections

Direction of arrow - Direction to obtain the front view.

Example 1

In the following Figs. from 6.51 to 6.72 the isometric projection of some solids and machine components are shown for which the three orthgraphic views are given in first angle projection.

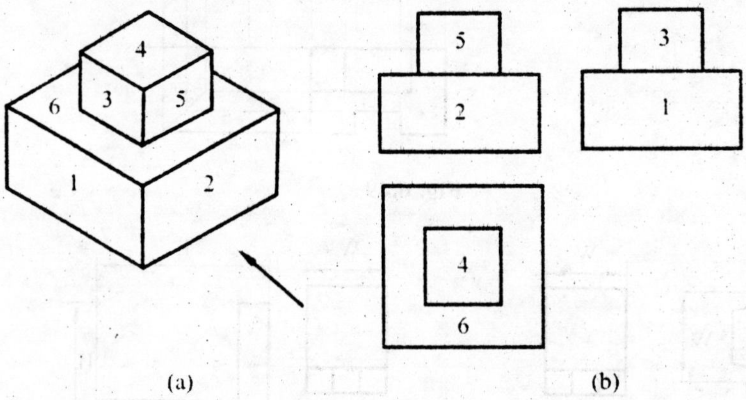

Fig. 6.51

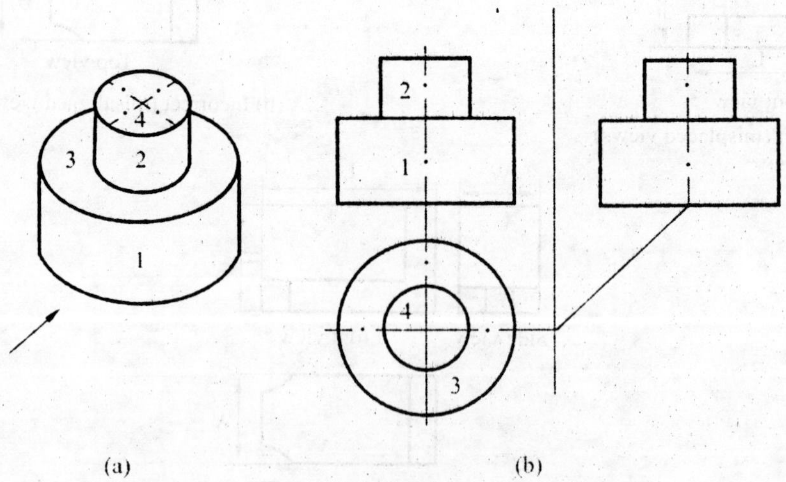

Fig. 6.52

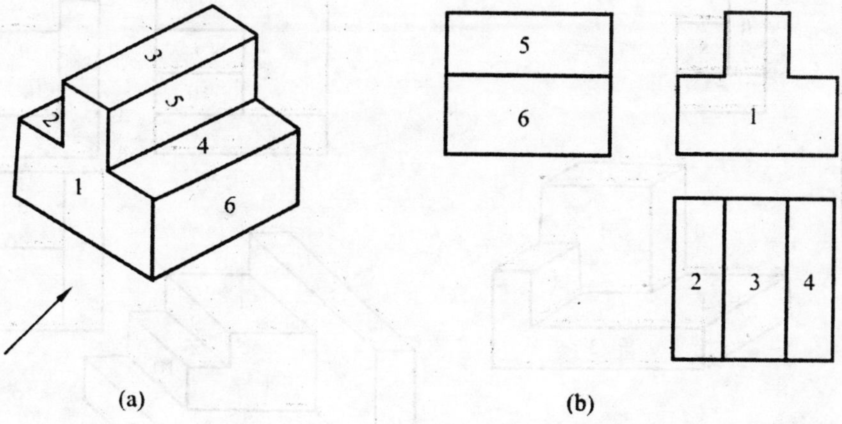

Fig. 6.53

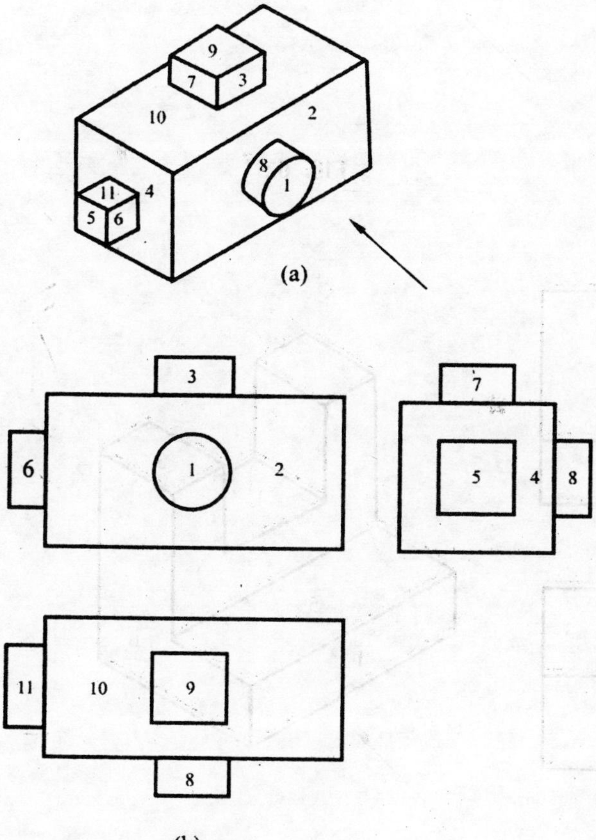

Fig. 6.54

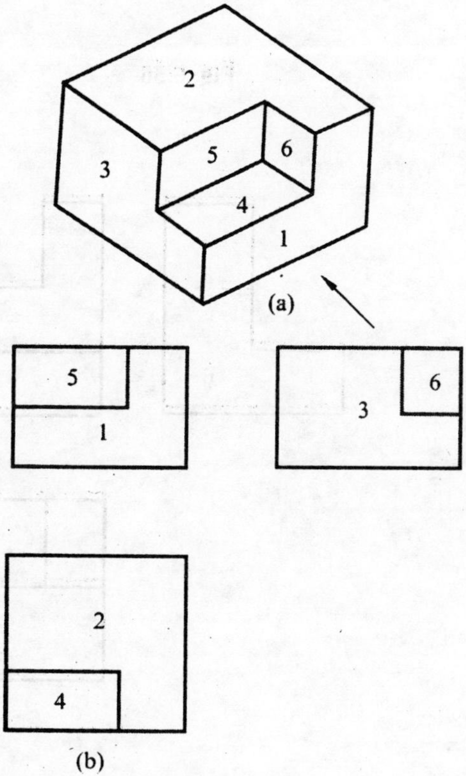

Fig. 6.55

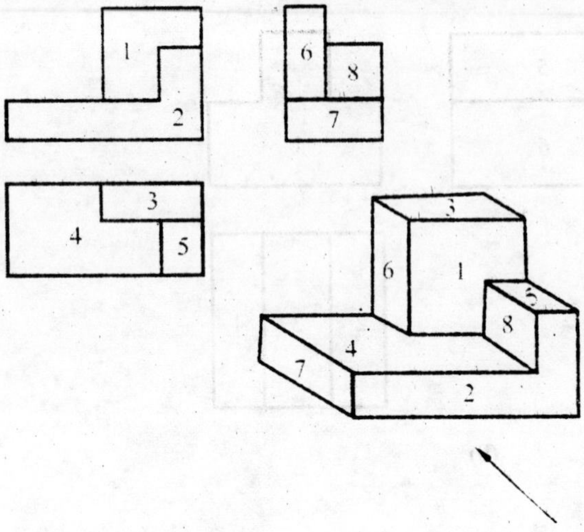

Fig. 6.56

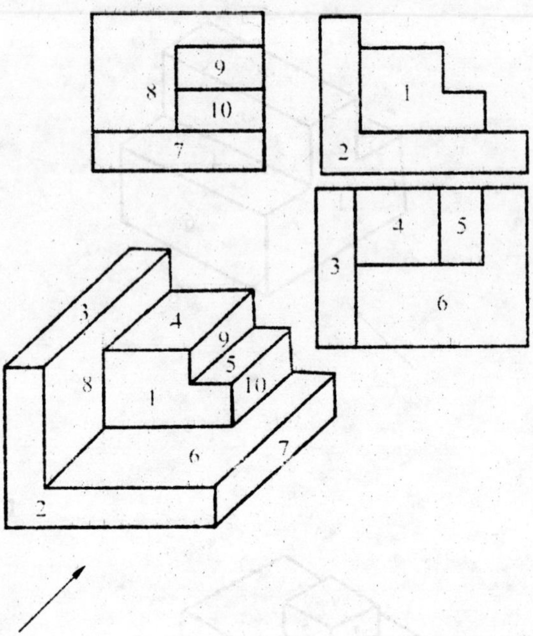

Fig. 6.57

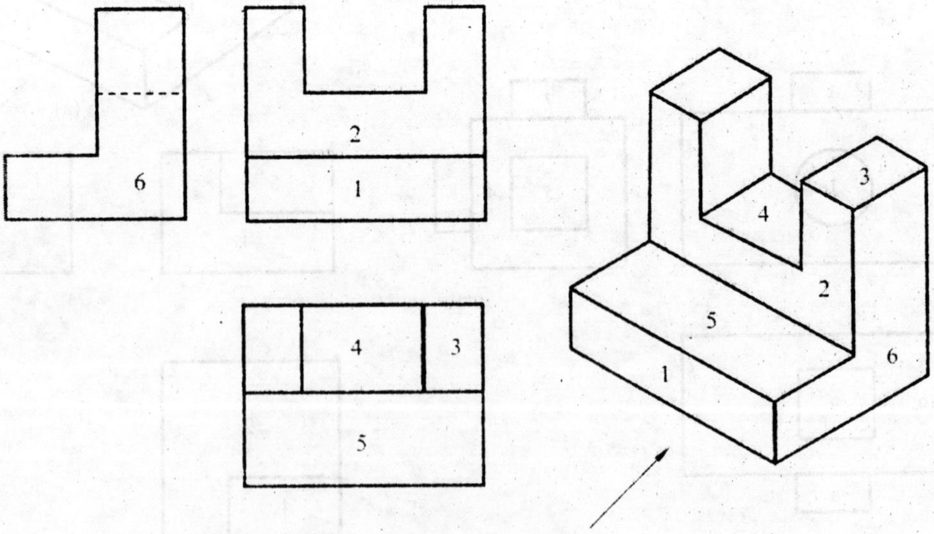

Fig. 6.58

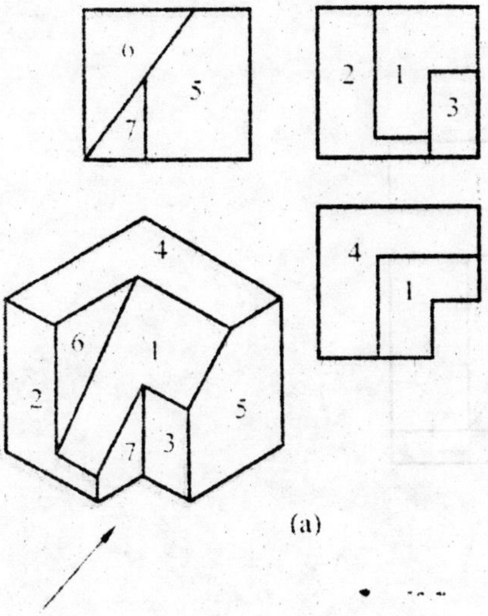

Fig. 6.59

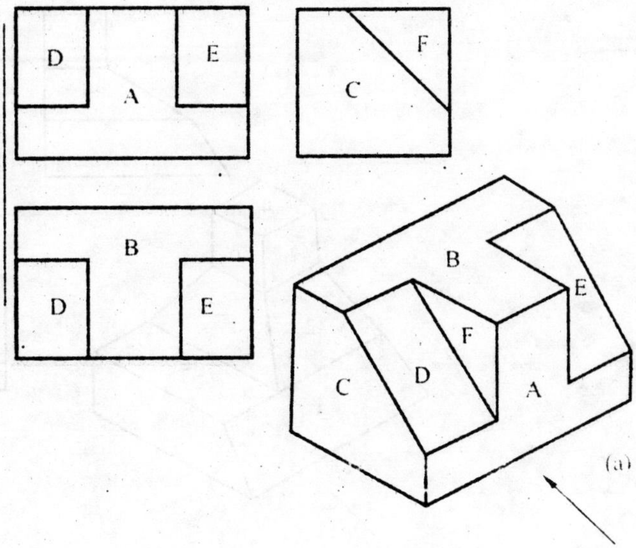

Fig. 6.60

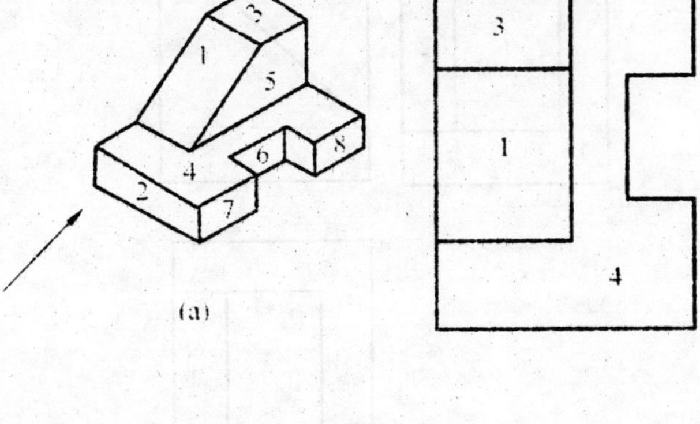

Fig. 6.61

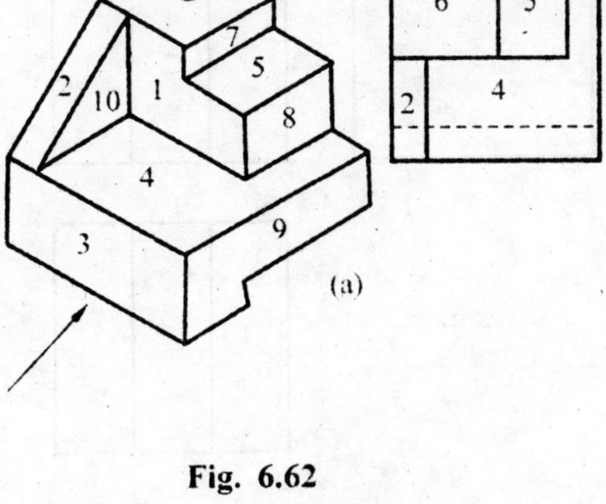

Fig. 6.62

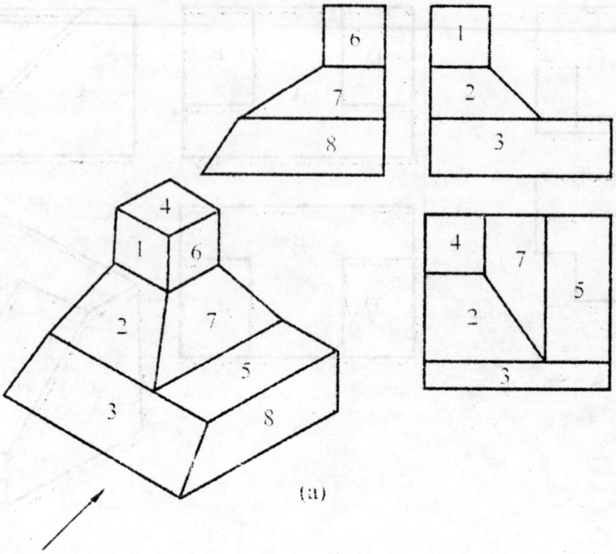

Fig. 6.63

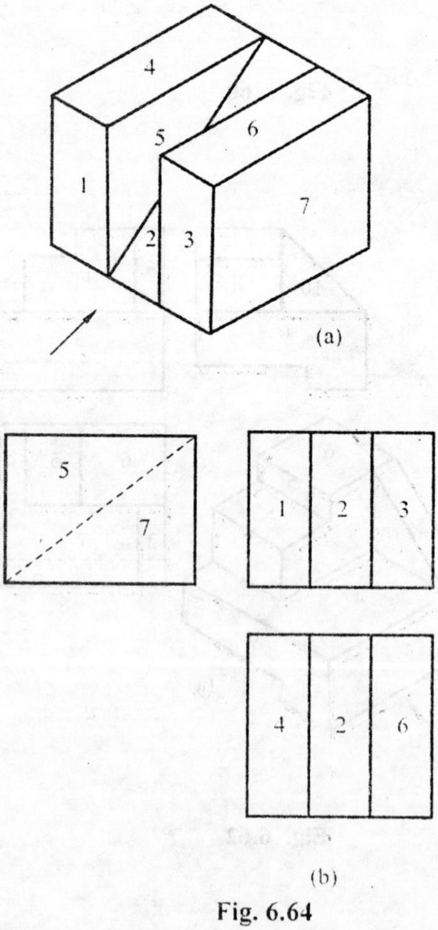

Fig. 6.64

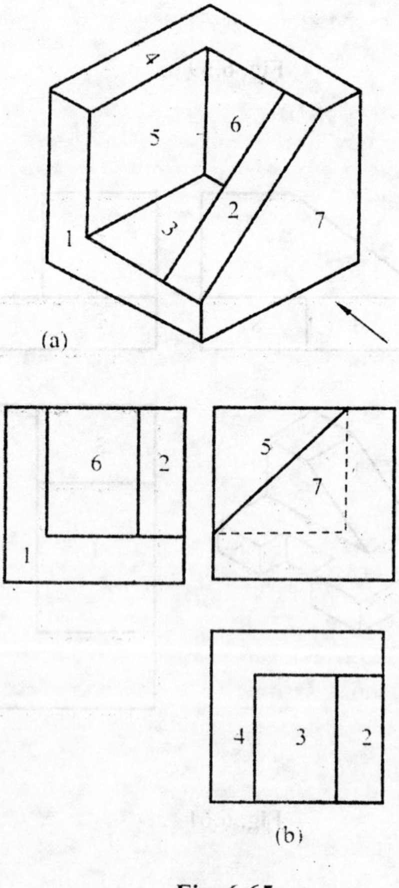

Fig. 6.65

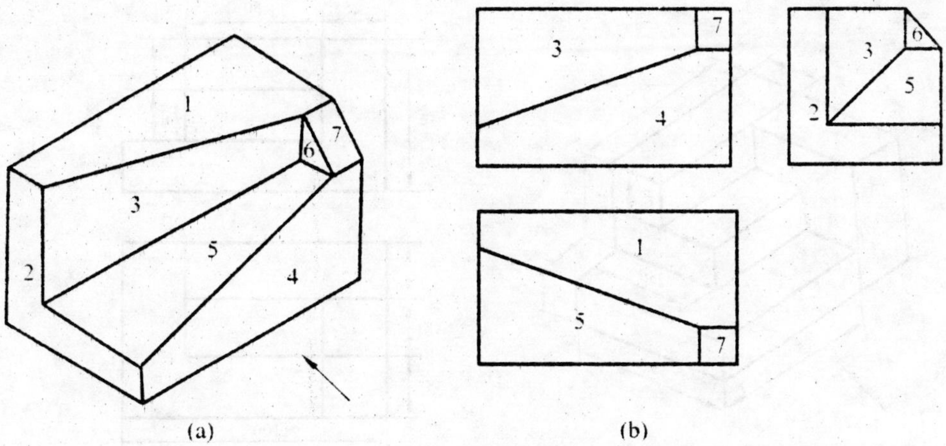

(a) (b)

Fig. 6.66

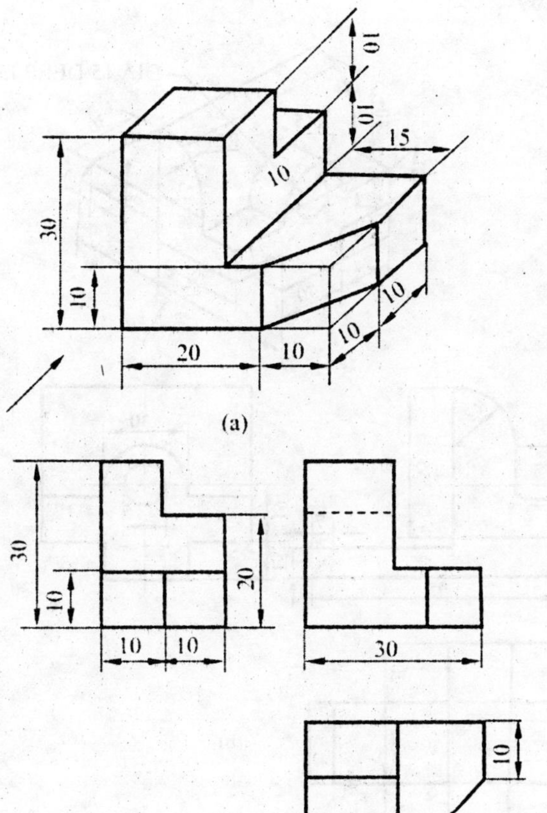

(a) (b)

Fig. 6.67

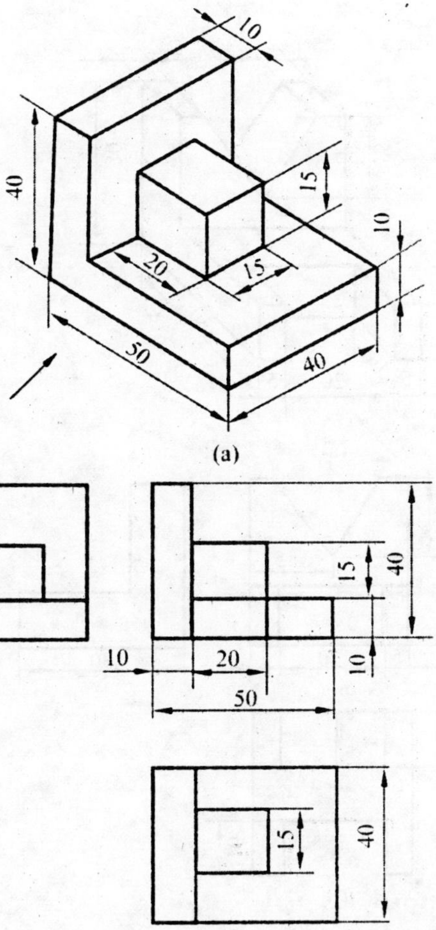

(a) (b)

Fig. 6.68

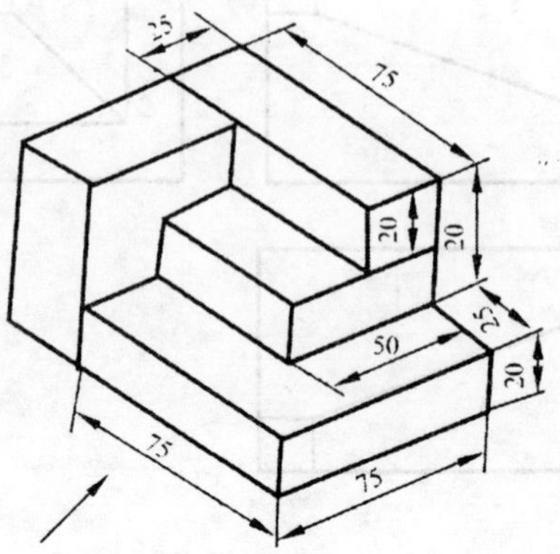

(a)

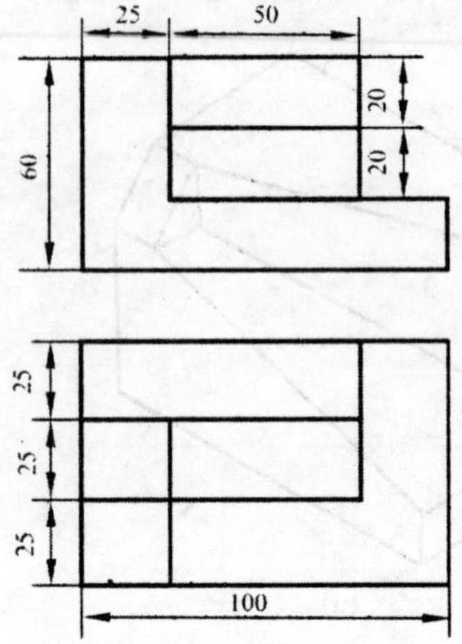

(b)

Fig. 6.69

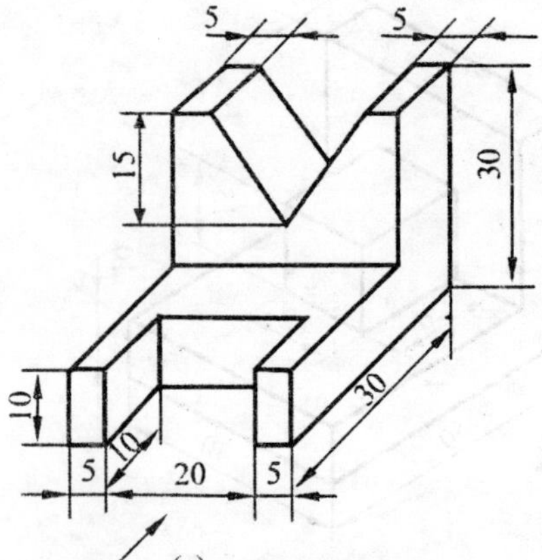

(a)

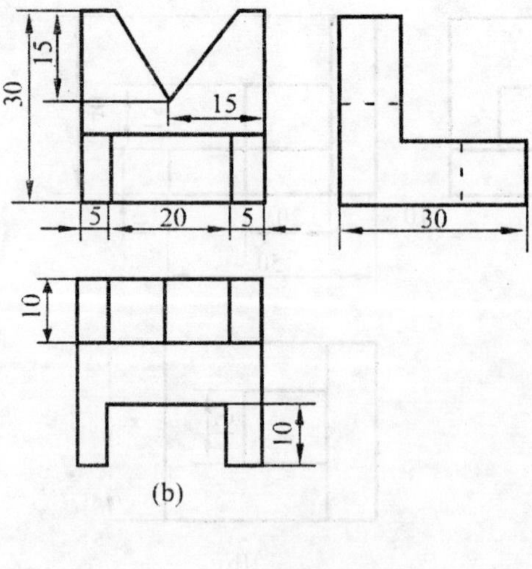

(b)

Fig. 6.70

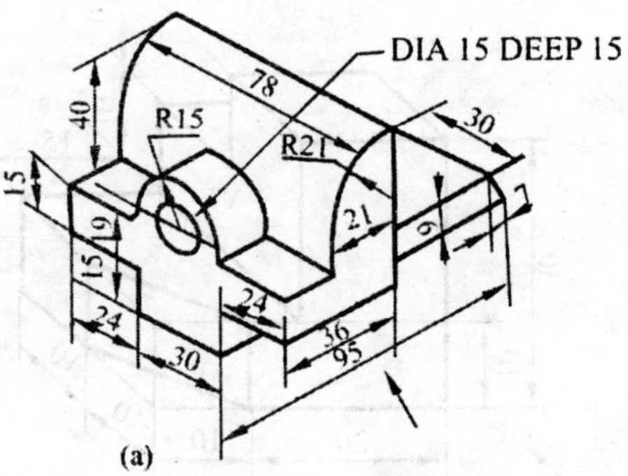

(a)

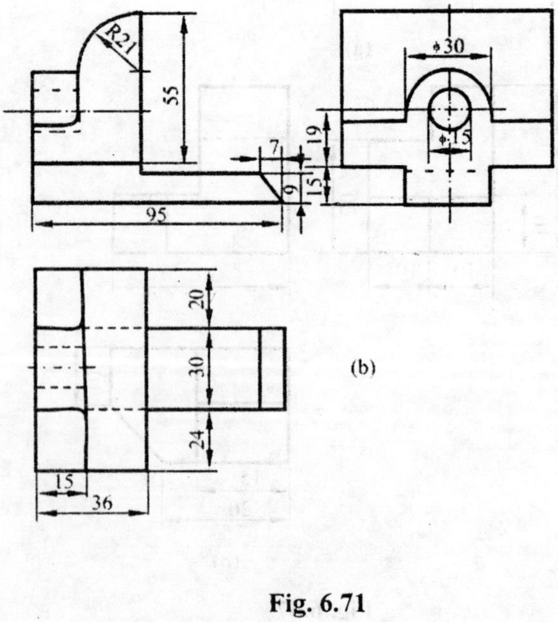

(b)

Fig. 6.71

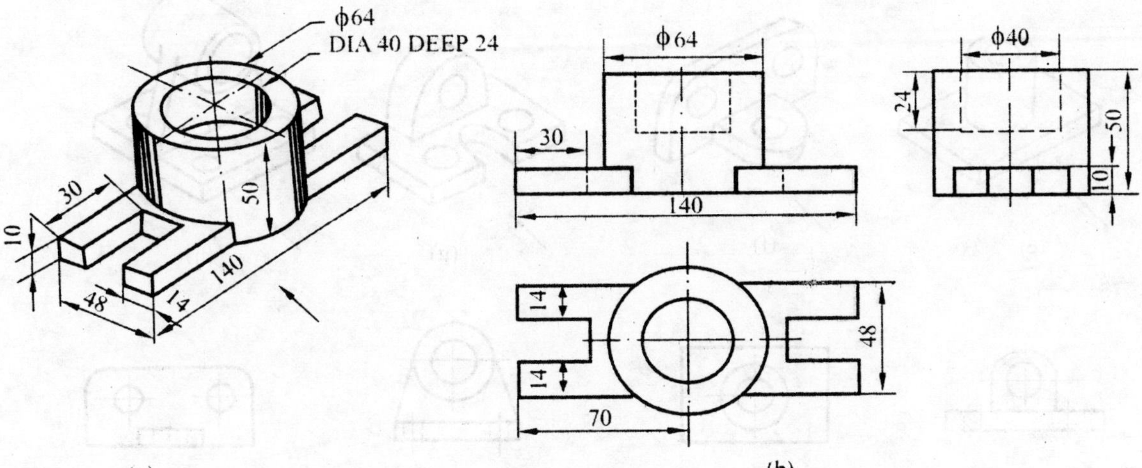

(a)

(b)

Fig. 6.72

Exercise 2

Figs. 6.73 and 6.76 show the isometric views of certain objects A to H along with their orthographic views. Identify the front, top or side views of the objects and draw the third view.

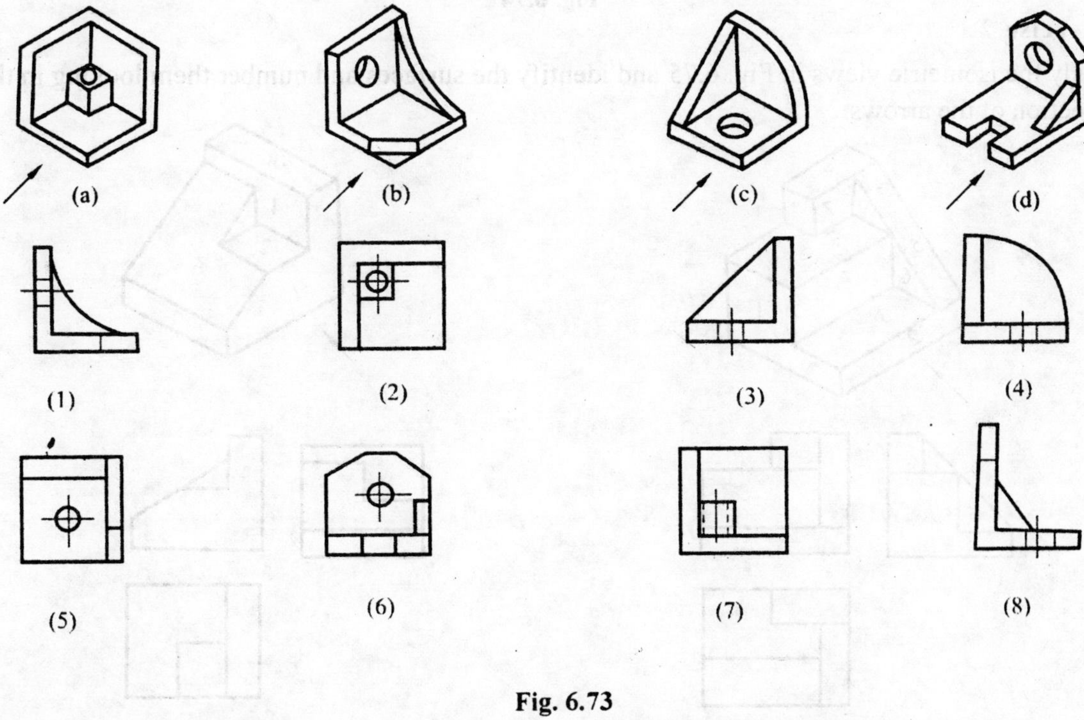

Fig. 6.73

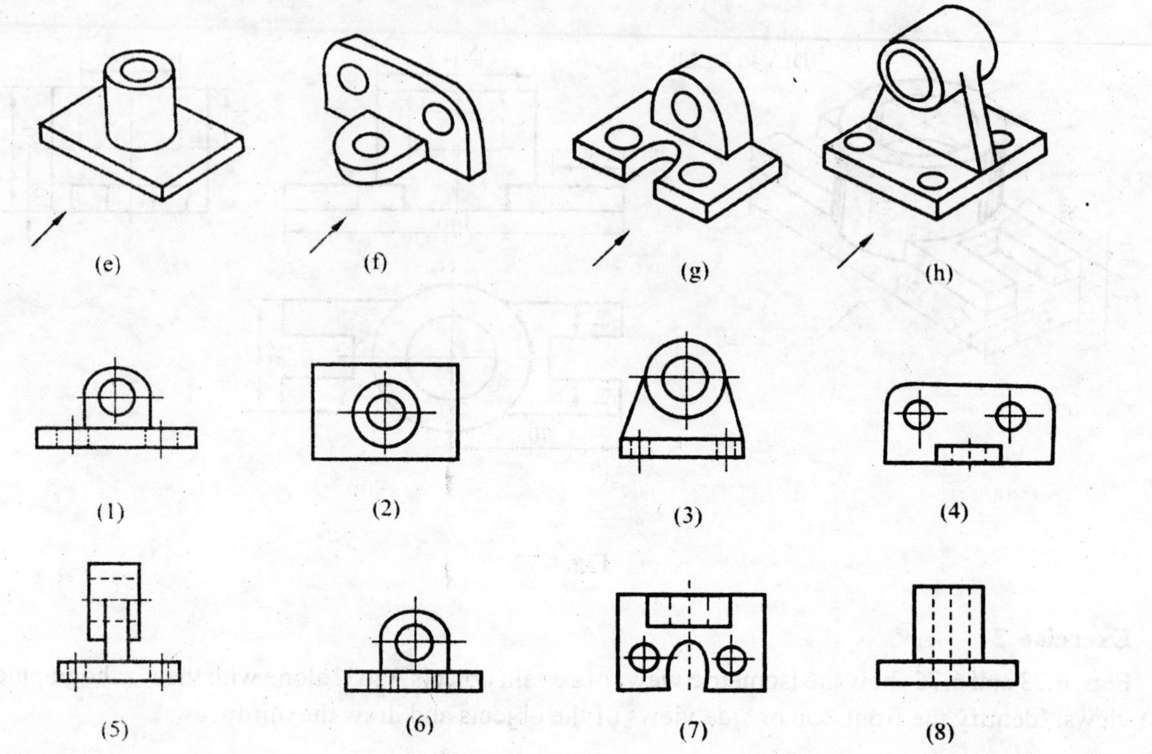

Fig. 6.74

Exercise 2

Study the isometric views in Fig. 6.75 and identify the surfaces and number them looking in the direction of the arrows.

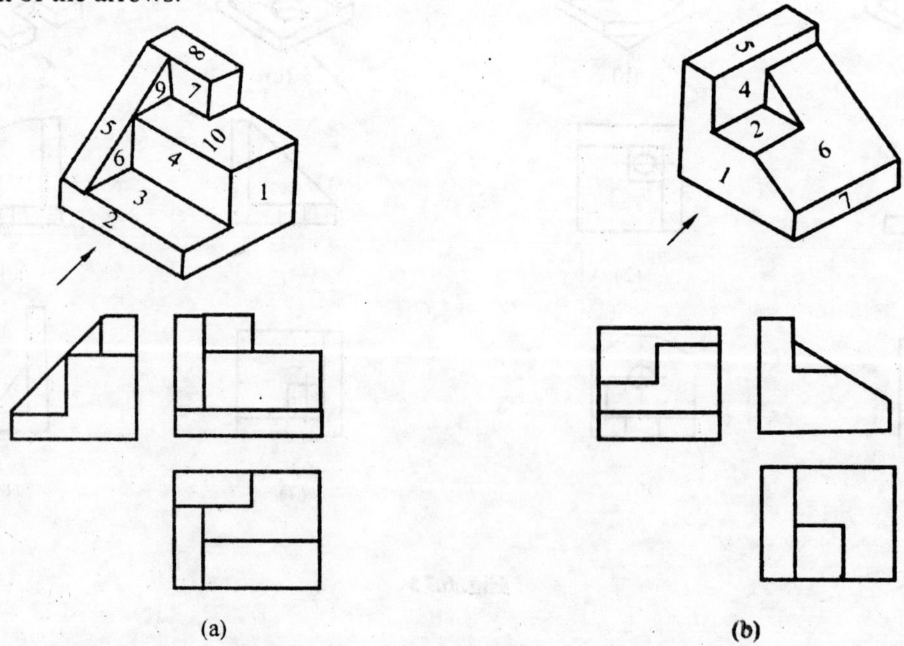

Fig. 6.75

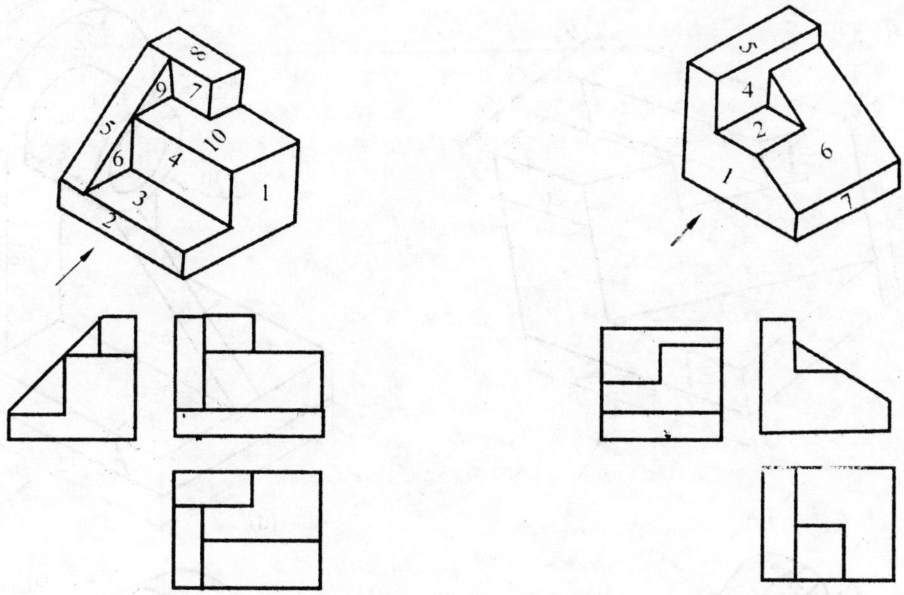

(c) (d)

Fig. 6.75

Exercise 3

Study the isometric views in Figs. 6.76(a) to (h) and draw the orthographic views looking in the direction of the arrows and number the surfaces.

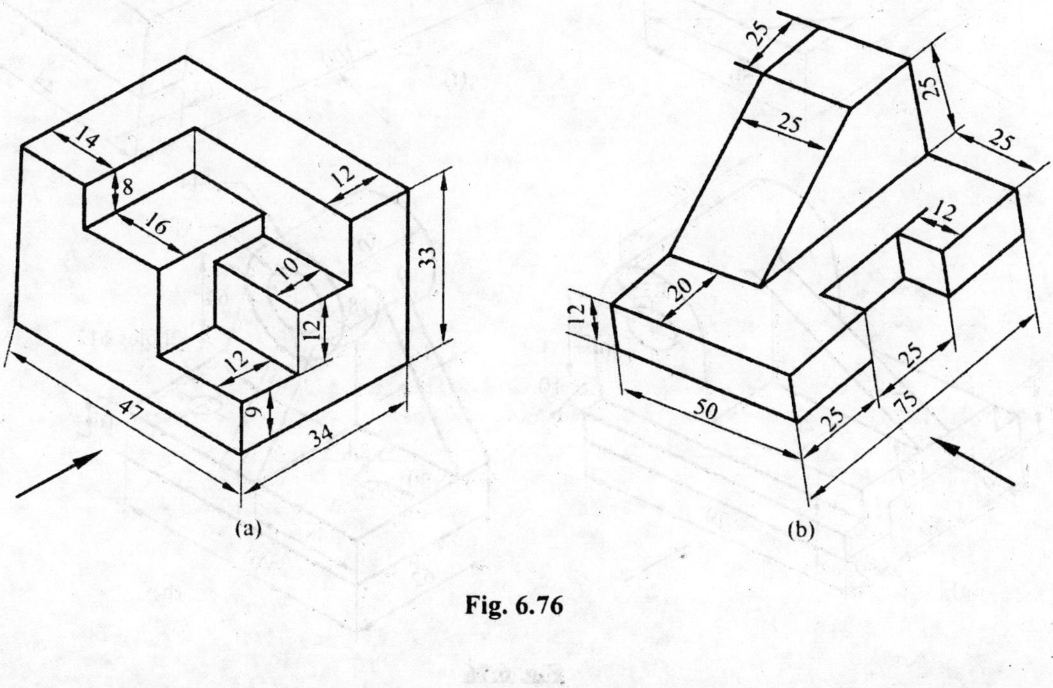

(a) (b)

Fig. 6.76

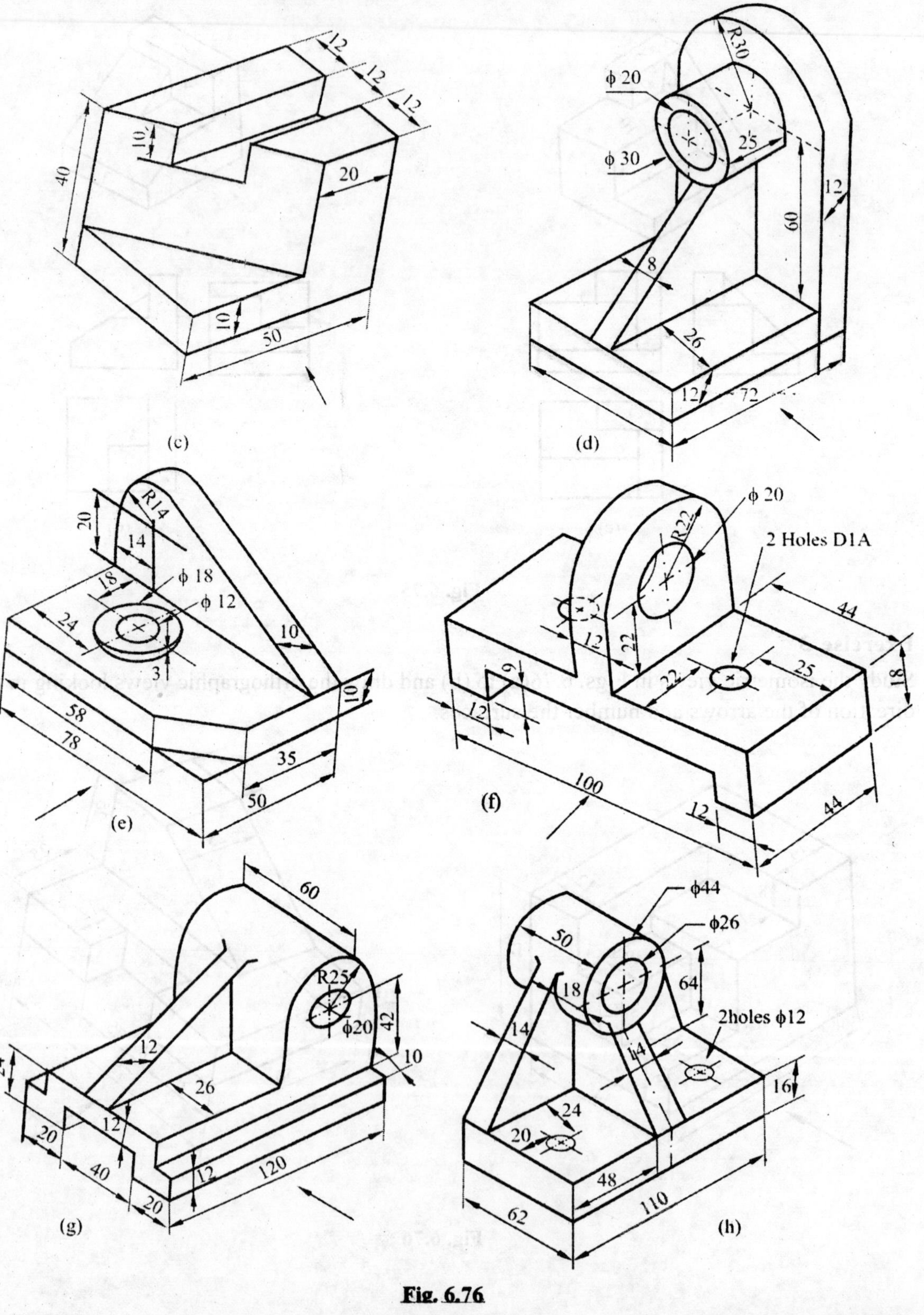
(c)
(d)
φ 20
φ 30
R30
(e)
R14
φ 18
φ 12
(f)
φ 20
R22
2 Holes D1A
(g)
R25
φ20
(h)
φ44
φ26
2holes φ12

Fig. 6.76

Exercise 4

Study the isometric views in Fig. 6.77(a) to (d) and draw the orthographic views.

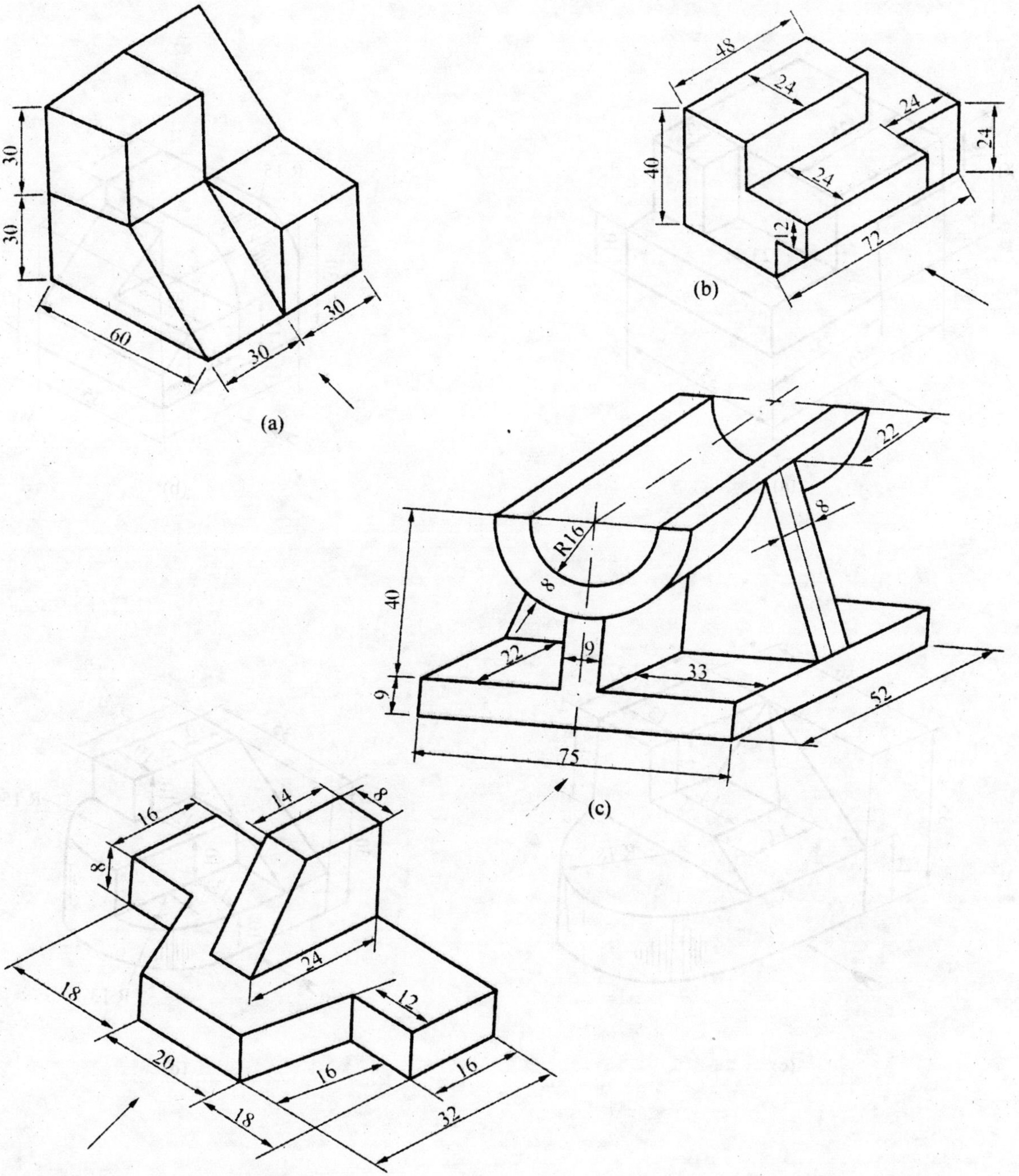

Fig. 6.77

Exercise 5

Study the isometric views in Fig. 6.78(a) to (d) and draw the three orthographic views.

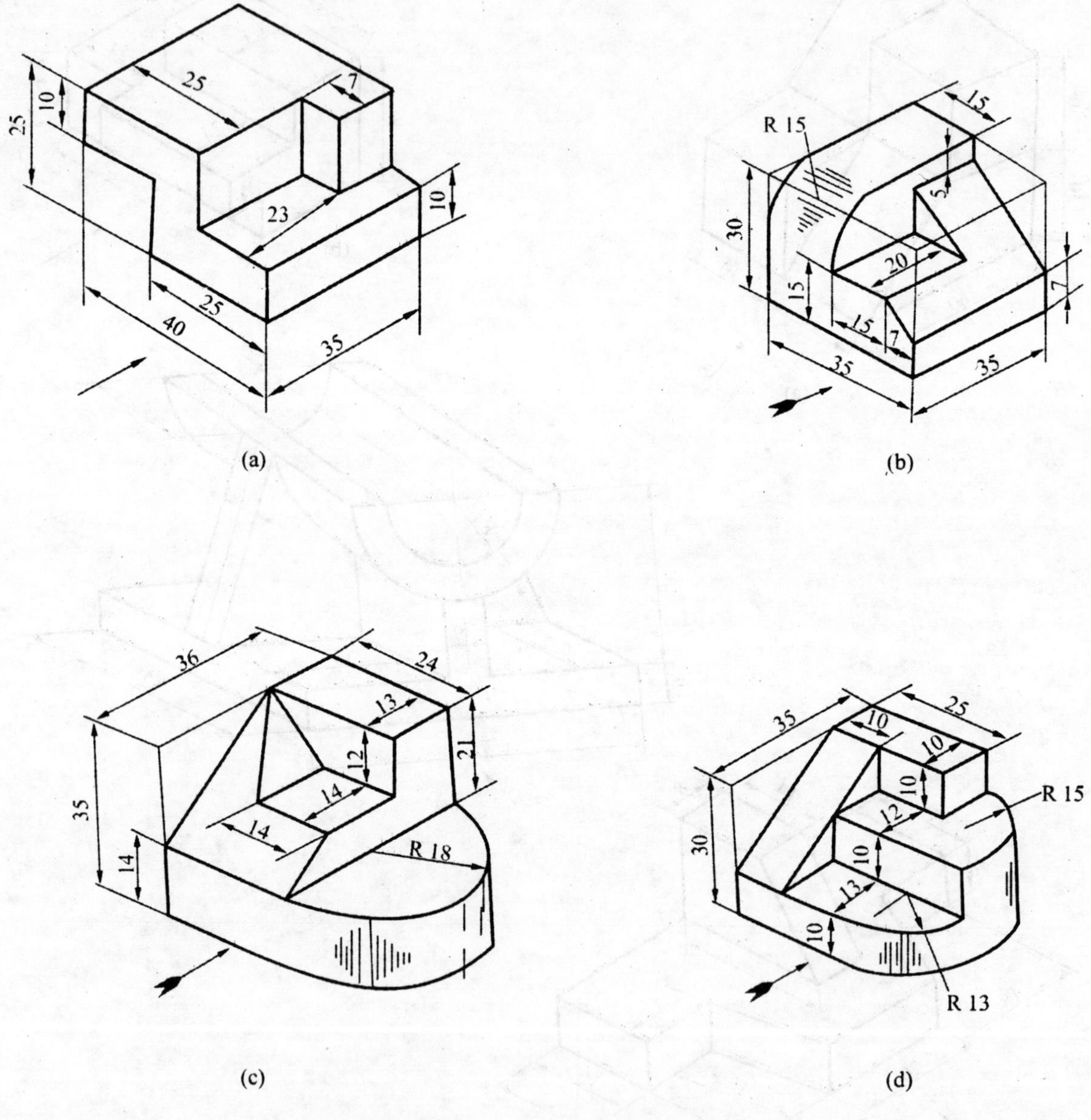

Fig. 6.78

Solutions to University Questions

Conversion of Isometric views to orthographic views.

Problem : Draw the three orthographic views of the object shown in the figure below.

(JNTU – CSE – May/June 2008)

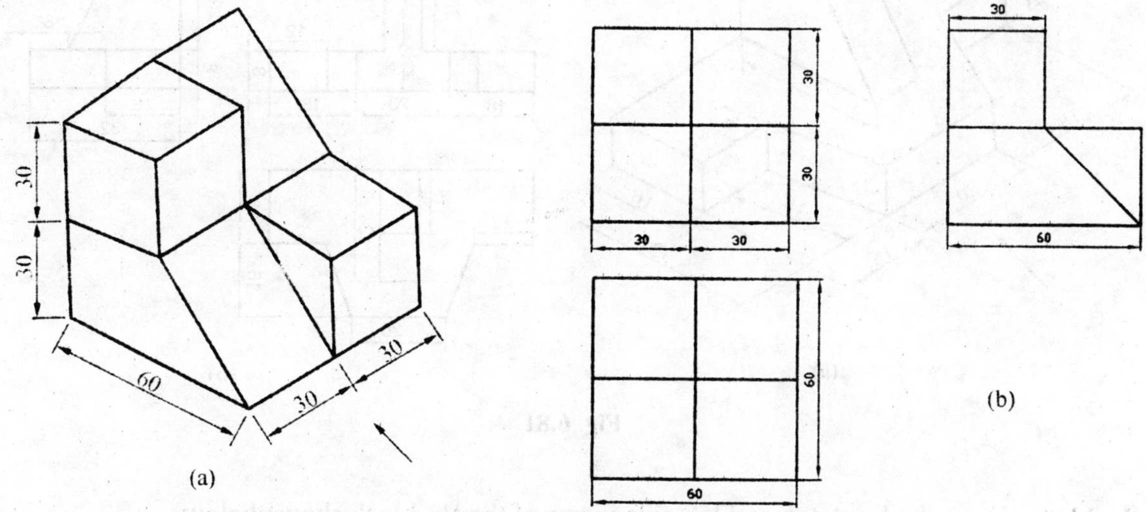

Fig. 6.79

Problem : Draw the three views of the object shown in the figure below.

(JNTU – CSE – May/June 2008)

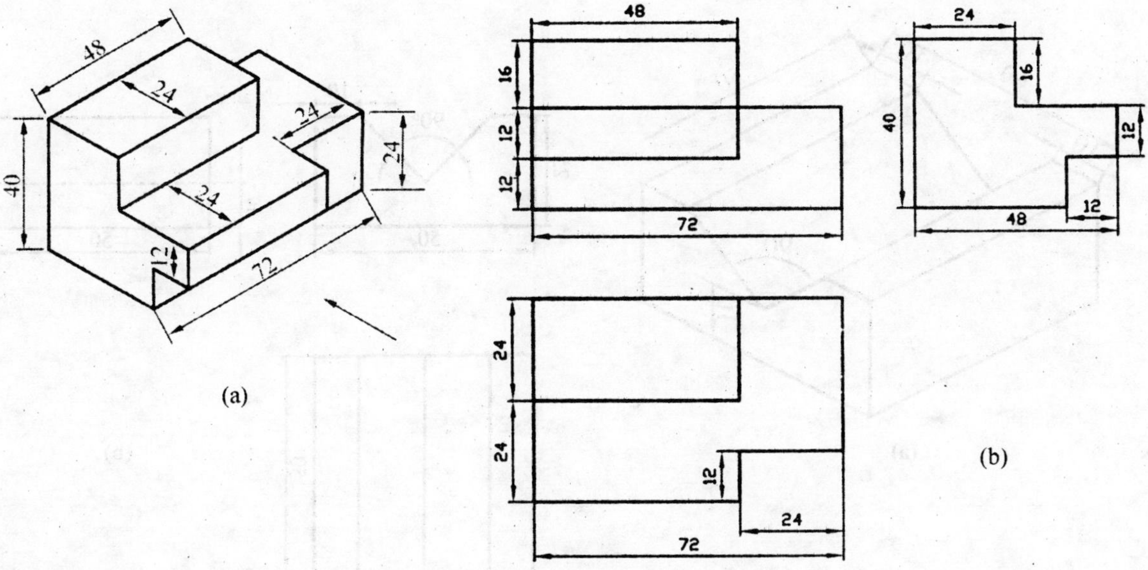

Fig. 6.80

Problem : Draw the front, top and side views of the block shown below.

(JNTU – EEE – May/June 2008)

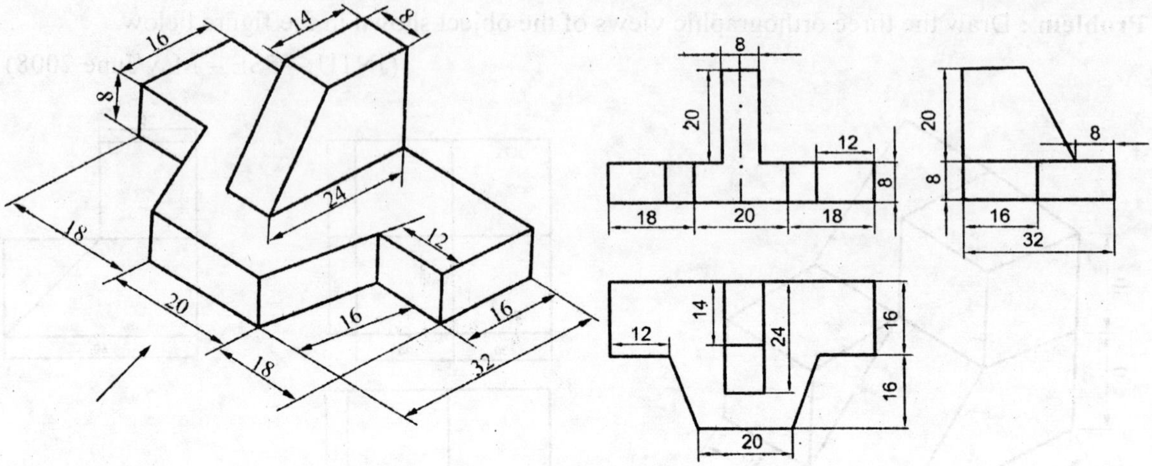

(a) (b)

Fig. 6.81

Problem : Draw the front, top and left side views of the V - block shown below.

(JNTU – CSE – May/June 2008)

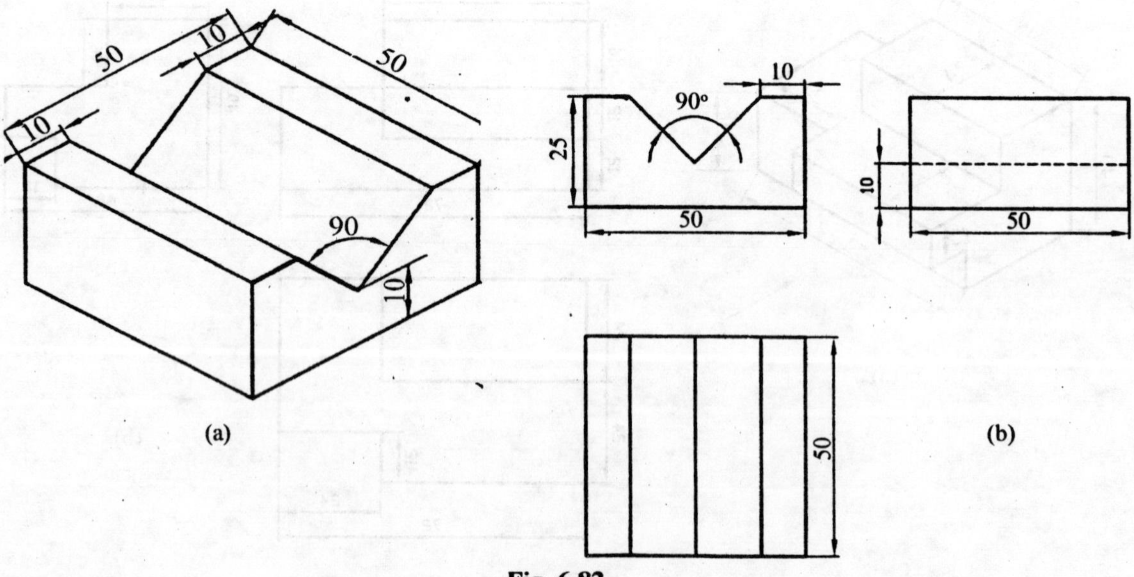

(a) (b)

Fig. 6.82

Problem : Draw the front, top and left side views of the part shown below.

(JNTU – EEE – May/June 2008)

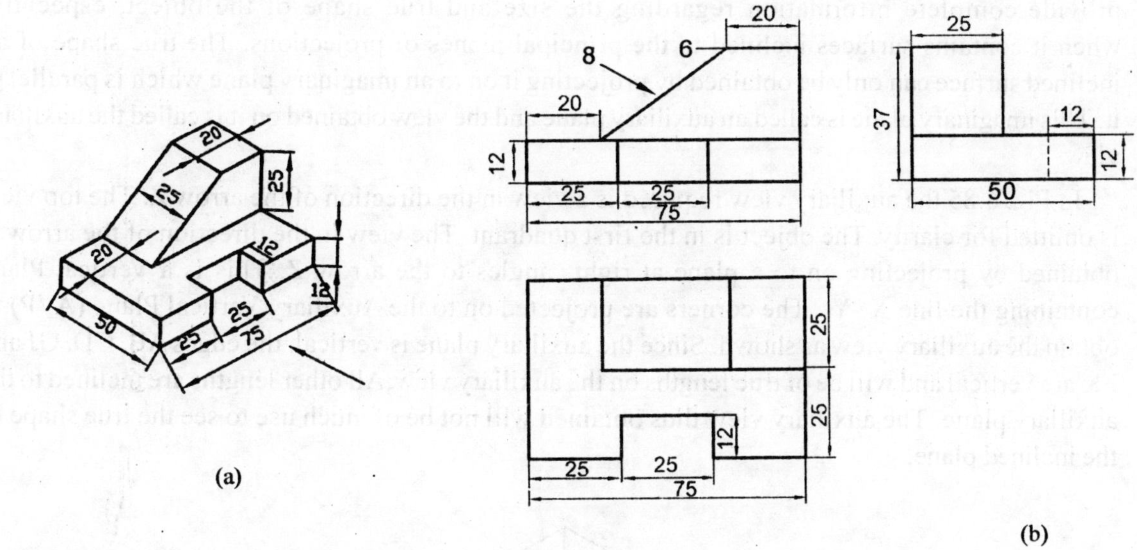

Fig. 6.83

Problem : Draw the orthographic projections for the isometric view shown below.

(JNTU – EEE – May/June 2008)

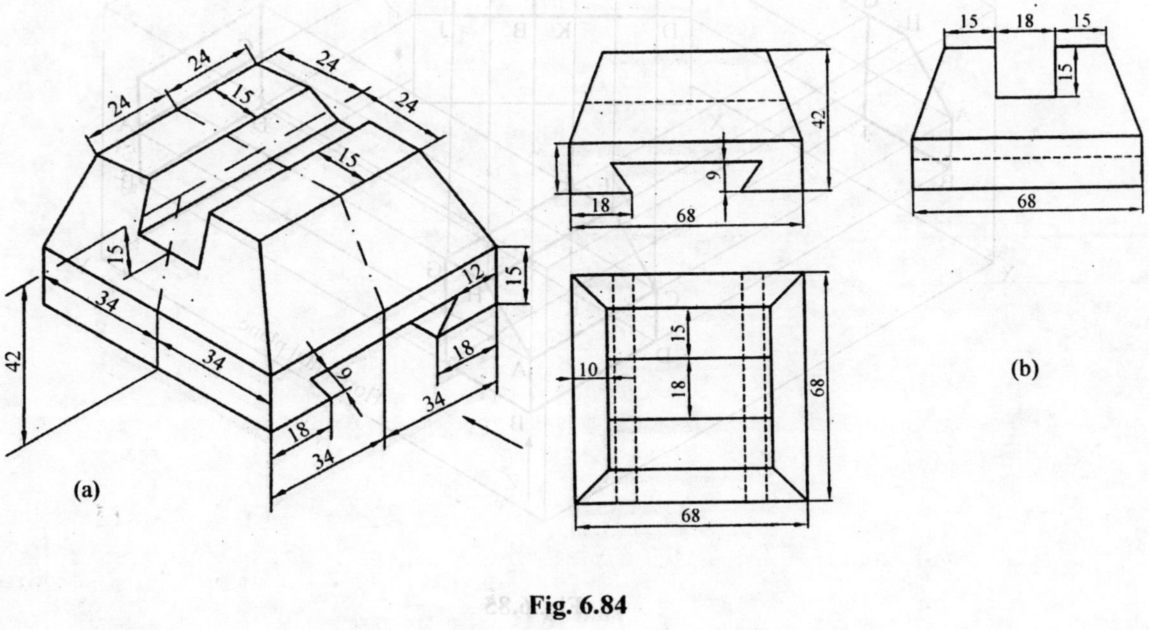

Fig. 6.84

6.10 Auxiliary Projections

The conventional orthographic views, viz, front, top and side views may not be sufficient always to provide complete information regarding the size and true shape of the object, especially when it contains surfaces inclined to the principal planes of projections. The true shape of an inclined surface can only be obtained by projecting it on to an imaginary plane which is parallel to it. This imaginary plane is called an auxiliary plane and the view obtained on it is called the auxiliary view.

In Fig. 6.85 the auxiliary view required is a view in the direction of the arrow Z. The top view is omitted for clarity. The object is in the first quadrant. The view in the direction of the arrow is obtained by projecting on to a plane at right angles to the arrow Z. This is a Vertical Plane containing the line X_1-Y_1. The corners are projected on to the Auxiliary Vertical Plane (AVP) to obtain the auxiliary view as shown. Since the auxiliary plane is vertical, the edges AB, CD, GJ and FK are vertical and wiil be of true lengths on the auxiliary view. All other lengths are inclined to the auxiliary plane. The auxiliary view thus obtained will not be of much use to see the true shape of the inclined plane.

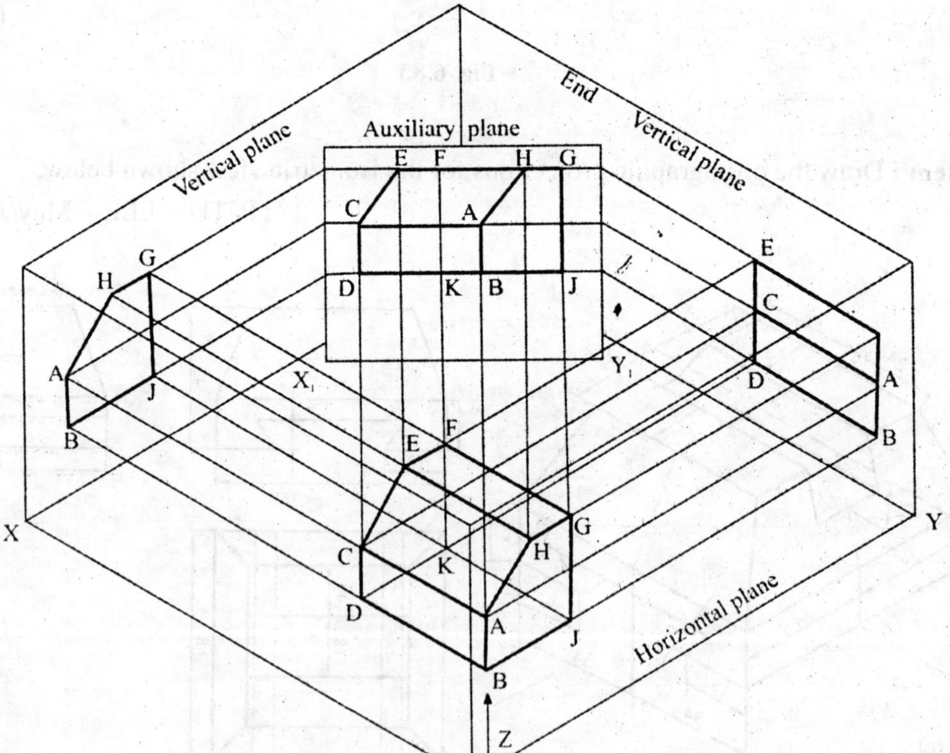

Fig. 6.85

The advantage of auxiliary projections to see the true shape of the inclined surfaces is illustrated in Fig. 6.86(a).

As the auxiliary view only shows the true shape and details of the inclined surface or feature, a partial auxiliary view pertaining to the inclined surface only is drawn. Drawing all other features lead to confusion of the shape discription (Fig. 6.86(b)).

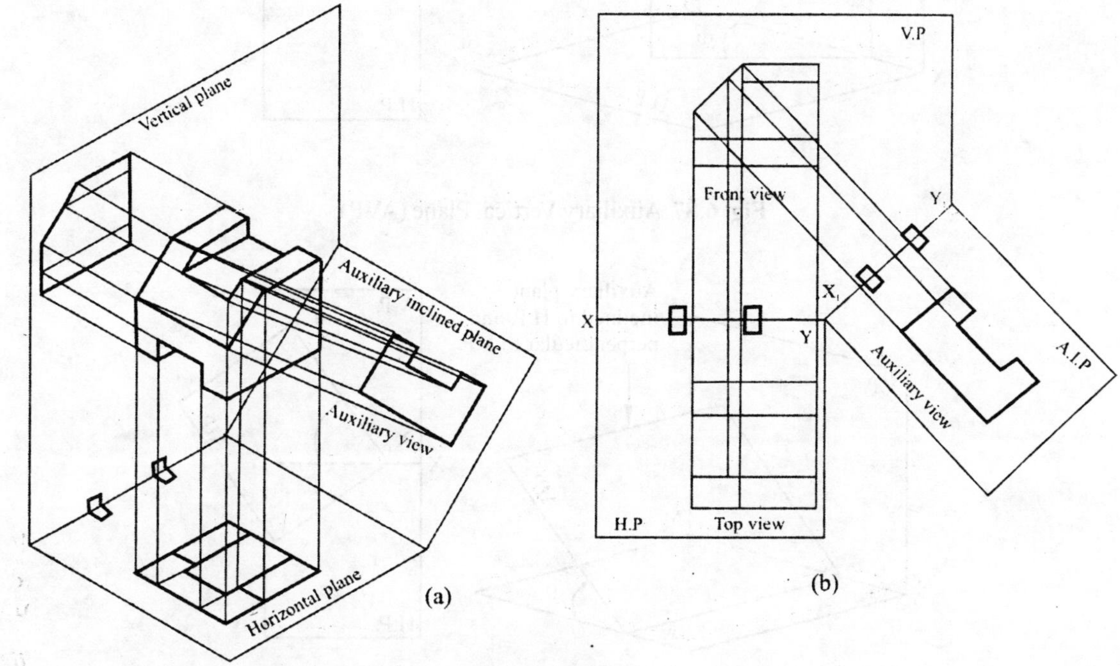

Fig. 6.86

6.11 Types of Auxiliary Planes

The auxilliary planes are of two types

(i) Auxiliary Vertical Plane (AVP)

(ii) Auxiliary Inclined Plane (AIP)

6.11.1 Auxiliary Vertical Plane

An Auxiliary vertical plane is perpendicular to HP and inclined to the V.P. The projection obtained on this AVP is called **auxiliary front view**.

Auxiliary views may be classified, based on the relation of the inclined surface of the object with respect to the principal planes of projections.

6.11.2 Auxiliary Inclined Plane (AIP)

It is perpendicular to the V.P but inclined to the H.P. The projection on this AIP is called **auxiliary top view**.

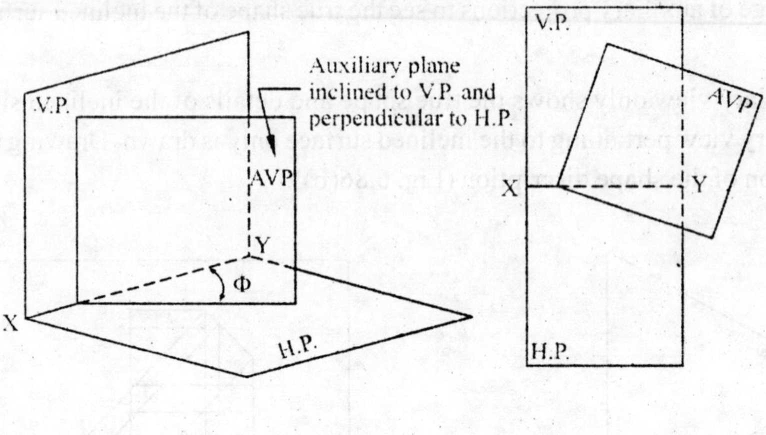

Fig. 6.87 Auxiliary Vertical Plane (AVP).

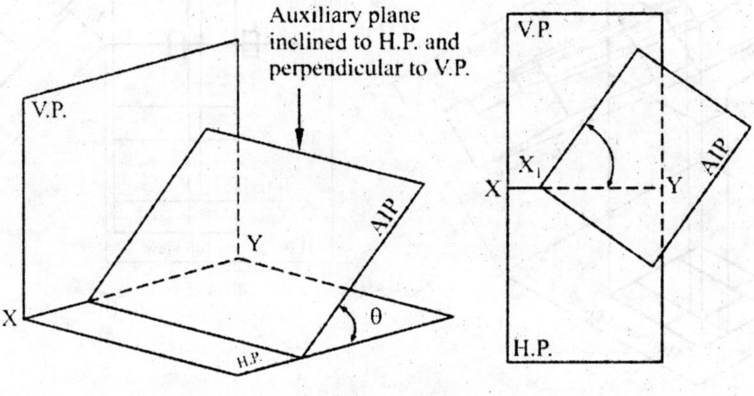

Fig. 6.88 Auxiliary Inclined Plane (AIP).

Steps to draw auxiliary projections of solid

The following table illustrates the steps to draw the auxiliary projection of solids.

S. No.	Position of Solid	Steps
1.	Axis of the solid is inclined to H.P. and parallel to V.P.	1. Draw the simple position of solid (The axis is perpendicular to H.P. parallel to V.P) 2. Draw the reference line (X_1Y_1) in front view at given angle to the axis of the front view to get the final auxiliary top view.
2.	Axis of the solid is inclined to V.P. and parallel to H.P	1. Draw the simple position of solid. (The axis is perpendicular to V.P. parallel to H.P) 2. Draw the reference line (X_1Y_1) in top view at given angle to the axis of the top view to get the final auxiliary front view.

Contd

S. No.	Position of Solid	Steps
3.	Axis of the solid is inclined to both H.P. and V.P.	1. Draw the simple position of solid. (The axis is perpendicular to H.P. parallel to V.P) 2. Draw the reference line (X_1Y_1) at given angle to the axis of the front view to get the auxiliary top view. 3. Now draw another reference line (X_2Y_2) at given angle to the axis of the first auxiliary top view to get the final auxiliary front view .

Auxiliary Front View

Figure 6.89 shows the auxiliary front view of a cube, projected on an Auxiliary Vertical Plane (AVP), inclined to VP and perpendicular to HP. Here, the auxiliary front view is projected from the top view, and its height is same as the height of the front view.

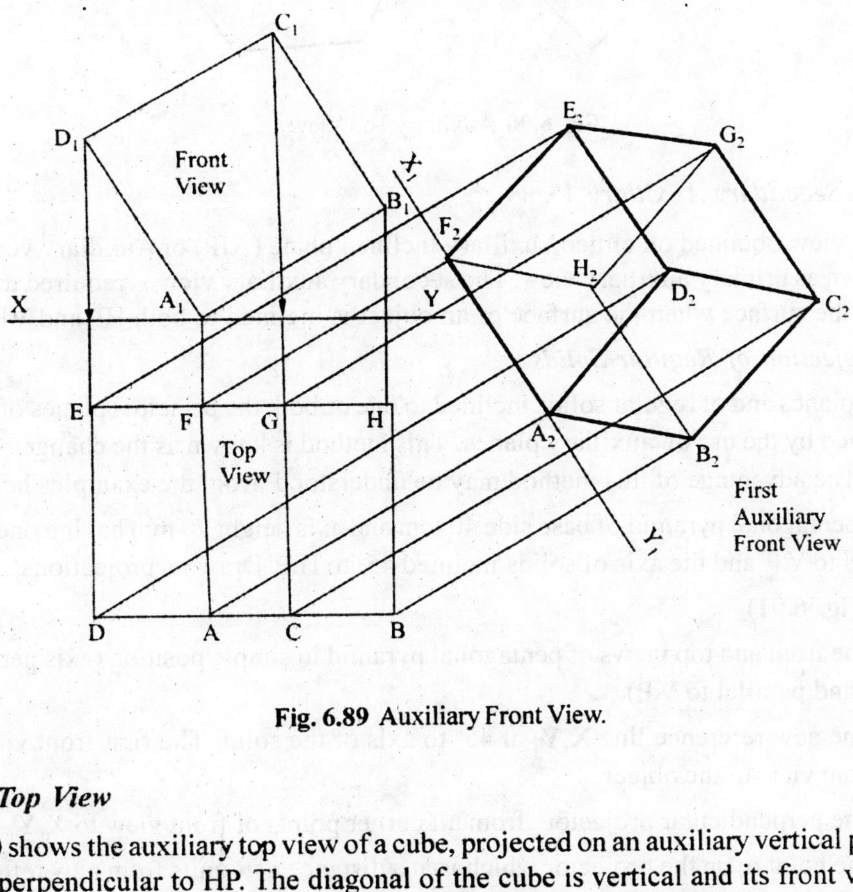

Fig. 6.89 Auxiliary Front View.

Auxiliary Top View

Figure 6.90 shows the auxiliary top view of a cube, projected on an auxiliary vertical plane inclined to VP and perpendicular to HP. The diagonal of the cube is vertical and its front view is given. The auxiliary top view is projected from the front view and its depth is the same as the depth of the top view.

The auxilairy top view is projected from front view and its depth is the same as the depth of the top view.

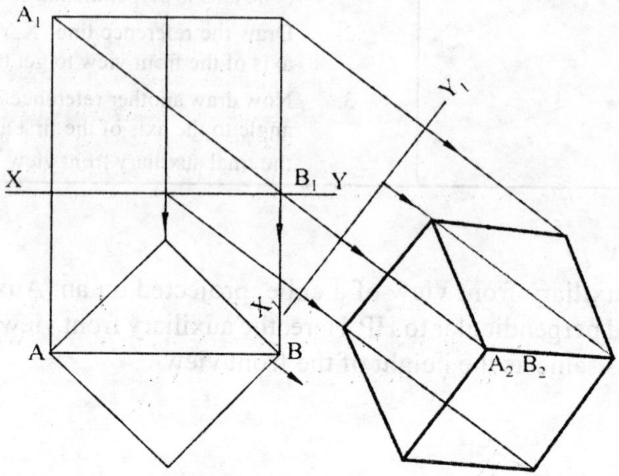

Fig. 6.90 Auxiliary Top View.

Primary and Secondary Auxiliary Views

The auxiliary view obtained on either Auxiliary inclined plane (AIP) or Auxiliary vertical plane (AVP) is known as primary auxiliary view. The secondary auxiliary view is required to obtain the true shape of the surface when the surface of an object is inclined to both HP and VP.

Auxiliary Projection of Regular Solids

Projection of planes and of regular solids inclined to one or both the principal planes of projection may be obtained by the use of auxiliary planes. This method is known as the change of reference line method. The advantage of this method may be understood from the examples below.

Problem : A pentagonal pyramid of base side 40 mm and axis height 70 mm having one of its base side is parallel to V.P. and the axis of solids inclined 45° to H.P. Draw its projections.

Solution : (Fig. 6.91)

1. Draw the front and top views of pentagonal pyramid in simple position (axis perpendicular to H.P and parallel to V.P).
2. Draw the new reference line $\mathbf{X_1Y_1}$ at 45° to axis of the solid. The first front view itself is final front view of the object.
3. Draw the perpendicular projectors from all corner points of front view to X_1Y_1.
4. Mark the point $\mathbf{o_1}$ on the projector which is at a distance of x units from new reference line, the **x** units is equal to the distance from **XY** line to **o** in the first plane.
5. Similarly mark remaining points on the projector with appropriate distances and get the final top view of the given pentagon.

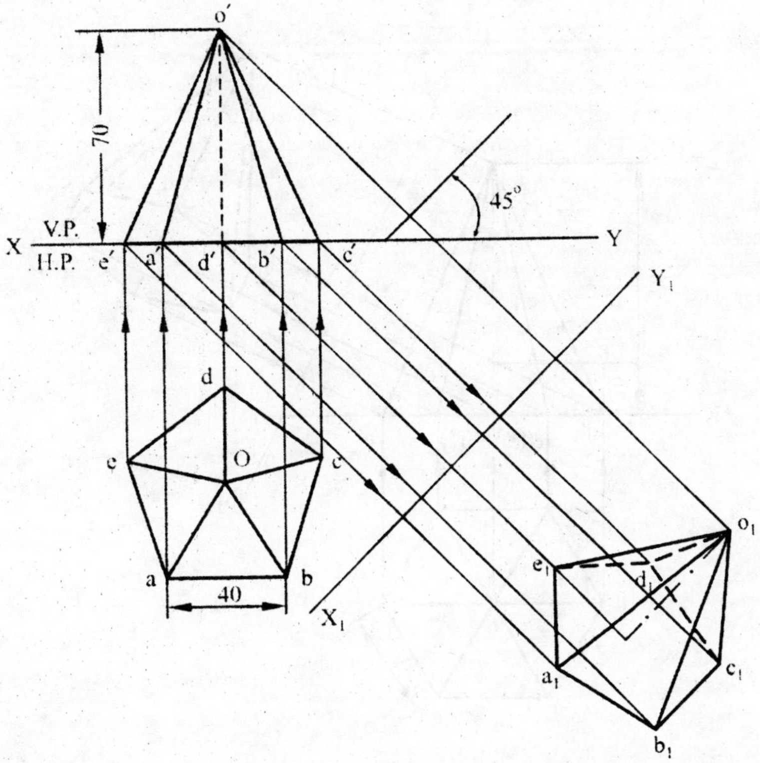

Fig. 6.91

Problem : A hexagonal pyramid of base side edge 30 mm and height 80 mm is freely suspended by one of the base corners with the axis parallel to V.P. Draw its projection.

Solution : (Fig. 6.92)

1. Draw the projection of hexagonal pyramid in simple position (axis perpendicular to H.P. and parallel to V.P).
2. The pyramid is freely suspended by one of its base corners. So the mass centre of gravity **g'** is 20 mm (one fourth of the height of the solid) above H.P.
3. Draw the new reference line $\mathbf{X_1Y_1}$ such that it is at 90° inclined to line **d'g'** .
4. Measure the distance of each point in plane from **XY** line and transfer those points on to corresponding projectors with respect to $\mathbf{X_1Y_1}$.
5. Join all the points in proper sequence to get the final front view of the given solid.

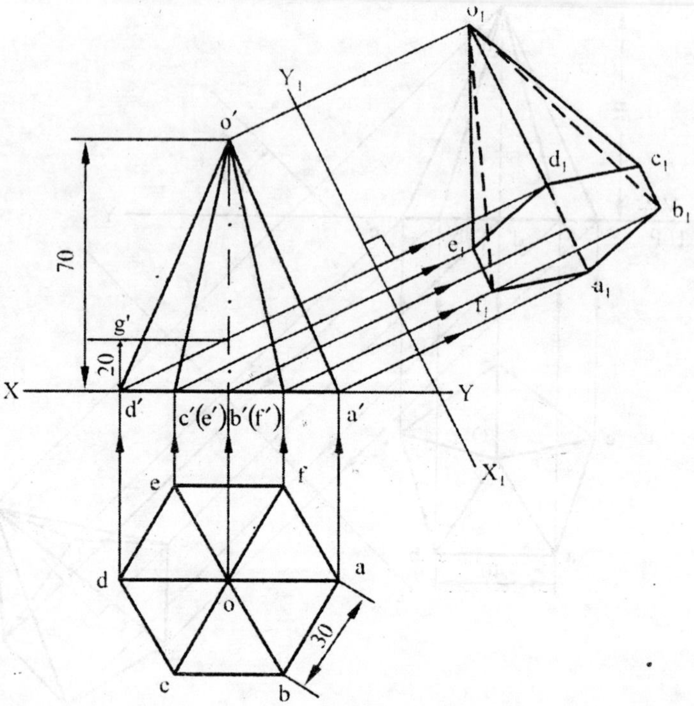

Fig. 6.92

Note:

1. For cone and pyramid, the centre of gravity lies in the axis at a distance 1/4 of height from base.
2. When a pyramid is suspended freely from one of its base corners, the line joining that corner and the CG (a'g') should be perpendicular to XY line.

Problem : A cone of base diameter 30 mm and axis 50 mm long, has the axis parallel to V.P and inclined at 30° to H.P. Draw the projection by auxiliary projection method.

Solution : **(Fig. 6.93)**

1. Draw the top and front views of cone with diameter 30 mm and height of 50 mm.
2. Here instead of tilting the solid, draw a new reference line $\mathbf{X_1Y_1}$ such that it is inclined at 30° to the axis of solid in front view. The first front view is itself the second front view. For practical convenience draw the reference line parallel to $\mathbf{X_1Y_1}$.
3. Draw the perpendicular line to new reference line X_1Y_1 from all the corner points in the front view.
4. Mark the point **o**, on the projector from **o'** at a distance of x units from new reference line the **x** units is equal to the distance from **XY** line to **o** in the first plane.

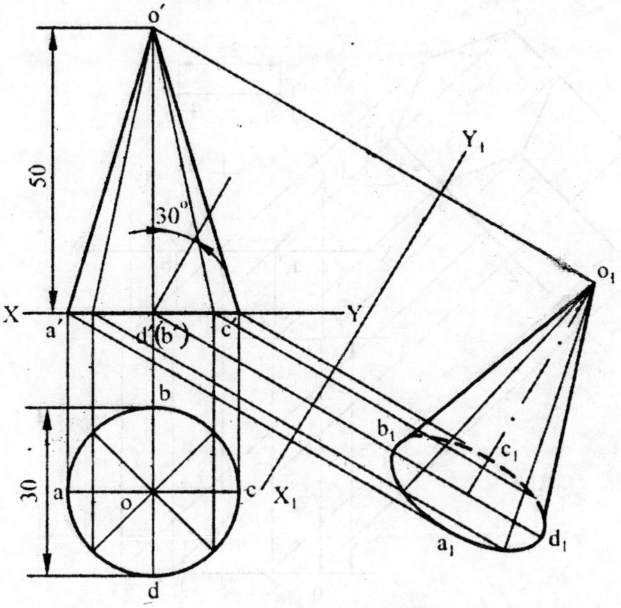

Fig. 6.93

5. Similarly mark the other points **a, b, c, d, e, f, g, h** on the corresponding projectors with appropriate distances.
6. Join these points to get the final top view of given cone.

Problem : A hexagonal prism with a side of base 25 mm and axis 60 mm long is resting on one of its rectangular faces on HP. Draw the projections of the prism when it is inclined at 45° to VP.

Solution : **(Fig. 6.94)**

1. Draw the projection of the prism assuming that the axis is perpendicular to VP, with one of its rectangular faces on HP.
2. Draw the reference line X_1Y_1 at any convenient location representing AVP and inclined at 45° to the axis of the initial top view.
3. Draw projectors perpendicular to X_1Y_1, from all the corners in the top view.
4. Measure the distances of the corners in the front view from XY, corresponding to the above corners and mark from X_1Y_1, along the above projectors.
5. Join the points in the order and complete the auxiliary front view.

The auxiliary view and the initial top view are the final views of the prism.

Problem : Draw the projection of a cylinder of dia 50 mm and axis length 65 mm. It is laying on the H.P. and its axis is inclined at 30° to V.P. and parallel to H.P.

Solution : **(Fig. 6.95)**

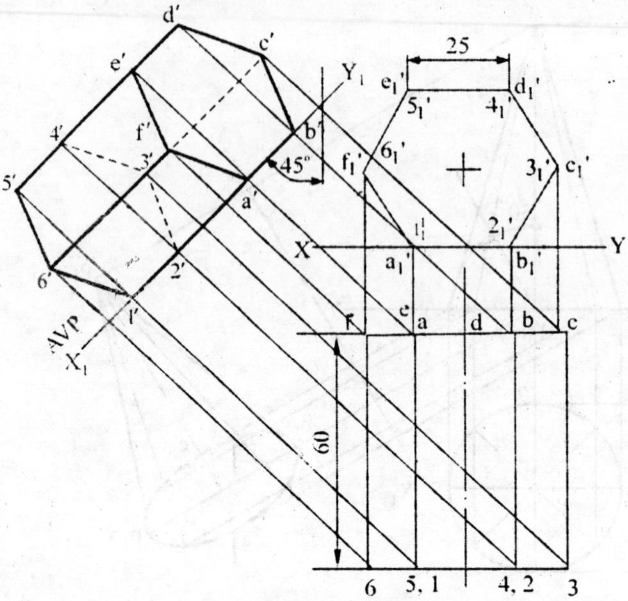

Fig. 6.94

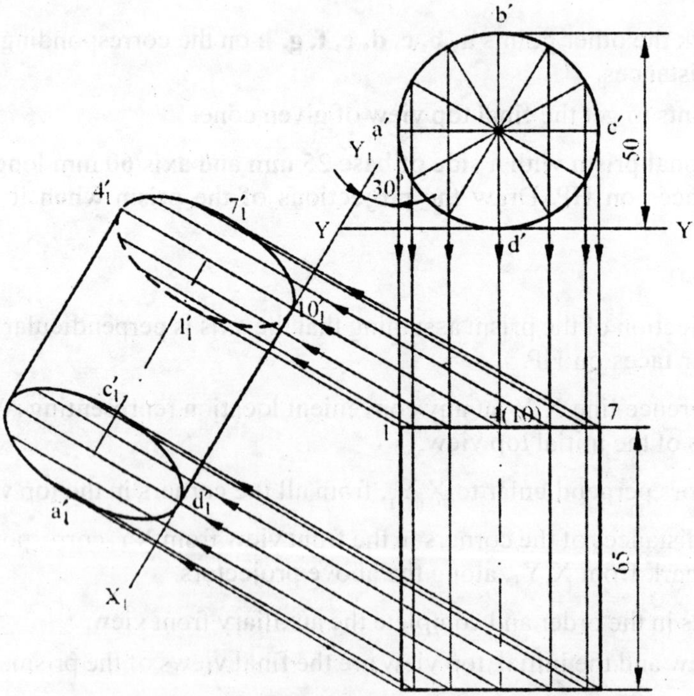

Fig. 6.95

1. Draw the projection of cylinder at simple position axis perpendicular to V.P.
2. Draw the new reference line X_1Y_1 at 30° to axis of cylinder.
3. Measure the distance between all corner points front view of from **XY** line and transfer those points in to corresponding projectors with respect to X_1Y_1.
4. Join all the corner points in proper sequence, to get front view the final of the object.

Problem : Draw the projection of a cone of dia 40 mm and height 60 mm when the base is perpendicular to H.P. and axis is inclined at 30° to V.P.

Solution : **(Fig. 6.96)**

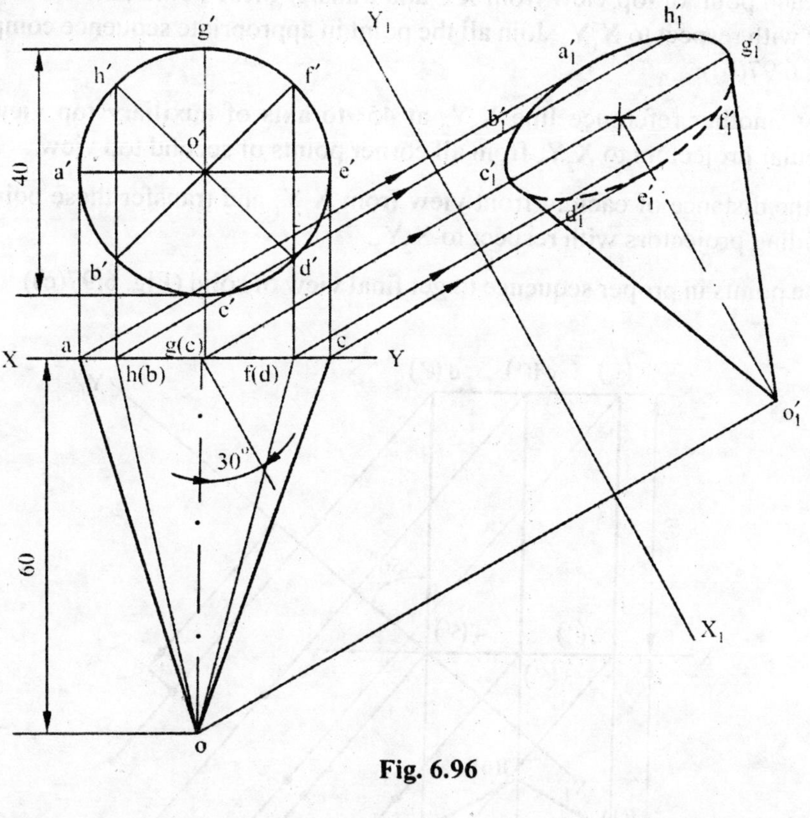

Fig. 6.96

1. Draw the top and front views of the cone in simple position with its axis perpendicular to V.P. and parallel to H.P.
2. Draw the reference line X_1Y_1 at 30° to axis of top view.
3. Draw perpendicular line to X_1Y_1 from all corners points in the top view.
4. Measure the distance of each points in front view from **XY** line and transfer those points into corresponding projectors with respect to X_1Y_1.
5. Join all the points in proper sequence and get the final front view of given solid.

Problem : A hexagonal prism of base side 25 mm and axis 60 mm long rests on the H.P. on one of its base edge and the axis is inclined at 50° to H.P. and the top view containing the axis is inclined at 45° to V.P. Draw its projections.

Solution : **(Fig. 6.97)**

1. Draw the top and front views of prism at simple position axis perpendicular to H.P. and parallel to V.P. (Fig. 6.97(a).
2. Tilt the reference line X_1Y_1 at 50° to first front view axis. For practical convenience draw the X_1Y_1 parallel to reference line. Draw the perpendicular projectors from all corner points front view to X_1Y_1 line.
3. Measure each point of top view from **XY** and transfer these points into the corresponding projectors with respect to X_1Y_1. Join all the point in appropriate sequence complete the top view (Fig. 6.97(b)).
4. Then draw another reference line X_2Y_2 at 45° to axis of auxiliary top view. Draw the perpendicular projectors to X_2Y_2 from all corner points of second top view.
5. Measure the distance of each in front view from X_1Y_1 and transfer these points on to the corresponding projectors with respect to X_2Y_2.
6. Join all the points in proper sequence to get final view of solid (Fig. 6.97(c)).

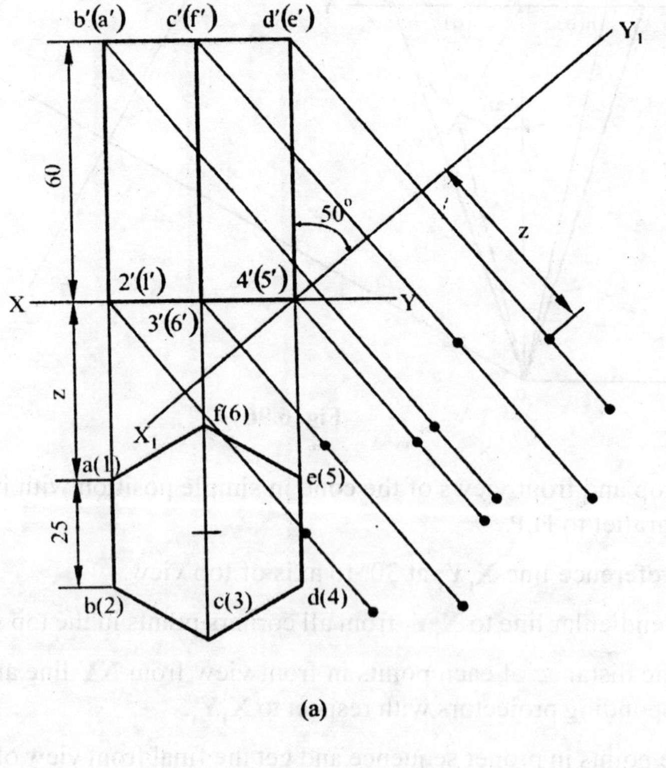

(a)

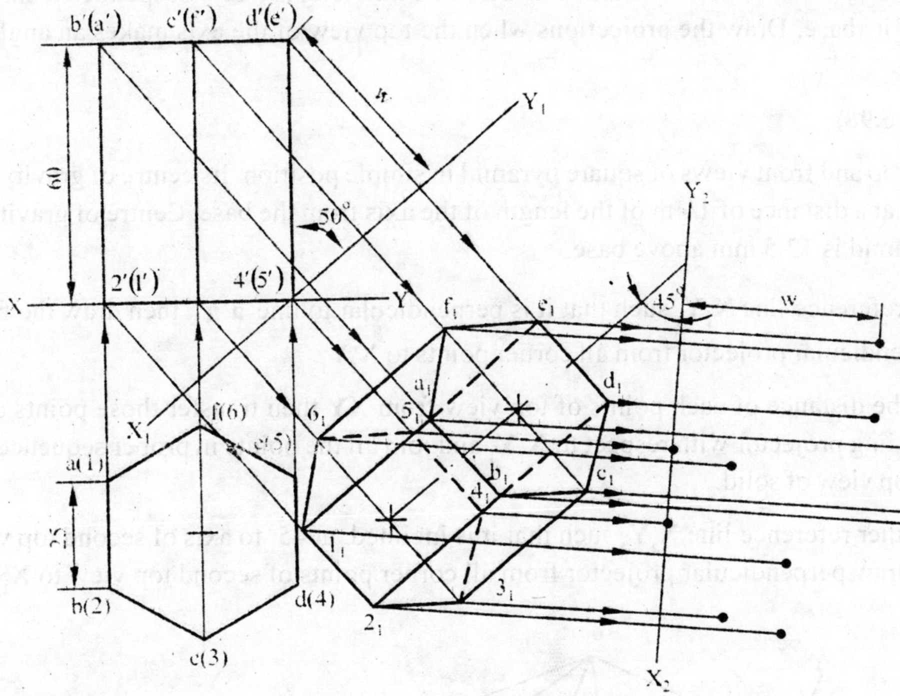
b'(a')
c'(f')
d'(e')
60
50°
Y1
Y2
2'(1')
4'(5')
X
Y
45°
w
f1
e1
a1
d1
X1
f(6)
e(5)
5
a(1)
b1
c1
41
25
11
31
b(2)
d(4)
21
c(3)
X2

(b)

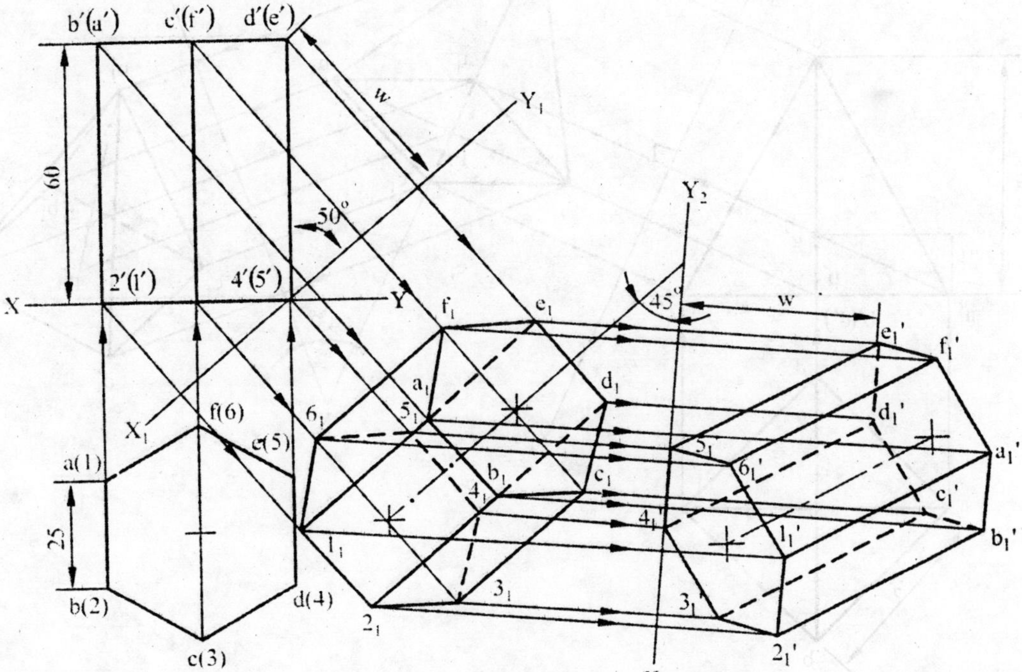
b'(a')
c'(f')
d'(e')
60
50°
Y1
Y2
2'(1')
4'(5')
X
Y
45°
w
f1
e1
e1'
f1'
a1
d1
d1'
X1
f(6)
e(5)
61
51
a1'
a(1)
b1
c1
61'
c1'
41
41'
11'
b1'
25
11
31
31'
b(2)
d(4)
21
21'
c(3)
X2

(c)

Fig. 6.97

Problem : A square pyramid base 38 mm side and axis 50 mm long is freely suspended from one of the corners of its base. Draw the projections when the top view of the axis makes an angle of 45° to V.P.

Solution : **(Fig. 6.98)**

1. Draw the top and front views of square pyramid in simple position. Its centre of gravity lies on its axis at a distance of 1/4th of the length of the axis from the base. Centre of gravity of given pyramid is 12.5 mm above base.
2. Draw the reference line $\mathbf{X_1Y_1}$ such that it is perpendicular to line **a'g'**, then draw the front view perpendicular projector from all corner points to $\mathbf{X_1Y_1}$.
3. Measure the distance of each points of top view from **XY** then transfer those points onto corresponding projector with respect to $\mathbf{X_1Y_1}$ and join all the points in proper sequence get the final top view of solid.
4. Draw another reference line $\mathbf{X_2Y_2}$ such that it is inclined at 45° to axis of second top view and then draw perpendicular projector from all corner points of second top view to $\mathbf{X_2Y_2}$.

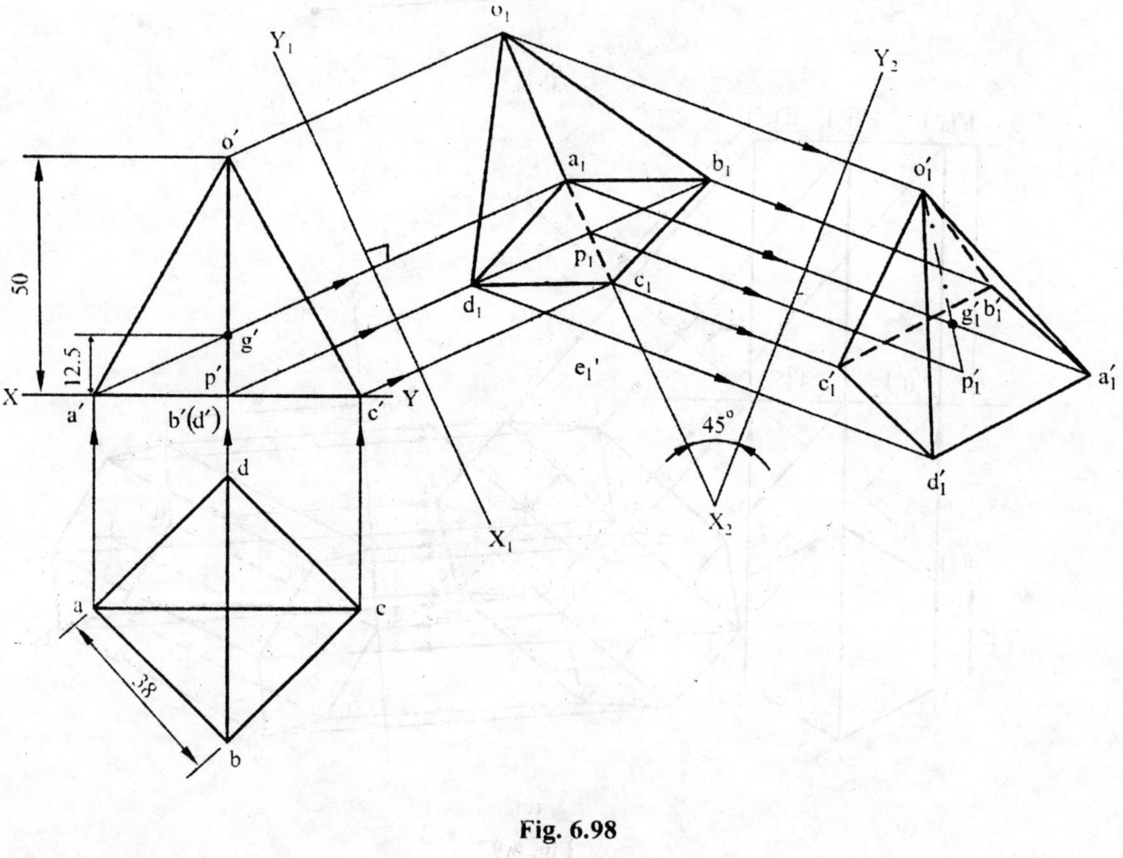

Fig. 6.98

5. Measure the distance of each point of front view from X_1Y_1 then transfer those points onto corresponding projector with respect to X_2Y_2 and join all points in proper sequence to get the final front view of solid.

Problem : A square headed bolt 20 mm diameter, 100 mm long and having a square neck of 20 mm height has its axis parallel to V.P. and inclined at 45° to the H.P. All faces of the square head are equally inclined to the V.P. Draw its projections neglecting the threads and chamfer.

Solution : **(Fig. 6.99)**

1. The object (bolt) is square head 20 mm diameter and 100 mm long. Draw the simple position of the object (top, front views). **Head of the bolt = 1.8 to 2 × dia of the bolt.**
2. Draw the new reference line X_1Y_1 such that it is inclined at 45° to axis of front view.
3. Draw the projectors from all corner points of front view to X_1Y_1. Measure distance of each corner points of from XY line and transfer those points in to corresponding projectors with respect to X_1Y_1.
4. Then join all visible and invisible edges of given bolt.

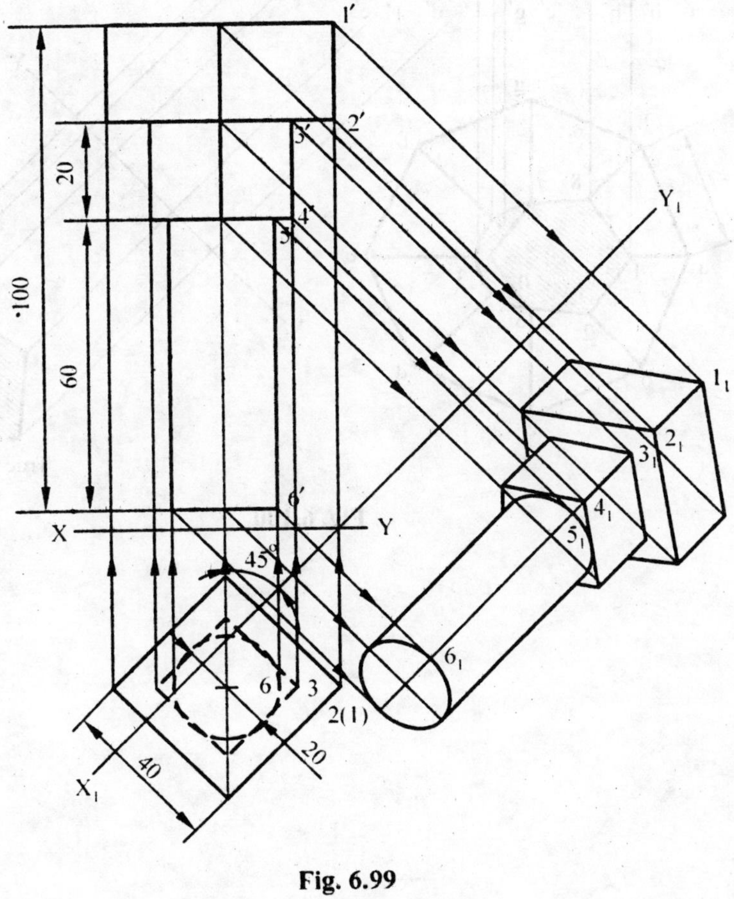

Fig. 6.99

Problem : To draw the true shape of a truncated octagonal pyramid.

Solution : **(Fig. 6.100)**

1. Draw the two views of the truncated pyramid.
2. Draw its auxiliary view as shown and get the true shape.

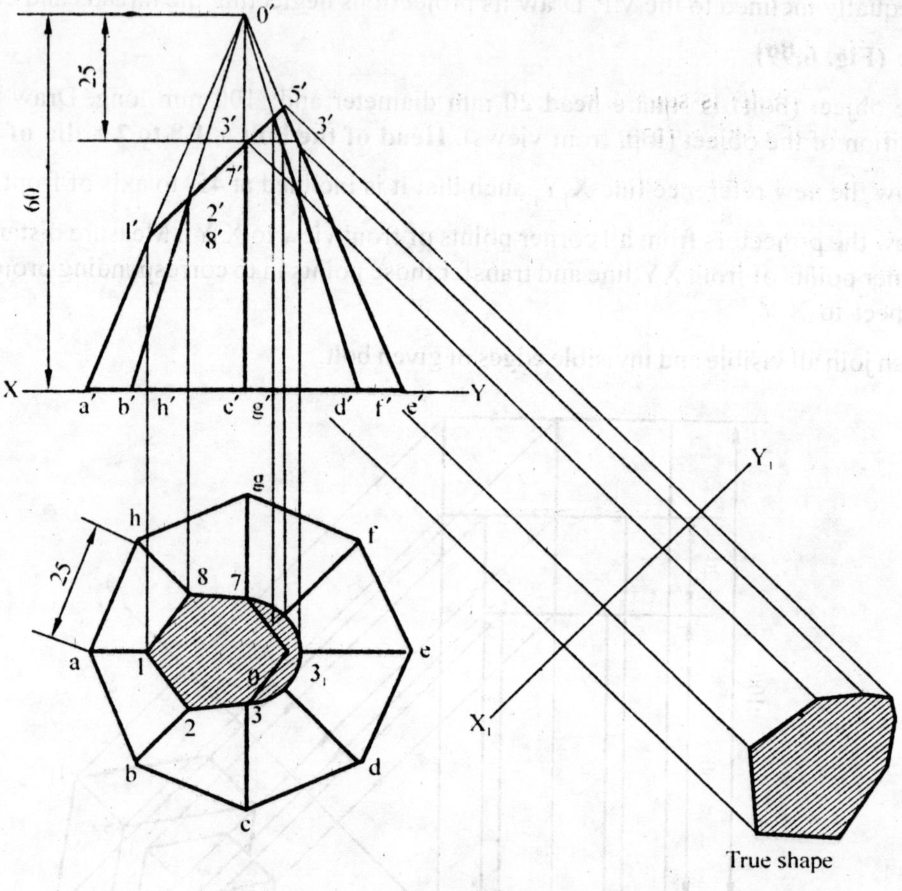

Fig. 6.100

EXERCISES

1. A cube of side 40 mm rests on its base in HP. It is then rotated such that one of its vertical faces makes an angle of 30° to VP. Draw the projection of the cube.
2. A pentagonal prism with side of base 25 mm and axis 50 mm long is lying on HP on one of its faces. Draw the projections of the prism, when the axis is parallel to VP.
3. A hexagonal pyramid of side 30 mm and height 60 mm is resting with its base on HP. One of the base edges is inclined at 60° to VP. Draw its projections.
4. A regular pentagonal prism of side 30 mm and height 60mm is resting on a corner such that one of its rectangular faces incline at 45° to the VP and perpendicular to HP. Draw its projection.
5. Draw the projection of cylinder with diameter of the base 40 mm and axis 70 mm long with its axis perpendicular to VP and 35 mm above HP; one end being 10 mm away from VP.
6. A pentagonal pyramid of base 30 mm and axis 60 mm long has its apex in the VP and the axis is perpendicular to VP. A corner of the base is resting on the ground and the side of the base contained by the corner is inclined at 30° to the ground. Draw its projections.
7. Draw the projections of hexagonal pyramid of base 25 mm and height 60 mm when one of its triangular faces lies on HP. and its base edge is at right angle to the VP and the axis of the pyramid is parallel to VP.
8. One of the body diagonals of a cube of 40 mm edge is parallel to HP and inclined at 60° to VP. Draw the projections of the cube.
9. Draw the projection of cylinder of 30 mm diameter and 50 mm long, lying on the ground with its axis inclined at 45° to the VP and parallel to the ground.
10. A cylinder of diameter 40 mm and axis 80 mm long is standing with its axis inclined at 30° to HP. Draw the projection.
11. Draw the projection of a right circular cone of 30 mm diameter and 50 mm height when a generator lines on HP making an angle of 30° with VP.
12. One of the body diagonals of a cube of 40 mm edge is parallel to HP and inclined at 60° to VP. Draw the projections of the cube.

CHAPTER 7

Development of Surfaces

7.1 Introduction

A layout of the complete surface of a three dimensional object on a plane is called the development of the surface or flat pattern of the object. The development of surfaces is very important in the fabrication of articles made of sheet metal.

The objects such as containers, boxes, boilers, hoppers, vessels, funnels, trays etc., are made of sheet metal by using the principle of development of surfaces.

In making the development of a surface, an opening of the surface should be determined first. Every line used in making the development must represent the true length of the line (edge) on the object.

The steps to be followed for making objects, using sheet metal are given below:

1. Draw the orthographic views of the object to full size.
2. Draw the development on a sheet of paper.
3. Transfer the development to the sheet metal.
4. Cut the development from the sheet.
5. Form the shape of the object by bending.
6. Join the closing edges.

Note: In actual practice, allowances have to be given for extra material required for joints and bends. These allowances are not considered in the topics presented in this chapter.

7.2 Methods of Development

The method to be followed for making the development of a solid depends upon the nature of its lateral surfaces. Based on the classification of solids, the following are the methods of development.

1. Parallel-line Development

It is used for developing prisms and single curved surfaces like cylinders in which all the edges / generators of lateral surfaces are parallel to each other.

2. Radial-line Development

It is employed for pyramids and single curved surfaces like cones in which the apex is taken as centre and the slant edge or generator (which are the true lengths) as radius for its development.

7.2.1 Development of Prism

To draw the development of a square prism of side of base 30 mm and height 50 mm

***Solution* :(Fig. 7.1)**

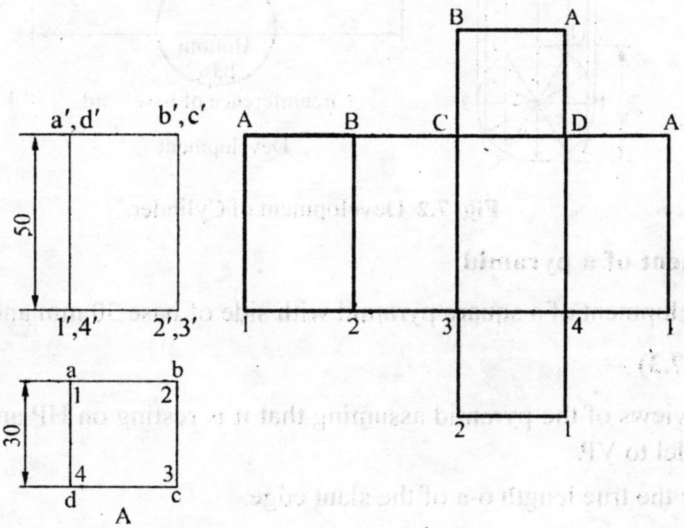

Fig. 7.1

1. Assume the prism is resting on its base on HP with an edge of the base pallel to VP and draw the orthographic views of the square prism.
2. Draw the stretch-out line 1-1 (equal in length to the circumference of the square prism) and mark off the sides of the base along this line in succesion ie 1-2, 2-3, 3-4 and 4-1.
3. Errect perpendiculars through 1,2,3 etc., and mark the edges (folding lines) 1-A, 2-B, etc., equal to the height of the prism 50 mm.
4. Add the bottom and top bases 1234 and ABCD by the side of any of the base edges.

7.2.2 Development of a Cylinder

***Solution* : (Fig. 7.2)**

Figure shows the development of a cylinder. In this the length of the rectangle representing the development of the lateral surface of the cylinder is equal to the circumference π d (here d is the diameter of the cylinder) of the circular base.

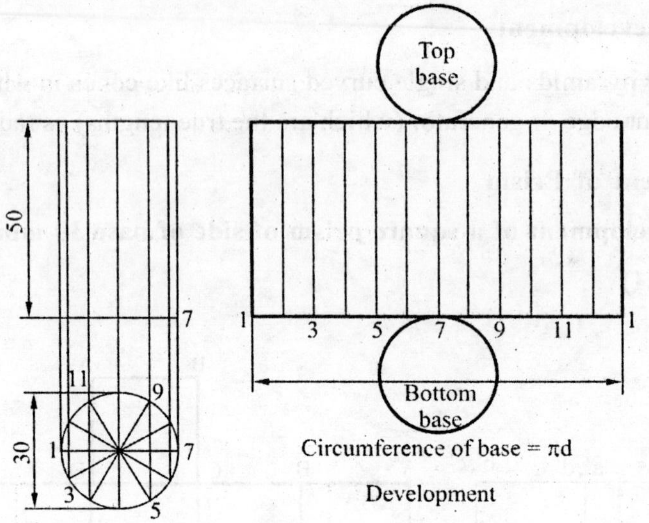

Fig. 7.2 Development of Cylinder.

7.2.3 Development of a pyramid

To draw the development of a square pyramid with side of base 30 mm and height 60 mm.

***Solution :* (Fig. 7.3)**

1. **Draw the views of the pyramid assuming that it is resting on HP and with an edge of the base parallel to VP.**
2. **Determine the true length o-a of the slant edge.**

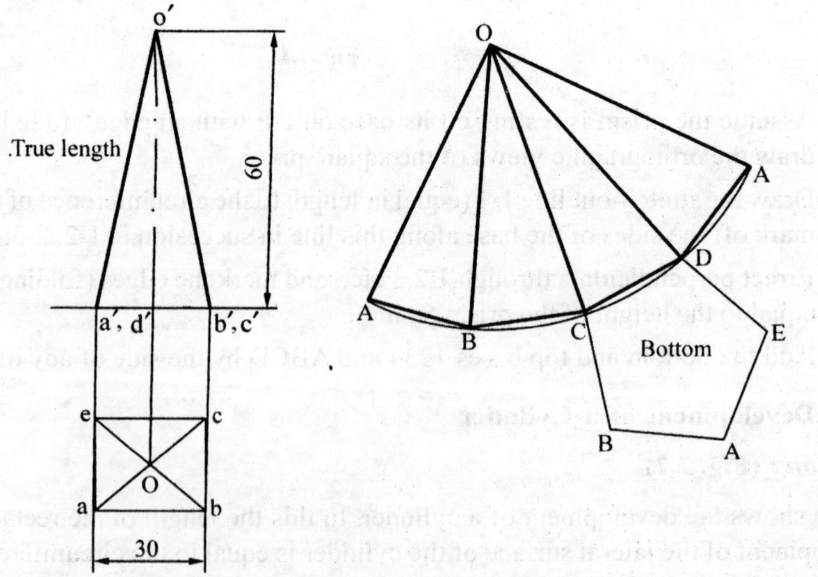

Fig. 7.3 Development of Square Pyramid.

Note:

In the orientation given for the solid, all the slant edges are inclined to both HP and VP. Hence, neither the front view nor the top view provides the true length of the slant edge. To determine the true length of the slant edge, say OA, rotate **oa** till it is parallel to XY to the position. oa_1. Through $\mathbf{a_1}$, draw a projector to meet the line XY at $\mathbf{a'_1}$. Then $\mathbf{o'_1\ a'_1}$ represents the true length of the slant edge OA. This method of determining the true length is also known as rotation method.

3. with centre O and radius **o'a'** draw an arc.
4. Starting from A along the arc, mark the edges of the base ie. AB, BC, CD and DE.
5. Join O to A,B,C, etc., representating the lines of folding and thus completing the development.

Development of Pentagonal Pyramid

To draw the development of a pentagonal pyramid with side of base 25 mm and height 60 mm.

***Solution :* (Fig. 7.4)**

1. Draw the orthgraphic views of the pyramid ABCDE with its base on HP and axis parallel to VP.
2. With centre **o** of the pyramid and radius equal to the true length of the slant edge draw an arc.
3. Mark off the edges starting from A along the arc and join them to **o** representing the lines of folding.
4. Add the base at a suitable location.

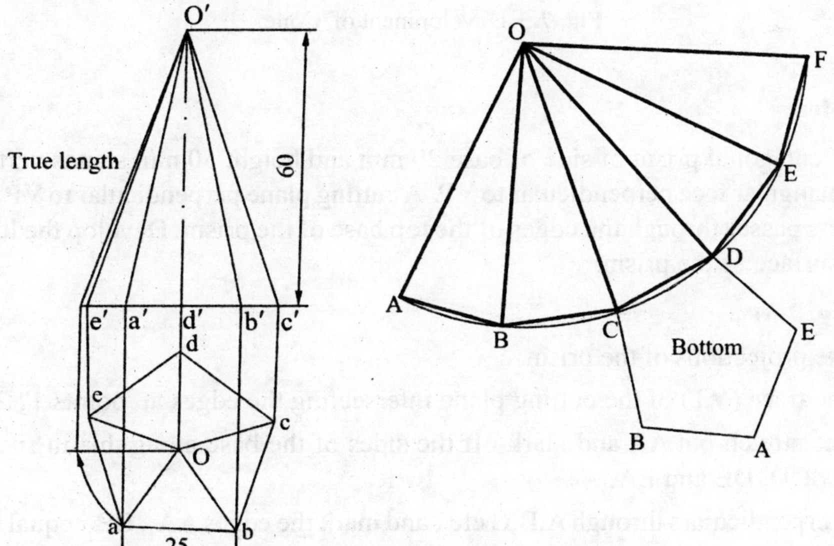

Fig. 7.4 Development of Pentagonal Pyramid.

7.2.4 Development of a Cone

Solution : **(Fig. 7.5)**

The development of the lateral surface of a cone is a sector of a circle. The radius and length of the arc are equal to the slant height and circumference of the base of the cone respectively. The included angle of the sector is given by (r / s) × 360°, where r is the radius of the base of the cone and s is the true length.

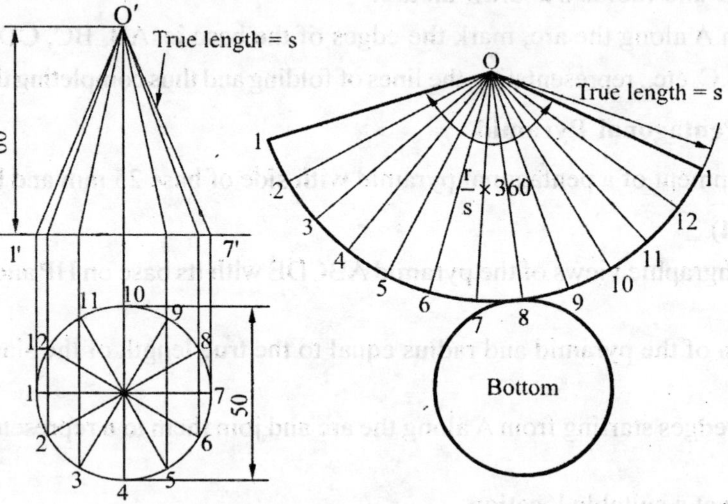

Fig. 7.5 Development of Cone.

7.2.5 Examples

Problem : A Pentagonal prism of side of base 20 mm and height 50 mm stands vertically on its base with a rectangular face perpendicular to VP. A cutting plane perpendicular to VP and inclined at 60° to the axis passes through the edges of the top base of the prism. Develop the lower portion of the lateral surface of the prism.

Solution : **(Fig. 7.6)**

1. Draw the projections of the prism.
2. Draw the trace (VT) of the cutting plane intersecting the edges at points 1', 2', 3', etc.
3. Draw the stretch-out AA and mark-off the sides of the base along this in succession i.e., AB, BC, CD, DE and EA.
4. Errect perpendiculars through A,B,C etc., and mark the edges AA_1, BB_1, equal to the height of the prism.
5. Project the points 1', 2', 3' etc., and obtain 1, 2, 3 etc., respectively on the corresponding edges in the development.

6. Join the points 1, 2, 3 etc.. by straight lines and darken the sides corresponding to the truncated portion of the solid.

Note

1. Generally, the opening is made along the shortest edge to save time and soldering.
2. Stretch-out line is drawn in-line with bottom base of the front view to save time in drawing the development.
3. AA_1-A_1A is the development of the complete prism.
4. Locate the points of intersectiion 1',2', etc., between VT and the edges of the prism and draw horizontal lines through them and obtain 1,2, etc.. on the corresponding edges in the devolopment
5. Usually, the lateral surfaces of solids are developed and the ends or bases are omitted in the developments. They can be added whenever required easily.

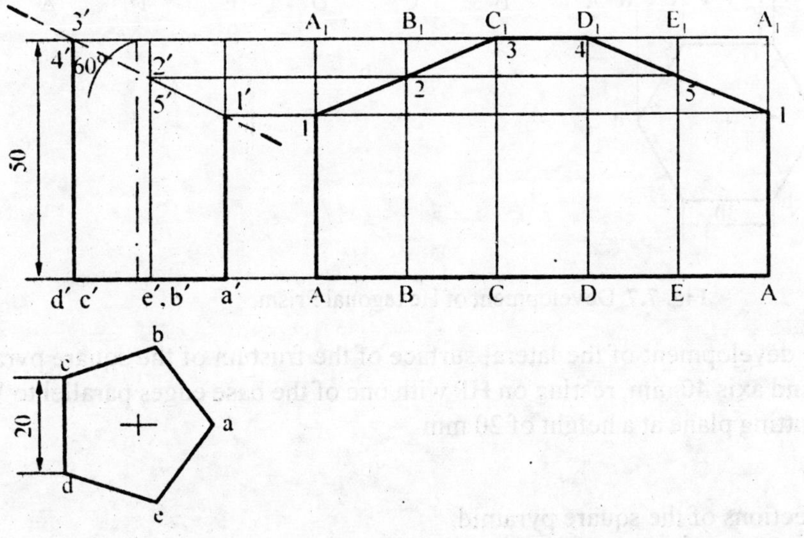

Fig. 7.6 Development of Pentagonal Prism.

Problem : A hexagonal prism of side of base 30 mm and axis 70 mm long is resting on its base on HP such that a rectangular face is parallel to VP. It is cut by a section plane perpendicular to VP and inclined at 30° to HP. The section plane is passing through the top end of an extreme lateral edge of the prism. Draw the development of the lateral surface of the cut prism.

***Solution :* (Fig. 7.7)**

1. Draw the projections of the prism.
2. Draw the section plane VT.
3. Draw the development AA_1-A_1A of the complete prism following the stretch out line principle.

4. Locate the point of intersectiion 1', 2' etc., between VT and the edges of the prism.
5. Draw horizontal lines thrugh 1',2' etc., and obtain 1,2, etc., on the corresponding edges in the development.
6. Join the points 1,2, etc., by straight lines and darken the sides corresponding to the retained portion of the solid.

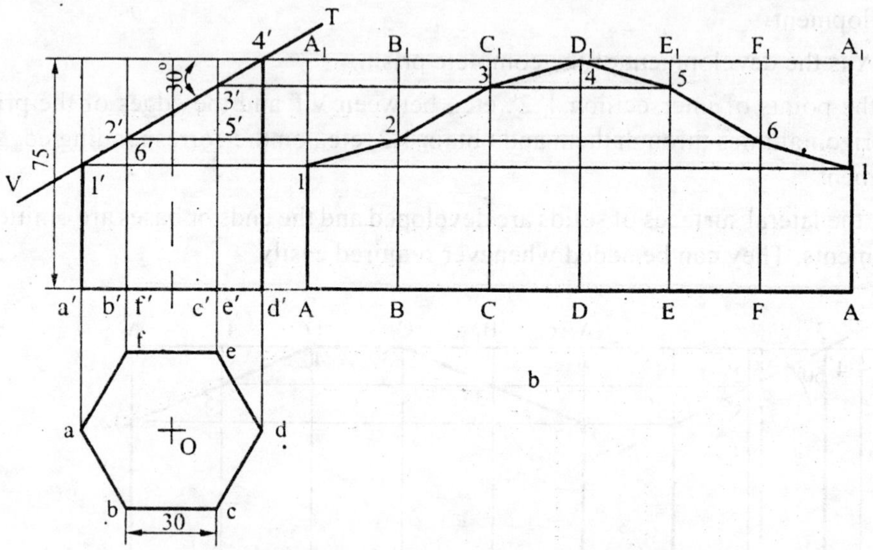

Fig. 7.7 Development of Hexagonal Prism.

Problem : Draw the development of the lateral surface of the frustum of the square pyramid of side of base 30 mm and axis 40 mm, resting on HP with one of the base edges parallel to VP. It is cut by a horizontal cutting plane at a height of 20 mm.

***Solution :* (Fig. 7.8)**

1. Draw the projections of the square pyramid.
2. Determine the true length. **o-a** of the slant edge.
3. Draw the trace of the cutting plane VT.
4. Locate the points of instersection of the cutting plane on the slant edges a'b'c'd' of the pyramid.
5. With any point **o** as centre and radius equal to the true length of the slant edge draw an arc of the circle.
6. With radius equal to the side of the base 30 mm, step-off divisions on the above arc.
7. Join the above division points 1,2,3 etc., in the order with the centre of the arc **o.** The full development of the pyramid is given by 0123410.
8. With centre **o** and radius equal to **oa** mark-off these projections at A, B, C, D, A. Join A-B, B-C etc. ABCDA-12341 is the development of the frustum of the square pyramid.

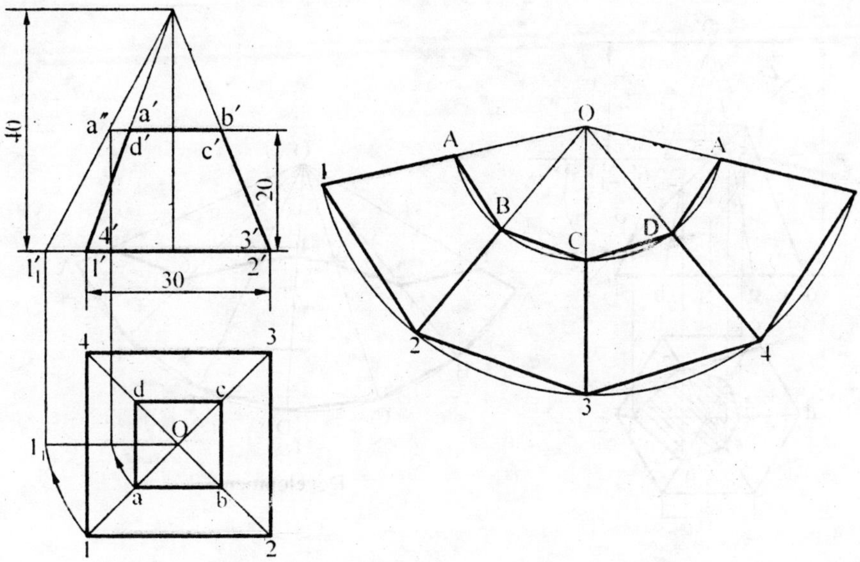

Fig. 7.8 Development of Frustum of Square Pyramid.

Problem : A hexagonal pyramid with side of base 30 mm and height 75 mm stands with its base on HP and an edge of the base parallel to VP. It is cut by a plane perpendicular to VP, inclined at 45^0 to HP and passing through the mid-point of the axis. Draw the (sectioned) top view and develop the lateral surface of the truncated pyramid.

Solution : **(Fig. 7.9)**

1. Draw the two views of the given pyramid and indicate the cutting plane.
2. Locate the points of interseciton 1',2',3',4',5' and 6' between the slant edges and the cutting plane.
3. Obtain the sectional top view by projecting the above points.
4. With **o** as centre and radius equal to the true length of the slant edge draw an arc and complete the total development by following construction as in Fig.7.8.
5. Determine the true length $o'2'_1$, $o'3'_1$, etc., of the slant edges o'2', o'3', etc.

Note

(i) To determine the true tength of the edge, say o'2', through 2' draw a line parallel to the base, meeting the true length line **o-a** at $2'_1$. The length $o'2'_1$ represents the true length of o'2'.

(ii) o'1' and o'4' represent the true lengths as their top views (o1, o4) are parallel to xy.

6. Mark 1,2,3 etc., along OA,OB,OC etc., corresponding to the true lengths o'1', o'2', o'3', etc., in the development.
7. Join 1,2,3 etc., by straight lines and darken the sides corresponding to the truncated portion of the solid.

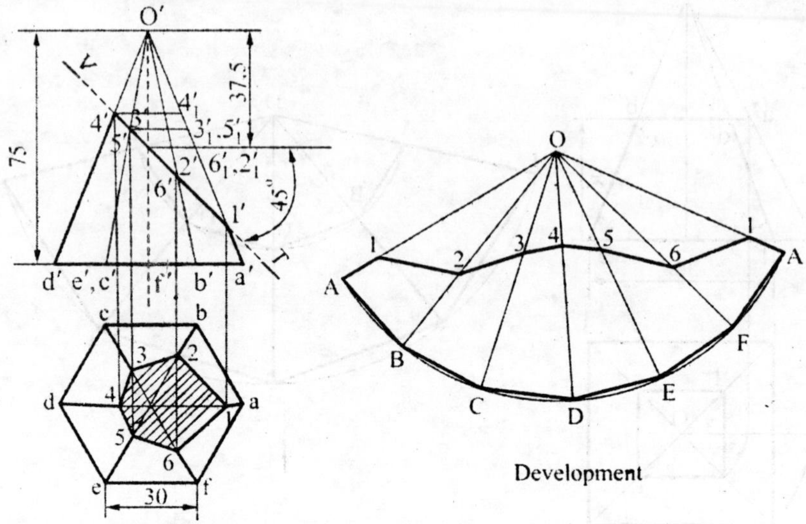

Fig. 7.9 Development of Frustum of Square Pyramid.

Problem : A cylinder of diameter of base 40 mm and height 50 mm is standing on its base on HP. A cutting plane inclined at 45° to the axis of the cylinder passes through the left extreme point of the top base. Develop the lateral surface of the truncated cylinder.

Solution : **(Fig. 7.10)**

1. Draw the views of the truncated cylinder.
2. Divide the circle (top view) into an equal number of parts.
3. Draw the genertors in the front view corresponding to the above division points.
4. Mark the points of intersection a',b',b'_1,c',c'_1, etc., between the truncated face and the generators.
5. Draw the stretch-out line of length equal to the circumference of the base circle.
6. Divide the stretch-out line into the same number of equal parts as that of the base circle and draw the generators through those points.
7. Project the points a,b,c, etc., and obtain A,B,C, etc., respectively on the corresponding generators 1,2,3 etc., in the development.
8. Join the points A,B,C etc., by a smooth curve.

Note

(i) The generators should not be drawn thick as they do not represent the folding edges on the surface of the cylinder.

(ii) The figure bounded by 1A-$A_1$1 represents the development of the complete cylinder.

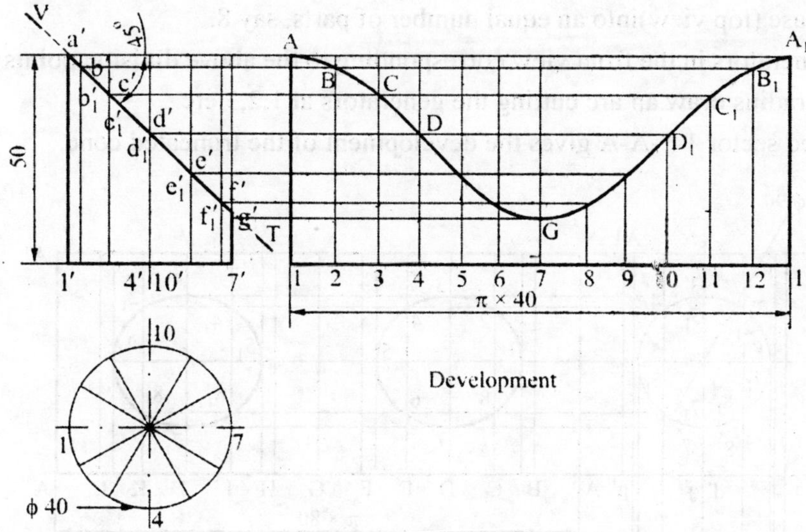

Fig. 7.10

Problem : A cylinder of base 120 mm and axis 160 mm long is resting on its base on HP. It has a circular hole of 90 mm diameter, drilled through centrally such that the axis of the hole is perpendicular to VP and bisects the axis of the cylinder at right angles. Develop the lateral surface of the cylinder.

Solution : **(Fig. 7.11)**

1. Draw the projections of the cylinder with the hole through it.
2. Divide the circle (top view) of the cylinder into 12 equal parts and locate the corresponding generators in the front view.
3. Obtain the complete development AA', A'A of the cylinder and locate the generatros on it.
4. Determine the points of intersection 1',2', etc and $1'_1$,$2'_1$, etc. between the hole and the generators in the front view.
5. Transfer these points to the development by projection, including the transition points $1'(1'_1)$ and $5'$ $(5'_1)$.
6. Join the points 1,2 etc., and 1_1,2_1, etc., by smooth curves and obtain the two openings in the development.

Problem : A cone of diameter of base 45 mm and height 60 mm is cut by horizontal cutting plane at 20 mm from the apex. Draw the devleopment of the truncated cone.

Solution : **(Fig.7. 12)**

1. Draw the two views of the given cone and indicate the cutting plane.
2. Draw the lateral surface of the complete cone by a sector of a circle with radius and arc length equal to the slant hight and circumference of the base respectively. The included anlgle of the sector is given by (360 × r/s), where r is the radius of the base and s is the slant height.

3. Divide the base (top view)into an equal number of parts, say 8.
4. Draw the generators in the front view corresponding to the above division points a,b,c etc.
5. With o'1' as radius draw an arc cutting the generators at 1,2,3 etc.
6. The truncated sector 1-1-A-A gives the development of the truncated cone.

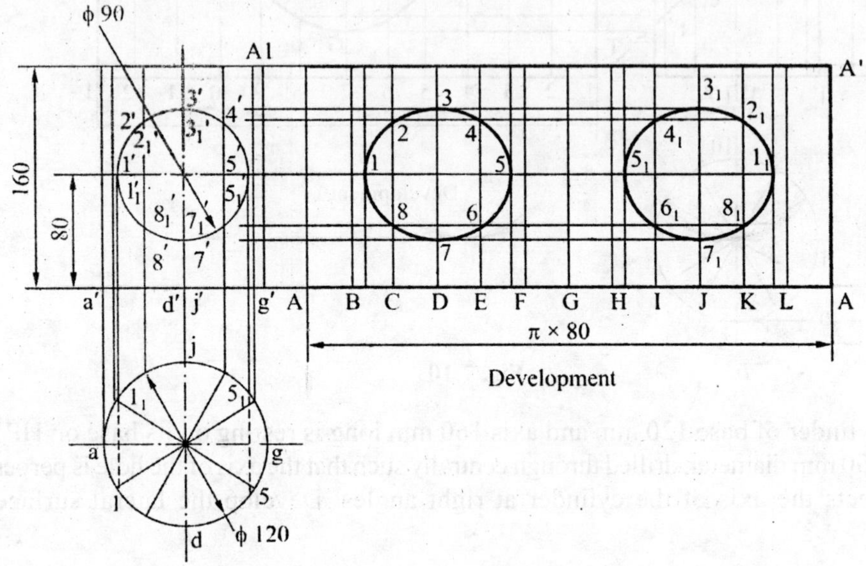

Fig. 7.11

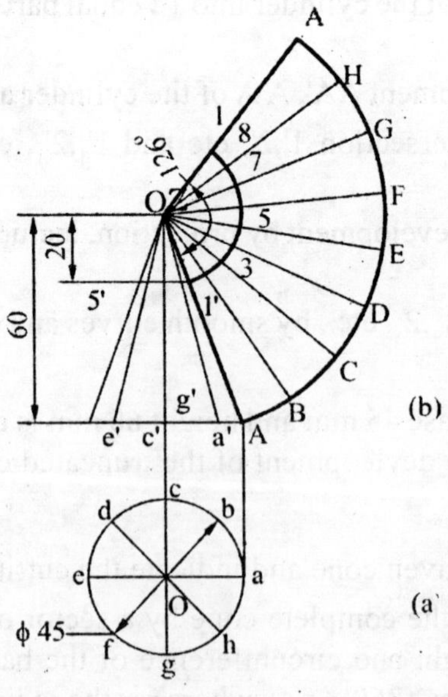

Fig. 7.12

Problem : A cone of base 50 mm diameter and height 60 mm rests with its base on HP and bisects the axis of the cone. Draw the development of the lateral surface of the truncated cone.

Solution : **(Fig. 7.13)**

1. Draw the two views of the given cone and indicate the cutting plane.
2. Draw the lateral surface of the complete cone.
3. Divide the base into 8 equal parts.
4. Draw the generators in the front view corresponding to the above divisions.
5. Mark the points of intersection 1, 2, 3 etc. between the cutting plane and the generators.
6. Trasfer the points 1, 2, 3 etc. to the development after finding the true distances of 1,2,3 etc from the apex **o** of the cone in the front view.

Note : To transfer a point say 4 on **od** to the development.

(i) Determine the true length of **o-4** by drawing a horizontal through 4 meeting **od** at 4.

(ii) On the generator OD, mark the distance **o-4** equal to **o-4**.

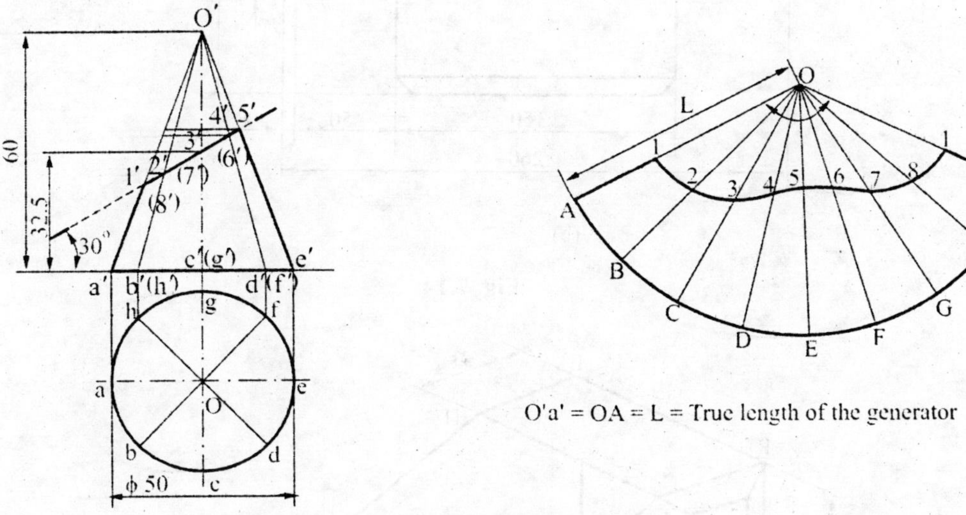

Fig. 7.13

Problem : Figure 7.14(a) shows a tools tray with an allowance for simple hem and lap-seam. Fig. 7.14(b) represents its development with dimensions.

Problem : Figure 7.15(a) shows a rectangualr scoop with allowance for lap-seam and Fig. 7.15(b) shows the development of the above with dimensions.

Problem : Figure 7.16(a) shows the pictorial view of a rectangular 90° elbow and Fig. 7.16(b) its development in two parts.

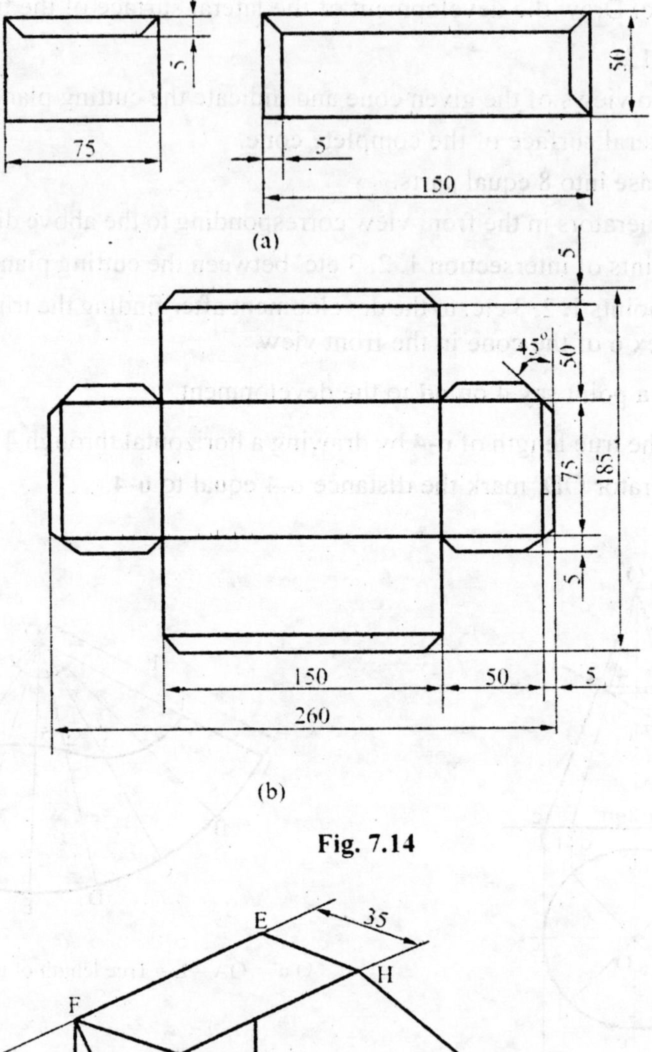

Fig. 7.14

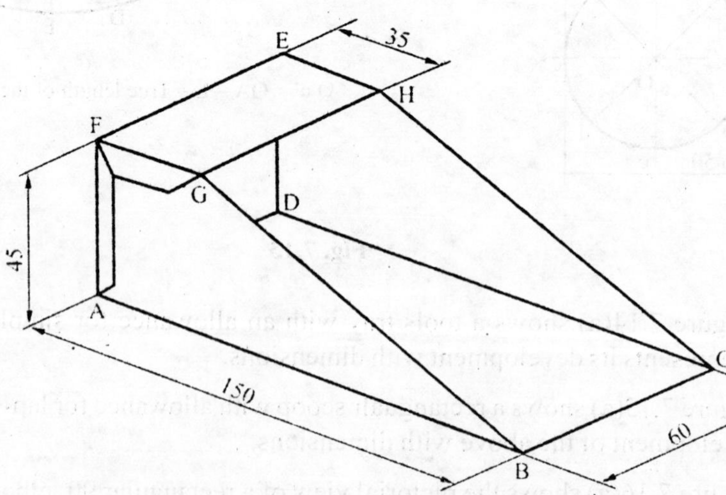

Fig. 7.15(a)

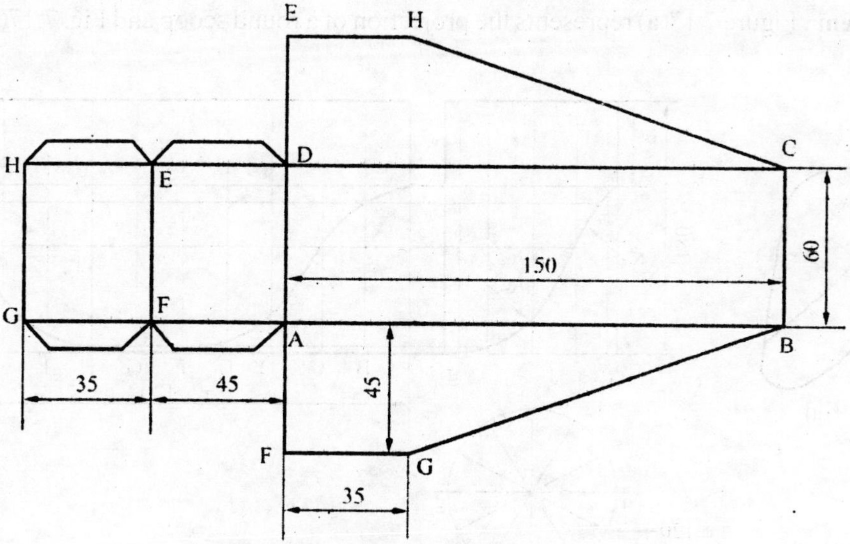

Fig. 7.15(b) Rectangular Scoop.

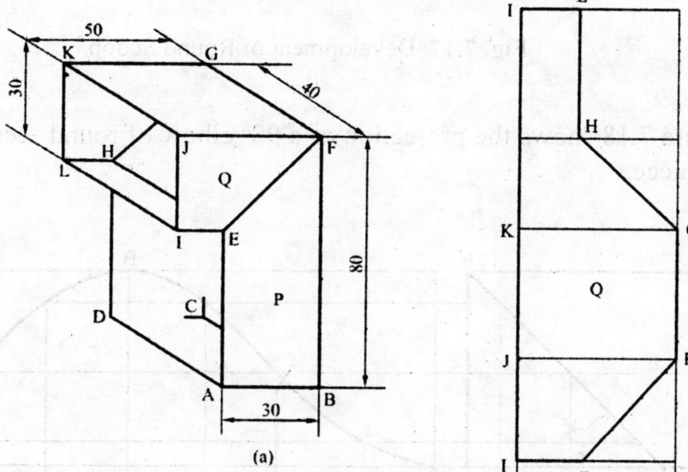

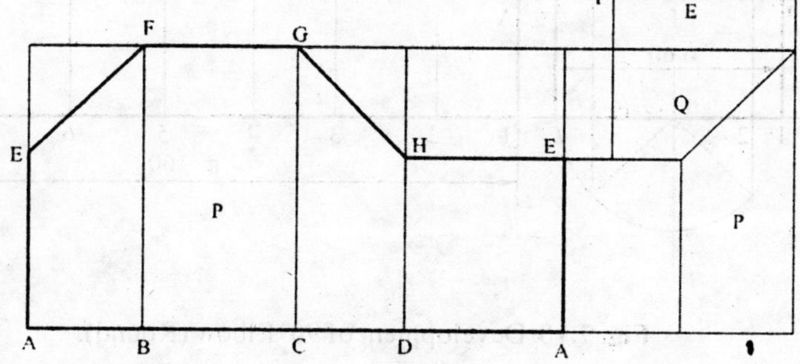

Fig. 7.16 Development of 90° Elbow (Rectangular).

Problem : Figure 7.17(a) represents the projection of a round scoop and Fig. 7.17(b) its development.

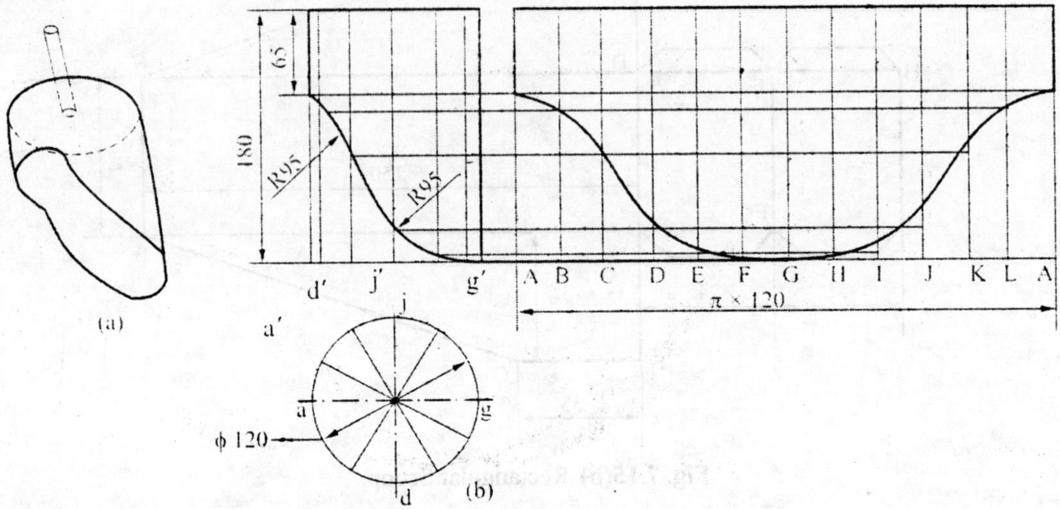

Fig. 7.17 Development of Round Scoop.

Problem : Figure 7.18 shows the projection of a 90° elbow of round section with development shown for one piece.

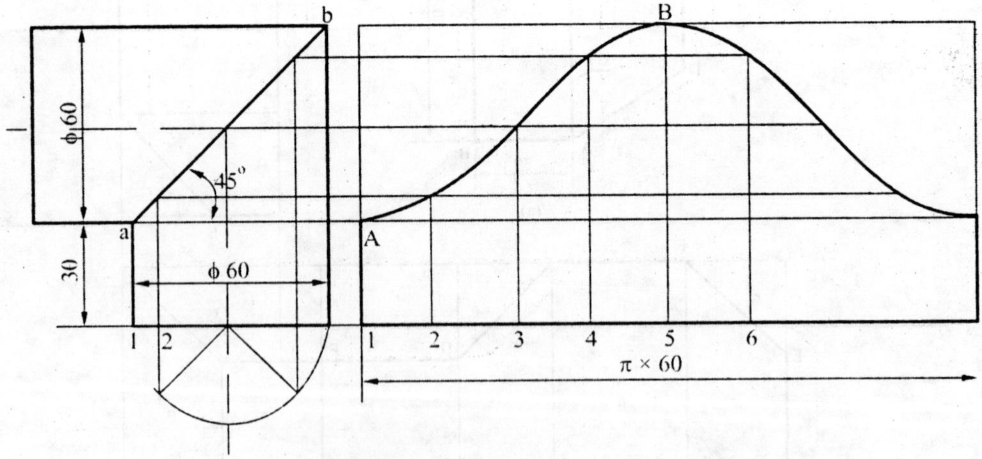

Fig. 7.18 Development of 90° Elbow (Round).

Problem : Figure 7.19 shows the orthographic projection and the development of parts of funnel.

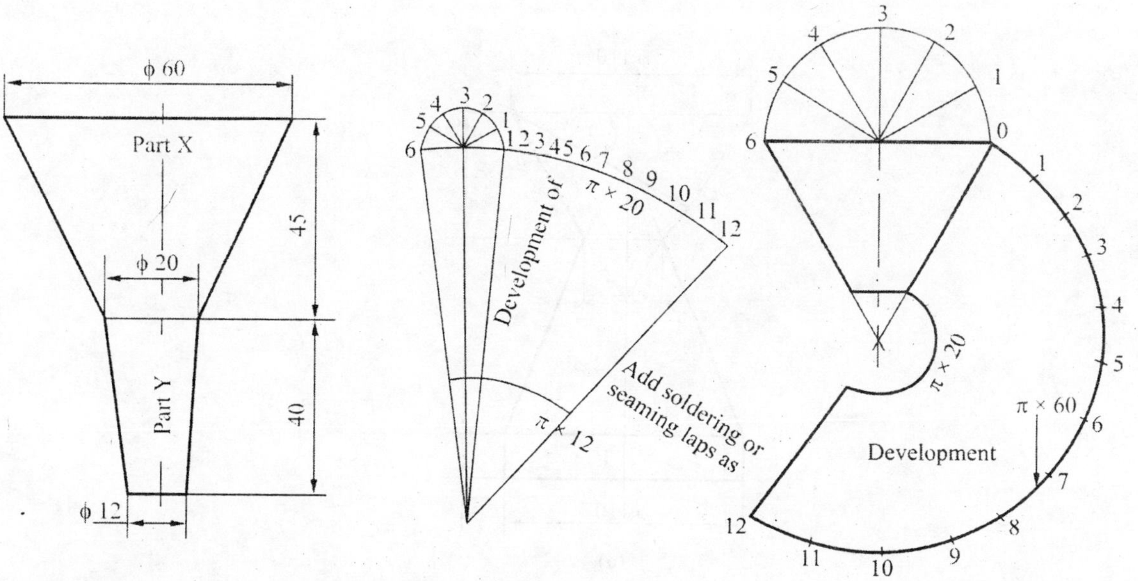

Fig. 7.19 Development of Funnel.

Problem : Figure 7.20 shows the orthographic projection of a chute and the development of the parts.

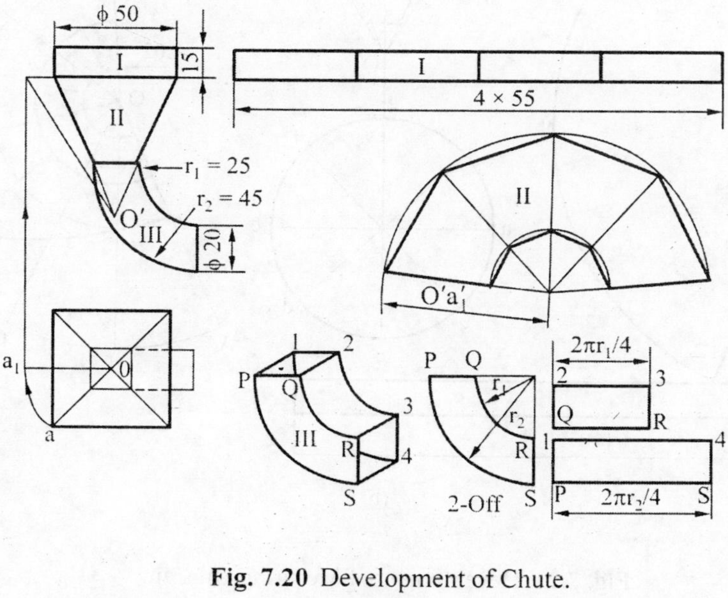

Fig. 7.20 Development of Chute.

Problem : Figure 7.21 shows the orthographic projection of measuring oil can and the development of its parts.

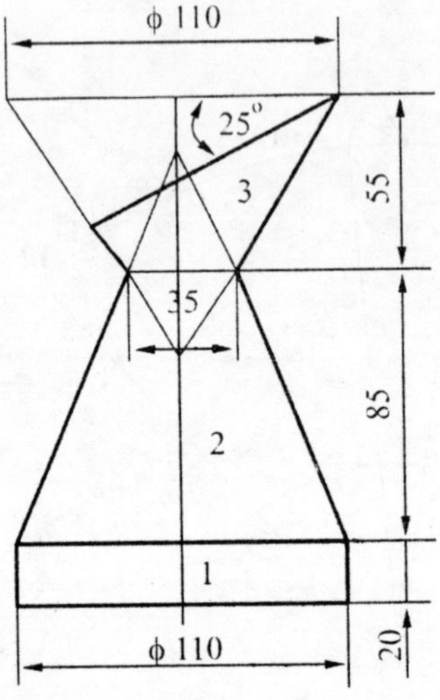

(a)

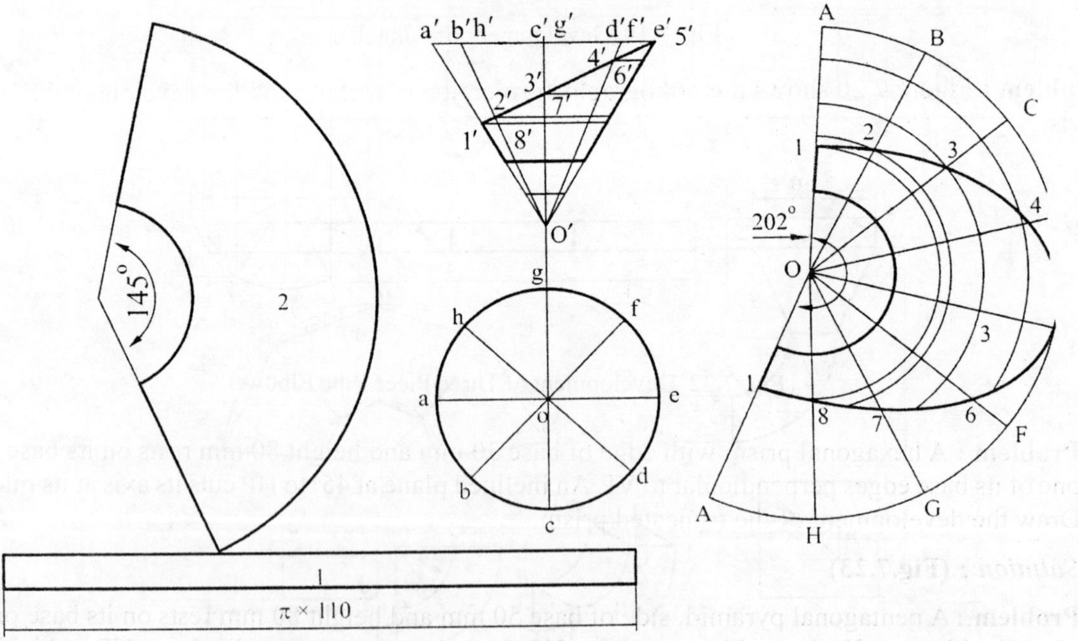

(b)

Fig. 7.21 Development of Mesuring Oil Can.

Problem : Figure 7.22 shows the development of a three piece pipe elbow.

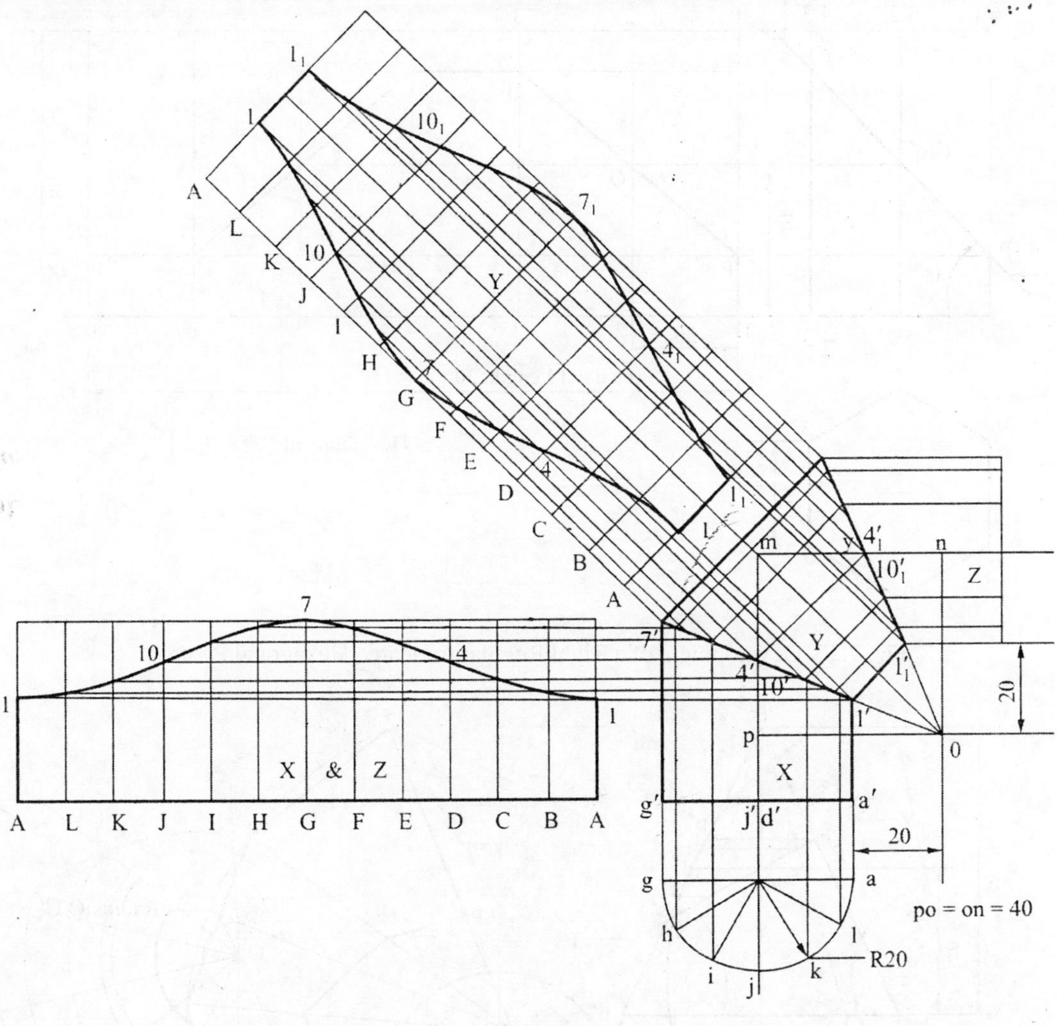

Fig. 7.22 Development of Three Piece Pipe Elbow.

Problem : A hexagonal prism with edge of base 30 mm and height 80 mm rests on its base with one of its base edges perpendicular to VP. An inclined plane at 45° to HP cuts its axis at its middle. Draw the development of the truncated prism.

Solution : **(Fig.7.23)**

Problem : A pentagonal pyramid, side of base 50 mm and height 80 mm rests on its base on the ground with one of its base sides parallel to VP. A section plane perpendicular to VP and inclined at 30° to HP cuts the pyramid, bisecting its axis. Draw the development of the truncated pyramid.

Solution : **(Fig.7.24)**

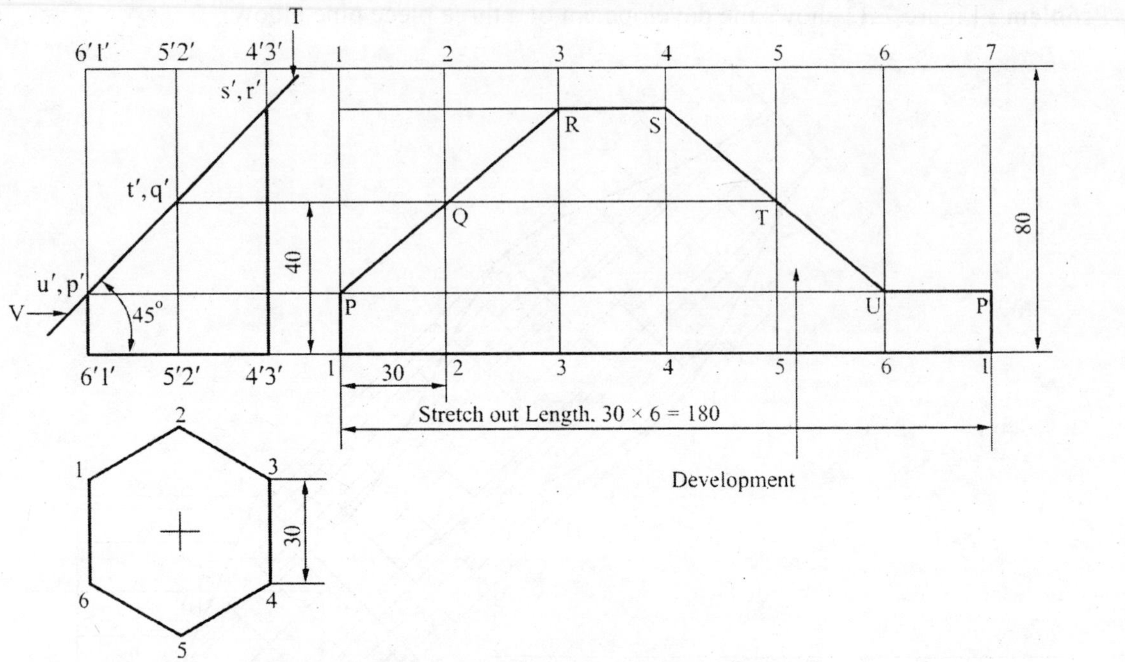

Fig. 7.23 Development of a Right Regular truncated Hexagonal Prism.

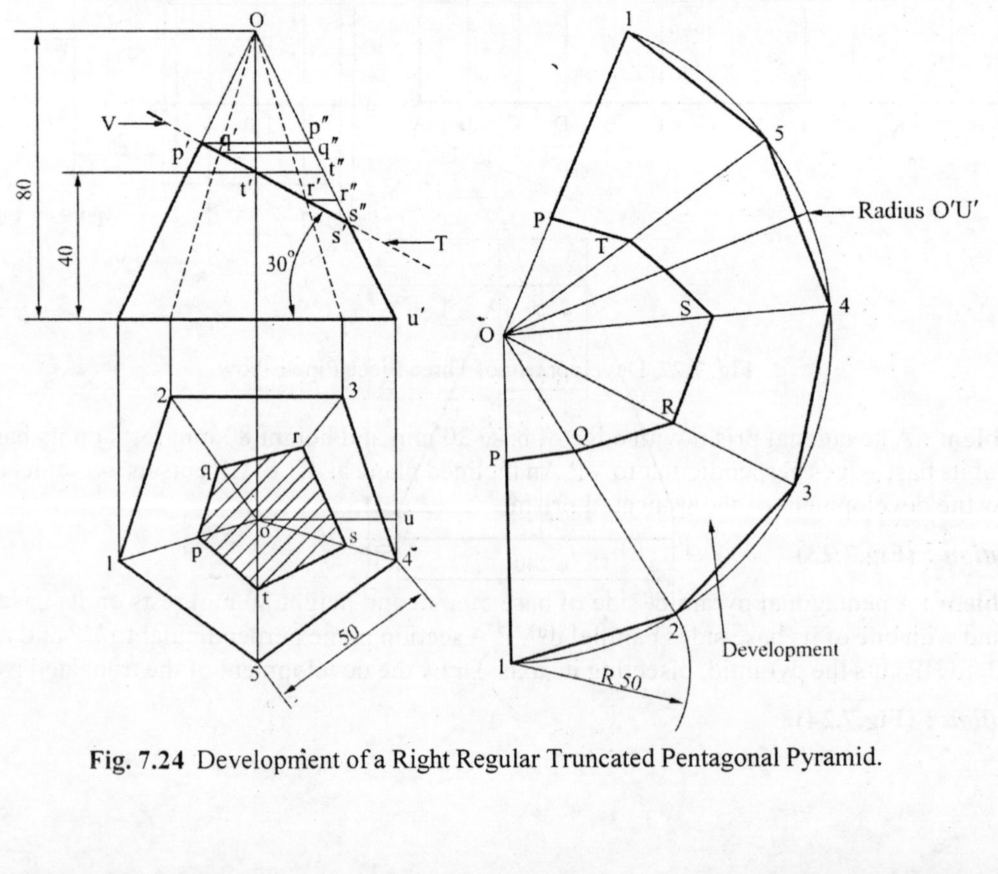

Fig. 7.24 Development of a Right Regular Truncated Pentagonal Pyramid.

Problem : Draw the development of a bucket shown in Fig.7.25(a)

Solution : **(Fig. 7.25(b))**

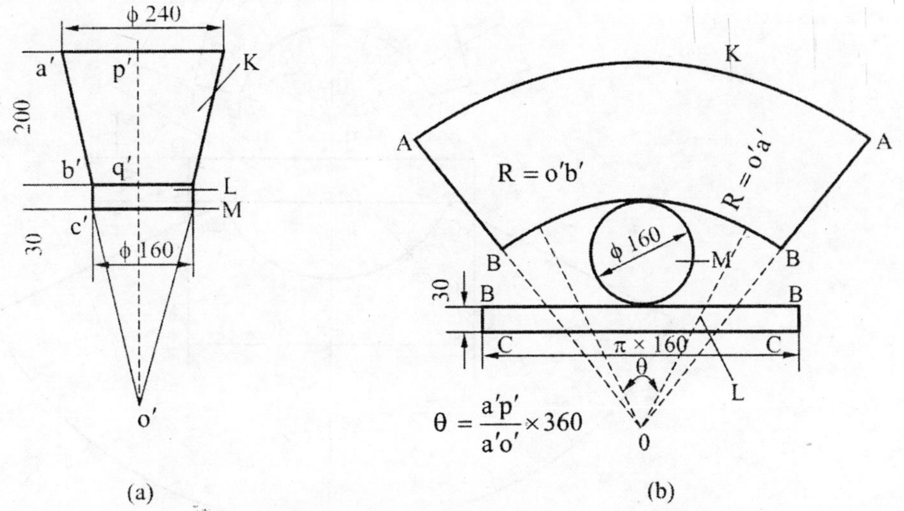

Fig. 7.25 Development of a Bucket.

Problem : Draw the development of the measuring jar shown in Fig. 7.26(a).

Solution : **(Fig. 7.26(b))**

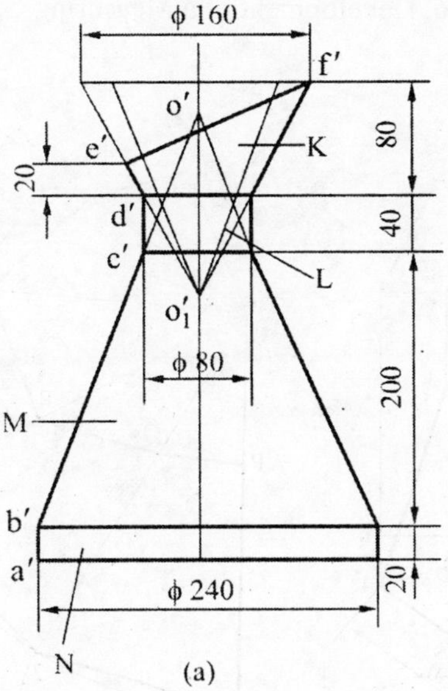

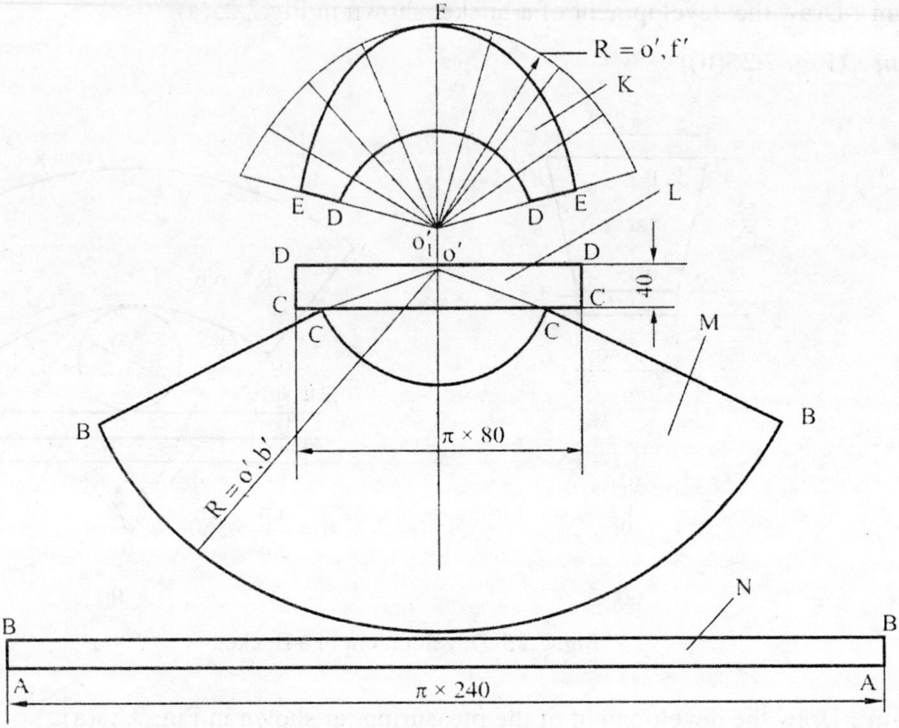

(b)

Fig. 7.26 Development of a Measuring Can.

EXERCISES

Development of Surfaces

1. A frustrum of a square pyramid has its base 50 mm side, top 25 mm side and height 60 mm. It is resting with its base on HP, with two of its sides parallel to VP. Draw the projections of the frustrum and show the development of its lateral surface.
2. A cone of diameter 60 mm and height 80 mm is cut by a section plane such that the plane passes through the mid-point of the axis and tangential to the base circle. Draw the development of the lateral surface of the bottom portion of the cone.
3. A cone of base 50 mm diameter and axis 75 mm long, has a through hole of 25 mm diameter. The centre of the hole is 25 mm above the base. The axes of the cone and hole intersect each other. Draw the development of the cone with the hole in it.
4. A transition piece connects a square pipe of side 25 mm at the top and circular pipe of 50 mm diameter at the bottom, the axes of both the pipes being collinear. The height of the transition piece is 60 mm. Draw its development.
5. Fig. 7.27 shows certain projections of solids. Draw the developments of their lateral surfaces.

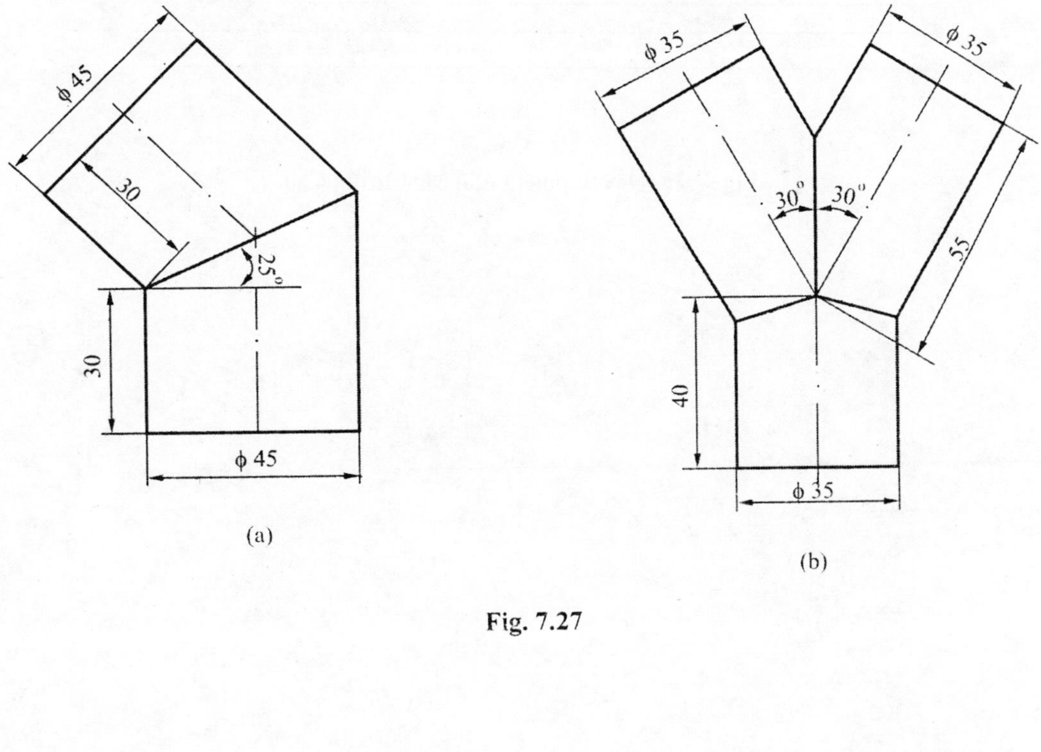

(a) (b)

Fig. 7.27

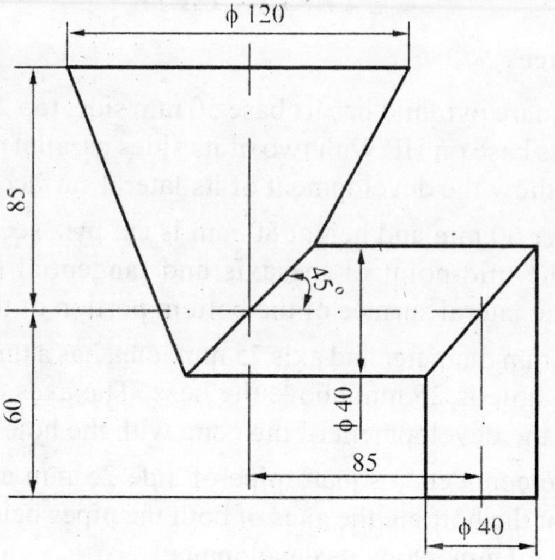

(c)

Fig. 7.27

CHAPTER 8

Intersection of Surfaces

8.1 Introduction

Ducts, pipe joints, smoke stacks, boilers, containers, machine castings etc., involve intersection of surfaces. Sheetmetal work required for the fabrication of the above objects necessiate the preparation of the development of the joints/ objects. Orthographic drawings of lines and curves of intersection of surfaces must be prepared first for the accurate development of objects. Methods of obtaining the lines and curves of intersection of surfaces of cylinder and cylinder, prism and prism are shown to introduce the subject. Figure 8.1 shows intersection of two cylinders.

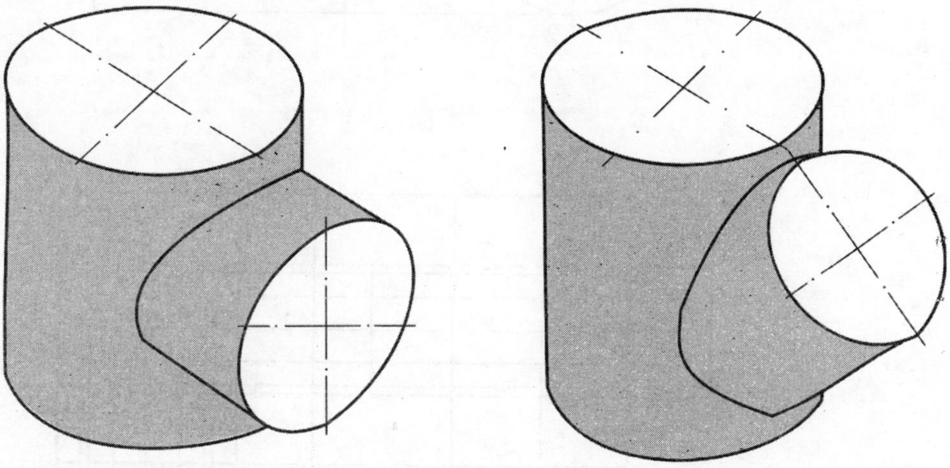

(a) (b)

Fig. 8.1

8.2 Intersection of Cylinder and Cylinder

Case 1

Problem : A horizontal cylinder of diameter 35 mm penetrates into a vertical cylinder of diameter 50 mm. The axes of the cylinders intersect at right angles. Draw the curves of intersection when the axis of the horizontal cylinder is parallel to the VP.

Solution : (Fig. 8.2)

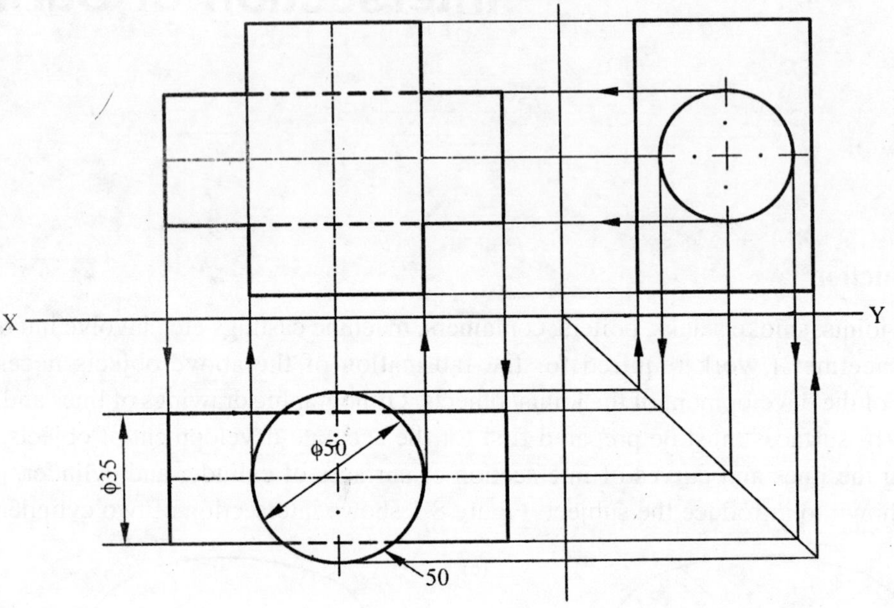

(a)

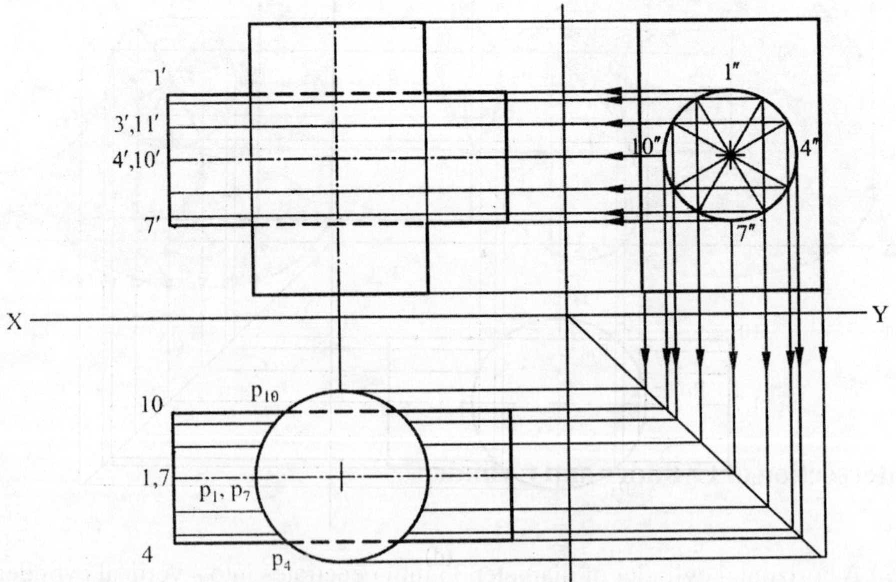

(b)

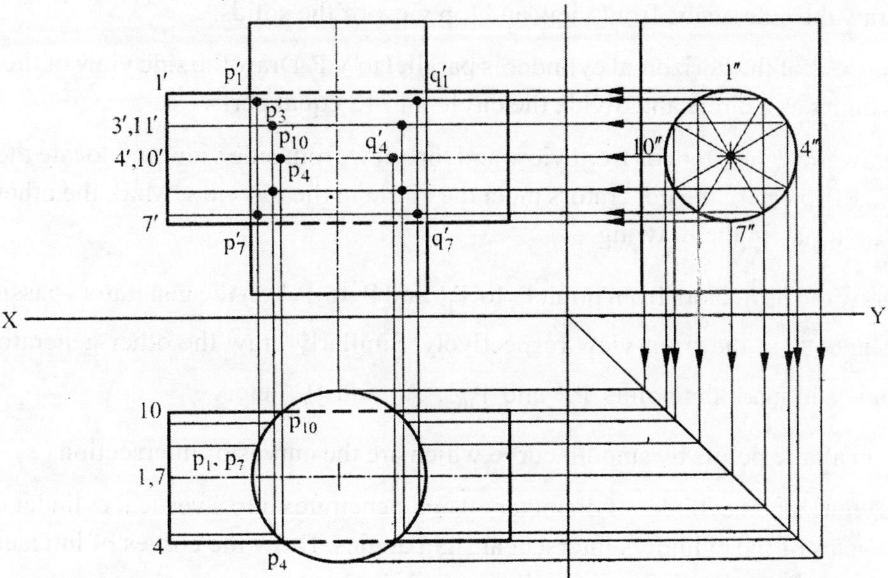

(c)

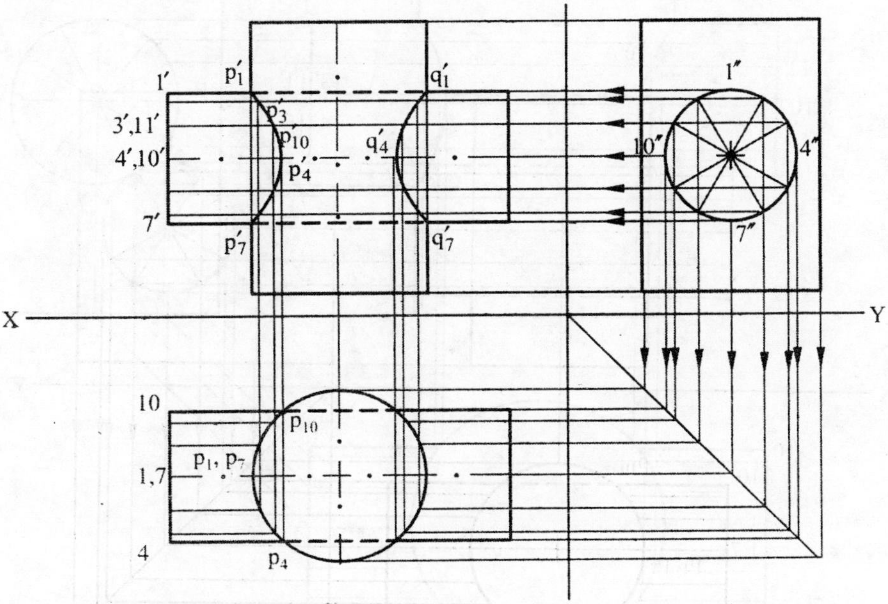

(d)

Fig. 8.2 Intersection of Unequal Cylinders.

Step 1: Draw the side view, front view and top view of the solids.

Step 2: The axis of the horizontal cylinder is parallel to V.P. Draw the side view of the horizontal cylinder as a circle and divide the circle into 12 equal parts.

Step 3: Draw the generators of front view and top view from side view and locate the points P_1, P_2,....P_{12}, where the generators meet the circle in the top view. Mark the other points q_1, q_2... q_{12} as in the drawing.

Step 4: Draw the projectors from point P_1 to P_1' and P_1 to P_7' on the generators passing through 1' and 7' in the front view respectively. Similarly draw the other generators to front view and mark the points P_4' and P_{10}', P_2' and P_{12}' etc.

Step 5: Join all the points by smooth curve which are the curves of intersection.

Problem : A horizontal cylinder of diameter 40 mm penetrates into a vertical cylinder of diameter 60 mm. The axes of the cylinders intersect at right angles. Draw the curves of intersection when the axis of the horizontal cylinder is parallel to the V.P.

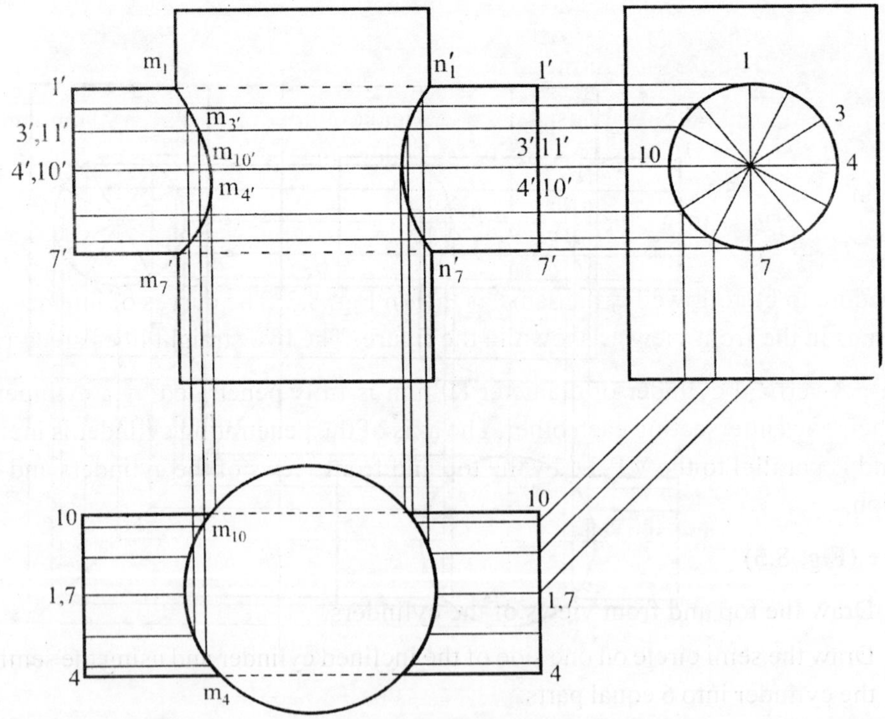

Fig. 8.3

1. Draw the top and front views of the cylinders.
2. Draw the left side view of the arrangement.
3. Divide the circle in the side view into number of equal parts say 12.
4. The generators of the horizontal cylinder are numbered in both front and top views as shown.
5. Mark point m_1, where the generator through 1 in the top view meets the circle in the top view of the vertical cylinder. Similarly mark m_2,...m_{12}.
6. Project m'_7 to m'_1 on the generator 1' 1' in the front view.
7. Project m^7 to m'_7 on $7'_7$. Similarly project all the point.
8. Draw a smooth curve through m'_1...m'_7.

   This curve is the intersection curve at the front. The curve at the rear through m'_4, m'_8 ...m'_{12} coincides with the corresponding visible curve at the front.

   Since the horizontal cylinder penetrates and comes out at the other end, similar curve of intersection will be seen on the right also.
9. Draw the curve through n'_1...n'_7 following the same procedure. The two curves m'_1-m'_7 and n'_1-n'_7 are the required curves of intersection.

Intersection of Cylinders of Same size

Problem : A T-pipe connection consists of a vertical cylinder of diameter 80 mm and a horizontal cylinder of the same size. The axes of the cylinders meet at right angles. Draw the curves of intersection.

Solution : (Fig. 8.4)

The procedure to be followed is the same as that in Fig. 8.2. The curves of intersection appear as straight lines in the front view as shown in the figure. The two straight lines are at right angles.

Problem : A vertical cylinder of diameter 80 mm is fully penetrated by a cylinder of diameter 50 mm, their axes intersecting each other. The axis of the penetrating cylinder is inclined at 30° to the HP and is parallel to the V.P. Draw the top and front views of the cylinders and the curves of intersection.

***Solution* : (Fig. 8.5)**

Step 1: Draw the top and front views of the cylinders.

Step 2: Draw the semi circle on one side of the inclined cylinder and using the semi circle divide the cylinder into 6 equal parts.

Step 3: Draw the projectors from these points obtained in bottom and top faces of inclined cylinder to top view and mark the points as **1, 2,...12** on both sides.

Step 4: Draw a semi circle in the top view of the inclined cylinder and divide it into six equal parts. Mark the intercepts obtained in the circle.

Step 5: From the intercept point, draw the generators to front view and mark the points as **1', 2' ... 12'**.

Step 6: Join these points by curves which are the required curves of intersection.

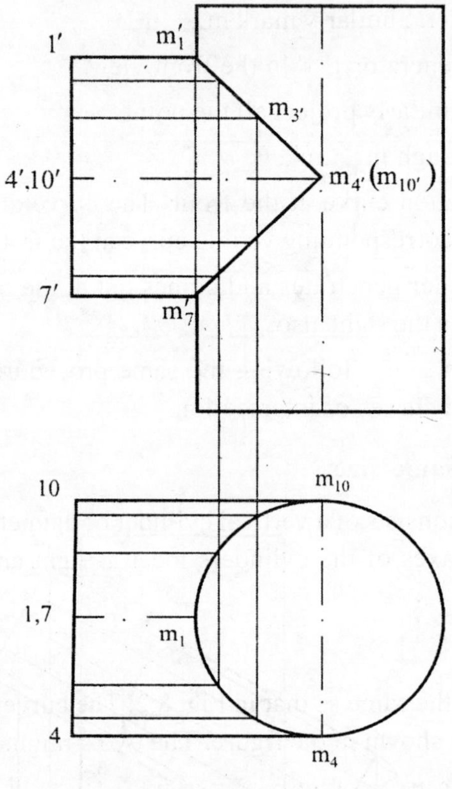

Fig. 8.4 Intersection of Cylinders of Same Size.

Problem : A vertical cylinder of diameter 120 mm is fully penetrated by a cylinder of diameter 90 mm, their axes intersecting each other. The axis of the penetrating cylinder is inclined at 30° to the HP and is parallel to the VP. Draw the top and front views of the cylinders and the curves of intersection.

Solution : **(Fig. 8.6)**

1. Draw the top and front views of the cylinders.
2. Following the procedure in Fig. 8.2, locate points m_1 in the top view. Project them to the corresponding generators in the inclined cylinder in the front view to obtain points m'_1, m'_2 etc.

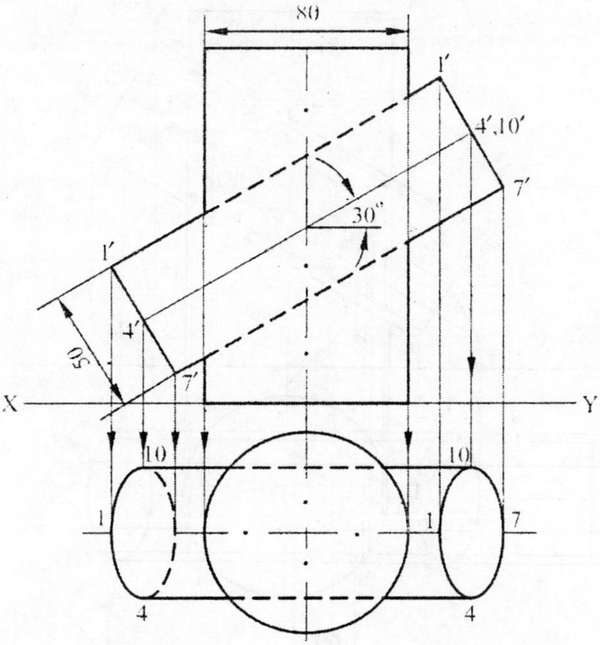

(a)

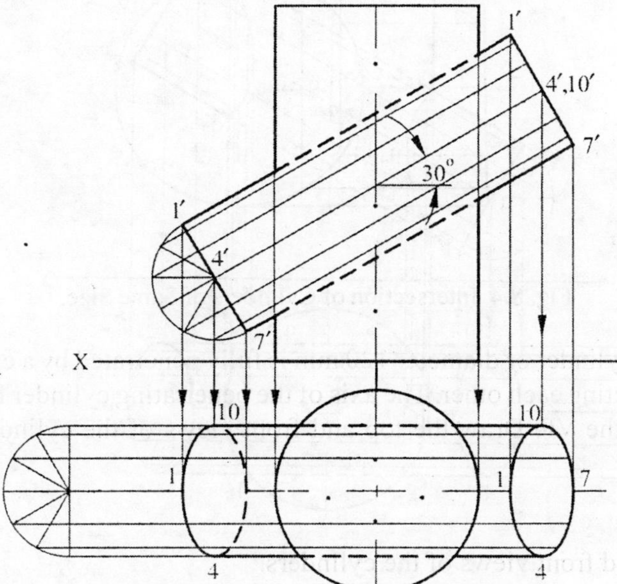

(b)

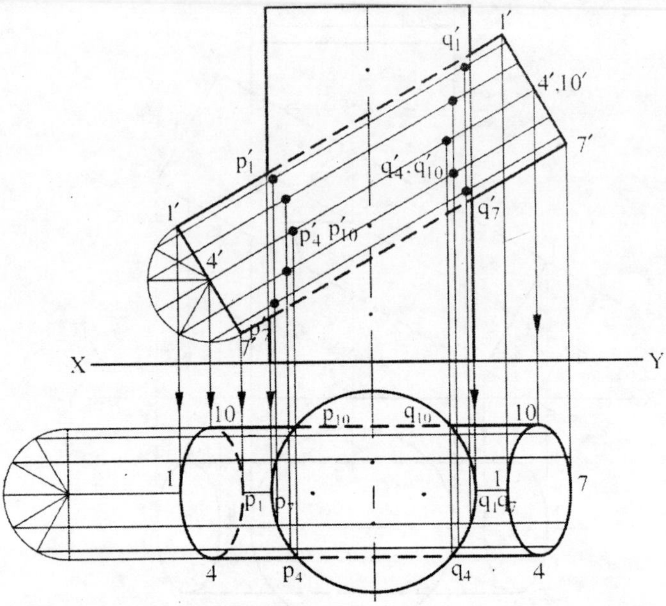

(c)

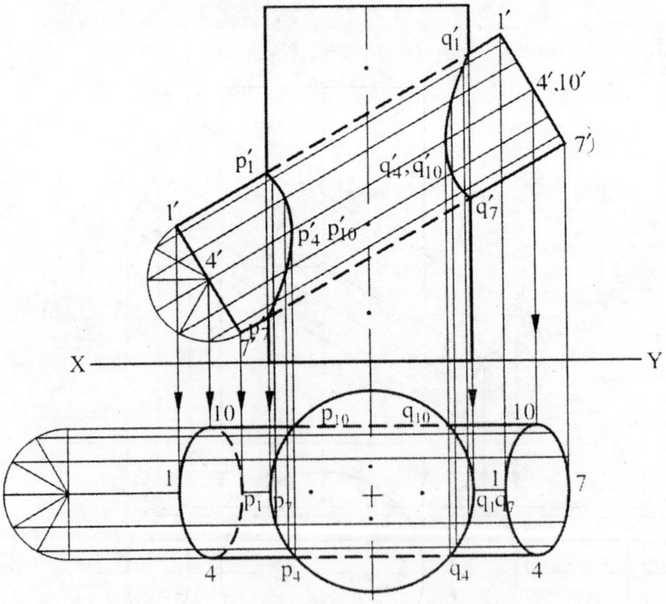

(d)

Fig. 8.5 Intersection of Cylinders of Inclined Axes.

3. Locate points n',.......n'$_{10}$ etc., on the right side using the same construction.
4. Draw smooth curves through them to get the required curve of intersection as shown in the Fig. 8.6.

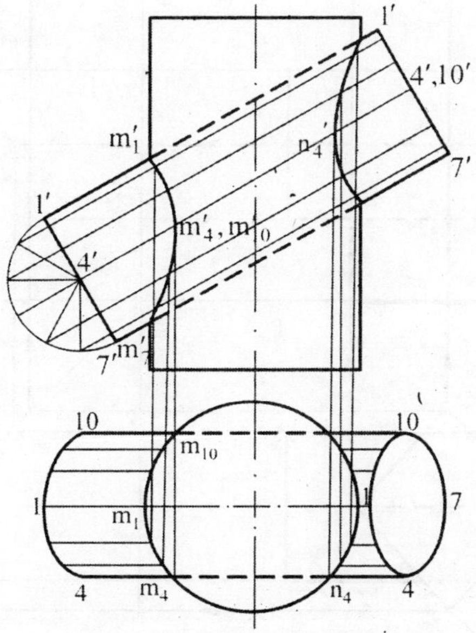

Fig. 8.6

8.3 Intersection of Prism and Prism

When a prism penetrates another prism, plane surface of one prism intersects the plane surfaces of another prism and hence the lines of intersection will be straight lines. In these cases, lines on the surface of one of the solids need not necessarily be drawn as it is done with cylinders. Instead, the points of intersections of the edges with the surface are located by mere inspection. These points are projected in the other view and the lines of intersection obtained.

Problem : A square prism of base side 65 mm rests on one of its ends on the H.P. with the base sides equally inclined to the V.P. It is penetrated fully by another square prism of base side 50 mm with the base side equally inclined to the H.P. The axes intersect at right angles. The axis of the penetrating prism is parallel to both the H.P and the V.P. Draw the projections of the prisms and show the lines of intersection.

Solution : **(Fig. 8.7)**

Step 1: Draw the top view, front view and side view of the prism.

Step 2: Draw the projectors from side view of the horizontal prism to front and top view and mark the intercepts in top view as $\mathbf{p_1, p_2, p_3, p_4}$ and $\mathbf{q_1, q_2, q_3, q_4}$.

Step 3: From intercept points on the top view, draw the projections to front view and mark the points.

Step 4: Join these points (p_1', p_2', p_3') and (q_1', q_2', q_3') which are the required lines of intersection.

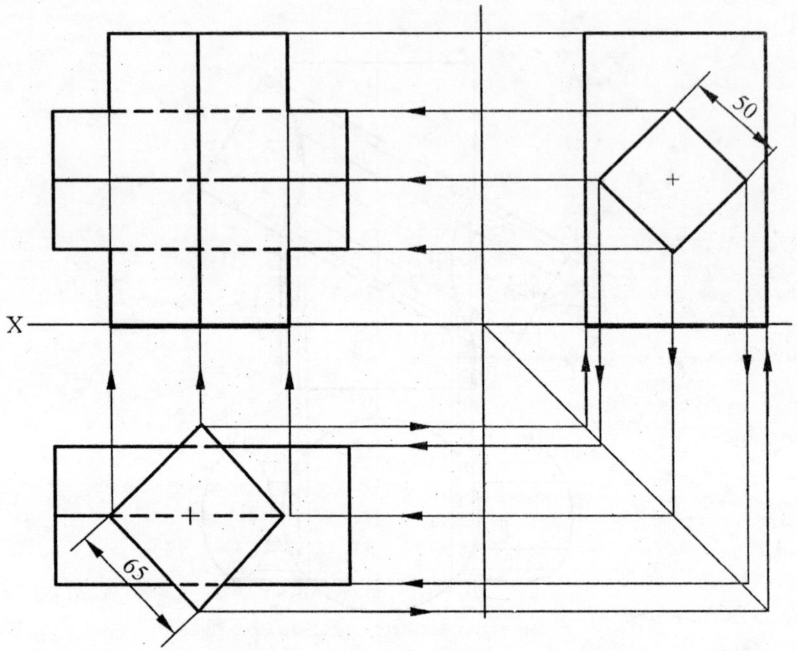

(a)

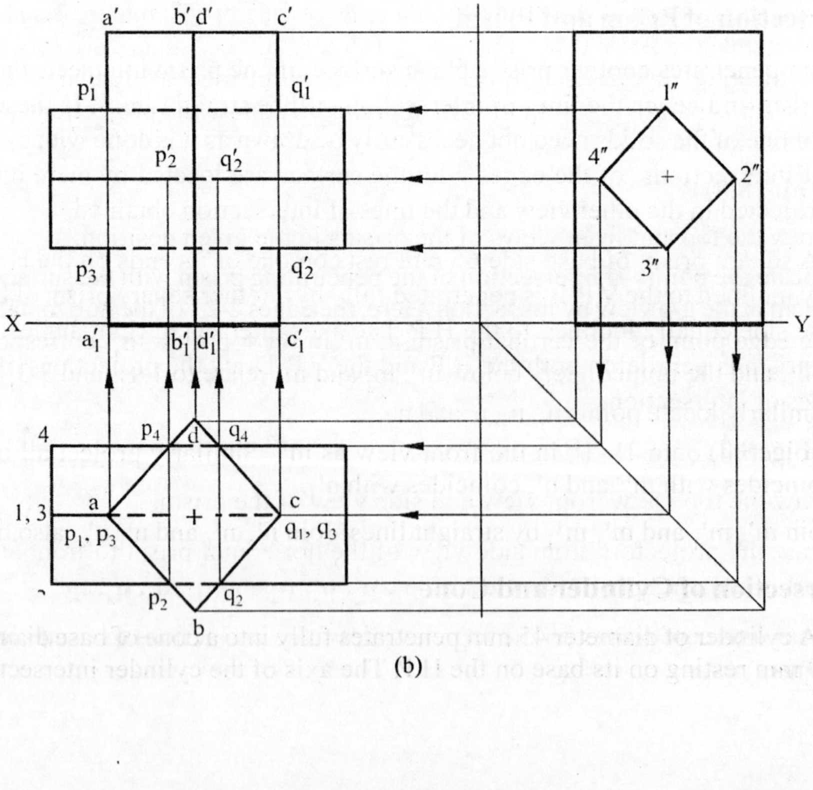

(b)

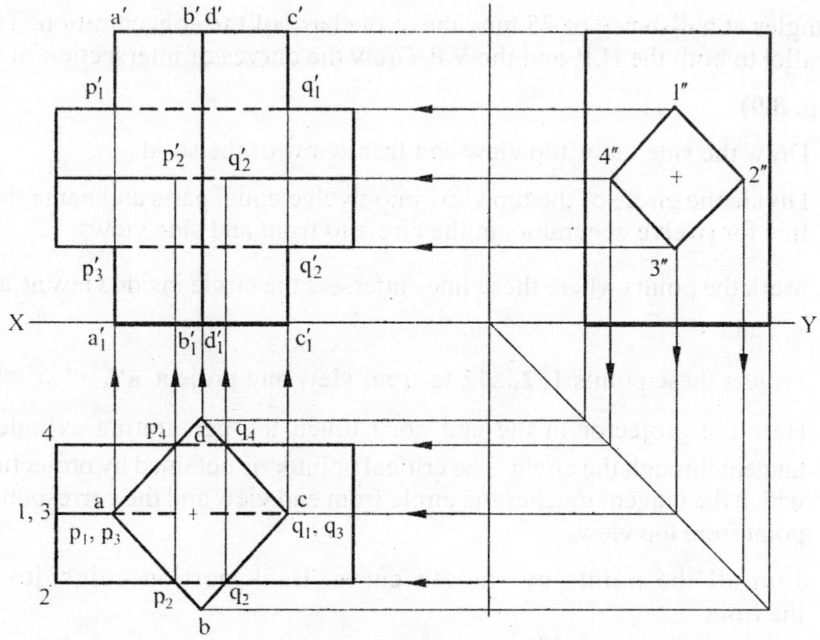

(c)

Fig. 8.7 Intersecting of Prism and Prism.

Problem : A square prism of base side 60 mm rests on one of its ends on the H.P. with the base sides equally inclined to the V.P. It is penetrated fully by another square prism of base side 45 mm with the base side equally inclined to the H.P. The axes intersect at right angles. The axis of the penetrating prism is parallel to both the H.P and the V.P. Draw the projections of the prisms and show the lines of intersection.

***Solution* : (Fig. 8.8)**

Step 1: Draw the top and front views of the prisms in the given position.

Step 2:. Locate the points of intersection of the penetrating prism with the surfaces of the vertical prism in the top view by inspection. Here, the edges 2-2_1, of the horizontal prism intersects the edge point of the vertical prism at m_2 in the top view. n_4 corresponds to the edge 4-4_1, and the immediately below m_2, m_1 and m_3 relate to 1-1_1, and 3-3_1 respectively.

Step 3: Similarly locate points n_1, n_2, n_3 and n_4.

Step 4: Project m_1 onto 1' -1', in the front view as m',. Similarly project all other points. m'_3 coincides with m'_1 and n'_3 coincides with n'_1.

Step 5: Join m'_2 m'_1 and m'_1 m'_4 by straight lines. Join n'_2 n'_1 and n'_1 n'_4 also by straight lines.

8.4 Intersection of Cylinder and Cone

Problem : A cylinder of diameter 45 mm penetrates fully into a cone of base diameter 75 mm and altitude 100 mm resting on its base on the H.P. The axis of the cylinder intersects the axis of the cone at right angles at a distance of 25 mm above the base of the concentration. The axis of the cylinder is parallel to both the H.P. and the V.P. Draw the curves of intersection of the solids.

***Solution :* (Fig. 8.9)**

Step 1: (i) Draw the side view, top view and front view of the solid.

(ii) Divide the circle of the top view into twelve equal parts and name them. Draw the line for twelve generators in the circle to front and side views.

(iii) Mark the points where these lines intersect the circle inside view at **a", b", c", d" e"** and **f"**

Step 2: (iv) Project these points **1, 2,...12** to front view and project **a", b" ...**

(v) Here the projector in the end don't touch the penetrating cylinder. So draw a tangent through the circle. The critical point **s'** is obtained by projecting the point at which the tangent touches the circle from end view and the corresponding intercept point from top view.

Step 3: (vi) Join all the points by smooth curve. Back portion coincides with that of the front.

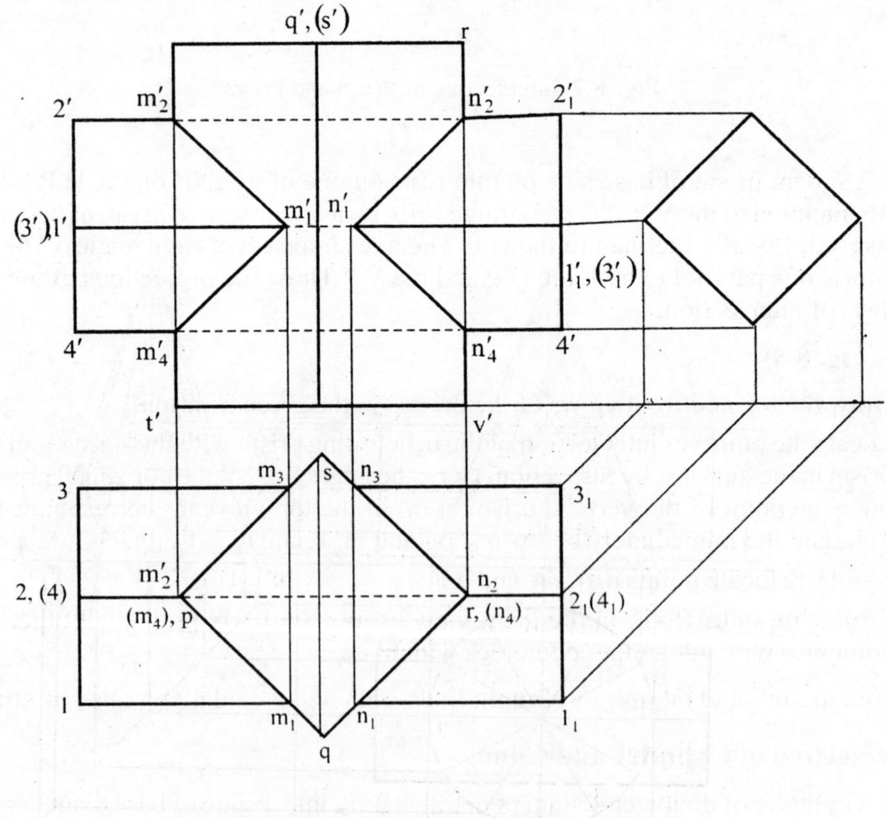

Fig. 8.8 Intersection of Square Prisms.

(vii) Mark the point at corresponding location where the projectors from top and side views meet.

Step 4: (viii) Join all these points by smooth curve. Hidden portions are marked by dotted lines in top views.

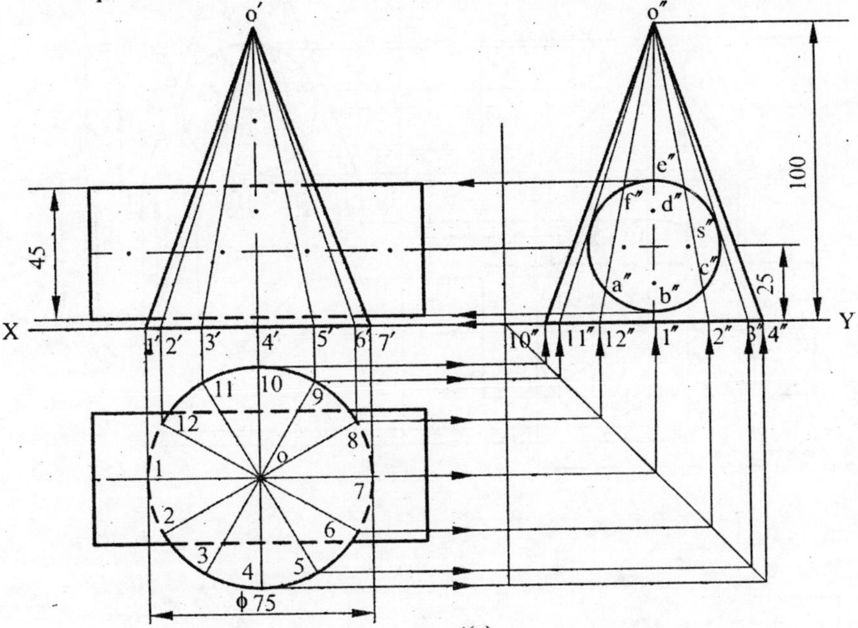

(a)

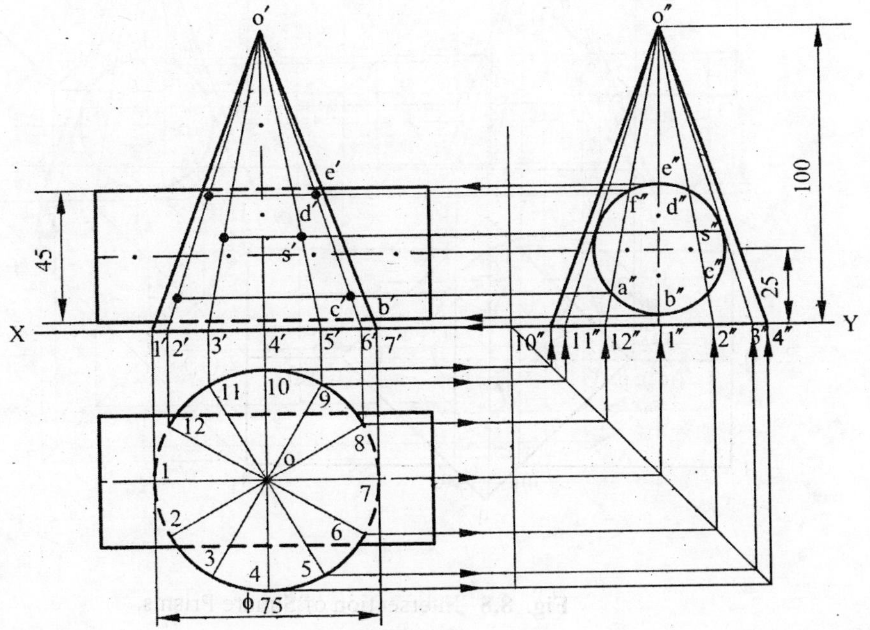

(b)

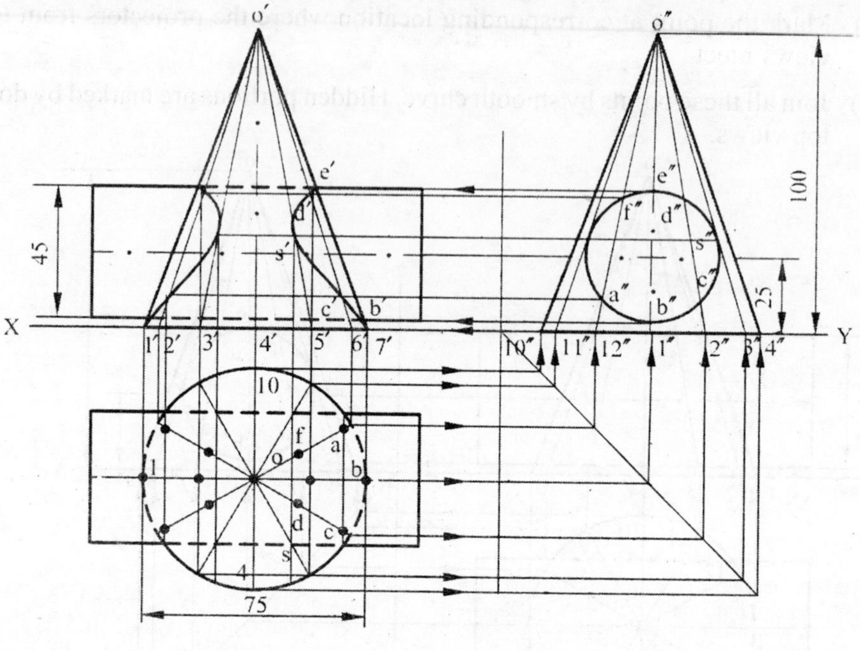

(c)

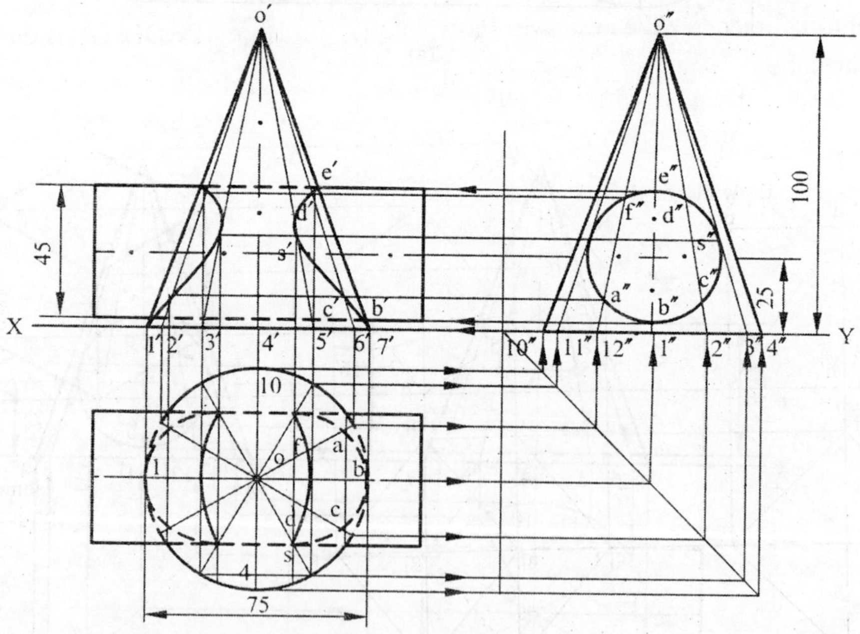

(d)

Fig. 8.9 Intersection of Cylinder and Cone.

CHAPTER 9

Isometric Projection

9.1 Introduction

Pictorial projections are used for presenting ideas which may be easily understood by persons even without technical training and knowledge of multi-view drawing. The pictorial drawing shows several faces of an object in one view, approximately as it appears to the eye.

9.2 Principle of Isometric Projections

It is a pictorial orthographic projection of an object in which a transparent cube containing the object is tilted until one of the solid diagonals of the cube becomes perpendicular to the vertical plane and the three axes are equally inclined to this vertical plane.

Insometric projection of a cube in steps is shown in Fig. 9.1. Here ABCDEFGH is the isometric projection of the cube.

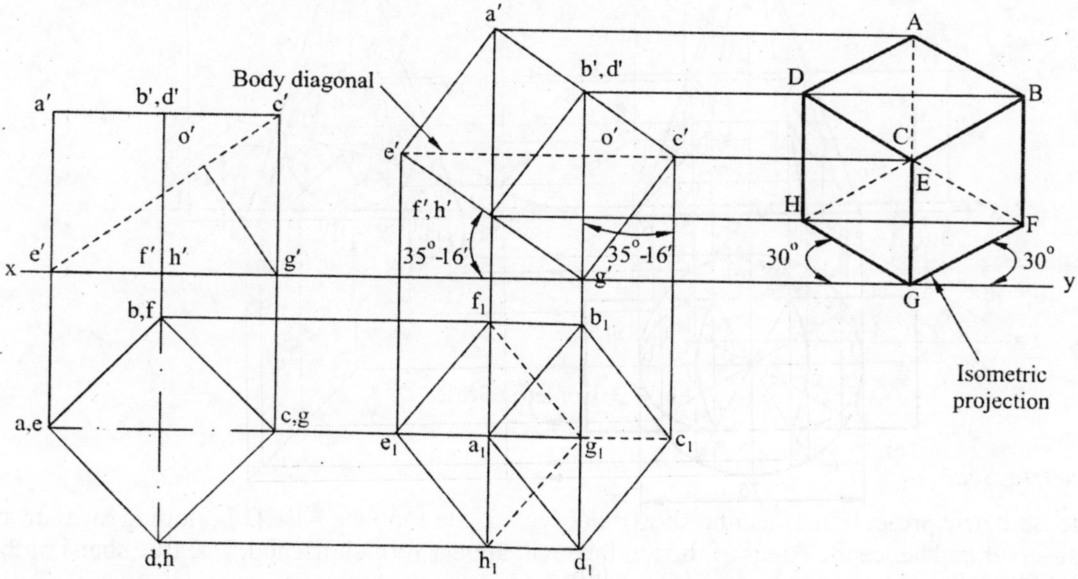

Fig. 9.1 Principle of Isometric Projection.

The front view of the cube, resting on one of its corners (G) is the isometric projection of the cube. The isometric projection of the cube is reproduced in Fig. 9.2.

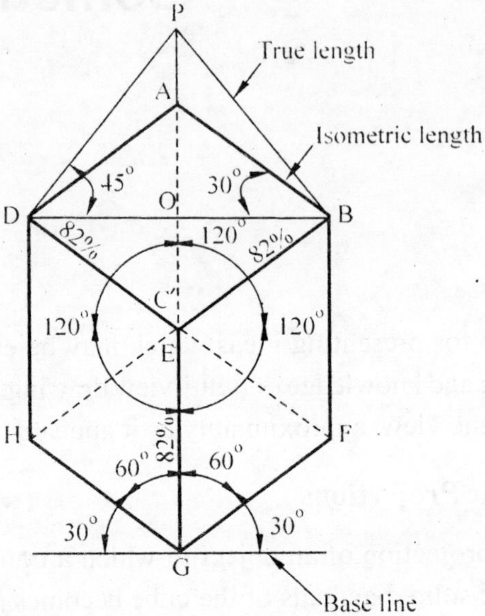

Fig. 9.2 An isometric Cube.

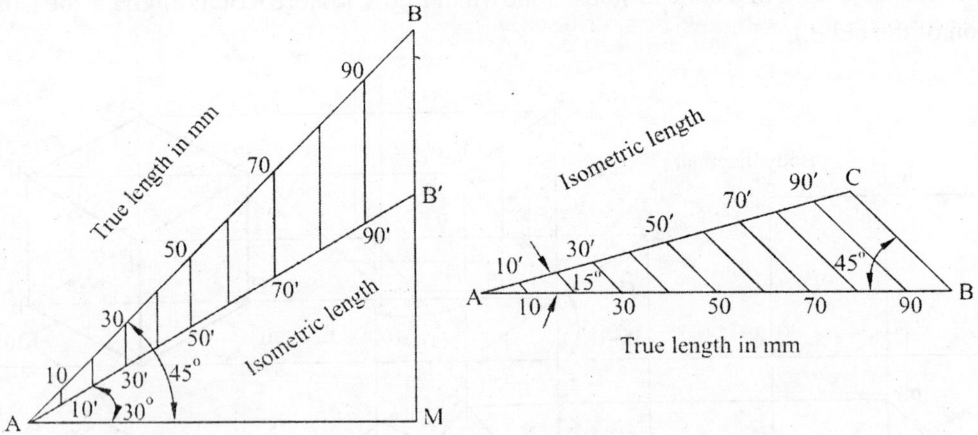

Fig. 9.3 Isometric Scale.

Isometric Scale

In the isometric projection of a cube shown in Fig. 9.2, the top face ABCD is sloping away from the observer and hence the edges of the top face will appear fore-shortened. The true shape of the triangle DAB is represended by the triangle DPB.

The extent of reducti'on of an sometric line can be easily found by construction of a diagram called **isometric scale**. For this, reproduce the triangle DPA as shown in Fig. 9.3. Mark the devisions of true length on DP. Through these divisions draw vertical lines to get the corresponding points on DA. The divisions of the line DA give dimensions to isometric scale.

From the triangle ADO and PDO in Fig. 9.2, the ratio of the isometric length to the true length, i.e., DA/DP = cos 45°/cos30° = 0.816

The isometric axes are reduced in the ratio 1:0.816 ie. 82% approximately.

9.2.1 Lines in Isometric Projection

The following are the relations between the lines in isometric projection which are evident from Fig. 9.2.

1. The lines that are parallel on the object are parallel in the isometric projection.
2. Vertical lines on the object appear vertical in the isometric projection.
3. Horizontal lines on the object are drawn at an angle of 30° with the horizontal in the isometric projection.
4. A line parallel to an isometric axis is called an isometric line and it is fore shortened to 82%.
5. A line which is not parallel to any isometric axis is called non-isometric line and the extent of fore-shoretening of non-isometric lines are different if their inclinations with the vertical planes are different.

9.2.2 Isometric Projection

Fig. 9.4(a) shows a rectangular block in pictorial form and Fig. 9.4(b), the steps for drawing an isometric projection using the isometric scale.

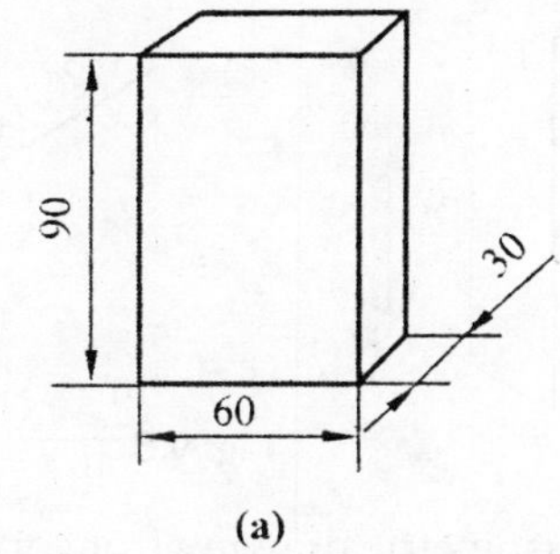

(a)

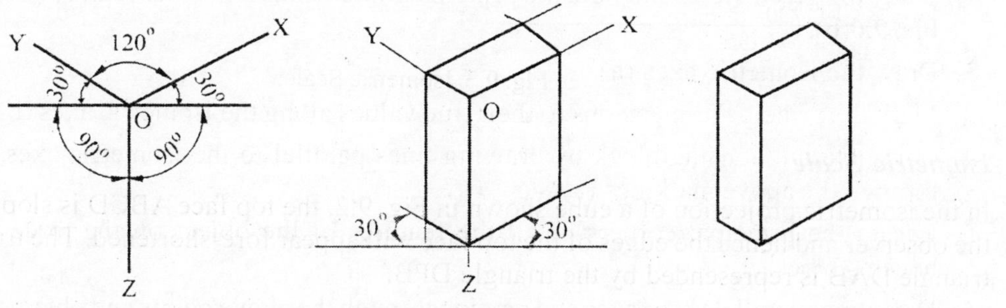

(b)

Fig. 9.4 Developing Isometric Projection.

9.2.3 Isometric Drawing

Drawing of objects are seldom drawn in true isometric projections, as the use of an isometric scale is inconvenient. Instead, a convenient method in which the foreshortening of lengths is ignored and actual or true lengths are used to obtain the projections, called isometric drawing or isometric view is normally used. This is advantageous because the measurement may be made directly from a drawing.

The isometric drawing of figure is slightly larger (approximaely 22%) than the isometric projection. As the proportions are the same, the increased size does not affect the pictorial view of the representation and at the same time, it may be done quickly. Fig. 9.5 shows the difference between the isometric drawing and isometric projection.

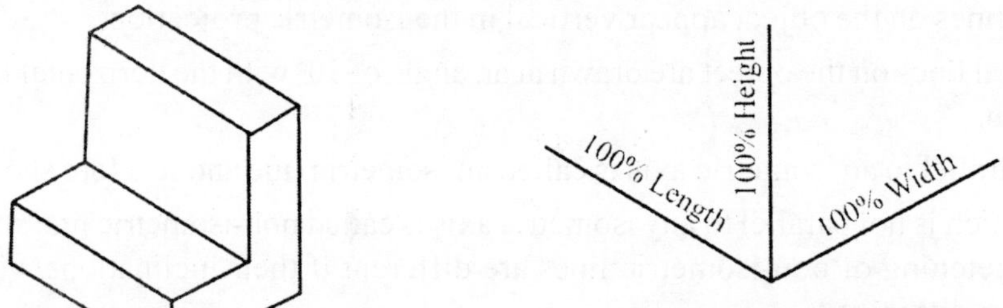

(a) Isometric Drawing

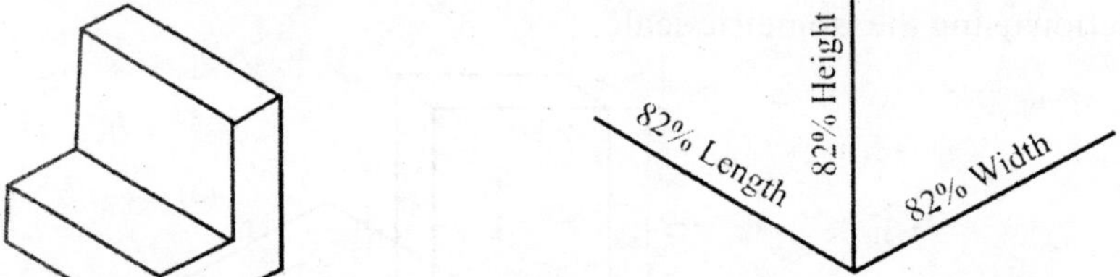

(b) Isometric Projection

Fig. 9.5

Steps to be followed to make isometric drawing from orthographic views are given in (Fig. 9.6).

1. Study the given views and note the principal dimensions and other features of the object in Fig. 9.6(b).
2. Draw the isometric axes (a).
3. Mark the principal dimensions to their true values along the isometric axes (b).
4. Complete the housing block by drawing lines parallel to the isometric axes and passing through the above markings (c).
5. Locate the principal corners of all the features of the object on the three faces of the housing block (d).
6. Draw lines parallel to the axes and passing through the above points and obtain the isometric drawing of the object by darkening the visible edges (e).

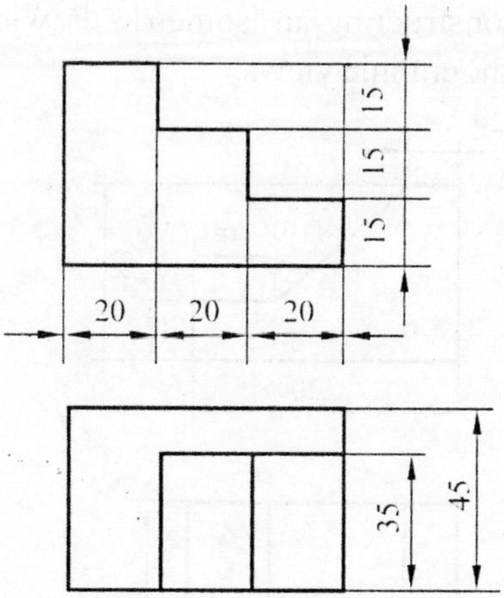

Fig. 9.6(a) Otrhographic View.

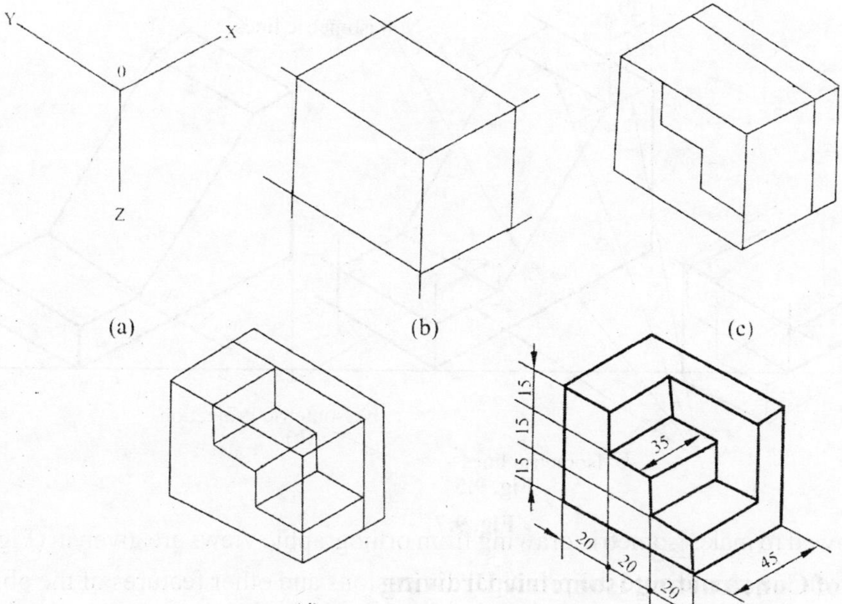

Fig. 9.6(b) Isometric View.

9.2.4 Non-Isometric Lines

In an isometric projection or drawing, the lines that are not parallel to the isometric axes are called non-isometric lines. These lines obviously do not appear in their true length on the drawing and can not be measured directly. These lines are drawn in an isometric projection or drawing by locating their end points.

Fig. 9.7 shows the steps in constructing an isometric drawing of an object containing non-isometric lines from the given orthographic views.

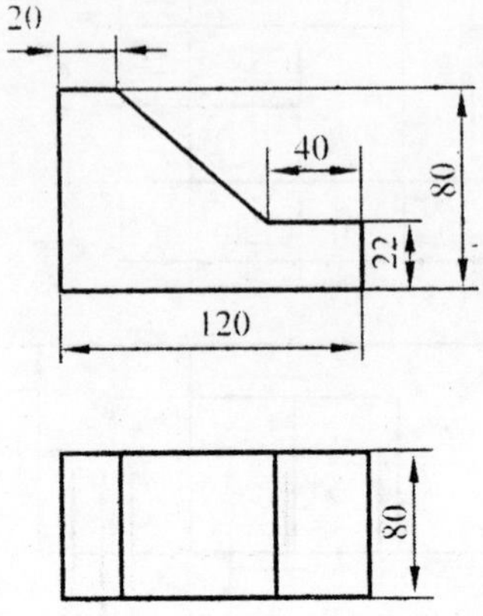

(a)

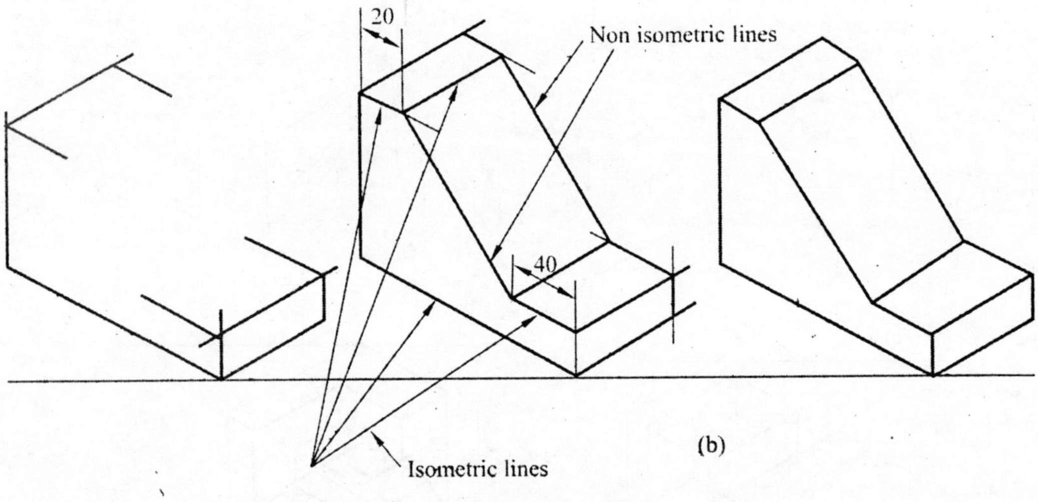

(b)

Fig. 9.7

9.3 Methods of Constructing Isometric Drawing

The methods used are :

1. Box method.
2. Off-set method.

9.3.1 Box Method (Fig. 9.8)

When an object contains a number of non-isometric lines, the isometric drawing may be conveniently constructed by using the box method. In this method, the object is imagined to be enclosed in a rectrangular box and both isometric and non-isometric lines are located by their respective points of contact with the surfaces and edges of the box.

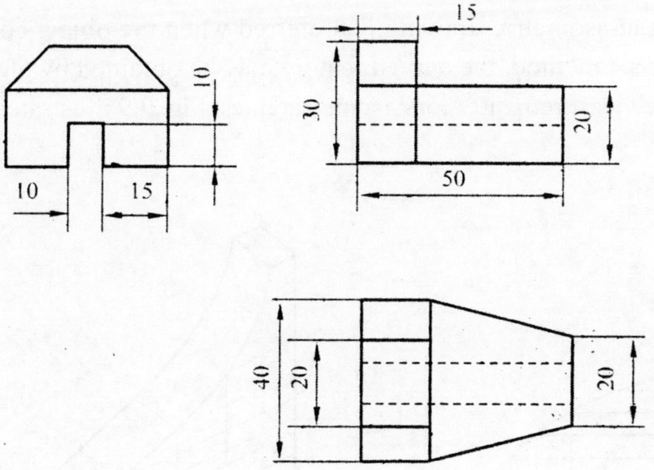

(a)

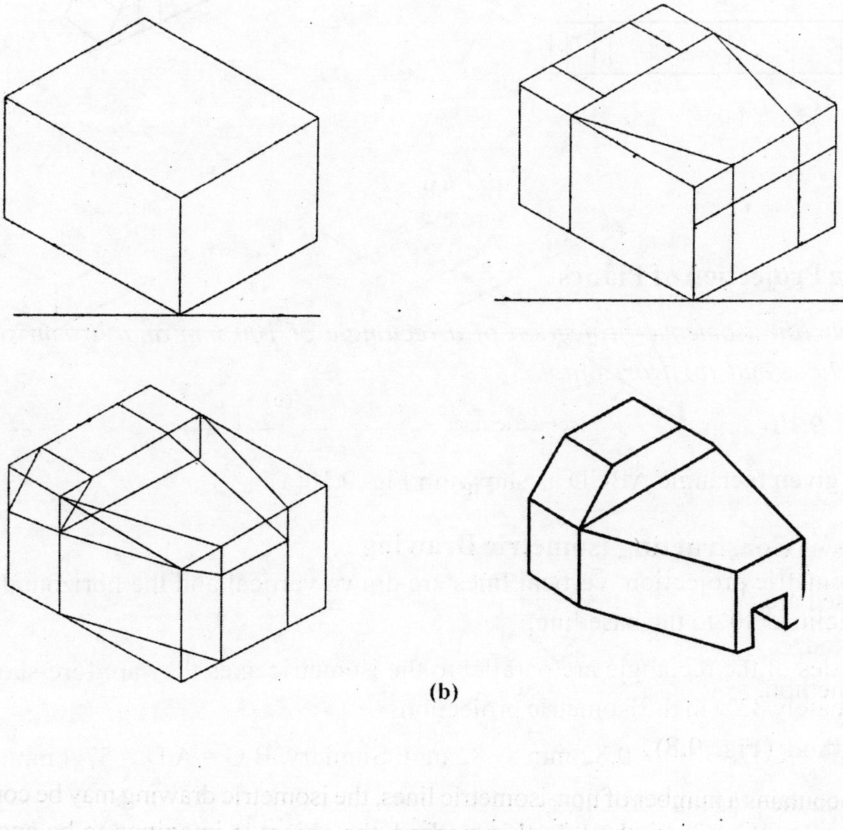

(b)

Fig. 9.8

9.3.2 Off-set Method

Off-set method of making an isometric drawing is preferred when the object contains irregular curved surfaces. In the off-set method, the curved feature may be obtained by plotting the points on the curve, located by the measurements along isometric lines. Fig. 9.9 illustrates the application of this method.

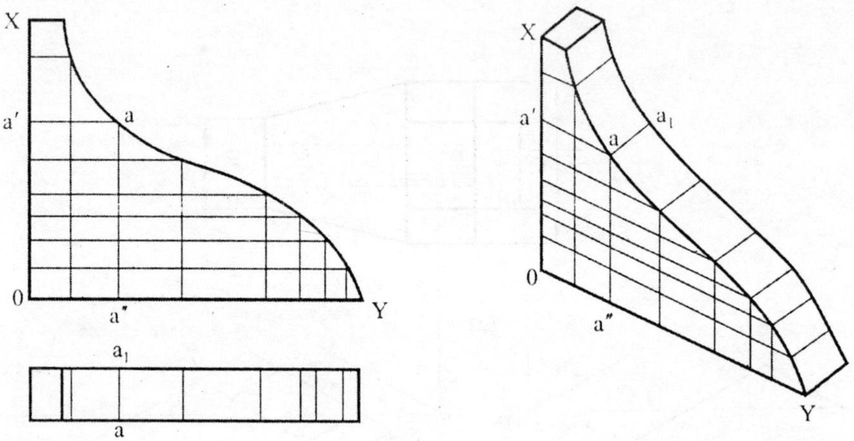

(a) **(b)**

Fig. 9.9

9.4 Isometric Projection of Planes

Problem : *Draw the isometric projection of a rectangle of 100 mm and 70 mm sides if its plane is (a) Vertical and (b) Horizontal.*

Solution : **(Fig. 9.10)**

1. Draw the given rectangle ABCD as shown in Fig. 9.10(a).

Note :

(i) In the isometric projection, vertical lines are drawn vertical and the horizontal lines are drawn inclined 30° to the base line.

(ii) As the sides of the rectangle are parallel to the isometric axes they are fore-shortened to approximately 82% in the isometric projections.

Hence AB = C D = 100 × 0.82 mm = 82 mm. Similary, B C = A D = 57.4 mm.

(a) When the plane is vertical:

2. Draw the side A D inclined at 30° to the base line as shwon in Fig.9.10b and mark A D = 57.4 mm.

3. Draw the verticals at A and D and mark off A B = D C = 82 mm on these verticals.
4. Join B C which is parallel to A D.

   A B C D is the required isometric projection. This can also be drawn as shown in Fig.9.10(c). Arrows show the direction of viewing.

   (b) When the plane is horizontal.

5. Draw the sides AD and DC inclined at 30° to be base line and complete the isometric projection A B C D as shown in Fig. 9.10(d). Arrow at the top shows the direction of viewing.

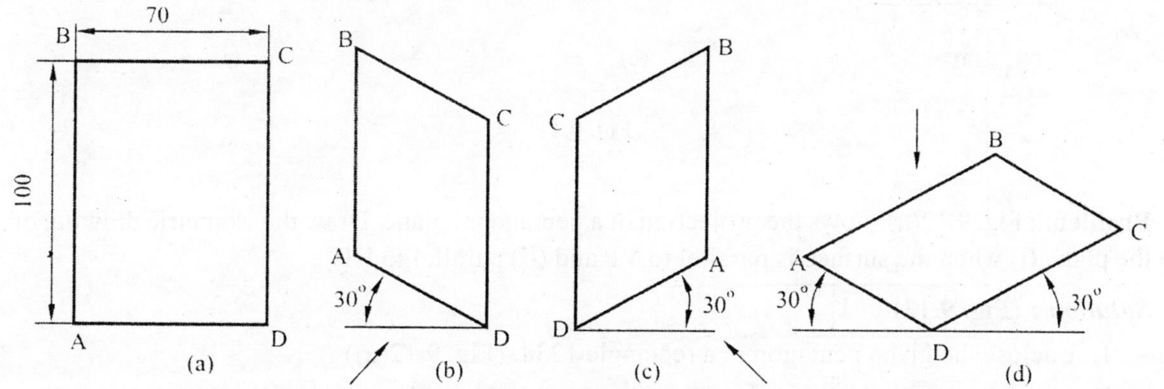

Fig. 9.10

To draw the isometric projection of a square plane. (Fig. 9.11(a))

***Solution :*(Fig. 9.11)**

Case 1 Vertical plane (Fig. 9.11(b))

1. Draw a line at 30° to the horizontal and mark the isometric length on it.
2. Draw verticals at the ends of the line and mark the isometric length on these parallel lines.
3. Join the ends by a straight line which is also inclined at 30° to the horizontal.

There are two possible positions for the plane.

Case II Horizontal plane (Fig. 9.11(c))

1. Draw two lines at 30° to the horizontal and mark the isometric length along the line.
2. Complete the figure by drawing 30° inclined lines at the ends till the lines intersect.

Note

(i) The shape of the isometric projection or drawing of a square is a **Rhombus**.

(ii) While dimensioning an isometric projection or isometric drawing true dimensional values only must be used.

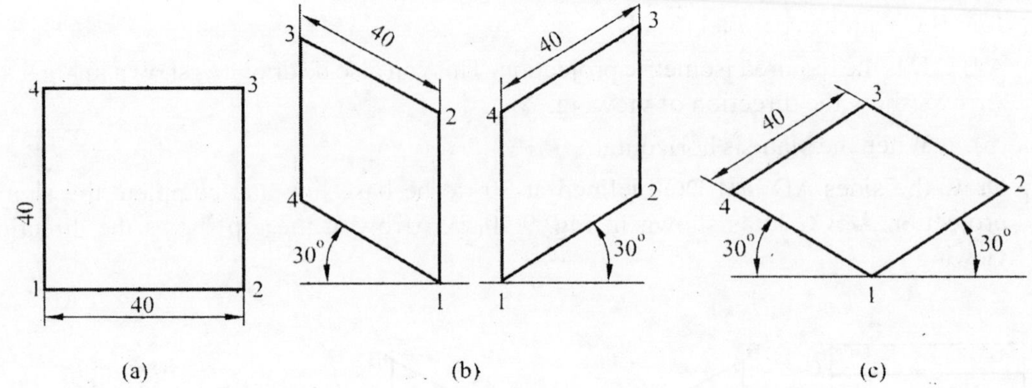

Fig. 9.11

Problem : Fig. 9.12(a) shows the projection of a pentagonal plane. Draw the isometric drawing of the plane (i) when the surface is parallel to VP and (ii) parallel to HP.

Solution : **(Fig. 9.12)**

1. Enclose the given pentagon in a rectangle 1234. (Fig. 9.12(a))
2. Make the isometric drawing of the rectangle 1234 by using true lengths.
3. Locate the points A and B such that 1a = 1A and 1b = 1B.
4. Similarly locate point C, D and E such that 2c = 2C, 3d = 3D and e4 = E4.
5. ABCDE is the isometric drawing of the pentagon (Fig. 9.12(b)).
6. Following the above princple of construction, 9.12(c) can be drawn

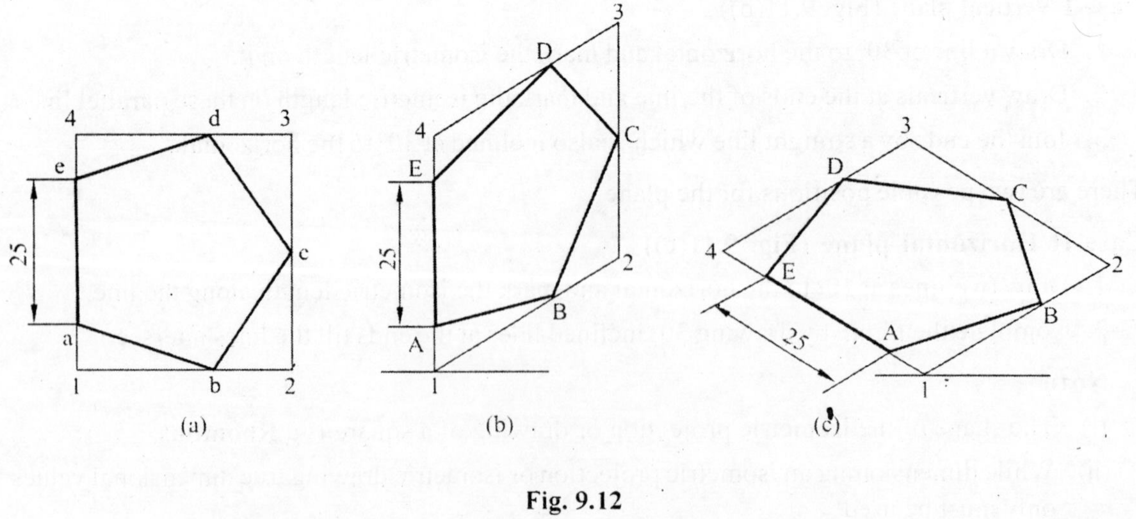

Fig. 9.12

Problem : Draw the isometric view of a pentagonal plane of 30 mm side when one of its sides is parallel to HP. (a) When it is horizontal and (b) vertical.

***Solution :* (9.13)**

1. Draw the pentagon ABCDE and enclose it in a rectangle 1-2-3-4 as shown in Fig.9.13(a).
   (a) When it is horizonta, the isometric view of the pentagon can be represented by ABCDE as shown in Fig.9.13(b).
   (b) When the plane is vertical it can be represented by ABCDE as shown in Fig.9.13(c) or (d).

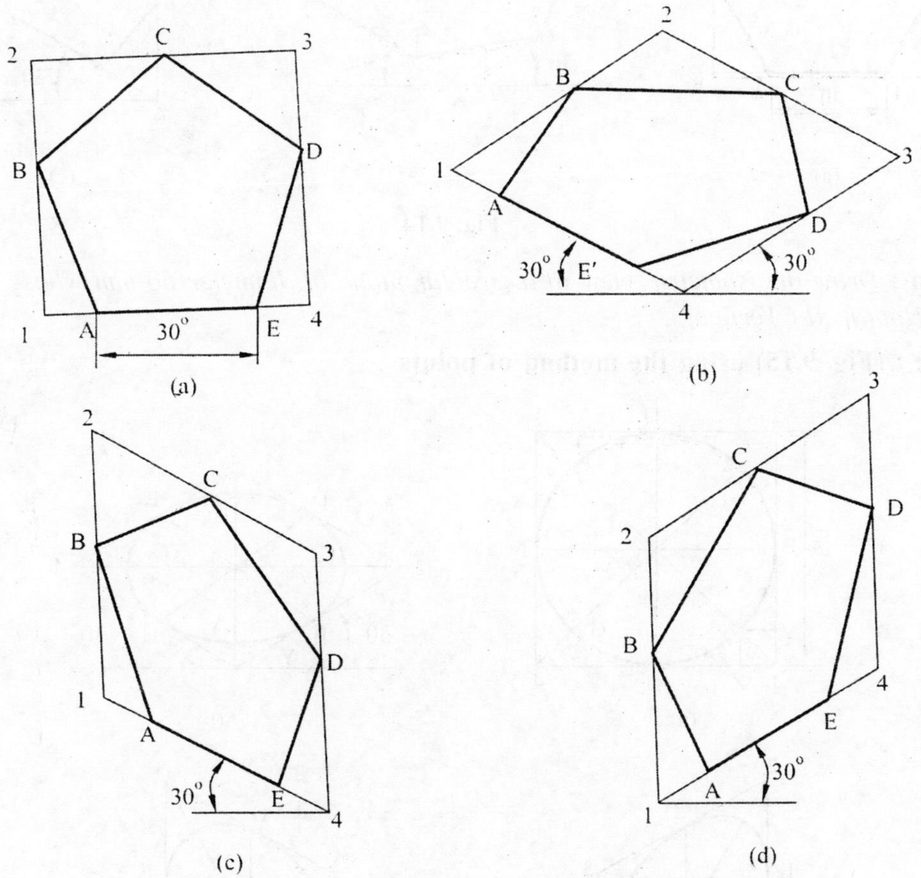

Fig. 9.13

Note : It may be noted that the point A on the isometric view can be marked after drawing the isometric view of the rectangle 1-2-3-4 for this, mark 1A' = 1A and so on.

Problem : Fig. 9.14(a) shows the orthographic view of a hexagonal plane of side 30 mm. Draw the isometric drawing (view) of the plane keeping it (a) horizontal and (b) vertical.

Solution : **(Fig. 9.14)**

Following the principle of construction of Fig. 9.13, obtain the Fig. 9.14(b) and 9.14(c) respectively for horizontal and vertical position of the hexagonal plane.

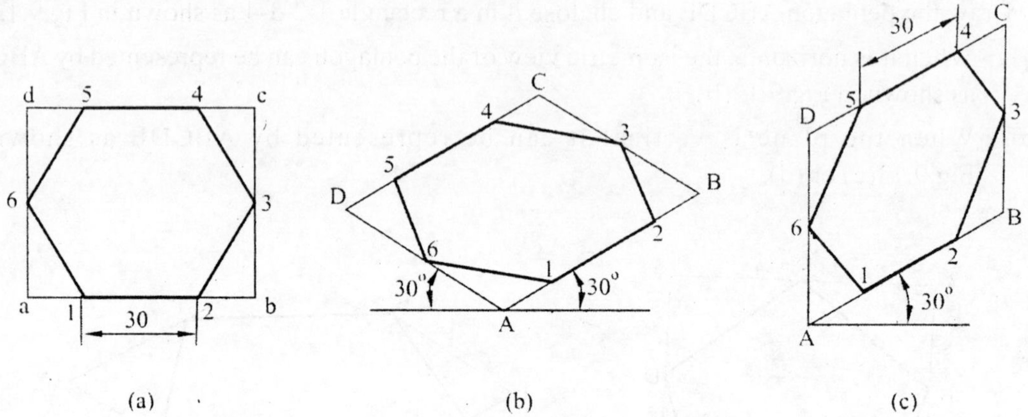

Fig. 9.14

Problem : *Draw the isometric view of a circular plane of diameter 60 mm whose surface is (a) Horizontal, (b) Vertical.*

Solution : **(Fig. 9.15) using the method of points**

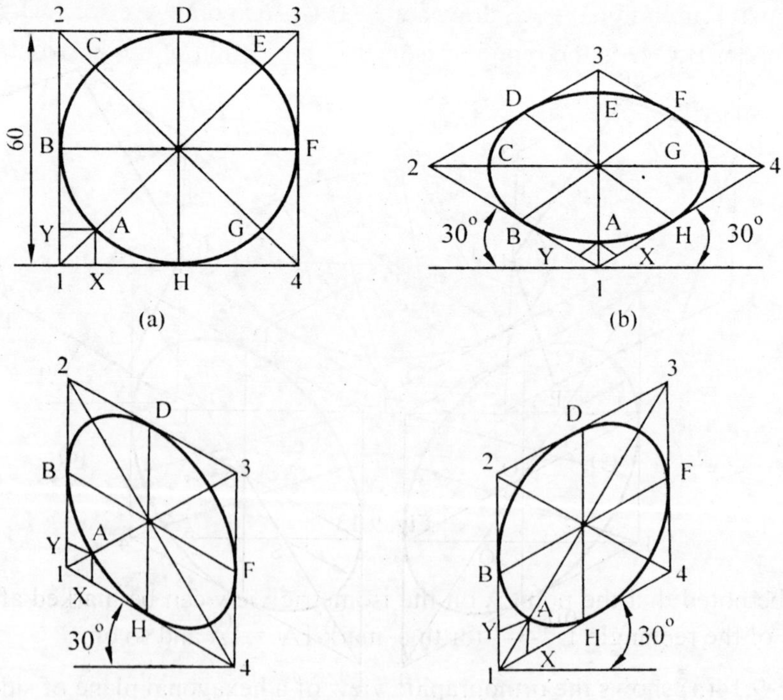

Fig. 9.15

1. Enclose the circle in a square 1-2-3-4 and draw diagonals, as shown in Fig. 9.15(a). Also draw lines YA horizontallly and XA vertically.

   To draw the isometeric view of the square 1-2-3-4 as shown in Fig. 9.15(b).
2. Mark the mid points of the sides of the square as B D F and H.
3. Locate the points X and Y on lines 1-4 and 1-2 respectively.
4. Through the point X, draw A X parallel to line 1-2 to get point A on the diagonal 1-3. The point A can be obtained also by drawing Y A through the point Y and parallel to the line 1-4.
5. Similarly obtain other points C, E and G.
6. Draw a smooth curve passing through all the points to obtain the required isometric view of the horizontal circular plane.
7. Similarly obtain isometric view of the vertical circular plane as shown in Fig. 9.15(c) and (d).

Problem : *Draw the isometric projection of a circular and semicircular plane of diameter 60 mm whose surface is (a) Horizontal and (b) Vertical; use four-centre method.*

Solution : **(Fig. 9.16)**

1. Draw the isometric projection of the square 1-2-3-4 (rhombus) whose length of side is equal to the isometric length of the diameter of the circle = 0.82 × 60.
2. Mark the mid points A, B, C and D of the four sides of the rhombus. Join the points 3 and A. This line intersects the line 2-4 joining the point 2 and 4 at M. Similarly obtain the intersecting point N.
3. With centre M and radius = MA draw an arc A B. Also draw an arc C D with centre N.
4. With centre 1 and radius = 1C, draw an arc B C. Also draw the arc A D.
5. The ellipse A B C D is the required isometric projection of the horizontal circular plane (Fig.9.16(a)).

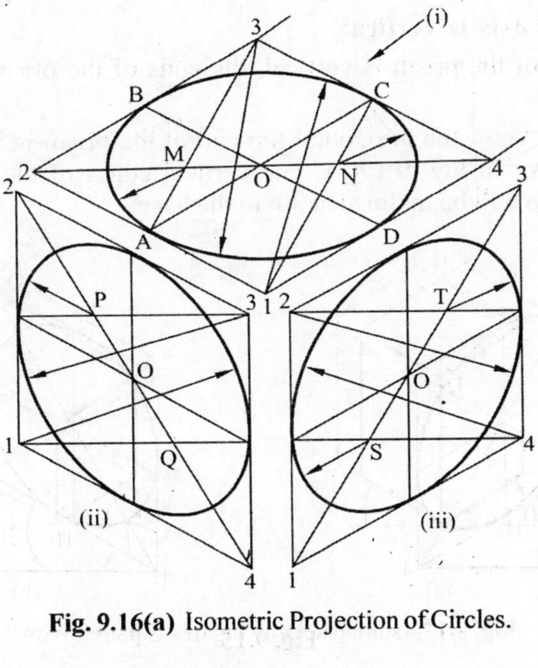

Fig. 9.16(a) Isometric Projection of Circles.

6. Similarly obtain the isometric projection in the vertical plane as shown.

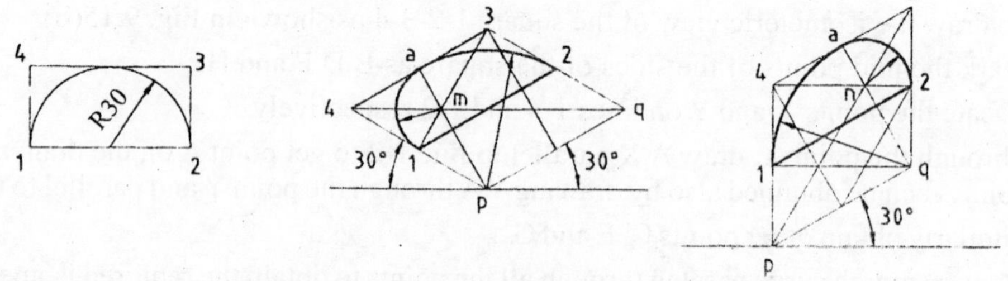

Fig. 9.16(b) Isometric porject of semicircle

To draw the isometric view of the semicircle with its surface parallel to VP

1. Construct a rectangle 1-2-3-4 enclosing the semicircle.
2. Draw one of the isometric axes at 30° to the horizontal and the other axis vertical. Considering it as a circle, draw the rhombus 4-*p*-*q*-3 and find the four centers as mentioned earlier.
3. With the center *n*, radius *n*2, draw an arc 2*a* and with *q* as center, radius *qa,* draw an arc *a*1.
4. Draw the line 1-2 as a dark line to complete the isometric view of the semicircle as in Fig. 9.16(b). The two other centers are not used to draw the arc as it is only a semicircle.

Note that, to draw isometric view/projection of a semicircle, it is always considered as a circle and the rhombus in isometric view/projection is constructed to get the four centres. Then the required part of the semicircle is drawn as a portion of the ellipse in isometric view/projection.

Problem : Draw the isometric view of square prism with a side of base 30 mm and axis 50 mm long when the axis is (a) vertical and (b) horizontal.

***Solution :* (Fig. 9.17)**

(a) Case 1 when the axis is vertical

1. When the axis of the prism is vertical, the ends of the prism which is square will be horizontal.
2. In an isometric view, the horizontal top end of the prism is represented by a rhombus ABCD as shown in Fig. 9.17(a). The vertical edges of the prism are vetical but its horizontal edges will be inclined at 30° to the base.

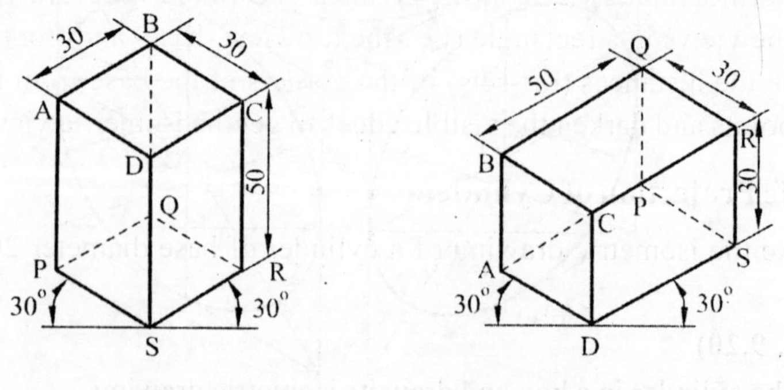

Fig. 9.17 Isometric Drawing of a Square Prism.

9.5 Isometric Projection of Prisms

Problem : Draw the isometric view of a pentagonal prism of base 60 mm side, axis 100 mm long and resting on its base with a vertical face perpendicular to VP.

Solution : **(Fig. 9.18)**

1. The front and top views of the prism are shown in Fig. 9.18(a).
2. Enclose the prism in a rectangular box and draw the isometric view as shown in Fig. 9.18(b) using the box method.

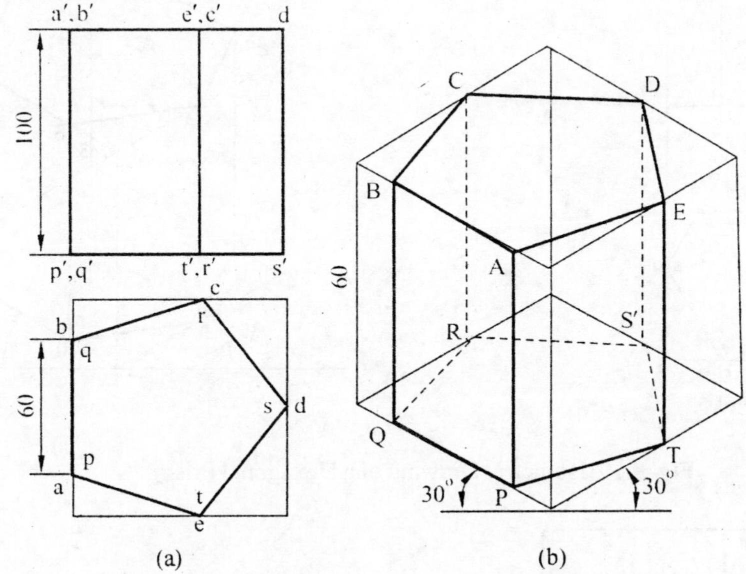

Fig. 9.18 Isometric Drawing of a Pentogonal Prism.

Problem : A hexagonal prism of base of side 30 mm and height 60 mm is resting on its base on HP. Draw the isometric drawing of the prism.

Solution : **(Fig. 9.19)**

1. Draw the orthographic views of the prism as shown in Fig.9.19(a).
2. Enclose the views in a rectangle (i.e., the top view -base- and front views).
3. Determine the distances (off-sets) of the corners of the base from the edges of the box.
4. Join the points and darken the visible edges to get the isometric view.

9.6 Isometric Projection of Cylinders

Problem : Make the isometric drawing of a cylinder of base diameter 20 mm and axis 35 mm long.

Solution : **(Fig. 9.20)**

1. Enclose the cylinder in a box and draw its isometric drawing.
2. Draw ellipses corresponding to the bottom and top bases by four centre method.
3. Join the bases by two common tangents.

***Solution* : (Fig. 9.20)**

1. Enclose the cylinder in a box and draw its isometric drawing.
2. Draw ellipses corresponding to the bottom and top bases by four centre method.
3. Join the bases by two common tangents.

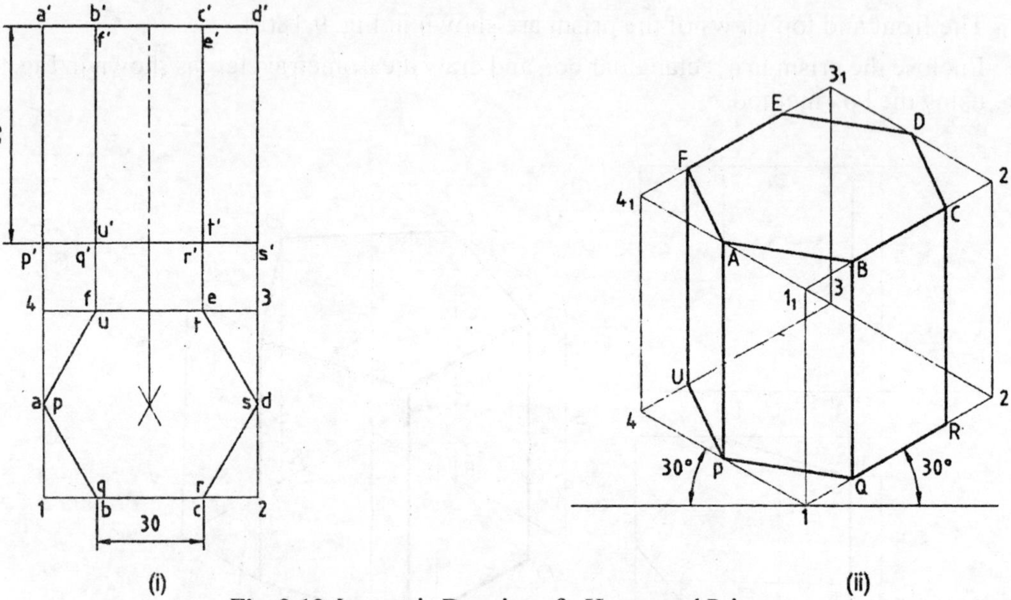

Fig. 9.19 Isometric Drawing of a Hexagonal Prism.

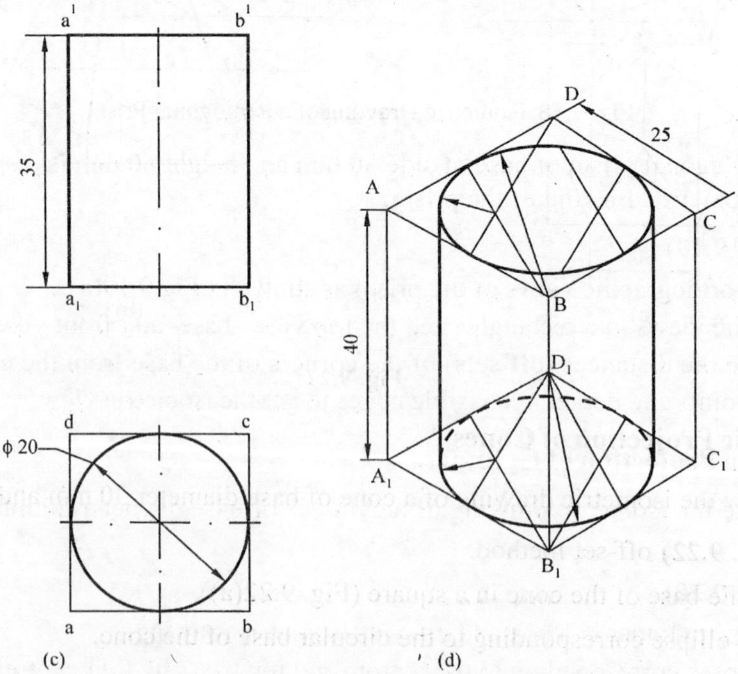

Fig. 9.20 Isometric Drawing of a Cylinder.

9.7 Isometric Projection of Pyramids

Problem : A pentagonal pyramid of side of base 30 mm and height 70 mm is resting with its base on HP. Draw the isometric drawing of the pyramid.

Solution : **(Fig. 9.21)**

1. Draw the projections of the pyramind (Fig.9.21(a)).
2. Enclose the top view in a rectangle abcd and measure the off-sets of all the corners of the base and the vertex.
3. Draw the isometric view of the rectangle ABCD.
4. Using the off-sets locate the corners of the base 1,2, etc. and the vertex o.
5. Join o-1, o-2, o-3, etc. and draken the visible edges and obtain the required view.

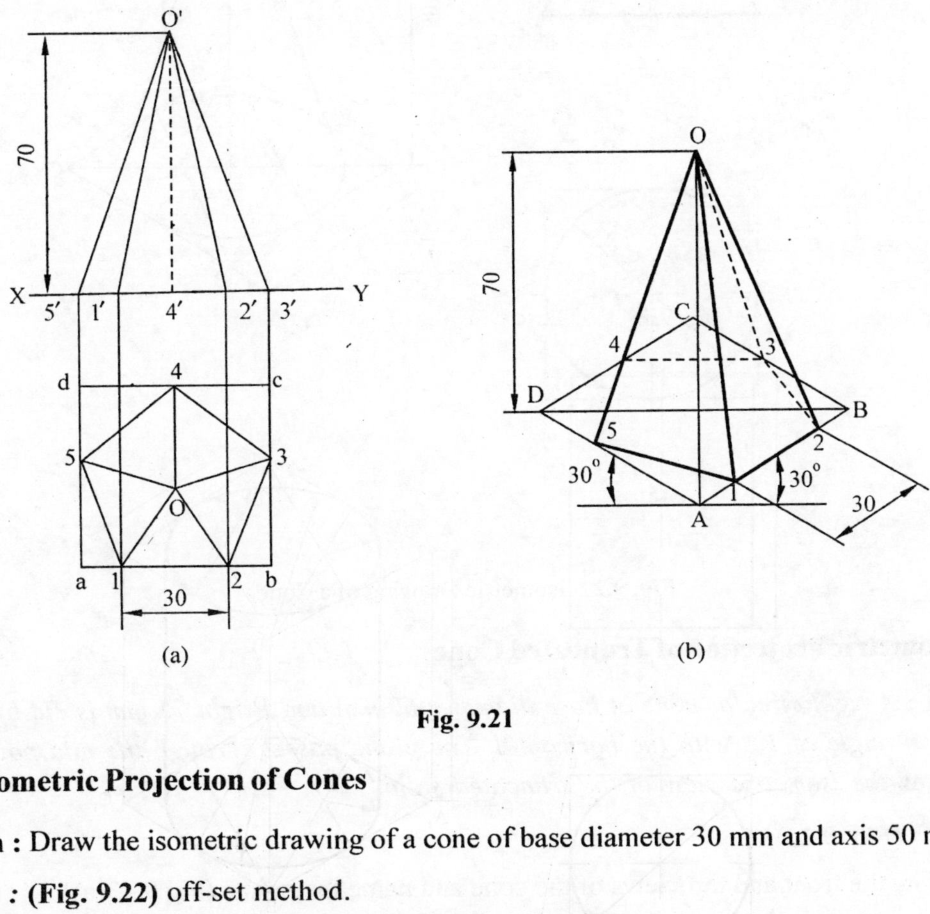

Fig. 9.21

9.8 Isometric Projection of Cones

Problem : Draw the isometric drawing of a cone of base diameter 30 mm and axis 50 mm long.

Solution : **(Fig. 9.22)** off-set method.

1. Enclose the base of the cone in a square (Fig. 9.22(a)).
2. Draw the ellipse corresponding to the circular base of the cone.

3. From the centre of the ellipse draw a vertical centre line and locate the apex at a height of 50 mm.
4. Draw the two outer most generators from the apex to the ellipse and complete the drawing.

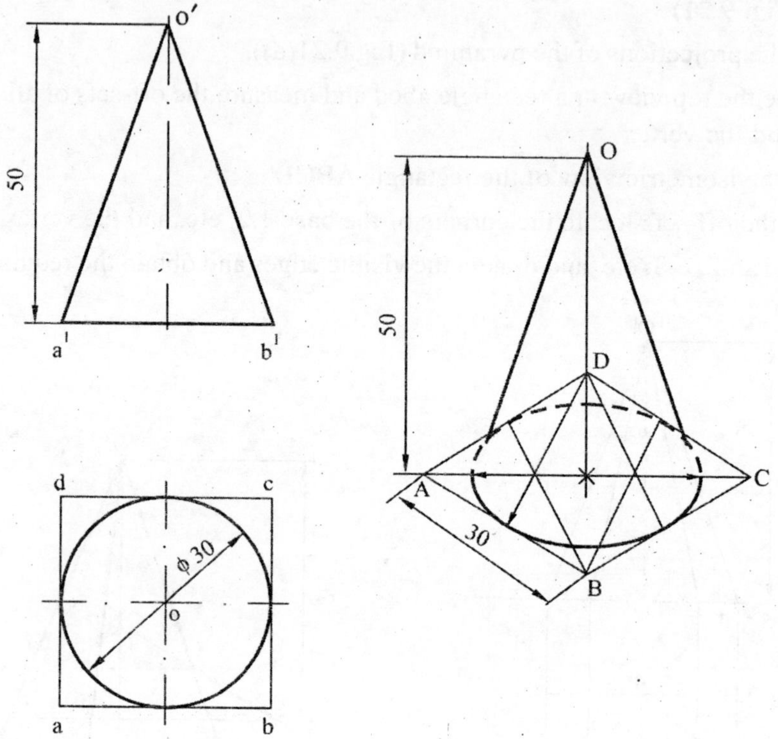

(a) (b)

Fig. 9.22 Isometric Drawing of a Cone.

9.9 Isometric Projection of Truncated Cone

Problem : *A right circular cone of base diameter 60 mm and height 75 mm is cut by a plane making an angle of 30° with the horizontal. The plane passes through the mid point of the axis. Draw the isometric view of the truncated solid.*

Solution : **(Fig. 9.23)**

1. Draw the front and top views of the cone and name the points (Fig.9.23(a)).
2. Draw a rectangular prism enclosing the complete pyramid.
3. Mark the plane containing the truncated surface of the pyramid. This plane intersects the box at p'p' in the front view and pppp in the top view

4. Draw the isometric view of the cone and mark the plane P P P P, containing the truncated surface of the pyramid as shown in Fig.9.23(b).
5. Draw the isometric view of the base of the cone which is an ellipse.
6. It is evident from the top view that the truncated surface is symmetrical about the line QQ. Hence mark the corresponding line Q Q in the isometric view.
7. Draw the line 1-1, 2-2, 3-3 and 4-4 passing through the points a_1, a_2, a_3, and a_4 in the top view. Mark the points 1, 2, 3, 4 on the corresponding edge of the base of the cone and transfer these points to the plane P P P P by drawing verticals as shown.
8. Point a_1 is the point of intersection of the lines qq and 1-1 in the top view. The point A_1 corresonding to the point a_1 is the point of intersection of the lines Q Q and 1-1 in the isometric view. Hence mark the the point A_1. Point Q lies on the line 2-2 in top view and its corresponding point in the isometric view is represented by A2 on the line 2-2 such that $2a_2 = 2A_2$. Similarly obtain the remaining points A_3 and A_4. Join these points by a smooth curve to get the truncated surface which is an ellipse.
9. Draw the common tangents to the ellipse to get the completed truncated cone.

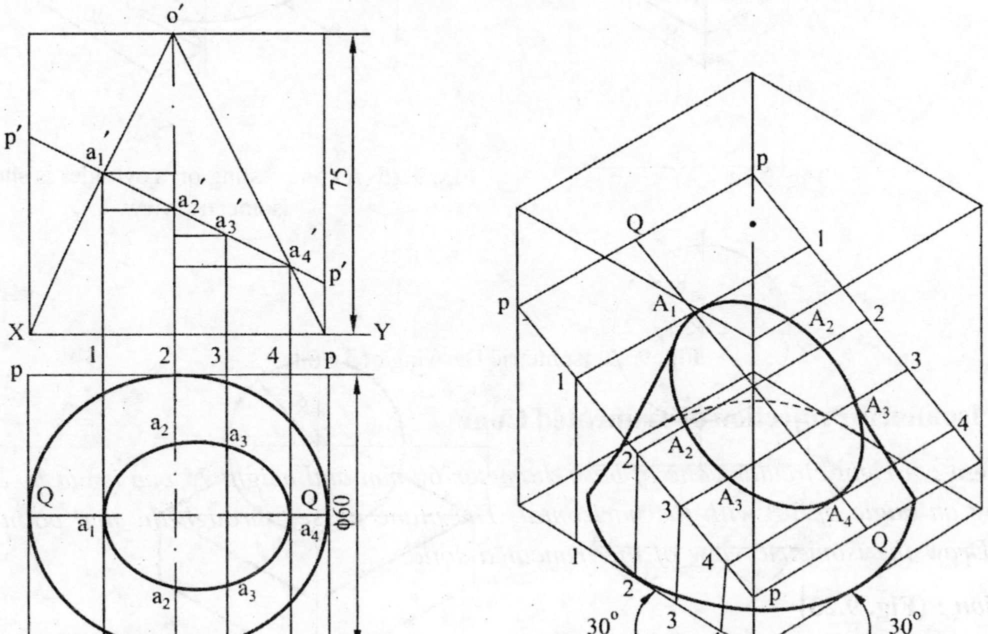

Fig. 9.23 Isometric View of a Trauncated Cone.

9.10 Examples

The orthographic projections and the isometric projections of some solids and machine components are shown from Fig.9.24 to 9.34.

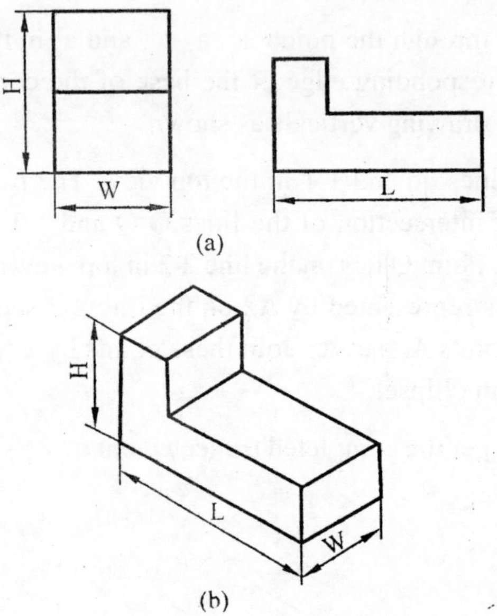

Fig. 9.24

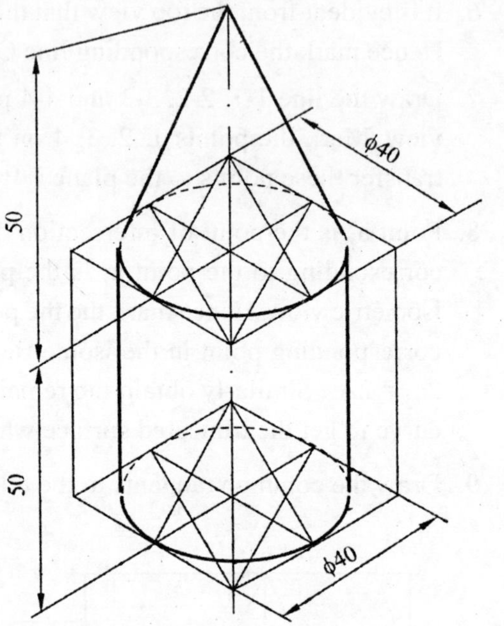

Fig. 9.25 A cone resting on a cylinder is shown in isometeric view.

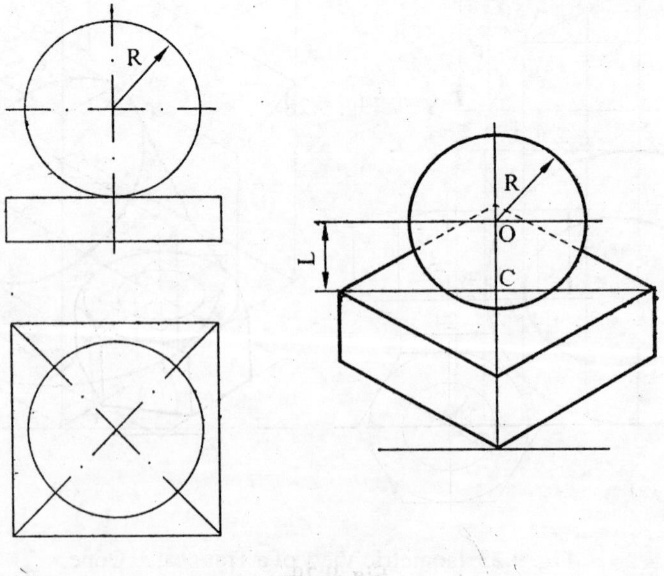

Fig. 9.26

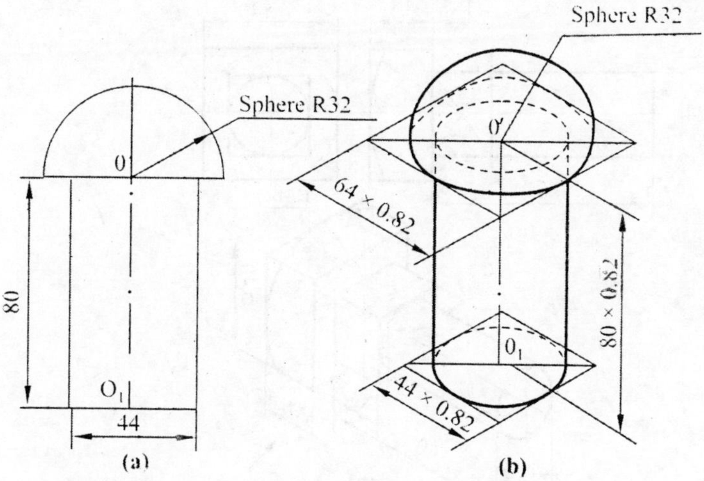

Fig. 9.27

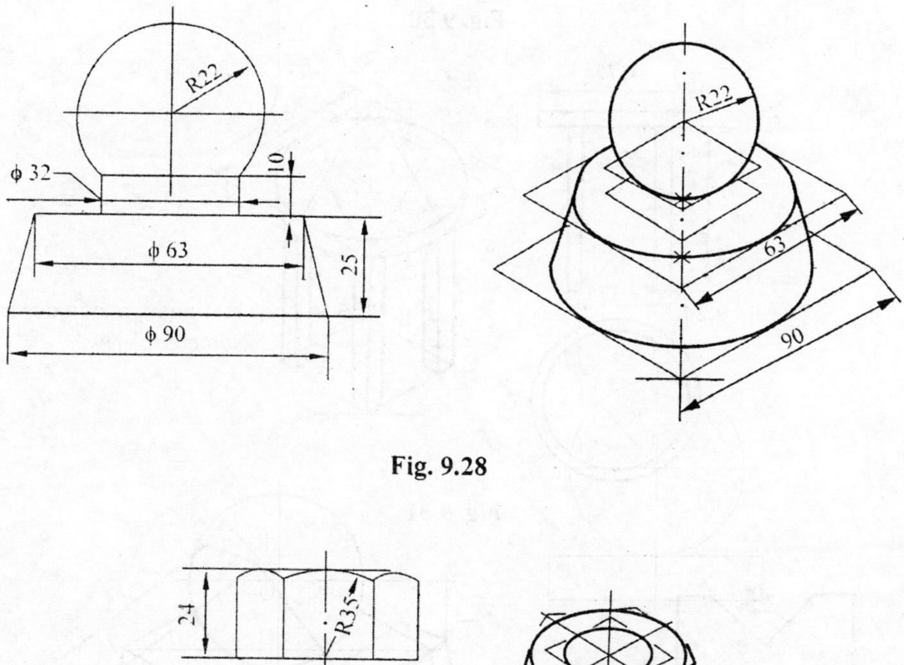

Fig. 9.28

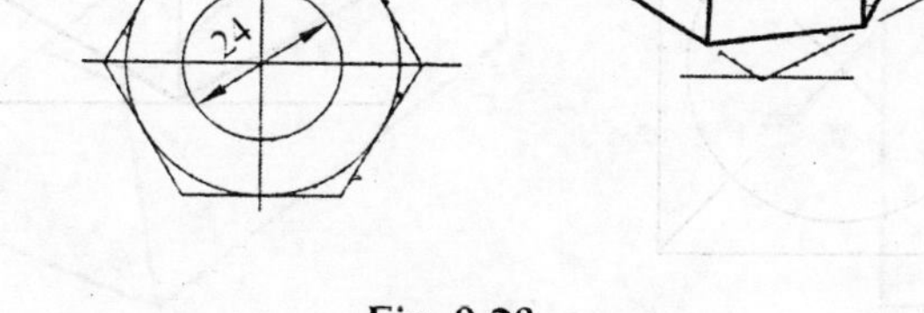

Fig. 9.29

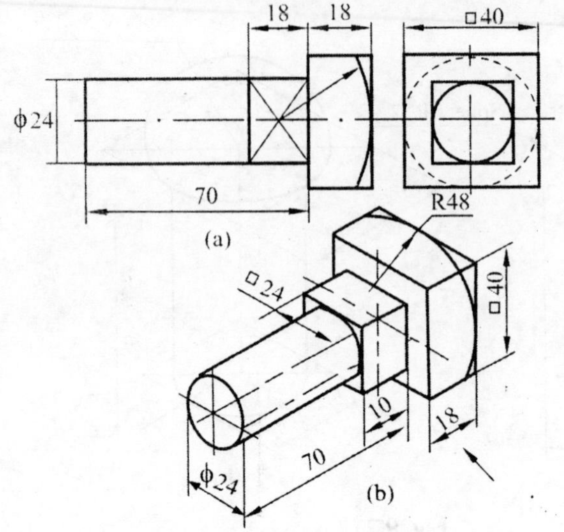

Fig. 9.30

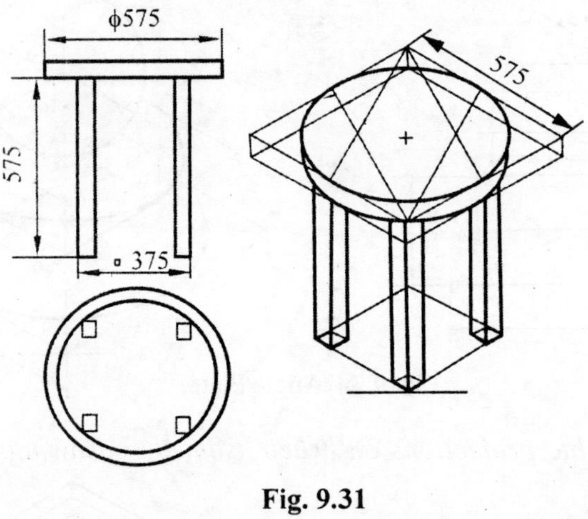

Fig. 9.31

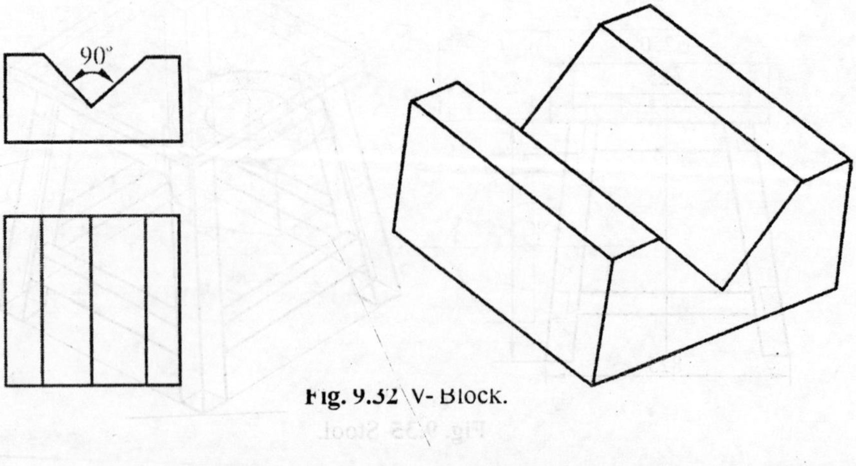

Fig. 9.32 V- Block.

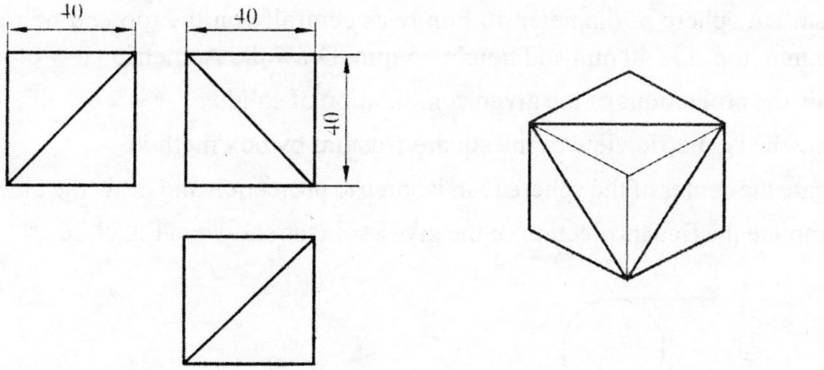

Fig. 9.33 Wedge Piece.

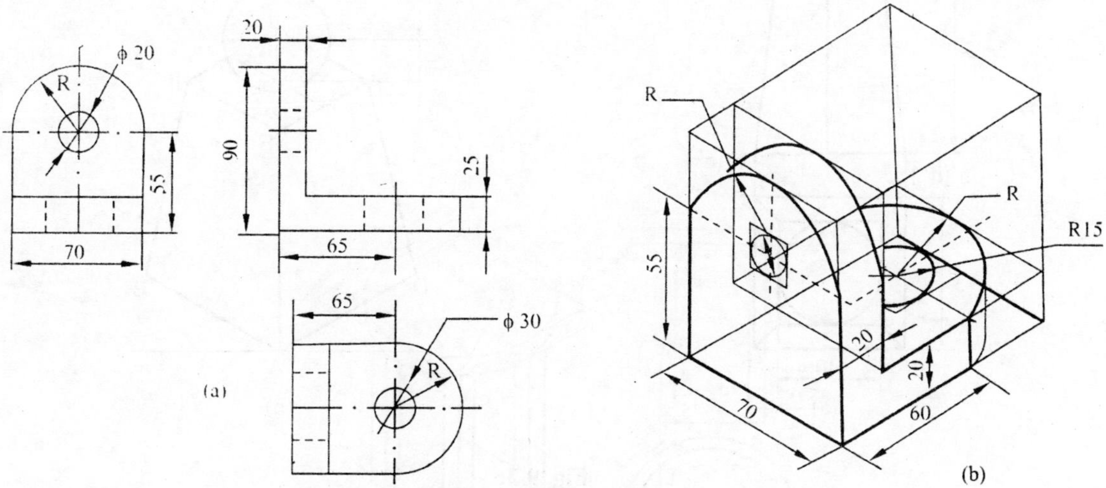

Fig. 9.34 Angle Plate.

Problem : *The orthographic projections and their isometric drawings of a stool is shown is Fig. 9.35.*

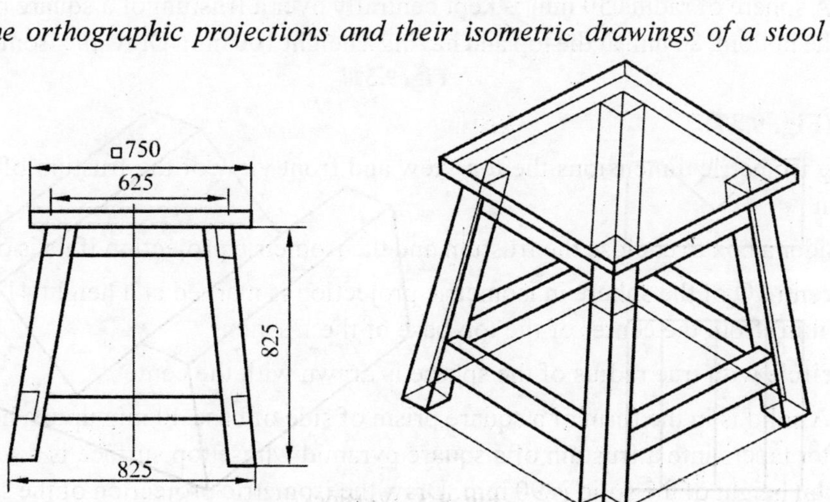

Fig. 9.35 Stool.

Problem : A sphere of diameter 40 mm rests centrally on the top end of a square frustum, base side 60 mm, top side 40 mm and height 75 mm. Draw the isometric view of the solids.

Draw the projections of the given combination of solids.

Draw the isometric view of the square frustum by box method

Locate the center of the sphere O in isometric projection and draw the circle of diamter 40 mm.

Complete the final projection of the given solid as shown in Fig. 9.36.

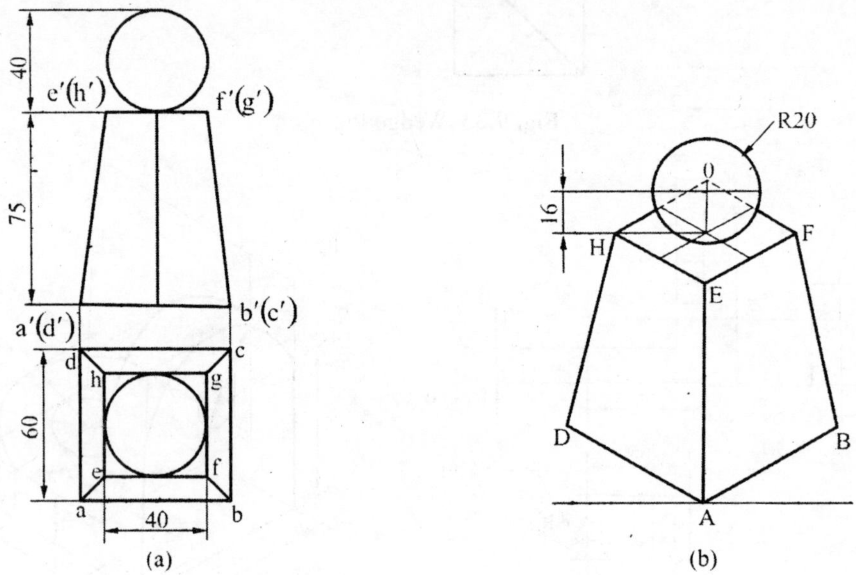

Fig. 9.36

Problem : A sphere of radius 50 mm is kept centrally over a frustum of a square pyramid of side 120 mm at the bottom, 80 mm at the top and having a height 100 mm. Draw the isometric projection of the solid.

Solution : **(Fig. 9.37)**

1. Using isometric dimensions the top view and front view of the frustum of the pyramid is drawn
2. Consider a box to enclose the frustum and the isometric projection if its is obtained.
3. The centre O of the sphere in isometric projection is marked at a height 41 mm (50 × 0.82 = 41 mm) from the center of the top base of the frustum.
4. The circle with true radius of the sphere is drawn with the center.

Problem : A solid is in the form of a square prism of side of base 40 mm upto a height of 60 mm and thereafter tapers into a frustum of a square pyramid whose top surface is a square of 20 mm side. The total height of the solid is 90 mm. Draw the isometric projection of the solid.

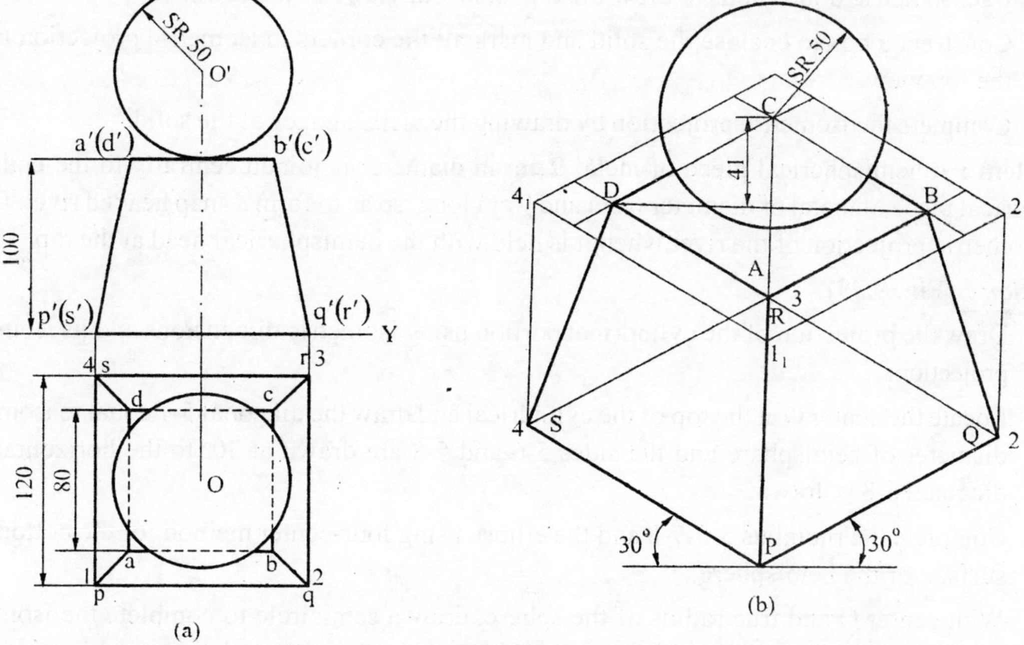

Fig. 9.37

***Solution :* (Fig. 9.38)**

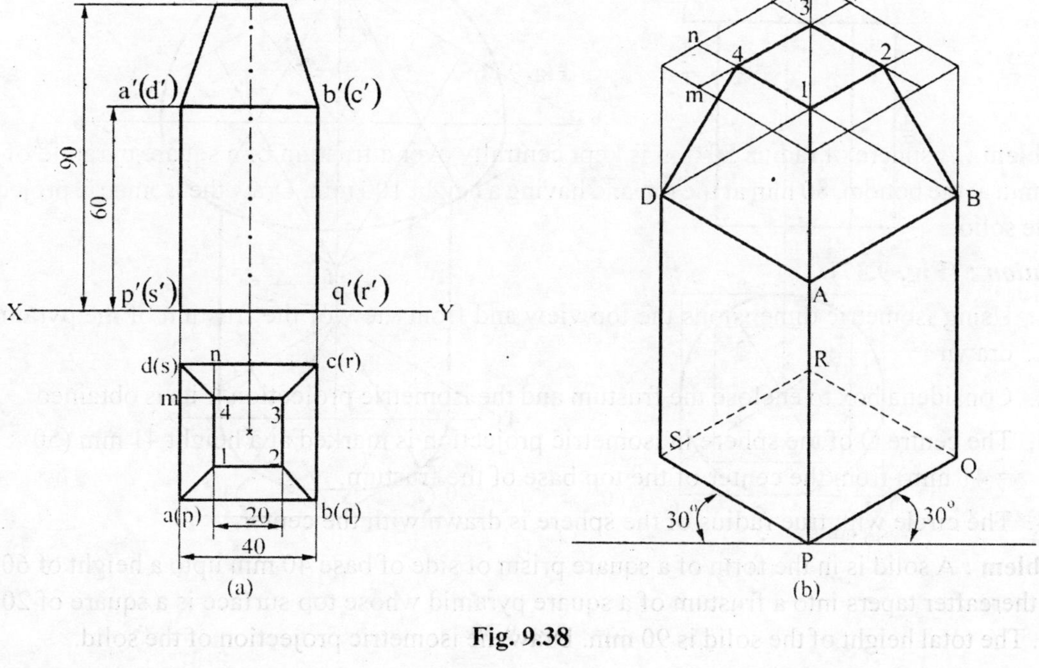

Fig. 9.38

1. Use isometric dimensions to draw the top and front views of the solid.
2. Construct a box to enclose the solid and mark all the corners in isometric projection using the top view.
3. Complete the isometric projection by drawing the visible edges of the solid.

Problem : A hemispherical piece of metal 9 cm in diameter is joined centrally to the end of a cylindrical piece of metal of diameter 6 cm and 9 cm long, so as to form a snap headed rivet. Draw the isometric projection of the rivet when it is held with the hemispherical head at the top.

***Solution :* (Fig. 9.39)**

1. Draw the projection of the cylindrical portion using isometric dimensions and its isometric projection.
2. Locate the center O at the top of the cylindrical and draw the diagonal 5-7 equal to isometric diameter of hemisphere and the sides 5-6 and 5-8 are drawn at 30° to the horizontal and diagonal 6-8 is drawn.
3. Complete the rhombus 5-6-7-8 and the ellipse using four-center method for the bottom flat surface of the hemisphere.
4. With center O and true radius of the sphere, draw a semicircle to complete the isometric projection and show the visible portion of the solid by drawing dark lines.

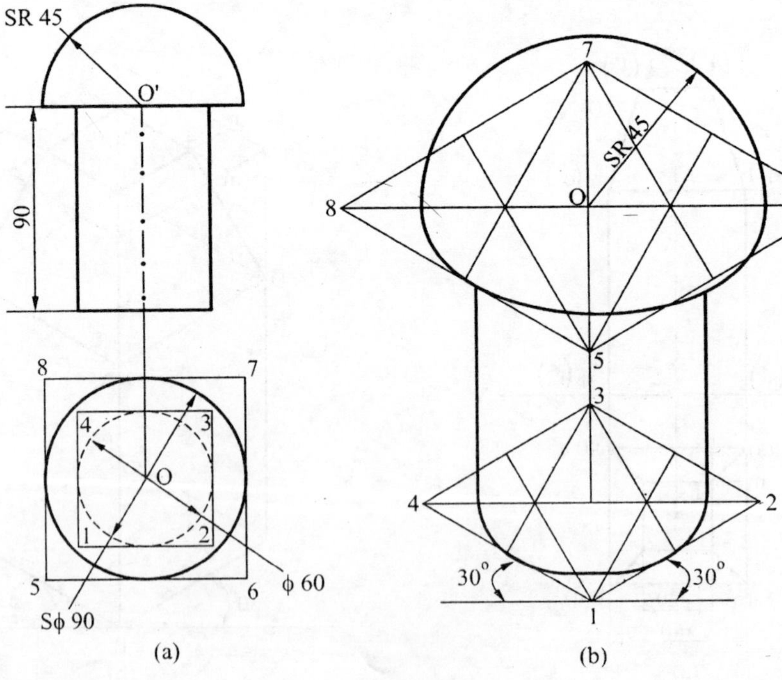

Fig. 9.39

Problem : A frustum of a cone having 25 mm as top diameter, 50 mm as bottom diameter and 50 mm axis length is placed vertically on a cylindrical block of 75 mm diameter and is 25 mm thick such that both the solids have the common axis. Draw the isometric projection of the combination of these solids.

1. Use isometric dimensions to draw isometric projections of the cylindrical slab by drawing the square box as discussed earlier, using the four-center method draw the ellipse for the bases.
2. Consider another square box to enclose the frustum of the cone and its bottom base is drawn as an ellipse using four-centre method.
3. Construct a square 9-10-11-12 to enclose the top base of the frustum of the cone which is marked in isometric projection. Using the four center method, the top base is also drawn as an ellipse.
4. Drawn tangents to the ellipses at the left and right extreme to complete the isometric projection of the cone. Similarly the cylindrical slab is also completed in isometric projection. Only the visible portion of the solids are shown.

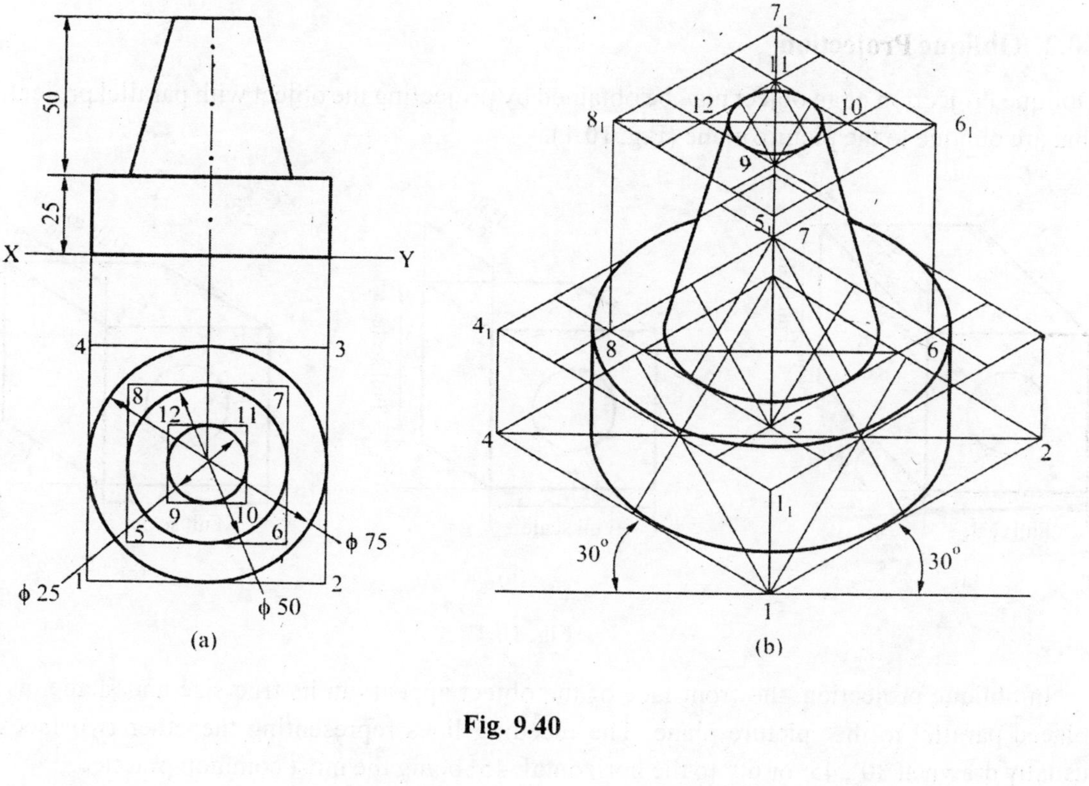

Fig. 9.40

CHAPTER 10

Oblique and Perspective Projections

10.1 Introduction

Pictorial projections are used for presenting ideas which may be easily understood by all without technical training. They show several faces of an object in one view, as it appears to the eye approximately. Among the pictorial projections, Isometric Projections are the most common as explained in previous chapter.

10.2 Oblique Projection

Oblique Projection of an object may be obtained by projecting the object with parallel projections that are oblique to the picture plane (Fig. 10.1).

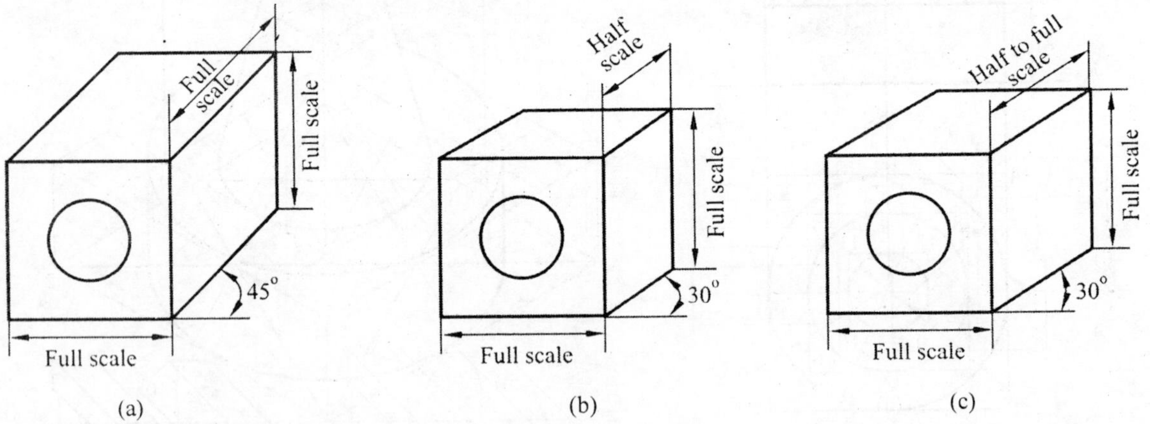

Fig. 10.1

In oblique projection, the front face of the object appears in its true size and shape, as it is placed parallel to the picture plane. The receding lines representing the other two faces are usually drawn at 30°, 45° or 60° to the horizontal, 45° being the most common practice.

As in the case of isometric projection, in oblique projection also, all lines that are parallel on the object appear parallel on the drawing and vertical lines on the object appear vertical.

10.3 Classification of Oblique Projection

Oblique projections are classified as cavalier, cabinet and general, depending on the scale of measurement followed along the receding lines, as shown in Fig 10.1. The oblique projection shown in Fig 10.1(a) presents a distorted appearance to the eye. To reduce the amount of distortion and to have a more realistic appearance, the length of the receding lines are reduced as shown, either in Fig. 10.1(b) or as in Fig 10.1(c). If the receding lines are measured to the true size, the projection is known as cavalier projection. If they are reduced to one half of their true lengths, the projection is called cabinet projection. In general, in oblique projections the measurement along the receding lines vary from half to full size.

Note: Oblique projection has the following advantages over isometric drawing:

1. Circular or irregular features on the front face appear in their true shape.
2. Distortion may be reduced by fore-shortening the measurement along the receding axis, and
3. A greater choice is permitted in the selection of the position of the axes.

10.4 Methods of Drawing Oblique Projection

The orthographic views of a V-block are shown in Fig. 10.2(a). The stages in obtaining the oblique projection of the same are shown in Fig. 10.2(b).

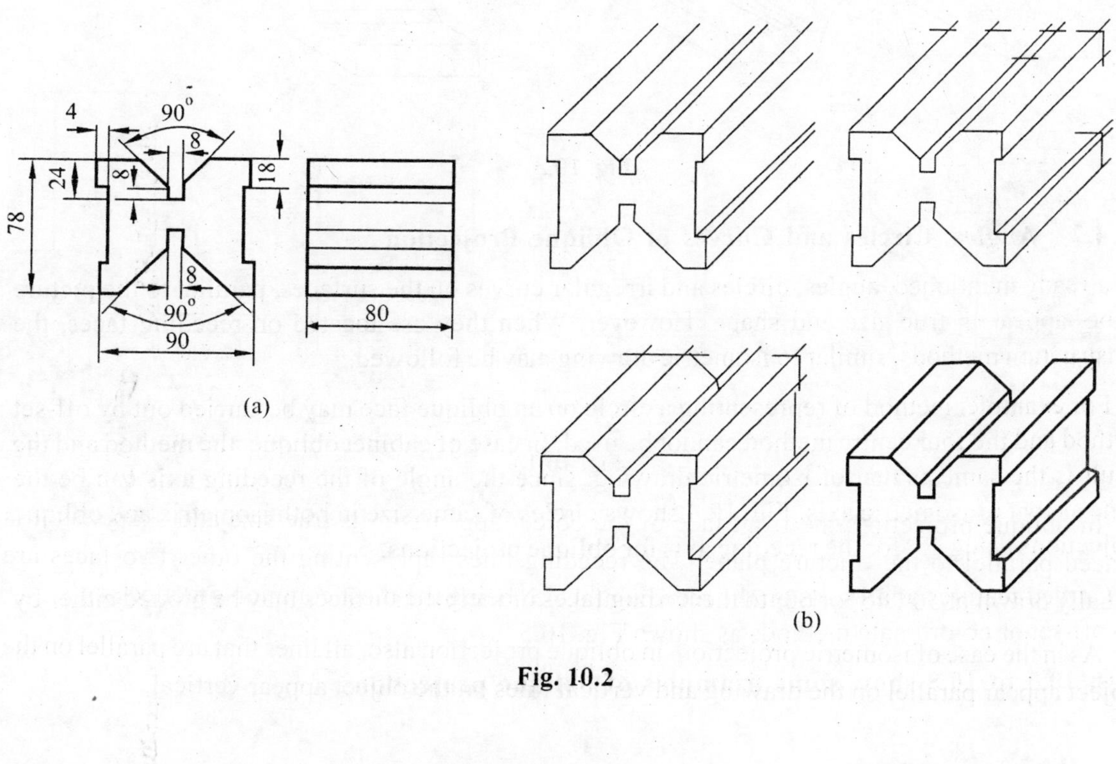

Fig. 10.2

1. After studying the views carefully, select the face that is either the most irregular one or the one with circular features if any. Make that face parallel to the picture plane to minimize distortion.
2. Draw the face to its tr'e size and shape
3. Draw the receding lines through all the visible corners of the front face.
4. Mark the length of the object along the receding lines and join these in the order.
5. Add other features if any on the top and side faces.

10.4.1 Choice of Position of the Object

For selecting the position of an object for drawing the oblique projection, the rules below are followed.

1. Place the most irregular face or the one with circular features parallel to the picture plane. This, simplifies the construction and minimizes distortion.
2. Place the longest face parallel to the picture plane. This results in a more realistic and pleasing appearance of the drawing (Fig. 10.3)

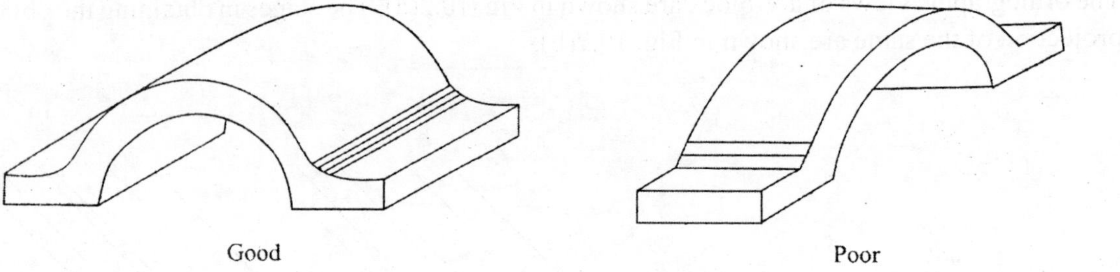

Fig. 10.3

10.4.2 Angles, Circles and Curves in Oblique Projection

As already mentioned, angles, circles and irregular curves on the surfaces, parallel to the picture plane, appear in true size and shape. However, When they are located on receding faces, the construction methods, similar to isometric drawing may be followed.

For example, method of representing a circle on an oblique face may be carried out by off-set method and the four centre method cannot be used. In case of cabinet oblique, the method and the result is the same as that of isometric drawing, since the angle of the receding axis can be the same as that of isometric axis. Fig. 10.4 shows circles of same size in both isometric and oblique projections using 45° for the receding axis for oblique projections.

Curved features of all sorts on the receding faces or inclined surfaces may be plotted either by the off-set or co-ordinate methods as shown Fig. 10.5

Figs. 10.6 to 10.8 show some examples of oblique projections.

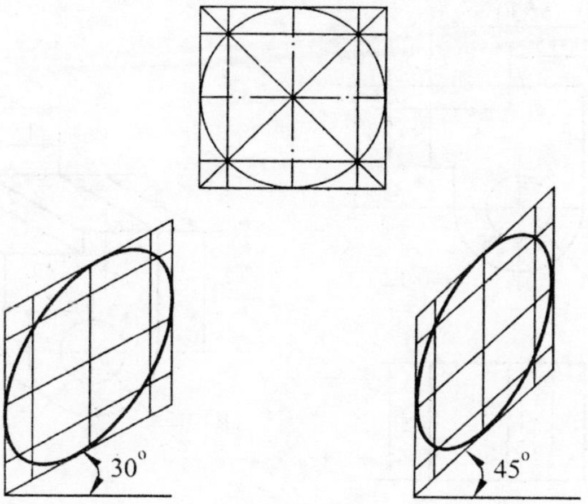

Fig. 10.4

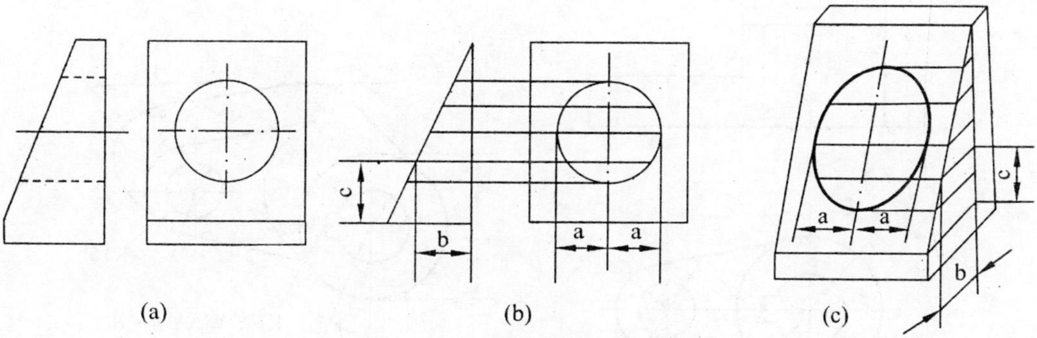

Fig. 10.5

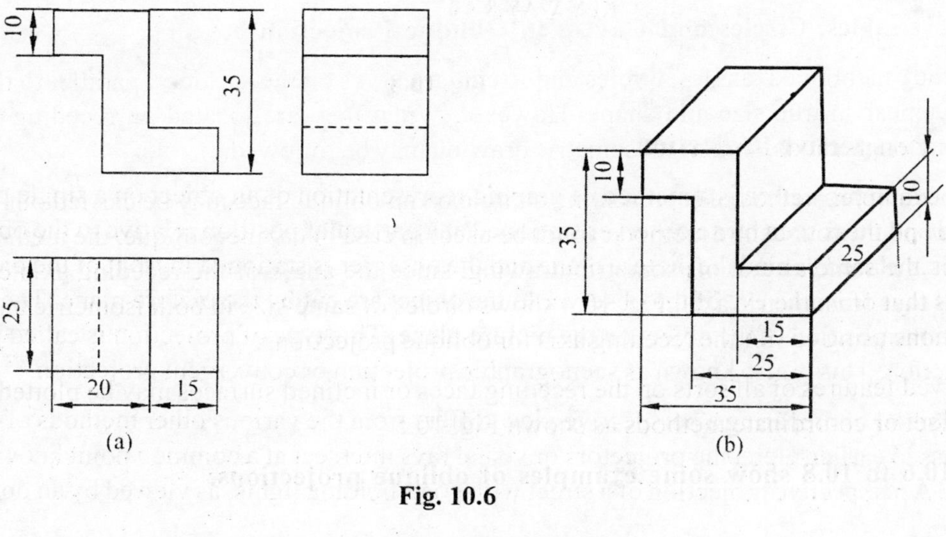

Fig. 10.6

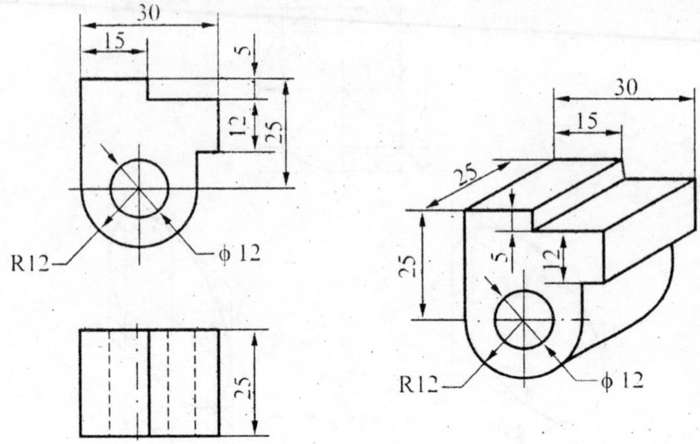

Fig. 10.7

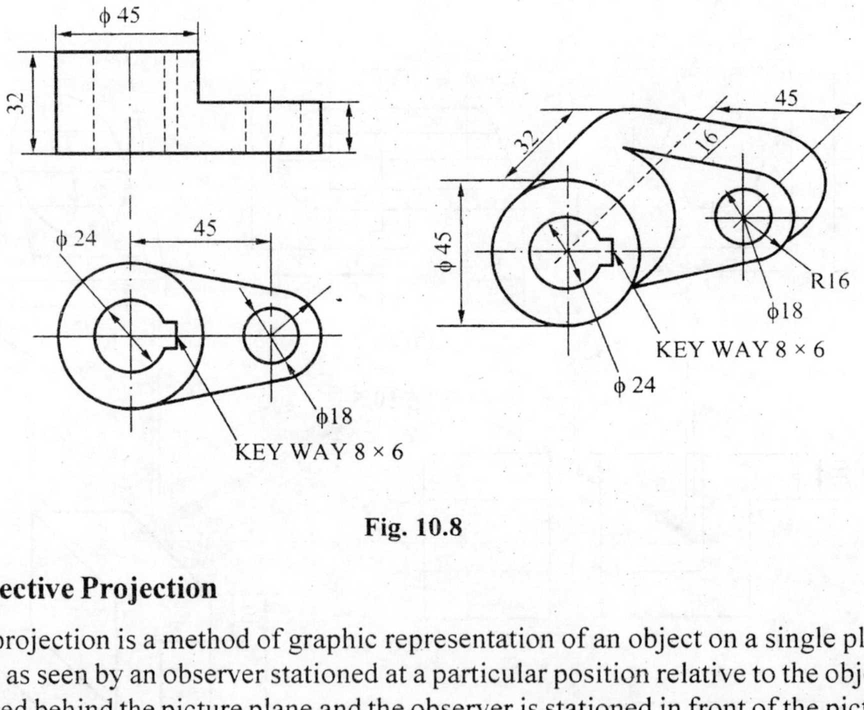

Fig. 10.8

10.5 Perspective Projection

Perspective projection is a method of graphic representation of an object on a single plane called picture plane as seen by an observer stationed at a particular position relative to the object. As the object is placed behind the picture plane and the observer is stationed in front of the picture plane, visual rays from the eye of the observer to the object are cut by the picture plane. The visual rays locate the position of the object on the picture plane. This type of projection is called perspective projection. This is also known as scenographic projection or convergent projection.

Method of preparing a perspective view differs from the various other methods of projections discussed earlier. Here, the projectors or visual rays intersect at a common point known as station point. A perspective projection of a street with posts holding lights, as viewed by an observer from a station point, is shown in Fig. 10.9. The observer sees the object through a transparent vertical plane called picture plane as shown in Fig. 10.9(a). The view obtained on the picture plane is shown in Fig. 10.9(b). In this view, the true shape and size of the street will not be seen as the object is viewed from a station point to which the visual rays converge. This method of projection is theoretically very similar to the optical system in photography and is extensively employed by architects to show the appearance of a building or by artist-draftman in the preparation of illustrations of huge machinery or equipment.

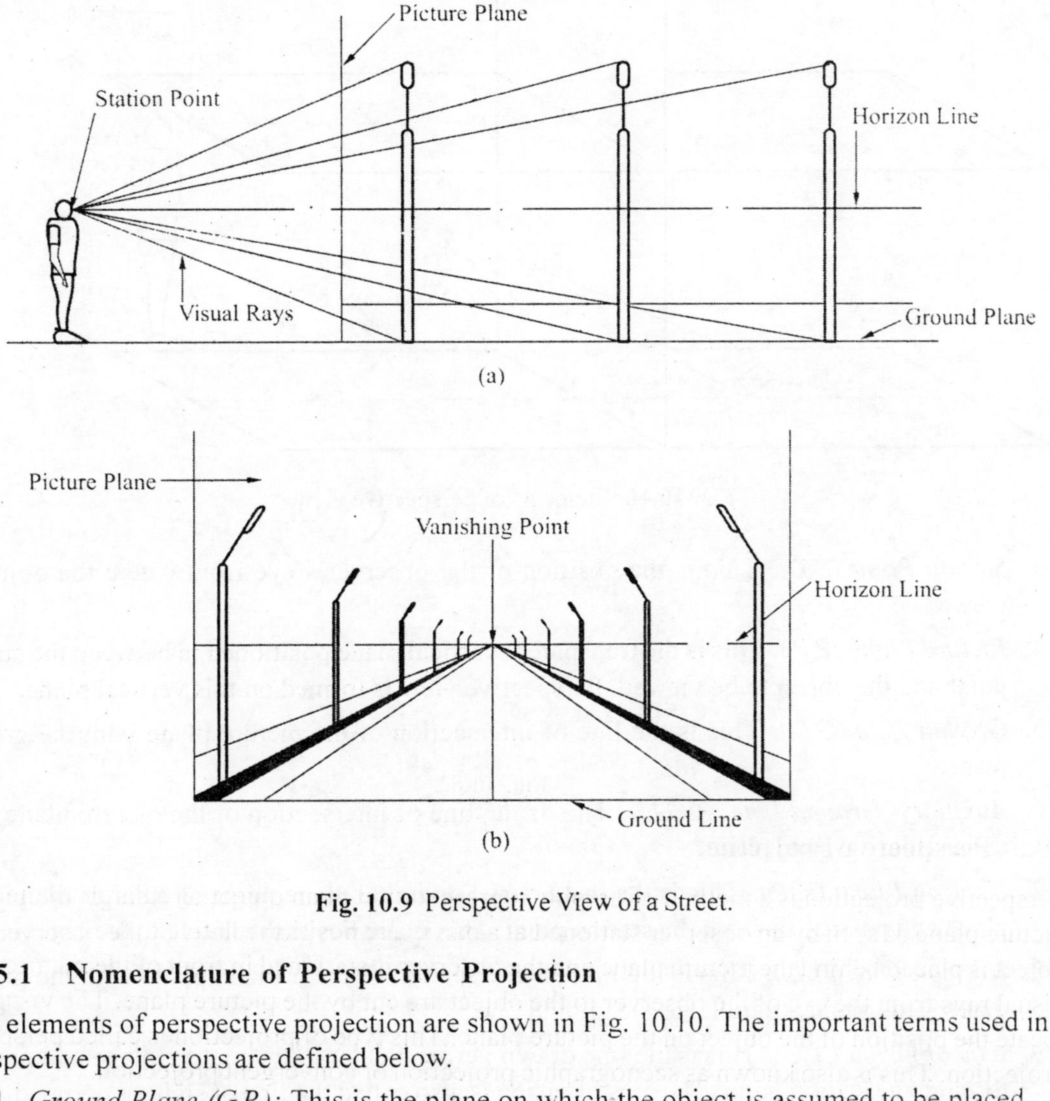

Fig. 10.9 Perspective View of a Street.

10.5.1 Nomenclature of Perspective Projection

The elements of perspective projection are shown in Fig. 10.10. The important terms used in the perspective projections are defined below.

1. *Ground Plane (G.P.):* This is the plane on which the object is assumed to be placed.
2. *Auxiliary Ground Plane (A.G.P):* This is any plane parallel to the ground plane (Not shown in Fig. 10.10).

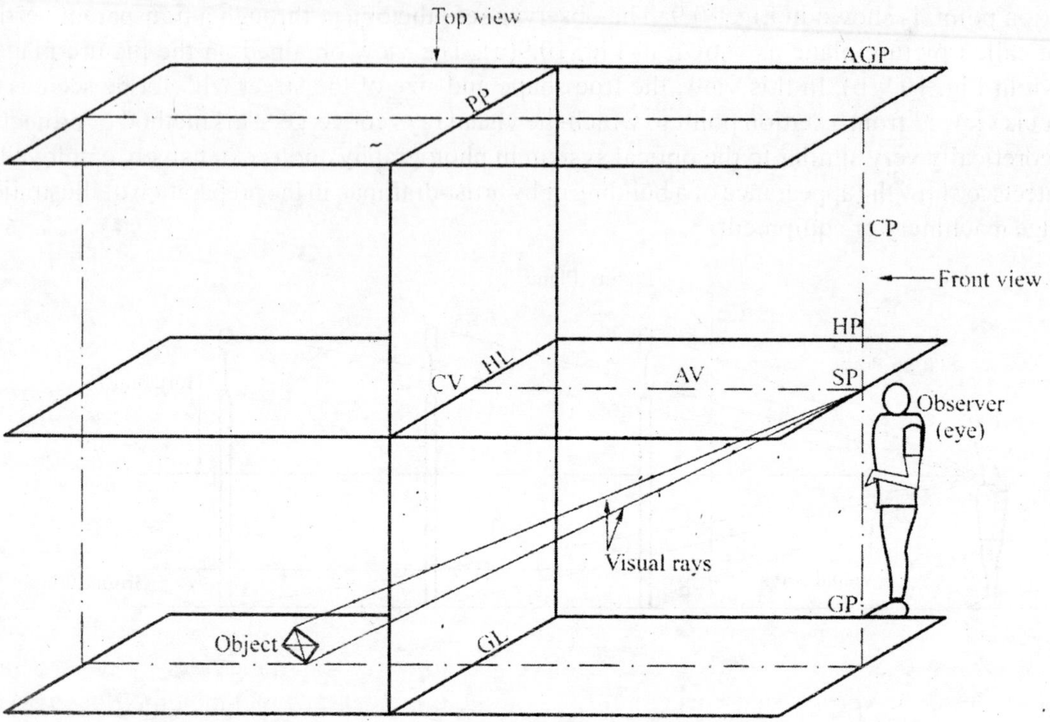

Fig. 10.10 Elements of perspective view

3. *Station Point (S.P.):* This is the position of the observer's eye from where the object is viewed.
4. *Picture Plane (P.P.):* This is the transparent vertical plane positioned in between the station point and the object to be viewed. Perspective view is formed on this vertical plane.
5. *Ground Line (G.L.):* This is the line of intersection of the picture plane with the ground plane.
6. *Auxiliary Ground Line (A.G.L.):* This is the line of intersection of the picture plane with the auxiliary ground plane.
7. *Horizon Plane (H.P.):* This is the imaginary horizontal plane perpendicular to the picture plane and passing through the station point. This plane lies at the level of the observer.
8. *Horizon Line (H.L.):* This is the line of intersection of the horizon plane with the picture plane. This plane is parallel to the ground line.
9. *Axis of Vision (A.V.):* This is the line drawn perpendicular to the picture plane and passing through the station point. The axis of vision is also called the line of sight or perpendicular axis.
10. *Centre of Vision (C.V.):* This is the point through which the axis of vision pierces the picture plane. This is also the point of intersection of horizon line with the axis of vision.

11. *Central Plane (C.P.):* This is the imaginary plane perpendicular to both the ground plane and the picture plane. It passes through the centre of vision and the station point while containing the axis of vision.
12. *Visual Rays (V.R.):* These are imaginary lines or projectors joining the station point to the various points on the object. These rays converge to a point.

10.5.2 Classification of Perspective Projections

Perspective projections can be broadly classified into three categories.

1. Parallel perspective or single point perspective.
2. Angular perspective or two point perspective.
3. Oblique perspective or three point perspective.

These perspective projections are based on the relative positions of the object with respect to the picture plane. All the three types of perspectives are shown in Fig. 10.11.

Parallel perspective or single point perspective

If the principal face of the object viewed, is parallel to the picture plane, the perspective view formed is called parallel perspective. Such a perspective view is shown in Fig. 10.11(a). In parallel perspective views, the horizontal lines receding the object converge to a single point called vanishing point (VP). But the vertical and horizontal lines on the principal face and the other faces of the object, do not converge, if these lines are parallel to the picture plane. Because the lines on the faces parallel to the picture plane do not converge to a point and the horizontal lines receding the object converge to a single vanishing point, the perspective projection obtained is called parallel or single point perspective. Single point perspective projection is generally used to present the interior details of a room, interior features of various components, etc.

Angular perspective or two-point perspective

If the two principal faces of the object viewed are inclined to the picture plane, the perspective view formed is called angular perspective. Such a perspective is shown in Fig. 10.11(b). In angular perspective views, all the horizontal lines converge to two different points called vanishing point left (V.P.L.) and vanishing point right (V.P.R.). But the vertical lines remain vertical. Because the two principal faces are inclined to picture plane and all the horizontal lines on the object converge to two different vanishing points, the perspective view obtained is called angular or two point perspective. Two point perspective projection is the most generally used to present the pictorial views of long and wide objects like buildings, structures, machines, etc.

Oblique perspective or three point perspective

If all the three mutually perpendicular principal faces of the object viewed, are inclined to the picture plane, the perspective view formed is called oblique perspective. Such a perspective view is shown in Fig. 10.11(c). In oblique perspective views, all the horizontal lines converge to two different points called vanishing point left (V.P.L.) and vanishing point right (V.P.R.) and all the vertical lines converge to a third vanishing point located either above or below the horizon line. Because all the three principal faces are inclined to the picture plane and all the horizontal and the vertical lines on the object converge to three different vanishing points, the perspective view obtained is called oblique or three point perspective.

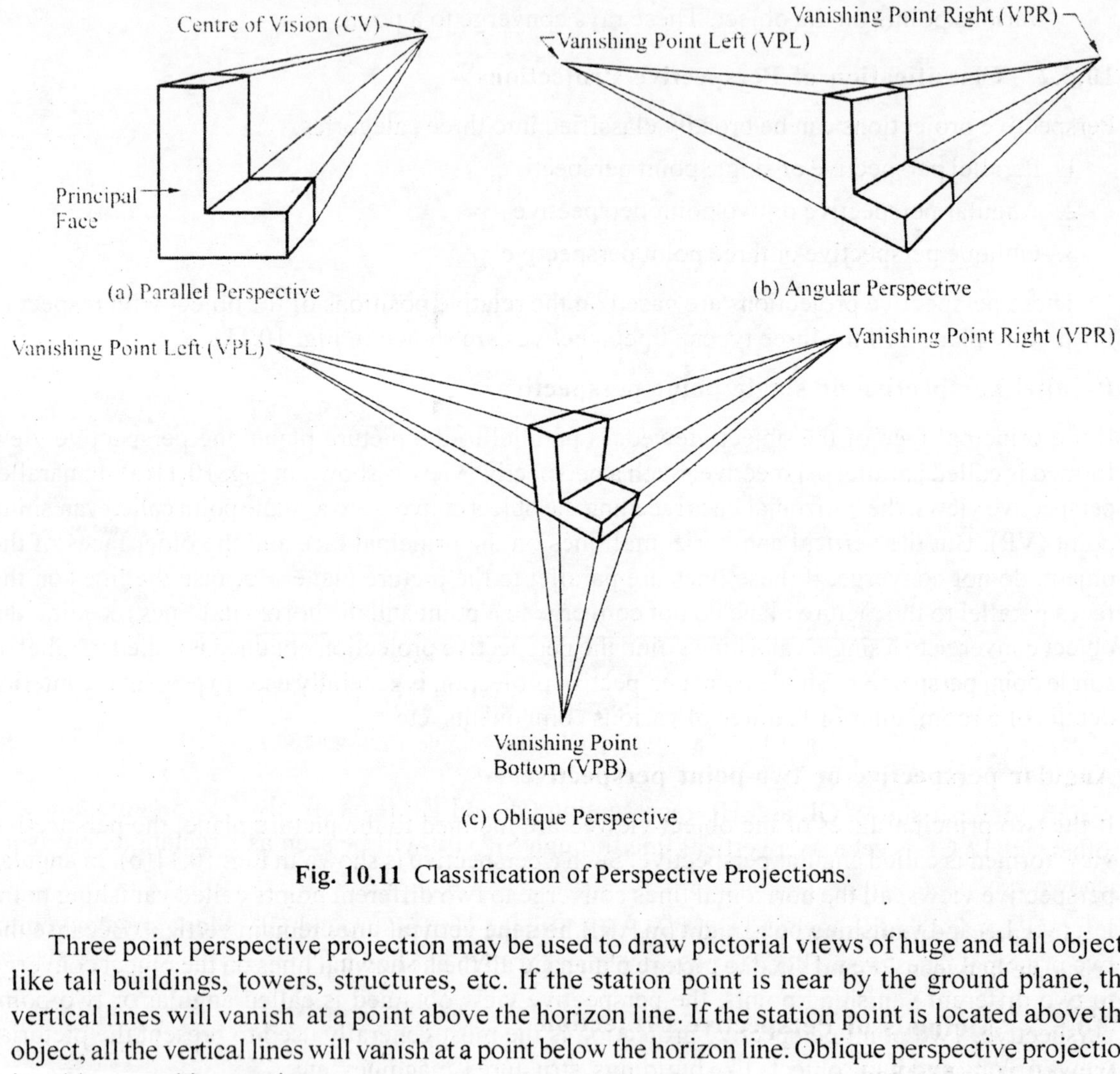

Fig. 10.11 Classification of Perspective Projections.

Three point perspective projection may be used to draw pictorial views of huge and tall objects like tall buildings, towers, structures, etc. If the station point is near by the ground plane, the vertical lines will vanish at a point above the horizon line. If the station point is located above the object, all the vertical lines will vanish at a point below the horizon line. Oblique perspective projection is seldom used in practice.

Orthographic Representation of Perspective Elements

Figure 10.12 shows orthographic views of the perspective elements in Third Angle Projection.

Top View: GP, HP and AGP will be rectangles, but are not shown. PP is seen as a horizontal line. Object is above PP. Top view SP of station point is below PP. Top view of center of vision is CV Line CV-SP represents the Perpendicular Axis CP

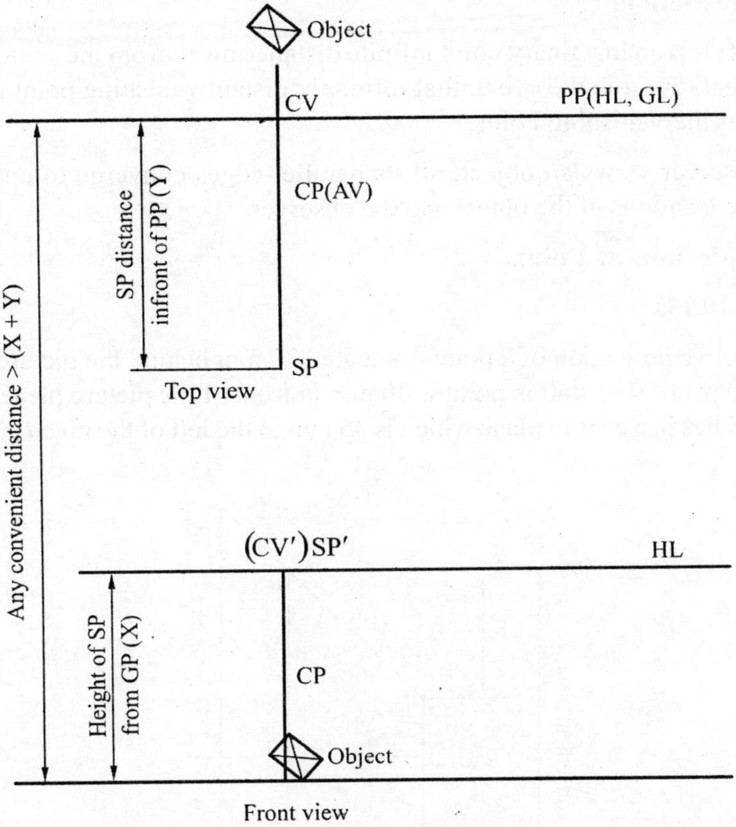

Fig. 10.12 Orthographic Representation (Third angle).

Front View: It shows GL and HL representing GP and HP respectively. CV, SP coincide each other on HL.CP is seen as a vertical line through SP[1]. PP will be seen as a rectangle, but is not shown.

Perspective projection, when drawn, will be seen above / around GL. Mark any convenient distance between PP and GL, i.e., greater than (X + Y) as shown.

10.5.3 Methods of Perspective Projection

Visual Ray Method

In this method, points on the perspective projection are obtained by drawing visual rays from SP to both top view and either front view or side view of the object. Top and side views are drawn in Third Angle Projection.

Perspective projection of a line is drawn by first marking the perspective projection of its ends (which are points) and then joining them. Perspective projection of a solid is drawn by first obtaining the perspective projection of each corner and then joining them in correct sequence.

Vanishing Point Method

Vanishing Point: It is an imaginary point infinite distance away from the station point. The point at which the visual ray from the eye to that infinitely distant vanishing point pierces the picture plane is termed as the Vanishing Point.

When the observer views an object, all its parallel edges converge to one/two/three points depending on the locations of the object and the observer.

Perspective Projection of Points

Problem : (Fig. 10.13)

Draw the perspective projection of a point A situated 20 mm behind the picture plane and 15 mm above the ground plane. The station point is 30 mm in front of the picture plane, 40 mm above the ground plane and lies in a central plane which is 35 mm to the left of the given pint.

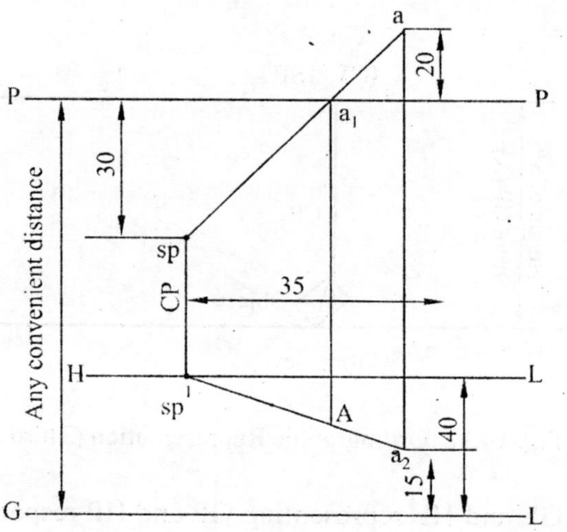

Fig. 10.13

Visual Ray Method

Top view

1. Draw a horizontal line pp to represent the top view of the picture plane.
2. The point A is 20 mm behind PP. Hence mark 'a' 20 mm above pp.
3. Station point SP lies in a central plane CP which is 35 mm to the left of point A. Therefore, draw a vertical line to represent the top of CP at 35 mm to the left of a.
4. SP is 30 mm in front of PP. Therefore on CP, mark SP 30mm below PP.
5. Join a and SP to represent the top view of the visual ray. It pierces the PP at a_1.

Front view

6. Draw a horizontal line GL at any convenient distance below PP to represent the ground line.
7. To avoid over lap of visual rays and get a clear perspective, select GL such that HL lies below sp'
8. sp' is 40 mm above GP. Therefore draw HL 40 mm above GL.
9. Further CP also represents front view of the CP. Hence mark sp' at the intersection of CP with HL.
10. Join a_2 sp', the front view of the visual ray.
11. From the piercing point a_1 erect vertical to intersect a_2 Sp' point at A, which is the required perspective projection of point A.

Perspective Projection of Straight Lines

In Visual Ray Method, perspective projection of a straight line is drawn by first marking the perspectives of its end points and then joining them.

Problem : (Fig 10.14)

Draw the perspective projection of a straight line AB 60 mm long, parallel to and 10 mm above the ground plane and inclined at 45° to PP. The end A is 20 mm behind the picture plane. Station point is 35 mm in front of the picture plane and 45 mm above the ground plane and lies in a central plane passing through the mid-point of AB.

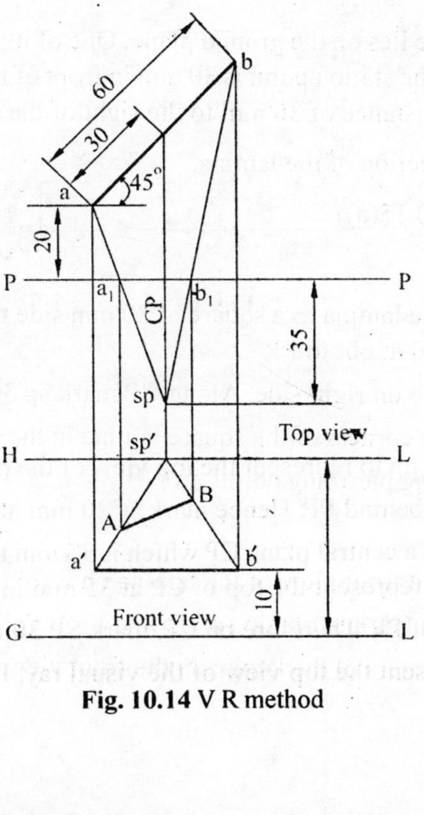

Fig. 10.14 V R method

Top View

1. Draw PP and mark a 20 mm above it.
2. Draw ab = 60 mm (True length of AB) inclined at 45° to PP
3. From the mid-point of ab erect a vertical line to represent the CP.
4. Along the central plane mark SP 35 mm below pp.
5. Join a and b with SP to represent the top view of the visual rays.
6. Mark the piercing points a and b on asp and bsp respectively.

Front View

7. Draw GL at any convenient distance below PP.
8. Draw a' b' parallel to and 10 mm above GL.
9. Draw HL 45 mm above GL.
10. Mark sp' at the intersection of CP & HL.
11. Join SP' with a' and b',
12. From a_1 and b_1 (piercing points) erect verticals to intersect sp' a' and sp' b' (the front view of the visual rays) at A and B respectively.
13. AB is the required perspective projection.

Perspective Projection of Plane Figures

Problem : (Fig. 10.15)

A square lamina of 30 mm side lies on the ground plane. One of its corners is touching the PP and edge is inclined at 60° to PP. The station point is 30 mm in front of PP, 45 mm above GP and lies in a central plane which is at a distance of 30 mm to the right of the corner touching the PP.

Draw the perspective projection of the lamina.

Visual Ray Method: (Fig. 10.15(a))

Top View

1. Draw the top view of the lamina as a square of 30 mm side that the corner b is touching PP and the edge bc inclined at 60° to PP.
2. Draw CP, 30 mm from b on right side. Along CP mark sp 30 mm below PP.
3. Join sp with all the four corners of the square lamina in the top view.
4. Obtain the corresponding pierrcing points on PP.

Front View

5. Draw GL and obtain the front view of the lamina on it ($a^1d^1b^1c^1$).
6. Draw HL 45 mm above GL and obtain sp' on it.
7. Joint sp' with all the corners of the lamina in the front view.

Perspective Projection

8. Since the corner b touches the picture plane, its perspective will be in its true position.
9. Since the lamina lies on the ground plane, b' is on GL and is also the perspective projection of B.
10. From a_1 draw vertical to intersect a' sp' at A.
11. Similarly obtain B, C and D.
12. Joint ABCD and complete the perspective projection.

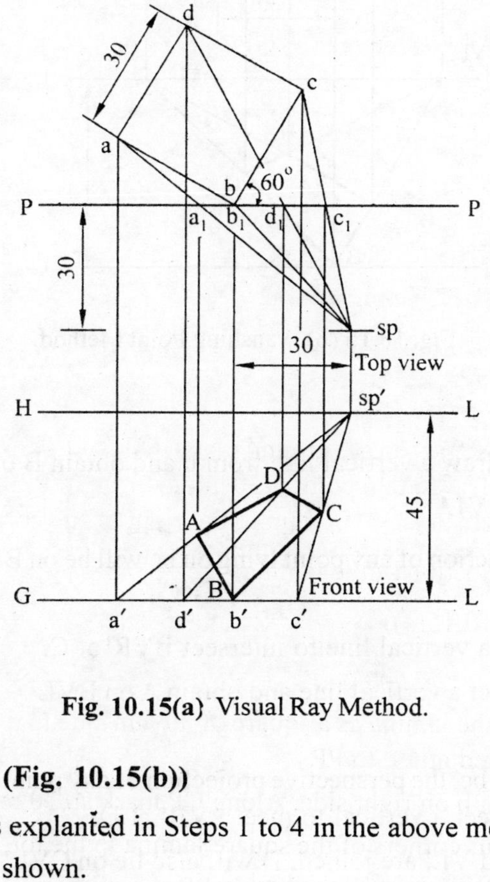

Fig. 10.15(a) Visual Ray Method.

Vanishing Point Method (Fig. 10.15(b))

1. Draw the top view as explanted in Steps 1 to 4 in the above method.
2. Draw GL and HL as shown.

Vanishing Points

3. From sp draw a line parallel to bc to intersect PP at VR
4. Erect vertical from VR to intersect HL at vanishing point VR'
5. Similarly from sp draw a line parallel to ba to cut PP at VL.
6. Erect vertical from VL and obtain the other vanishing point VL' on HL.

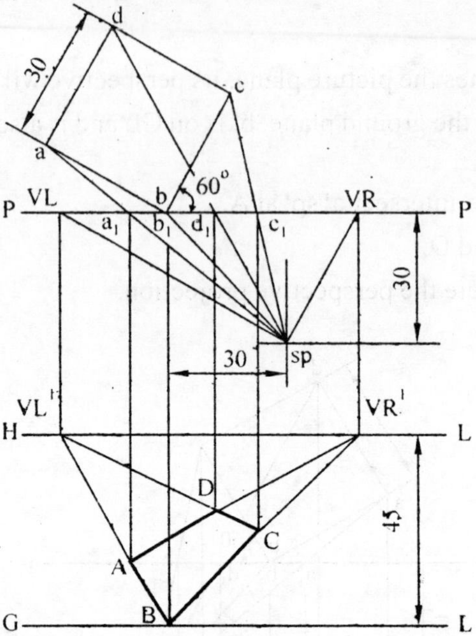

Fig. 10.15 (b) Vanshing Point Method.

Perspective Projection

7. Since b touches PP, draw a vertical line from b and obtain B on GL.
8. Join B with VR' and VL'

Note: The perspective projection of any point lying on bc will be on BVR' and any point on ba will be on BVL'.

9. Hence from c_1 erect a vertical line to intersect BVR' at C.
10. Similarly from a_1 erect a vertical line and obtain A on BVL^1.
11. Joint A with VR'.
12. Since ad is parallel to bc, the perspective projection of any point lying on ad will lie on AVR'. Therefore from d_1 erect a vertical to meet AVR' at D.
13. Note that when C and VL' are joined, D will also lie on CVL'.
14. Join ABCD and complete the perspective.

Problem : (Fig. 10.16)

A pentagonal lamina of 40 mm side lies on the ground. The corner which is nearest to PP is 15 mm behind it and an edge containing that corner is making 45° with PP. The station point is 40 mm in front of PP, 50 mm above GP and lies in a central plane which is at a distance of 70 mm to the left of the corner nearest to the PP.

Draw the perspective projection of the lamina.

Perspective projection is drawn by Visual Ray Method using top and front views

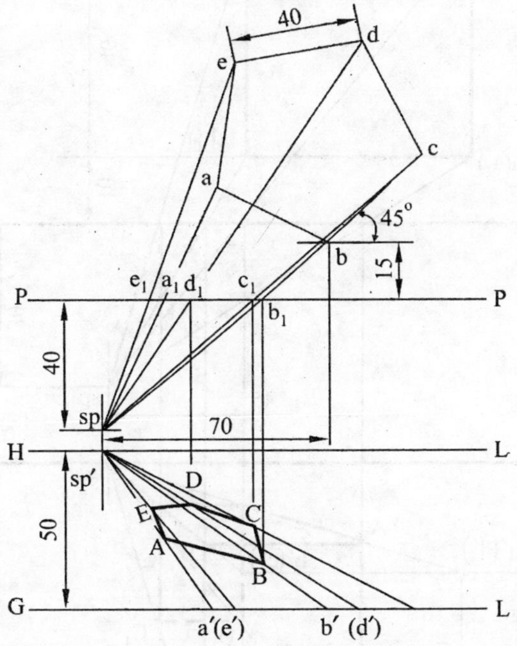

Fig. 10.16 Visual Ray Method.

Perspective Projection of Solids

Problem : (Fig. 10.17)

A square prism, side of base 40 mm and height 60 mm rests with its base on the ground such that one of its rectangular faces is parallel to and 10 mm behind the picture plane. The station point is 30 mm in front of PP, 80 mm above the ground plane and lies in a central plane 45 mm to the right of the centre of the prism.

Draw the perspective projection of the square prism.

Visual Ray Method (Fig. 10.17)

Top View

1. Draw the top view of the prism as a square of side 40 mm such that ab is parallel to and 10 mm above PP.
2. Locate sp and draw the top view of the visual rays.
3. Mark the piercing points.

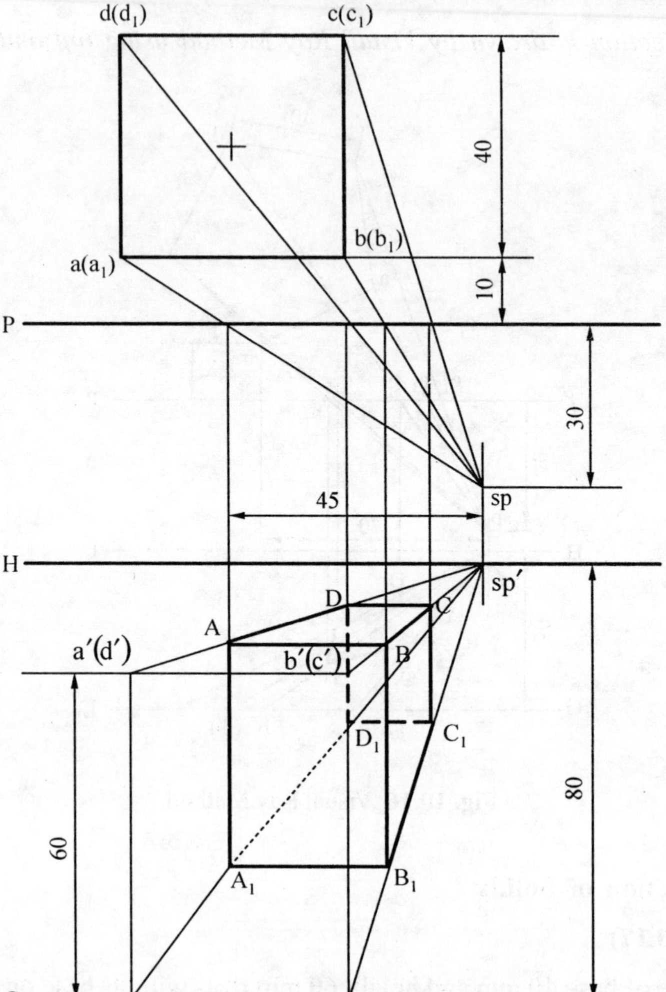

Fig. 10.17 Visual Ray Method.

Front View

4. Draw front view of the prism for given position.
5. Locate sp' and draw front view of the visual rays.
6. From piercing points erect vertical lines to cut the corresponding visual rays in the front view. Thus obtain all corners in the perspective projection.

To mark the visible and invisible edges in the perspective

7. Draw the boundary lines as thick lines.
8. The faces ab (b_1) (a_1) and bc (c_1) (b_1) are nearer to sp and visible. Hence draw BB_1, BA and BC as thick lines.
9. Edge d(d_1) is farther away from sp. Hence draw DD_1, D_1A_1 and D_1C_1 as dashed lines.

Problem : A square prism 30 mm side and 50 mm long is lying on the ground plane on one of its rectangular faces in such a way that one of its square faces is parallel to 10 mm behind the picture plane. The station point is located 50 mm in front of the picture plane and 40 mm above the ground plane. The central plane is 45mm away from the axis of the prism towards the left. Draw the perspective view of the prism.

The perspective view shown in Fig. 10.18 is developed from the top and front views.

1. Draw the top view of the picture plane (P.P.) and mark the ground line (G.L.) at a convenient distance from the line P.P. Draw the horizon line (H.L.) at a distance of 40 mm above the G.L.

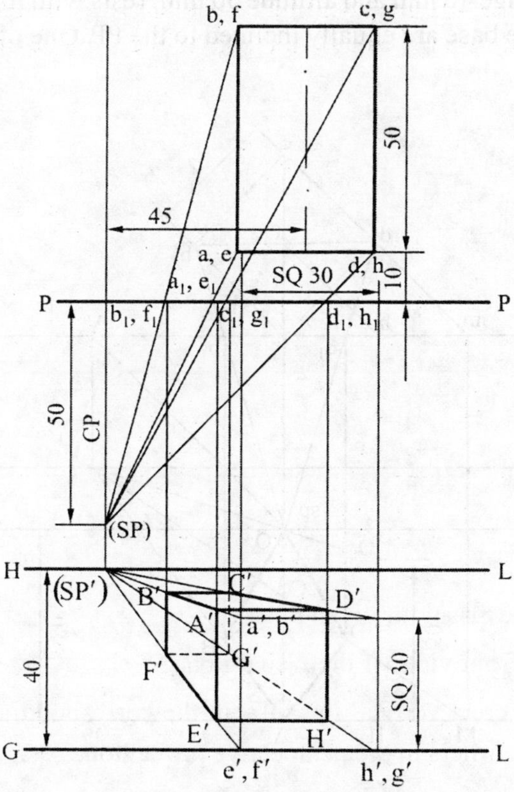

Fig. 10.18 Visual Ray Method of Parallel Perspective (from the top and front views).

2. Draw the top view of the square prism keeping the face adhe parallel to and 10 mm behind the P.P. Mark the central plane (C.P.) 45 mm away from the axis of the prism towards the left side. Locate the top view of the station point (S.P.) at a distance of 50 mm infornt of the P.P. and on C.P. Also mark the front view of the station point (S.P.) on the H.L.
3. Draw visual rays from (S.P.) to the various corners of the top view of the prism, piercing the P.P. at a_1, b_1, c_1, etc
4. Draw the front view of the prism a'd'h'e' on the G.L. and visual rays (V.R.) from (S.P.) to all corners of the front view.
5. Draw vertical lines from the points a_1, b_1, c_1, etc. to intersect the corresponding visual rays drawn from a', b', c', etc. from the front view to get the points A' B'C', etc. Join the points to get the required perspective.

Note: If the hidden edges are to be shown, they should be represented by short dashes. In the figure, F'G', C'G' and G'H' are hidden. If square faces of an object are parallel to P.P., in the perspective view these square faces will also be square but of reduced dimensions.

Problem : (Fig. 10.19)

A square pyramid of base edge 40 mm and altitude 50 mm, rests with its base on the ground plane such that all the edges of the base are equally inclined to the PP. One of the corners of the base is

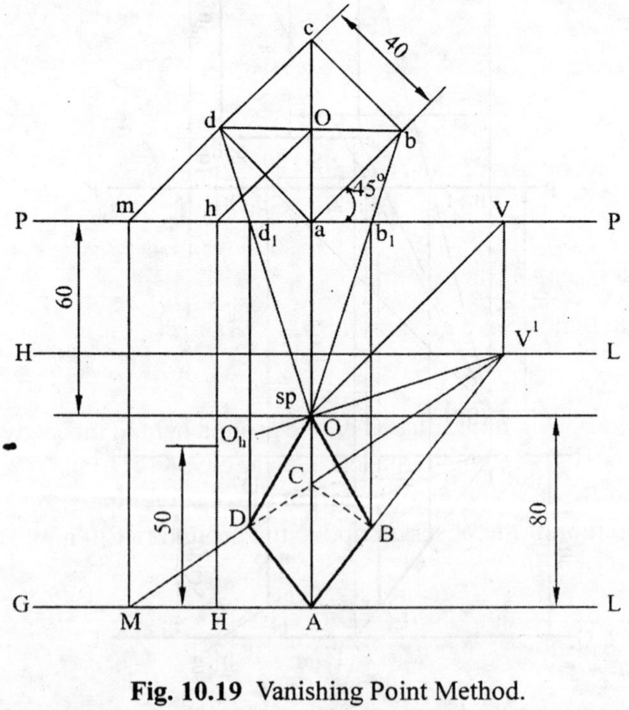

Fig. 10.19 Vanishing Point Method.

touching the PP. The station point is 60 mm in front of the PP, 80 mm above the ground plane and lies in a central plane which passes through the axis of the pyramid.

Draw the perspective projection.

Vanishing Piont Method

1. Draw the top view of the square pyramid and the visual rays.
2. From sp draw a line parallel to ab.
3. Obtain the vanishing point V on PP.

To obtain the perspective projection

4. Corner A is touching the PP and on the ground. Hence erect a vertical line from a and mark the perspective of A on GL.
5. Join A with V'.
6. As edge AB is on the ground, obtain the perspective of B on AV'.
7. To obtain the perspective or D or C extend cd to meet PP at m.
8. Draw the measuring line Mm.
9. Since the edge DC is on the ground, joint M with V'.
10. Obtain the perspective of D and C on this line M V'.
11. To mark the perspective of apex O, draw a line parallel to ab and passing through o to meet the picture plane at h.
12. Draw another measuring line Hh.
13. On this line mark the height of the apex as $O_hH = 50$ mm.
14. Join O_h with V'.
15. Obtain the perspective of O on this line.
16. Then complete the perspective as shown.

Problem

A rectangular prism 80 × 60 × 30 mm is placed on the ground behind the *PP* with the longest edges vertical and the shortest edges receding to the left at an angle of 40° to the *PP*. The nearest vertical edge is 10 mm behind the *PP* and 15 mm to the left of the observer who is at a distance of 150 mm in front of the *PP*. The height of the observer above the ground is 120 mm. Draw the perspective view of the prism.

Solution **(Fig. 10.20)**

To draw the perspective view

1. Draw the *PP* line and prism in top view and mark the station point.

2. Locate the vanishing points as discussed earlier and draw the visual rays joining various corners and *SP* in top view to mark the piercing points with *PP*.
3. Draw *GL* and *HL*, then project vanishing points V_L and V_R from top view to get V_L' and V_R' in front view.
4. The nearest vertical edge *ap* is behind *PP*, so the perspective of it on *PP* will be smaller than the true height of it. To mark the perspective of *ap*, produce it towards *PP* and from the meeting point of it with *PP*, draw a vertical line and mark a_2p_2 as the true height of *AP* which is 80 mm.
5. Join a_2 with V_L' and P_2 with V_L', project a_1 to get A' on a_2-V_L' and p_1 to get P' on P_2-V_L'.
6. Similarly get the perspective of other corners as discussed earlier and complete the perspective projection by showing the visible and invisible edges of the prism.

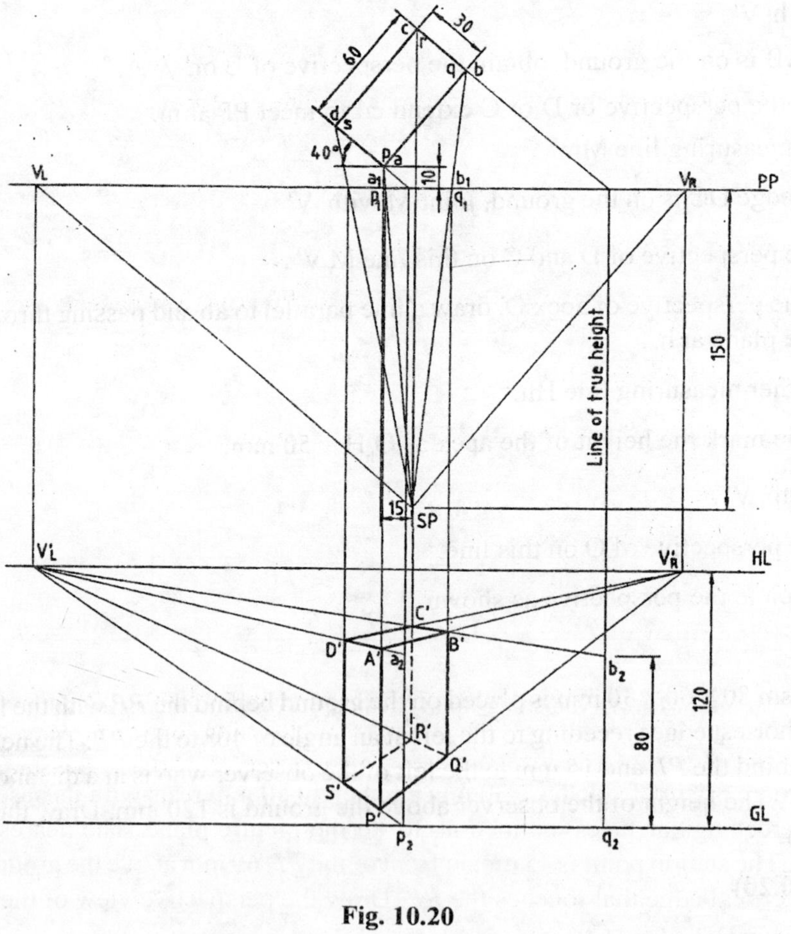

Fig. 10.20

Problem : Figure 10.21(a) shows isometric view of an object. Draw the angular perspective of it when the object is resting on the ground plane keeping the face F inclined 30° to and the edge QR 20 mm behind the picture plane. The station point is 120 mm in front of the picture plane, 80 mm above the ground plane and lies in the central plane which passes through the edge QR.

Solution: (Fig 10.21)

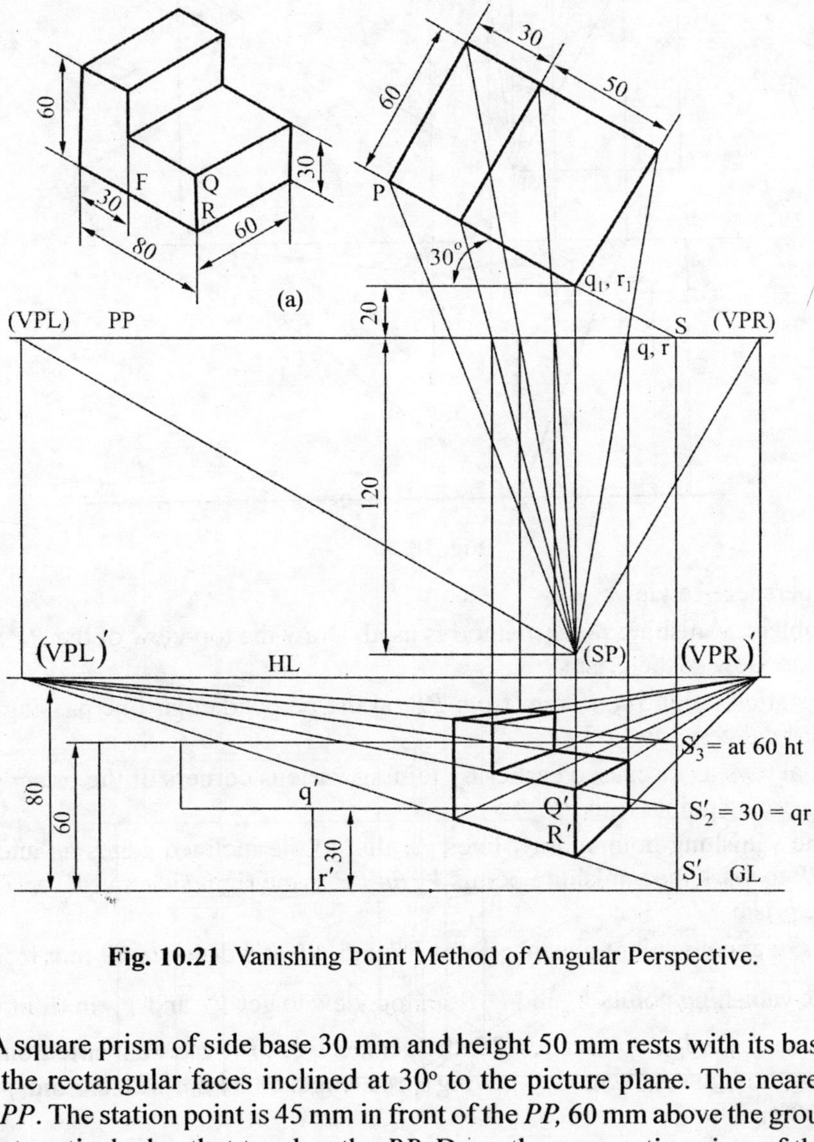

Fig. 10.21 Vanishing Point Method of Angular Perspective.

Problem: A square prism of side base 30 mm and height 50 mm rests with its base on the ground and one of the rectangular faces inclined at 30° to the picture plane. The nearest vertical edge touches the *PP*. The station point is 45 mm in front of the *PP*, 60 mm above the ground and opposite to the nearest vertical edge that touches the *PP*. Draw the perspective view of the prism.

Solution (Fig. 10.22)

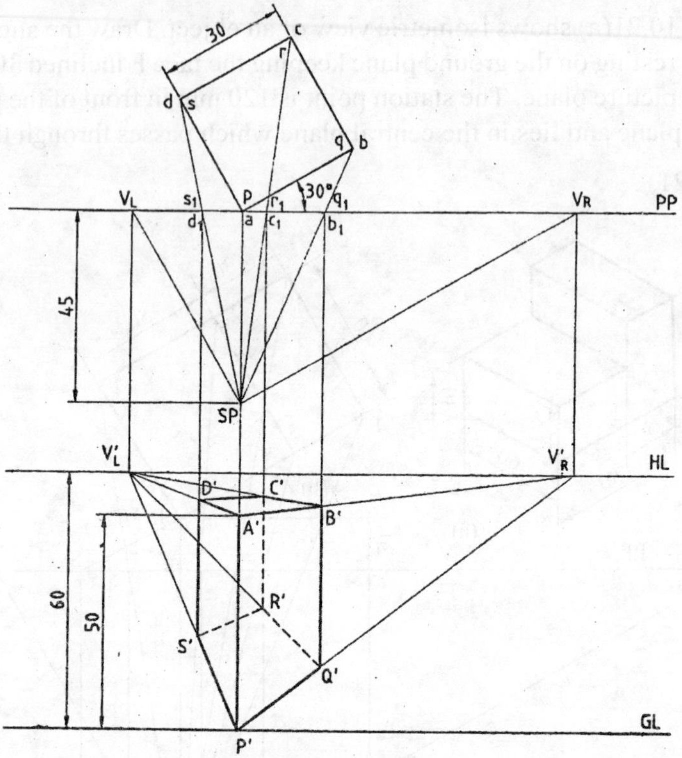

Fig. 10.22

To draw the perspective view

1. In this problem, vanishing point method is used. Draw the top view of the *PP* as a line and draw the top view of the prism.
2. Mark the station point *SP*, 45 mm from *PP* on the perpendicular line passing through the nearest vertical edge *ap*.
3. Draw visual rays as discussed earlier by joining various corners of the prism with *SP* and mark the piercing points with *PP*.
4. To find the vanishing points, draw lines parallel to the inclined edges *ap* and *ad* passing through *SP* to mark the vanishing points V_R *on PP* at the right side and V_L on *PP* at the left side of the prism.
5. Draw the *GL* at convenient distance from *PP* and *HL* at a distance 60 mm from *GL*.
6. Project the vanishing points V_R and V_L from top view to get V_R' and V_L' in front view on *HL*.
7. Since the vertical edge *ap* touches *PP*, the perspective of it can be marked on the vertical line passing through it in front view having a true length of 50 mm with the end *P'* on *GL* and *P' A'* = 50 mm.
8. Join *A'* with V_R', since the edge *ab* recedes towards V_R, by projecting the piercing point b_1 on to it, mark the perspective of *B'*.

9. Similarly by joining the perspective of the existing corners with the vanishing points V_L' or V_R' considering the edges, contains it recedes towards that vanishing point, mark the perspective of other corners as discussed in example 12.5.

10. Join A', B', etc. showing the visible and invisible edges to complete the perspective view of the prism.

Problem: A square prism of size 30 × 30 × 60 mm rests with the rectangular face on the ground. The nearest corner is 10 mm to the left of the central plane and 20 mm behind picture plane. Long edges are inclined at 45° to the picture plane and vanishes to the left. The station point is 75 mm above ground and 65 mm from picture plane. Draw the perspective projection.

Solution (Fig. 10.23)

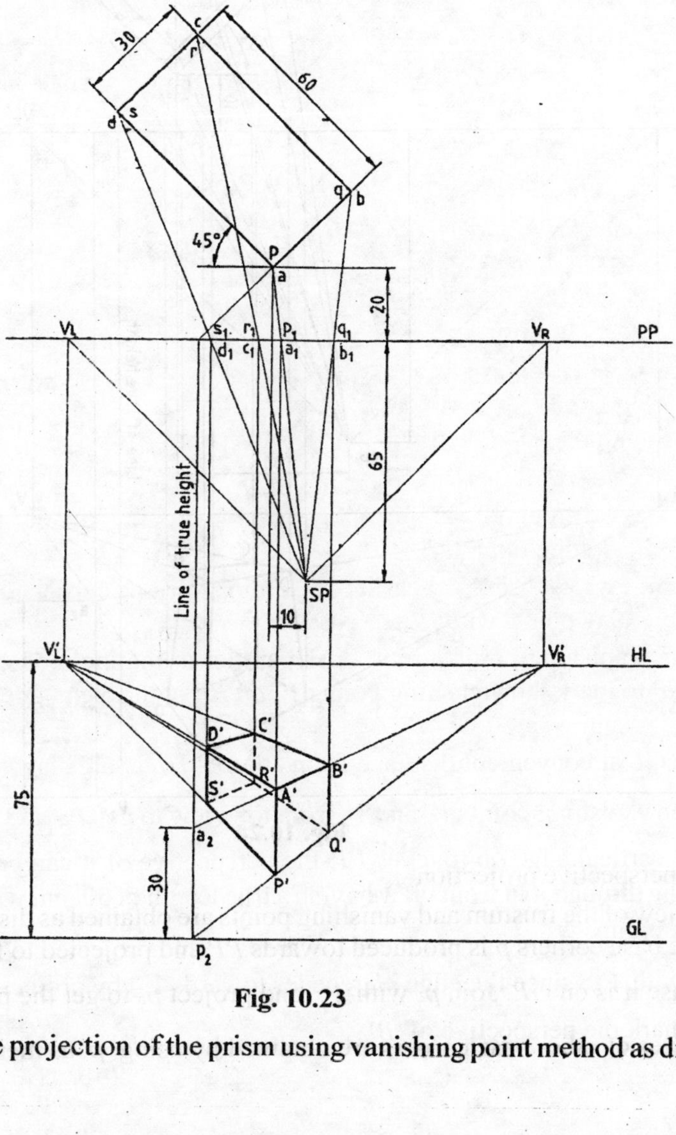

Fig. 10.23

Draw the perspective projection of the prism using vanishing point method as discussed earlier.

Problem A frustum of a square pyramid of base edge 26 mm and top edge 20 mm. The height of the frustum is 35 mm. It rests on its base on the ground, with the base edges equally inclined to *PP*. The axis of the frustum is 30 mm to the right of the eye. The eye is 55 mm in front of *PP* and 50 mm above the ground. The nearest base corner is 10 mm behind the *PP*. Draw the perspective projection of the frustum.

Solution (Fig. 10.24)

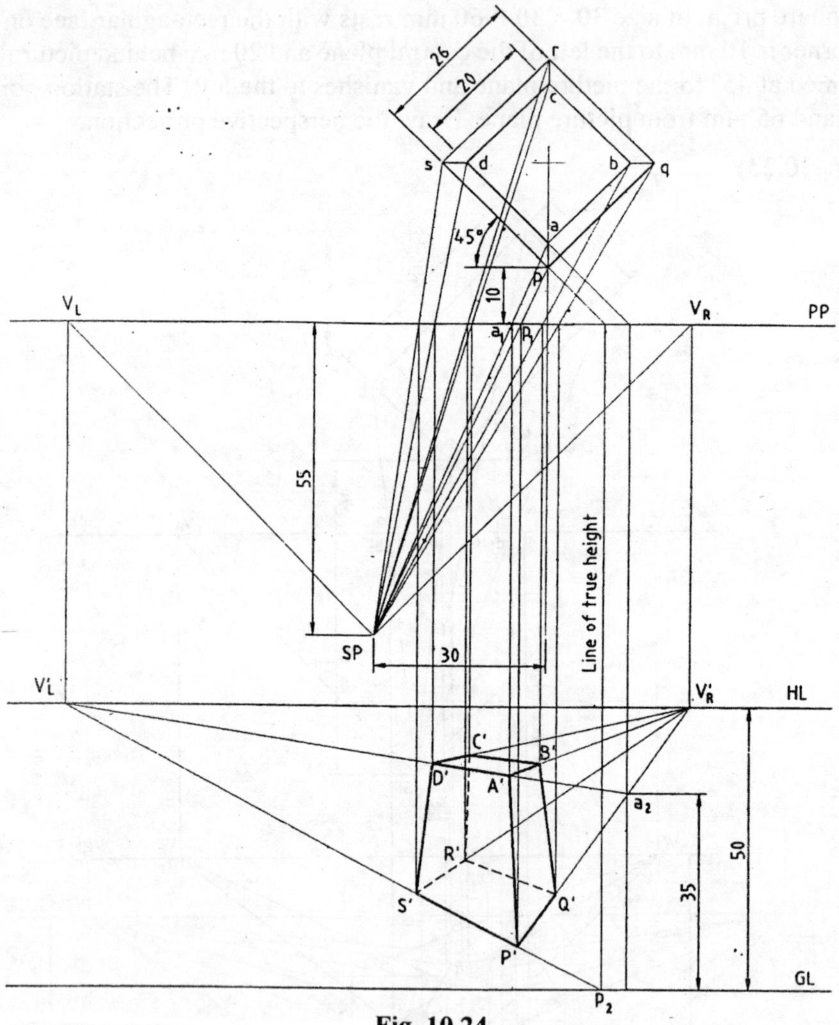

Fig. 10.24

To draw the perspective projection

1. The top view of the frustum and vanishing points are obtained as discussed earlier. Note that one of the base corners *p* is produced towards *PP* and projected to front view to mark p_2 on *GL* because it is on *GP*. Join p_2 with V_L' and project p_1 to get the perspective p'.
2. Perspectives of other base corners are obtained by using p' as mentioned earlier.

3. One of the top base corners a is produced towards PP and projected to front view to mark a_2 at height of 35 mm from GL.
4. Join a_2 with V_L' and project a_1 to get the perspective A'.
5. The perspective of other corners are obtained by using A' and P' as discussed earlier.
6. Draw the visible and invisible base edges and slant edges to complete the perspective projection of the frustum of the square pyramid.

Problem: A solid is in the form of a square prism of side of base 40 mm upto a height of 50 mm and thereafter tapers into frustum of a square pyramid whose top surface of 25 mm side. The total height of the solid is 70 mm. Draw the solid in perspective, given that one side of the base of the solid resting on the ground is inclined at 25° to the PP and the corner containing that side is 40 mm to the right of the eye and is touching the PP. The eye is 100 mm from PP and 90 mm above the ground.

Solution (Fig. 10.25)

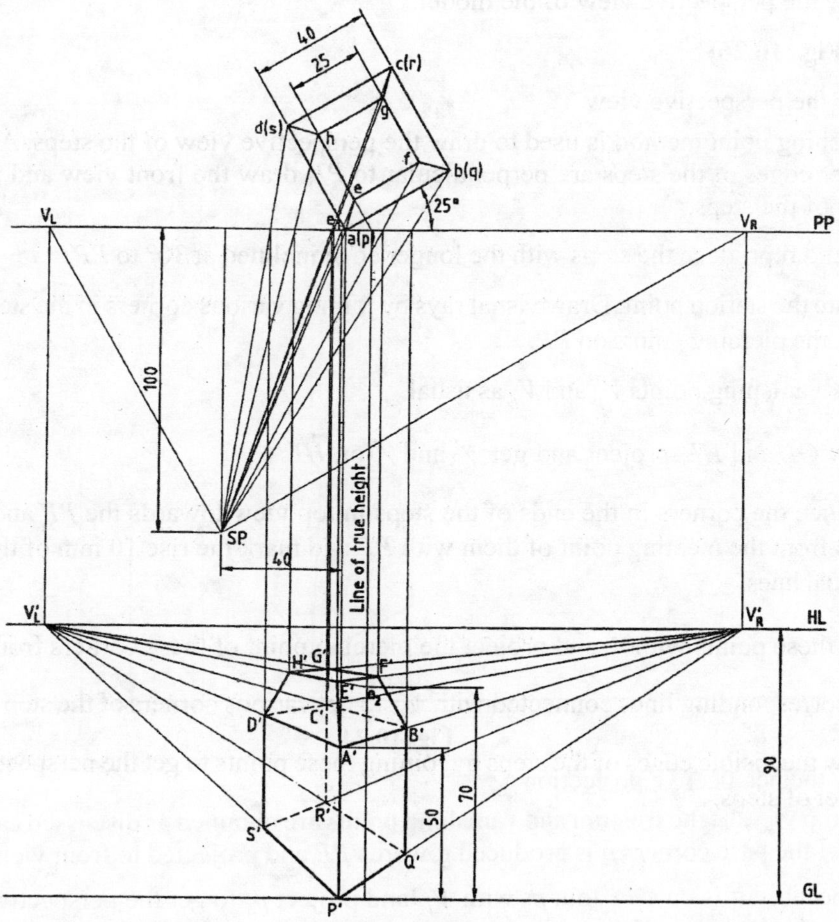

Fig. 10.25

To draw the perspective projection

1. Draw the perspective projection of the solid using vanishing point method as explained earlier.
2. Note that the top base corner of the frustum is produced towards *PP* and a vertical line is drawn from the meeting point of it with *PP* and the true height of it 70 mm is marked from *GL*.
3. Then e_2 is connected with V_L' and e_1 is projected to get the perspective E' on $e_2 - V_L'$.
4. Similarly other corners of the frustum are obtained as discussed earlier. Complete the perspective projection by showing the visible and invisible edges of the solid.

Problem: A model of steps has 3 steps of 15 mm tread and rise 10 mm. The steps measure 60 mm wide. The vertical edge of bottom step which is nearer to the picture plane is 25 mm behind *PP* and the width of steps recede to the left at an angle of 30° to *PP*. The station point is 100 mm in front of *PP* and 60 mm above the ground plane and 30 mm to the right of the vertical edge which is nearest to *PP*. Draw the perspective view of the model.

Solution (Fig. 10.26)

To draw the perspective view

1. Vanishing point method is used to draw the perspective view of the steps. Assume that the longer edges of the steps are perpendicular to *PP*, draw the front view and project the top view of the steps.
2. Tilt and reproduce the steps with the longer edge inclined at 30° to *PP*.
3. Locate the station point. Draw visual rays by joining various corners in the steps with *SP* and mark the piercing points on *PP*.
4. Mark vanishing points V_R and V_L as usual.
5. Draw *GL* and *HL*, project and get V_R' and V_L' on *HL*.
6. Produce the corners in the ends of the steps in top view towards the *PP* and draw vertical lines from the meeting point of them with *PP* and mark the rise 10 mm of the steps at both vertical lines,
7. Join these points with V_R' and project the piercing point of these corners from front view to the corresponding lines connected with V_R' to get various corners of the step in perspective.
8. Show the visible edges of the steps by joining these points to get the perspective view of the model of steps.

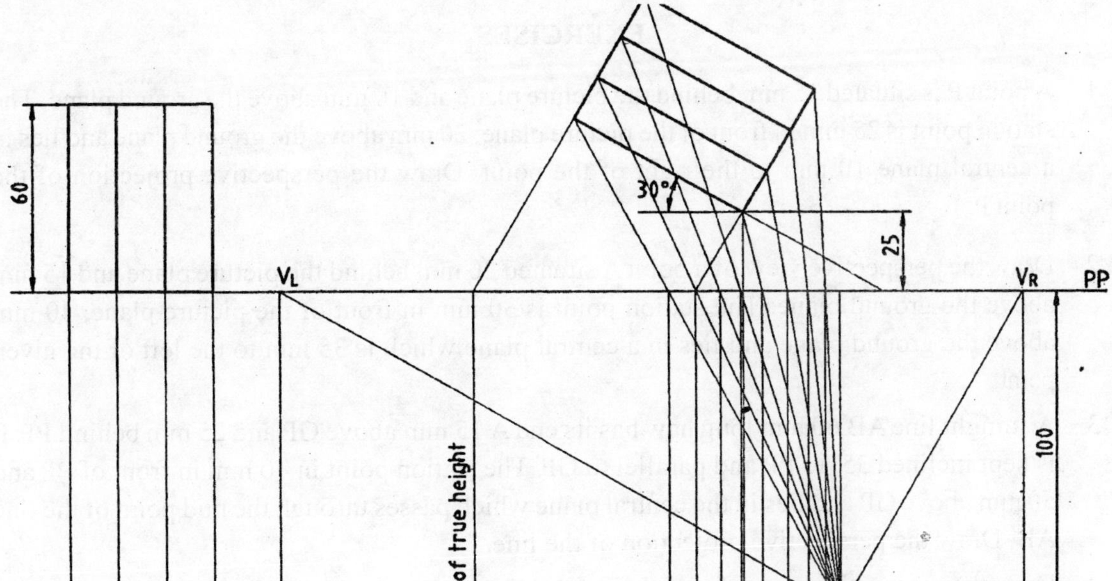

Fig. 10.26

EXERCISES

1. A point P is situated 15 mm behind the picture plane and 10 mm above the ground plane. The station point is 25 mm in front of the picture plane, 20 mm above the ground plane and lies in a central plane 10 mm to the right of the point. Draw the perspective projection of the point P.

2. Draw the perspective view of a point A situated 20 mm behind the picture plane and 15 mm above the ground plane. The station point is 30 mm in front of the picture plane, 40 mm above the ground plane and lies in a central plane which is 35 mm to the left of the given point.

3. A straight line AB 60 mm long haw has its end A 15 mm above GP and 25 mm behind PP. It is kept inclined 35° to PP and parallel to GP. The station point in 70 mm in front of PP and 50 mm above GP and lies in the central plane which passes through the mid point of the line AB. Draw the perspective projection of the line.

4. Draw the perspective view of straight line AB, 30 mm long, which is in the ground plane and inclined at 30° to the picture plane. End A is in picture plane. Central plane is 10 mm right of A. The station point is 30 mm in front of picture plane, 30 mm above the ground plane and lies in central plane.

5. A straight line AB, 35 mm long, which is parallel to both picture plane and ground plane. Line is 10 mm above ground plane and 20 mm behind the picture plane. Central plane is 10 mm left of A. The station point is 30 mm in front of picture plane, 25 mm above the ground plane and lies in central plane. Draw its perspective view.

6. A square lamina of 30 mm side rests on one of its sides on the ground touching the picture plane. The station point is 40 mm above the ground plane, 30 mm in front of picture plane and lies in a central plane 20 mm to the right of the center of the square. Draw the perspective projection of the square.

7. A hexagonal plane of 25 mm stands vertically on the ground plane and inclined at 40° to the picture plane. The corner nearest to picture plane is 15 mm behind it. The station point is 35 mm in front of the picture plane, 45 mm above the ground plane and lies in central plane which passes through the centre of the plane. Draw the perspective view of the plane.

8. A circular lamina of diameter 50 mm is lying on the ground plane touching the picture plane. The station point is 50 mm above the ground plane, 60 mm in front the picture plane and contained in the central plane which passes at a distance of 40 mm from the central of the circle. Draw the perspective projection of the circle.

9. Draw the perspective projection of a rectangular block of 300 mm × 200 mm × 100 mm resting on a horizontal plane with one side of the rectangular plane making an angle 45° with VP. The observer is at a distance of 600 mm from the picture. Assume eye level as 100 mm.

10. A square prism of 30 cm side and 50 cm length is lying on the ground plane on one of its rectangular faces, in such a way that one of its square faces is parallel to an 10 cm behind the picture plane. The station point is located 60 cm in front of the picture plane and 40 cm above the ground plane. The central plane is 50 cm away from the axis of the prism towards the left. Draw the perspective projection of the prism.
11. A cube of edge 30 mm rests with one of the faces on the ground plane such that a vertical edge touches the picture plane. The vertical faces of the cube are equally inclined to the PP and behind it. A station point is 40 mm in front of the PP, 50 mm above the ground plane and lies in a central plane 15 mm to the right of the axis of the cube. Draw the perspective projection of the cube.
12. A rectangular pyramid, sides of base 55 mm × 20 mm and height 60 mm rests with its base on the ground plane such that one of the longer edges of the base is parallel to and 20 mm behind the picture plane. The station point is 40 mm in front of the picture plane, 60 mm above the ground plane and lies in a central plane which passes through the axis of the pyramid. Draw the perspective projection of the cube.

CHAPTER 11

Conversion of Isometric Views to Orthographic Views and Vice Versa

11.1 Introduction

The following principles of orthographic views are considered in making the above drawings:

1. In first angle projection; the **Front view** on the above and the **Top view** at the bottom are always in line vertically.
2. The **front view** and the **side view** are always in line horizontally.
3. Each view gives two dimensions; usually the front view gives lengh and height, top view gives length and width and side view gives hight and width.
4. When the surface is parallel to a plane its projection on that plane will show its true shape and size.
5. When the surface is inclined, its projection will be foreshortened as shown in Fig.11.1(b).

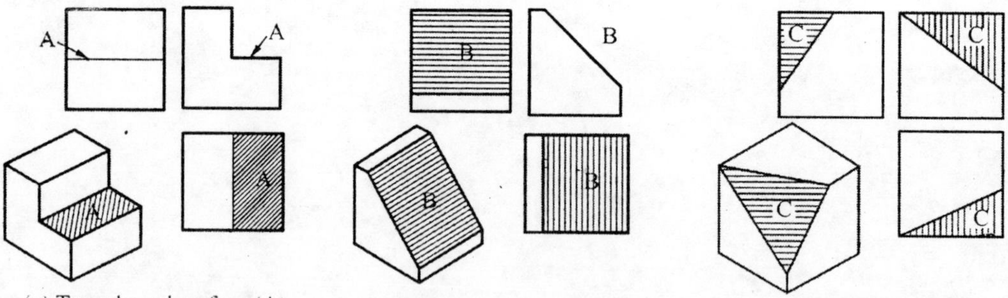

(a) True shaped surface (A) (b) Foreshortened surface (B) (c) Oblique surface (C)

Fig. 11.1 Representation of Surfaces.

Selection of Views

1. The number of orgthographic views required for clear description of the object is taken as the criteria to select the views. As far as possible least number of views are drawn.
2. While selecting the views, the object is placed in such a way that the number of hidden lines are kept to minimum.

3. Front view is drawn seeing the object in a direction is which its length is seen. It is also chosen such that the shape of the object is revealed. The direction of the view is indicated by arrows.

11.2 Conversion of Isometric Views to Orthographic Views

The isometric views of some objects and two or three of their orthographic views are shown from Fig. 11.2 to Fig. 11.22 in the following pages drawn as per the principles indicated above.

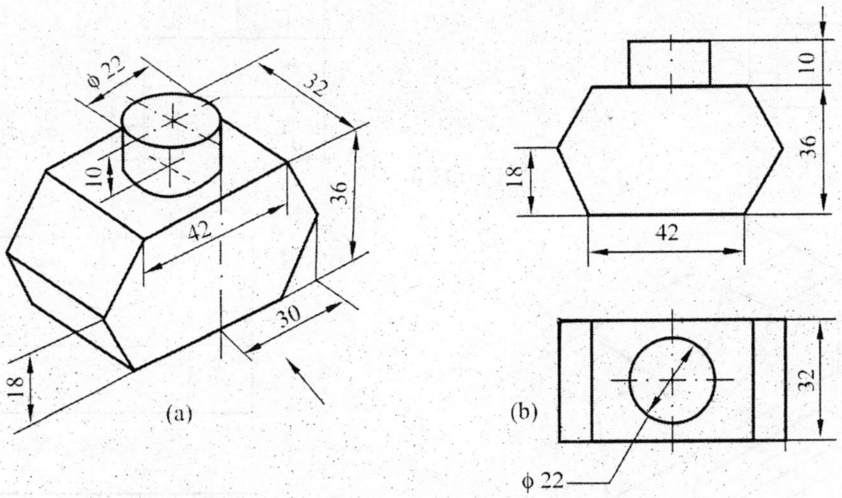

Fig. 11.2

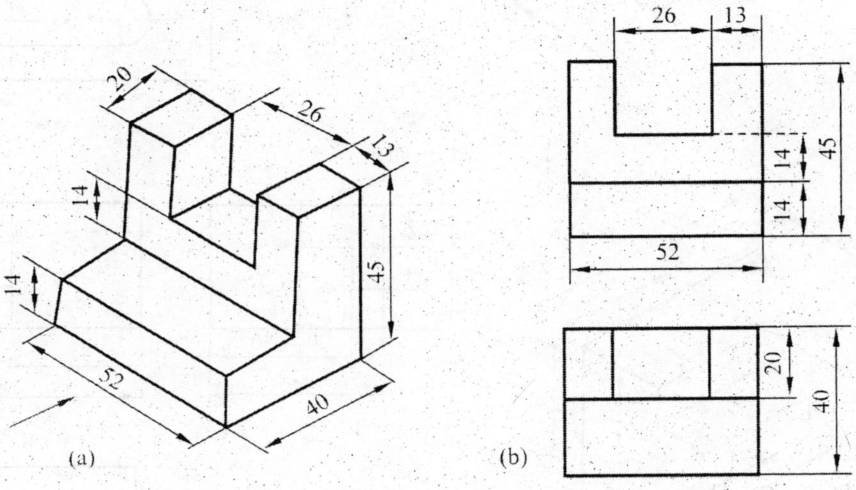

Fig. 11.3

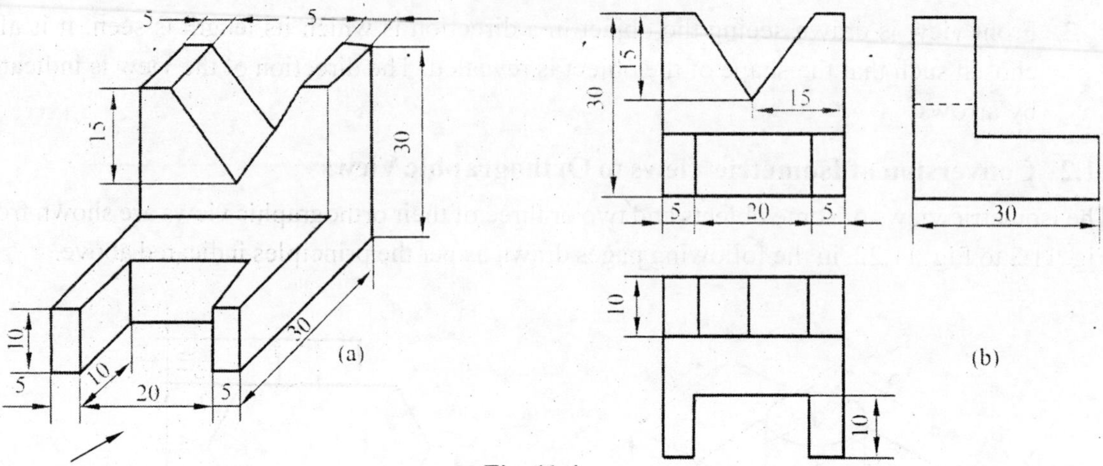

Fig. 11.4

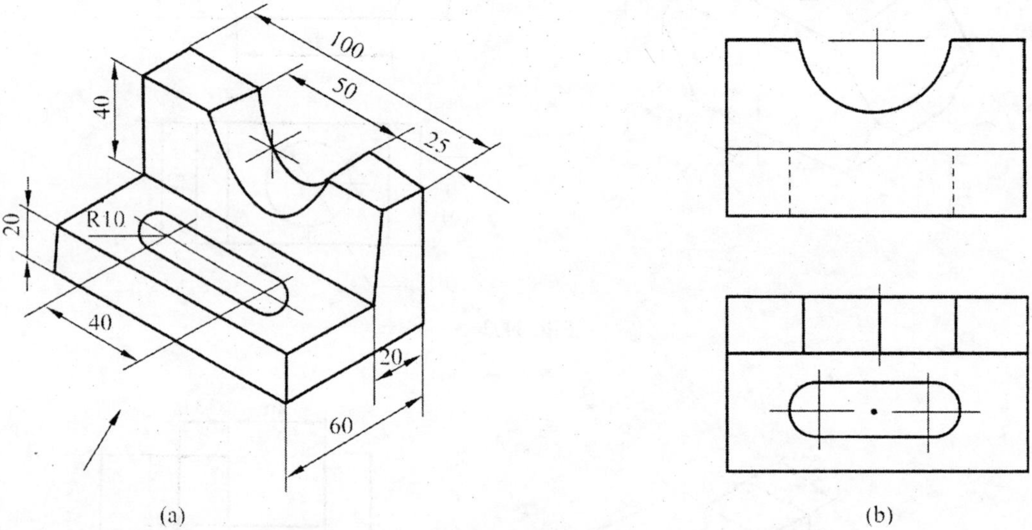

Fig. 115

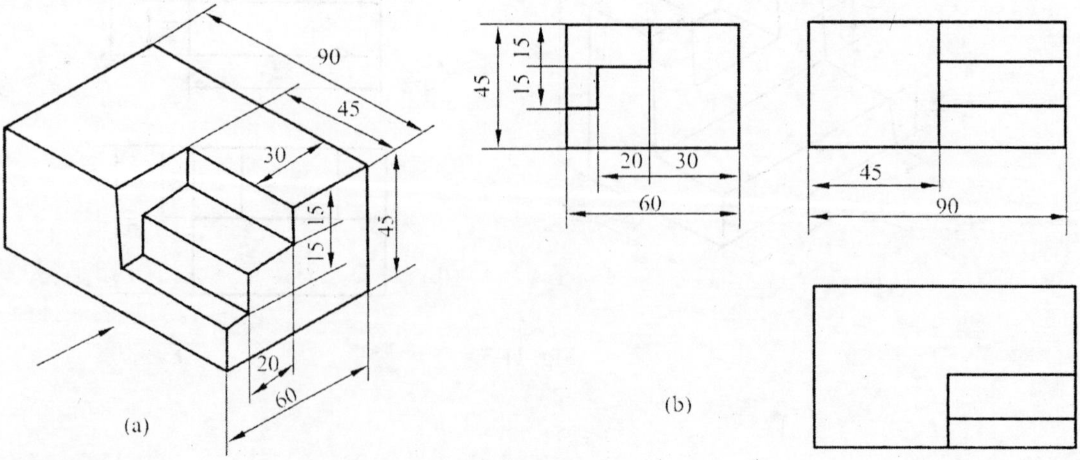

Fig. 11.6

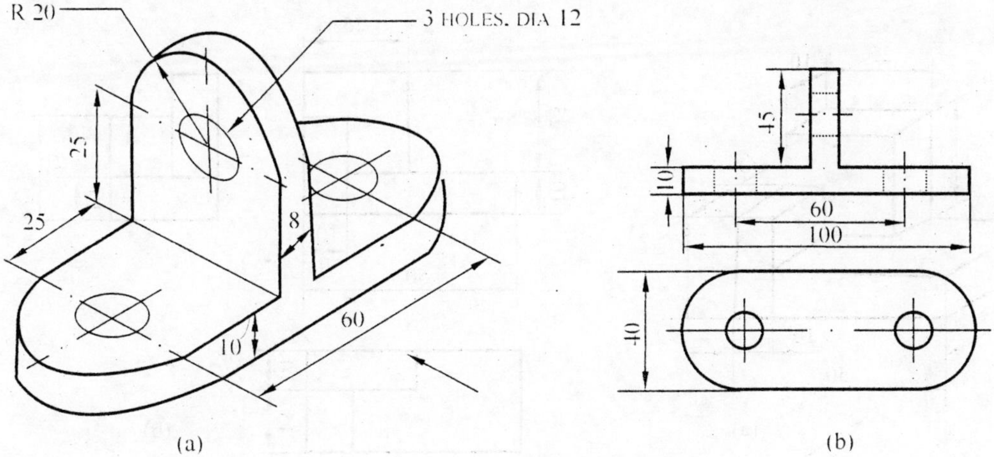

Fig. 11.7

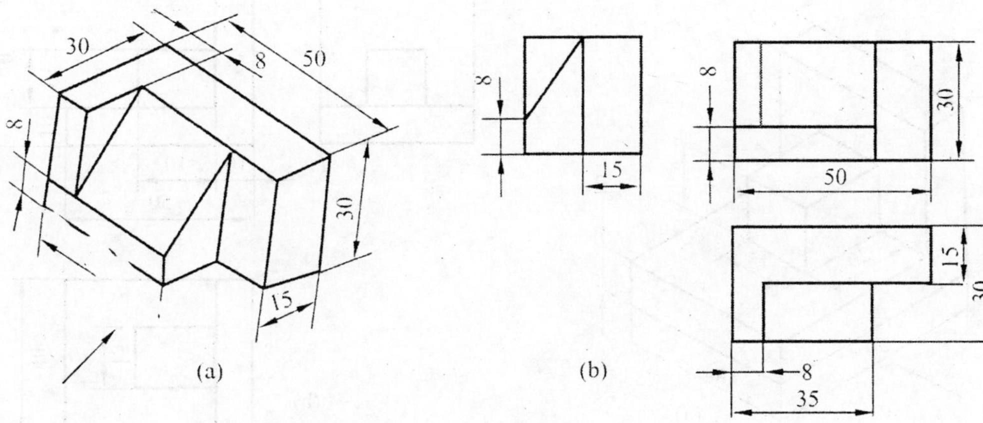

Fig. 11.8

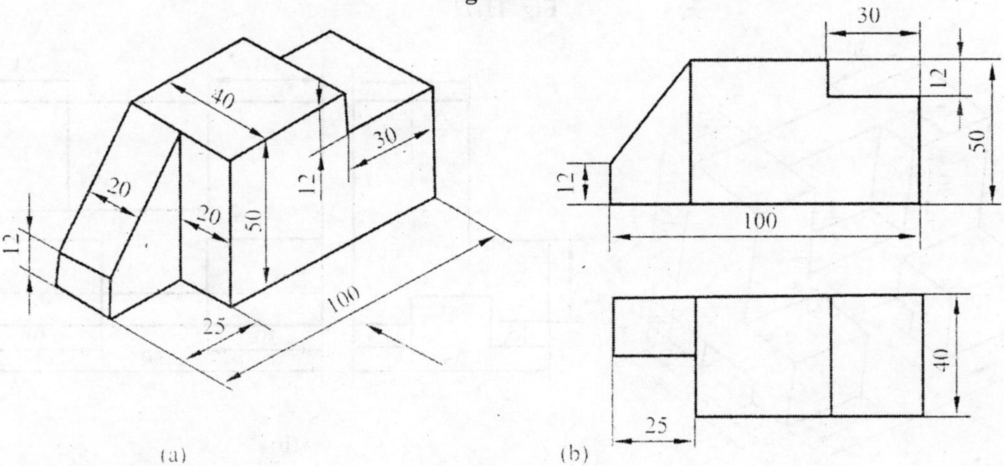

Fig. 11.9

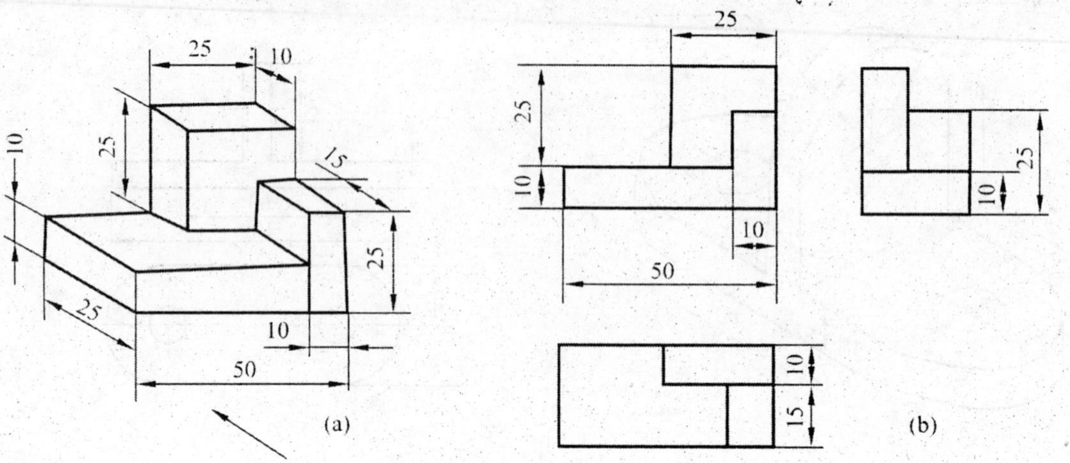

Fig. 11.10

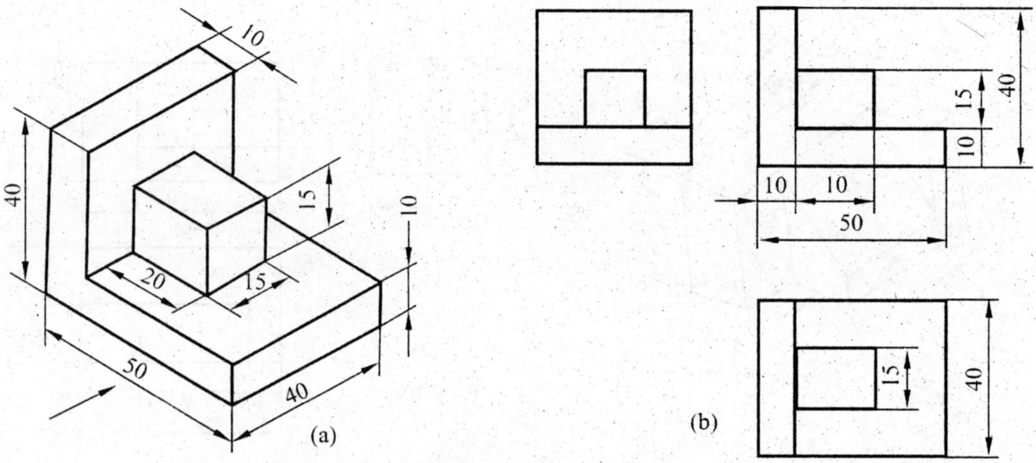

Fig. 11.11

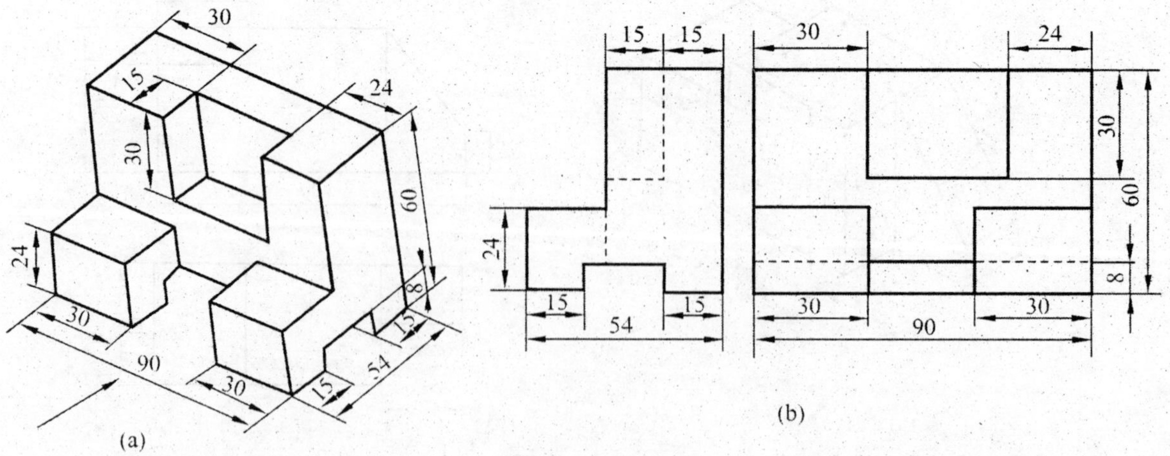

Fig. 11.12

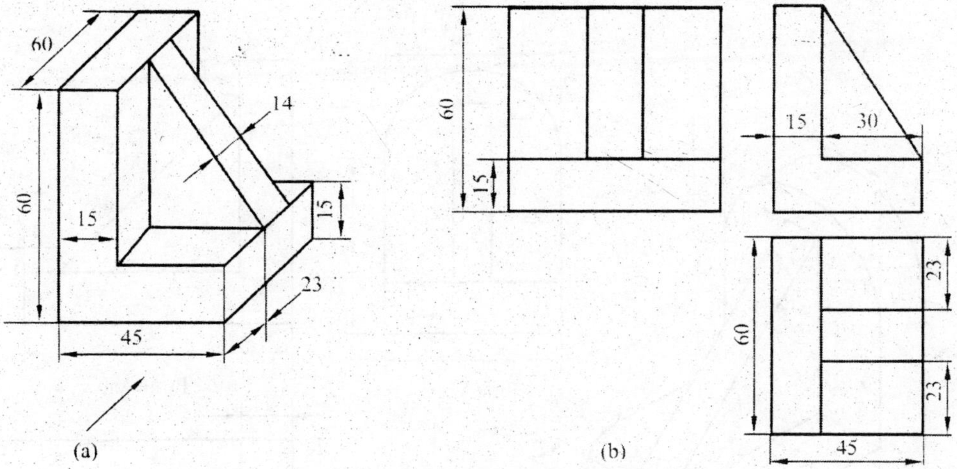

Fig. 11.13

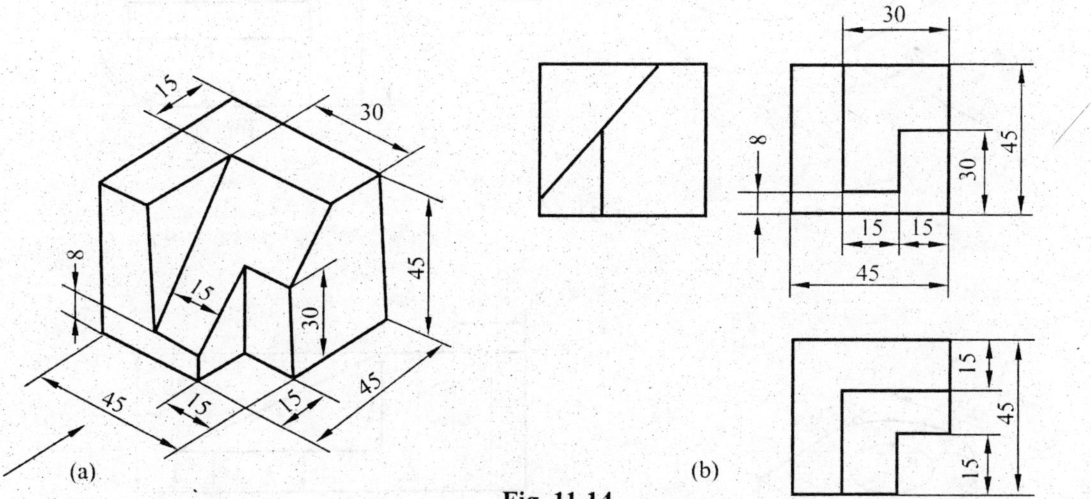

Fig. 11.14

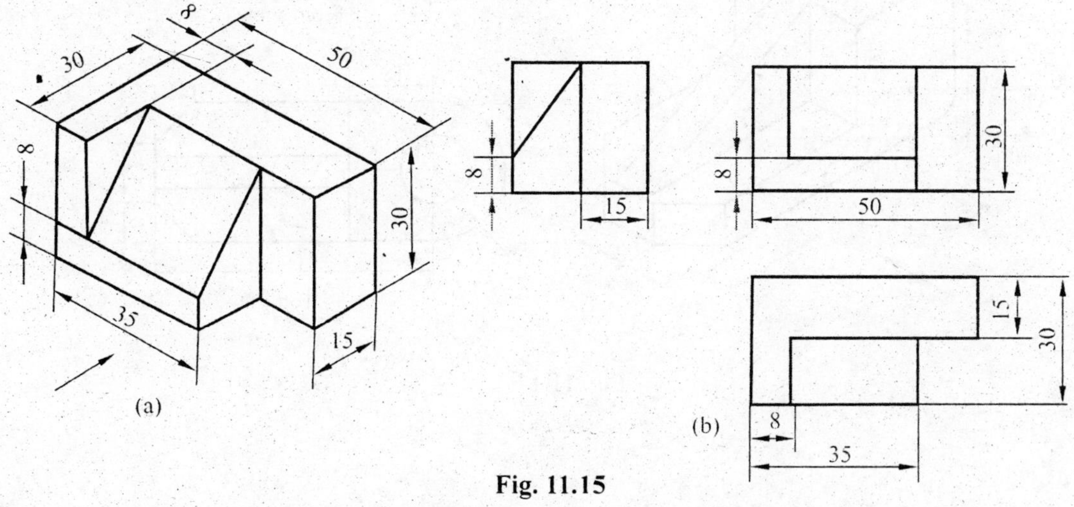

Fig. 11.15

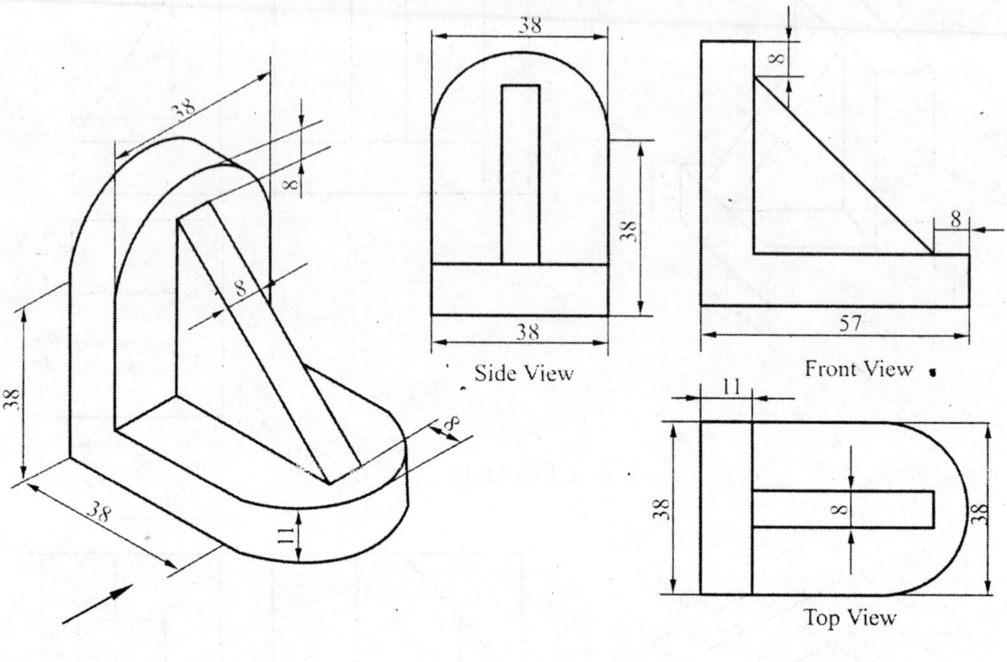

Fig. 11.16

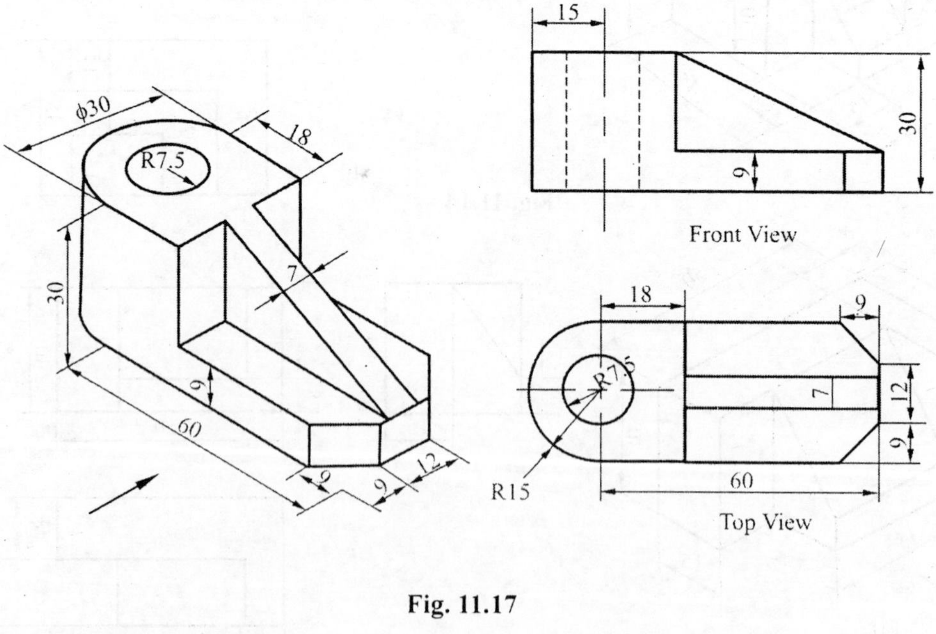

Fig. 11.17

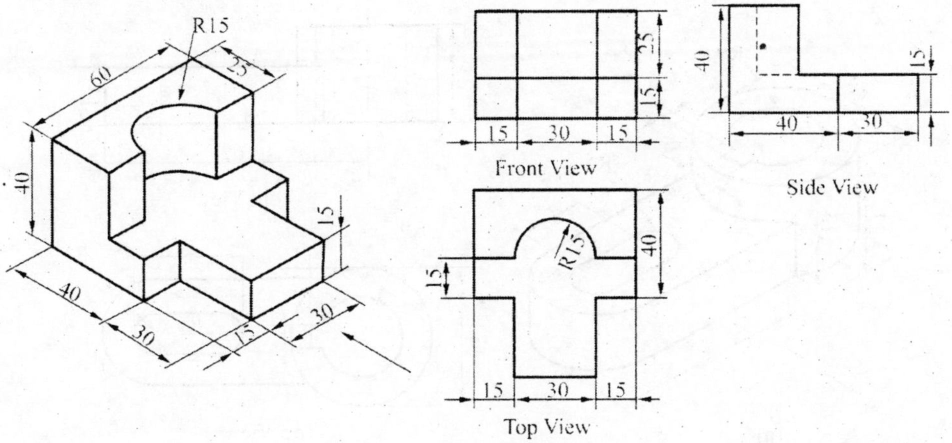

Fig. 11.18

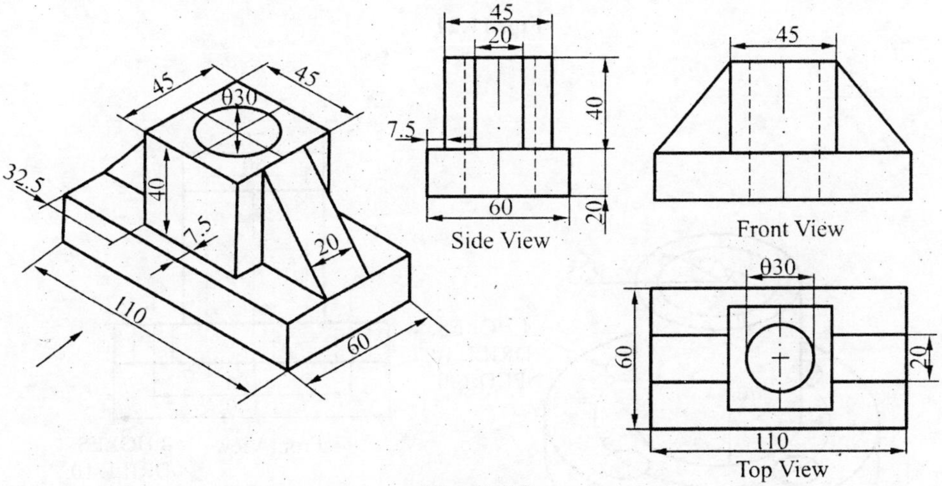

Fig. 11.19

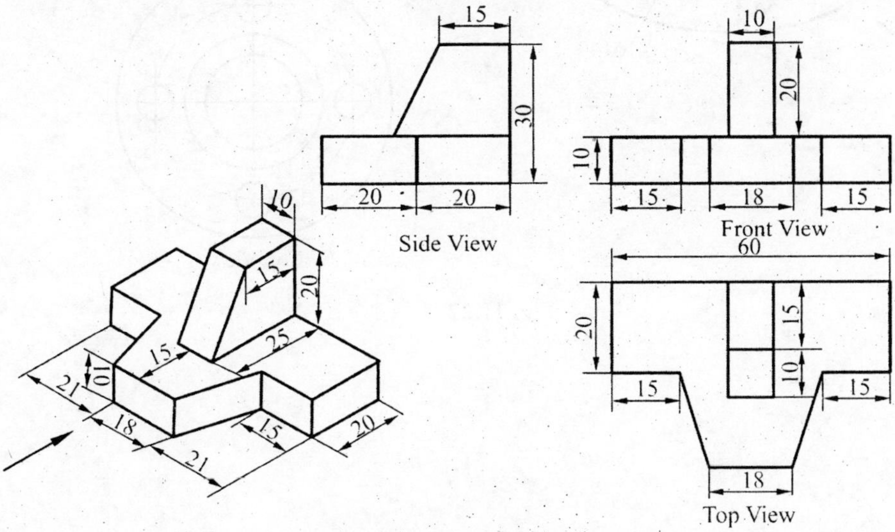

Fig. 11.20

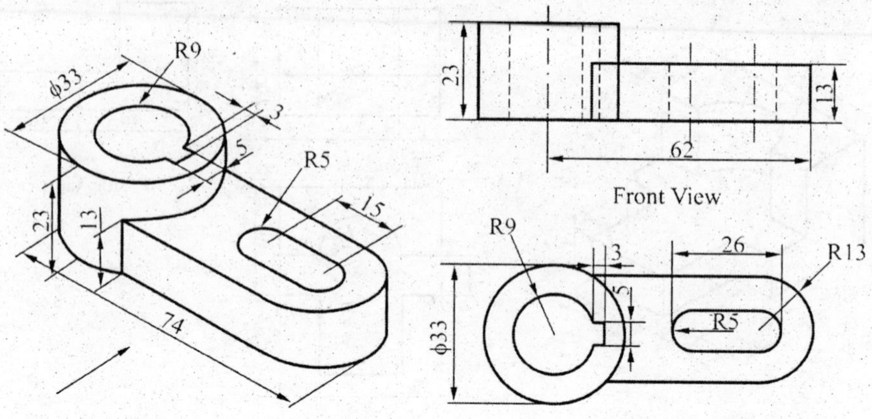

(a)

Fig. 11.21

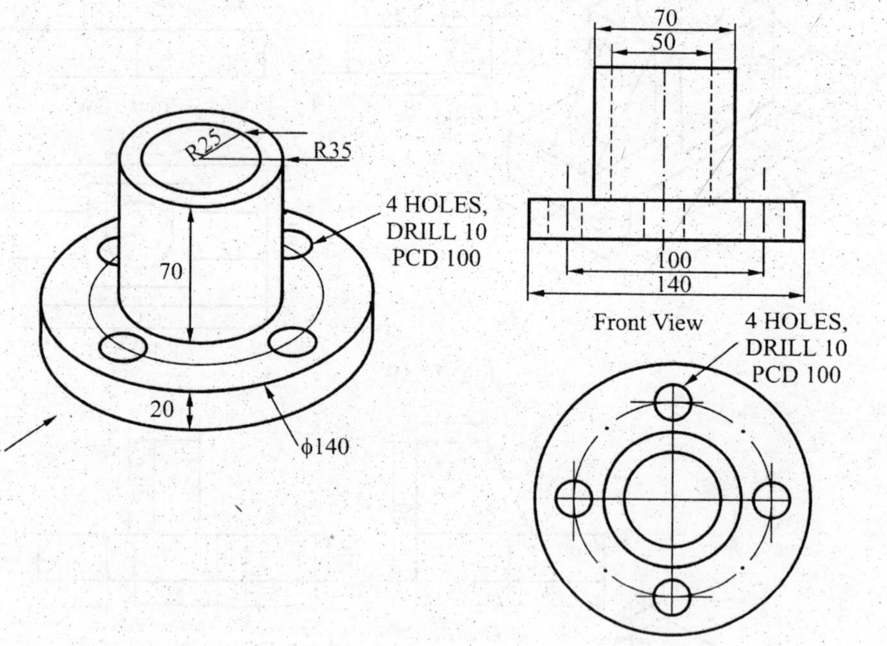

Fig. 11.22

Problem : Match the isometric views A to E with the orthographic views 1 to 5 shown in Fig. 11.23.

Fig. 11.23

11.3 Conversion of Orthographic Views to Isometric Views

Principles of conversion of orthographic views into isometric views are explained in chapter 7.

Examples

A few orthographic views and their isometric drawing are shown from Fig.11.24 to Fig.11.34.

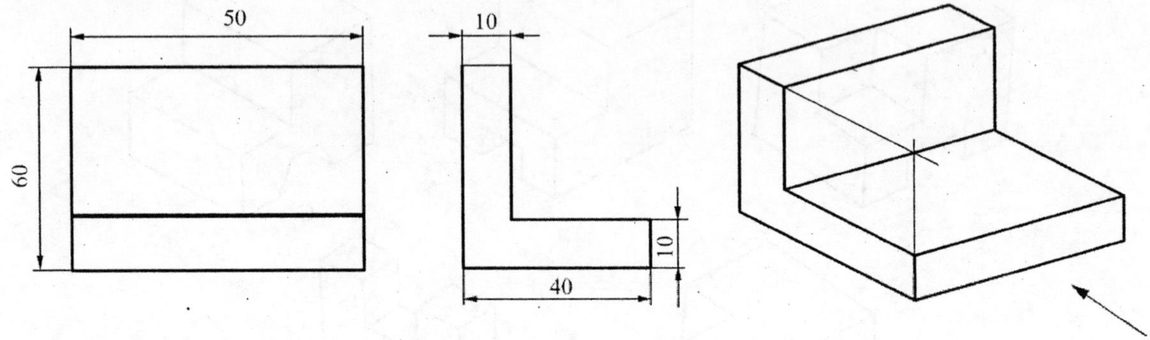

Fig. 11.24

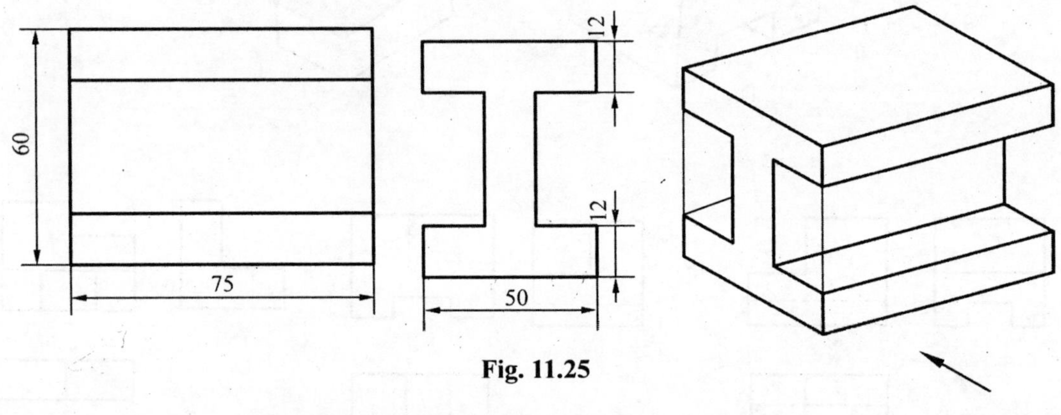

Fig. 11.25

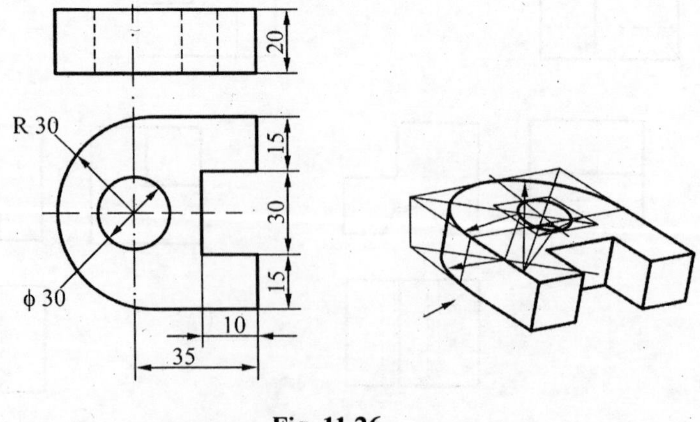

Fig. 11.26

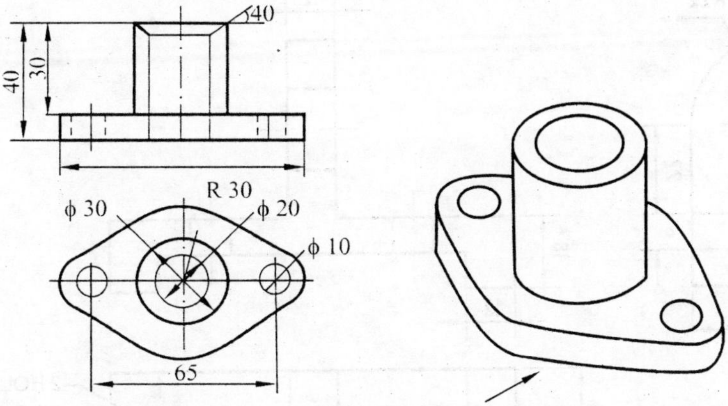

Fig. 11.27

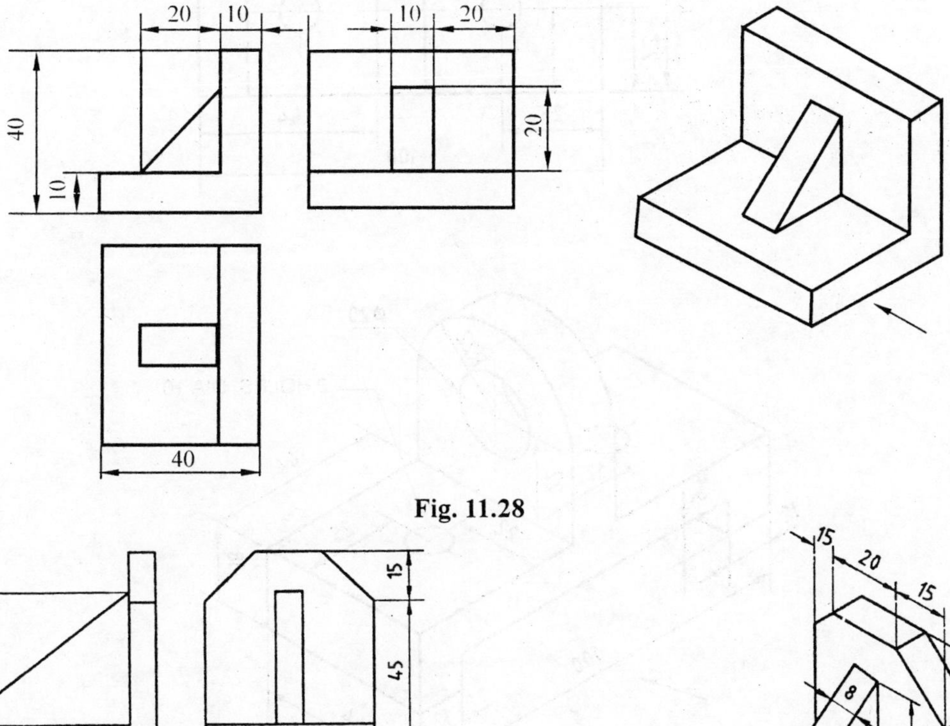

Fig. 11.28

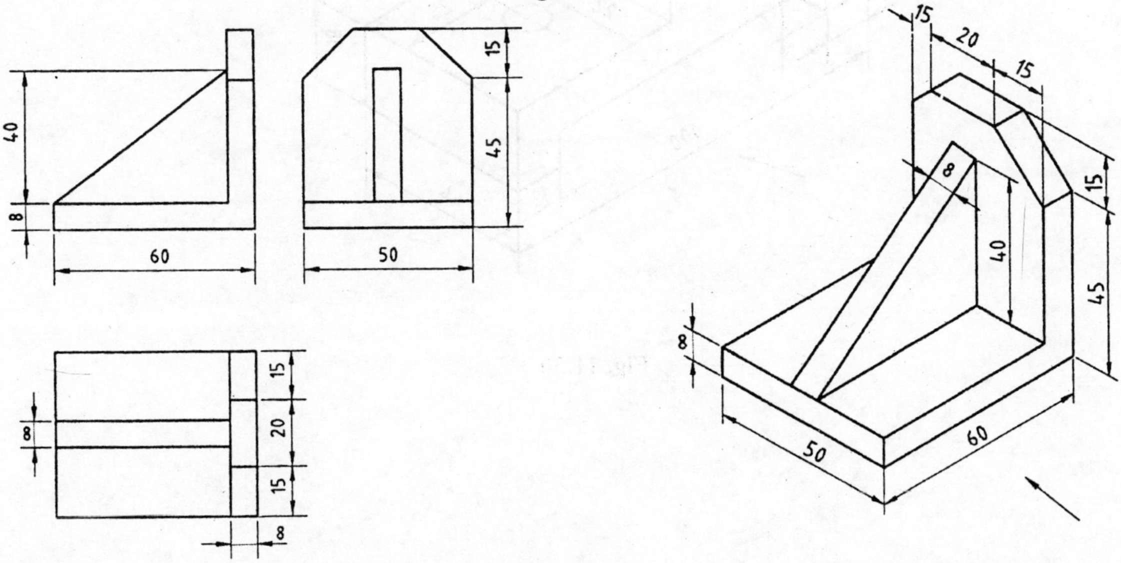

Fig. 11.29

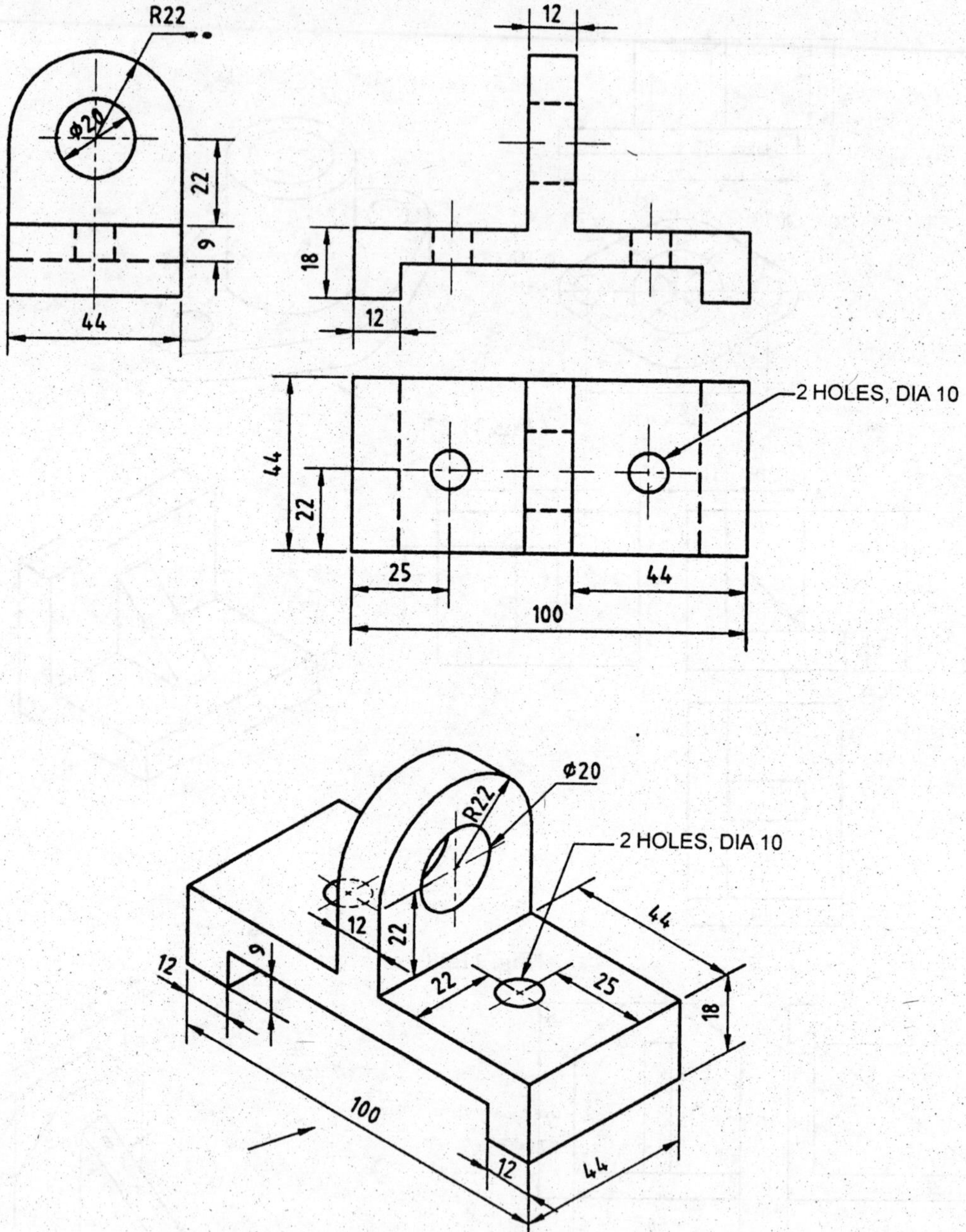
R22
ϕ20
22
9
44
12
18
12
2 HOLES, DIA 10
44
22
25
44
100
ϕ20
R22
2 HOLES, DIA 10
12
22
9
12
44
22
25
18
100
12
44

Fig. 11.30

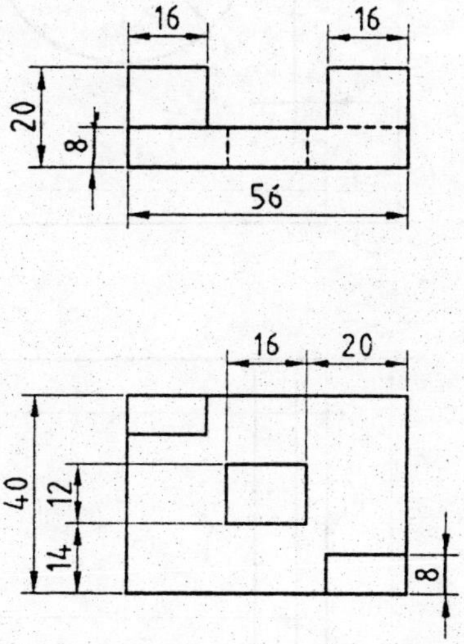

(a)

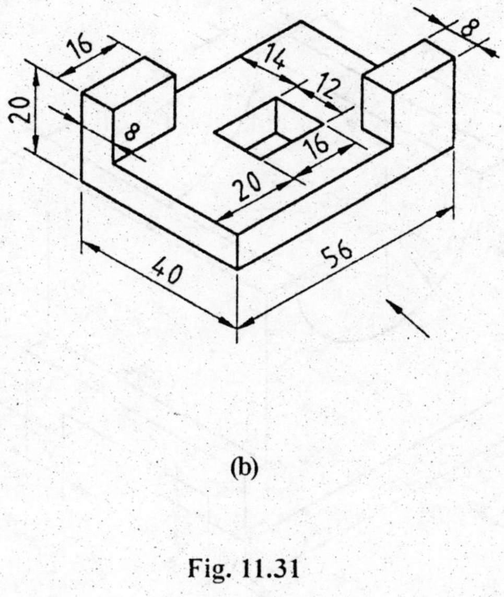

(b)

Fig. 11.31

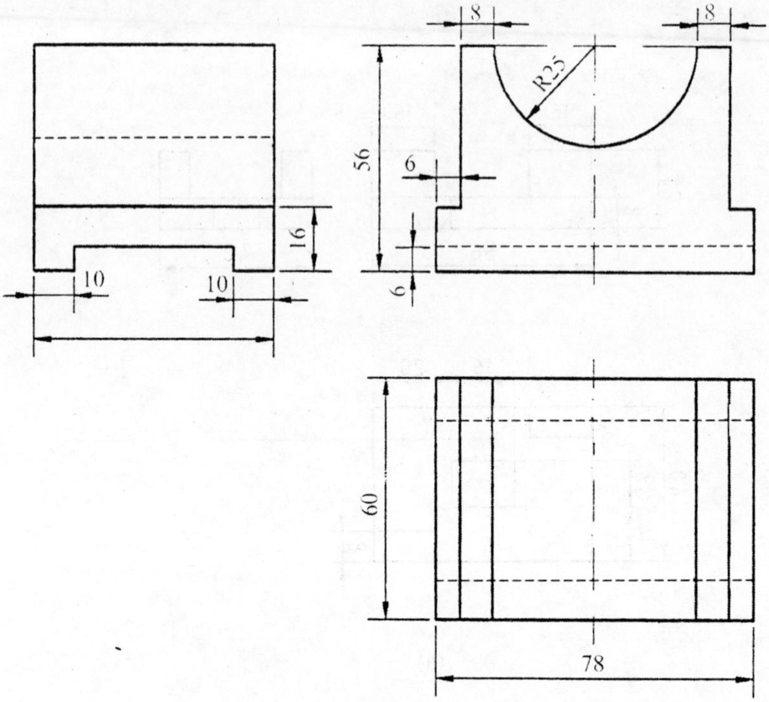

(a)

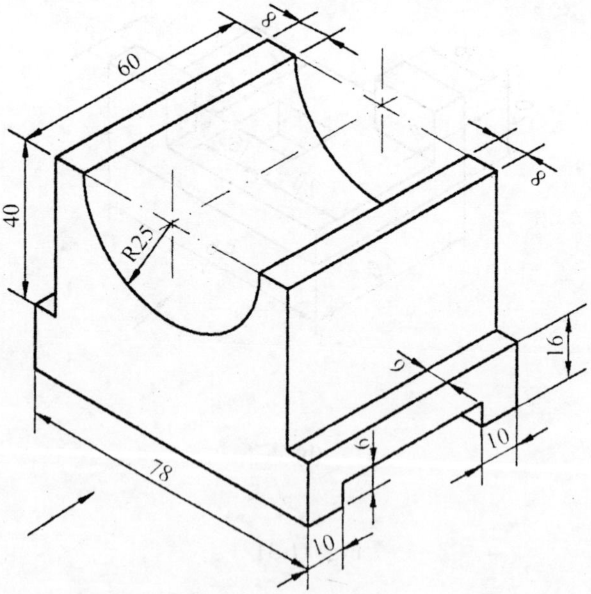

(b)

Fig. 11.32

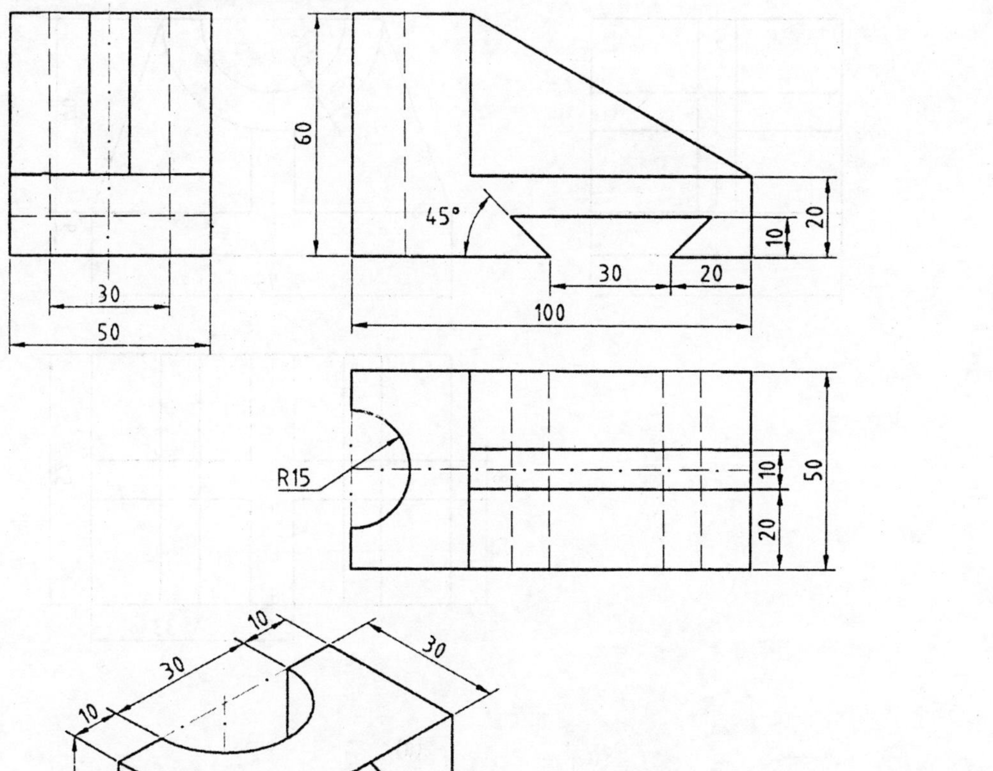
30
50
60
45°
10
20
30
20
100
R15
10
20
50
10
30
10
30
10
60
20
45°
10
30
100
20
50
20

(b)

Fig. 11.33

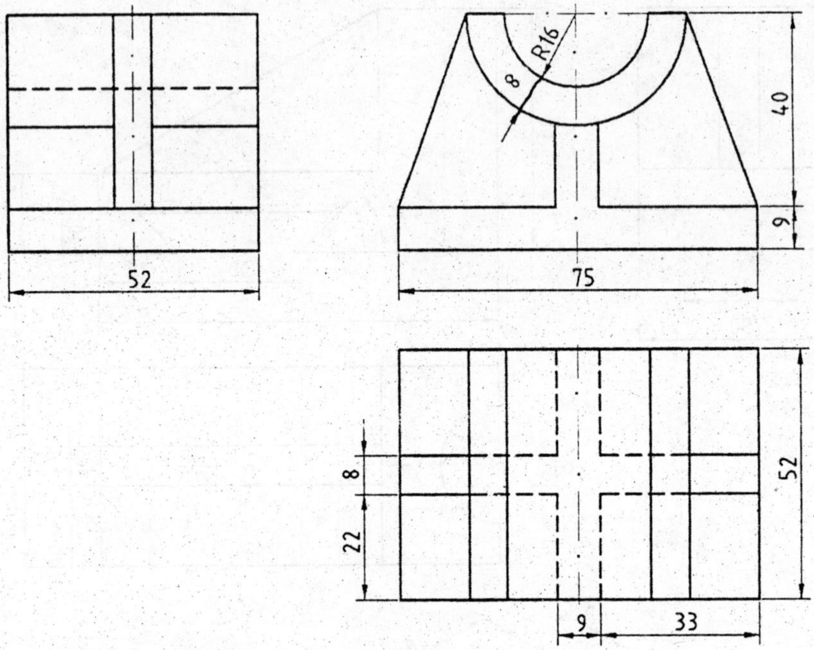

(a)

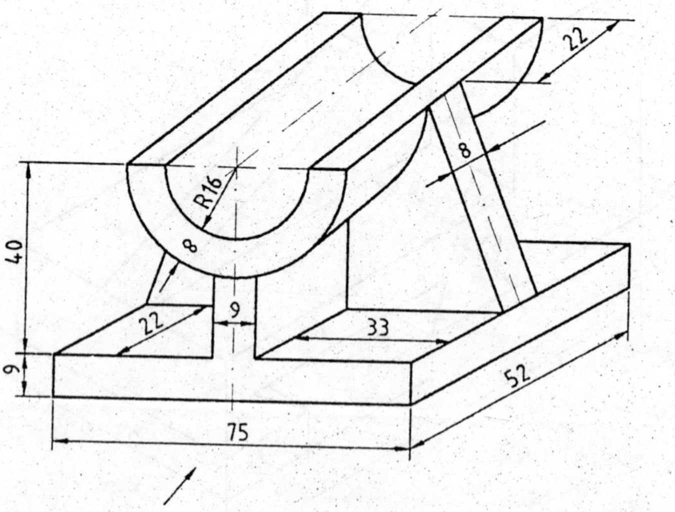

(b)

Fig. 11.34

CHAPTER 12

Sections of Solids

12.1 Sectioning of Solids

12.1.1 Introduction

Sections and sectional views are used to show hidden details more clearly. They are created by using a cutting plane to cut the object.

A section is a view of no thickness and shows the outline of the object at the cutting plane. Visible outlines beyond the cutting plane are not drawn.

A sectional view, displays the outline of the cutting plane and all visible outlines which can be seen beyond the cutting plane. It improves visualization of interior features.

Section views are used when important hidden details are in the interior of an object. These details appear as hidden lines in one of the orthographic principal views; therefore, their shapes are not very well described by pure orthographic projection.

12.1.2 Types of Section Views

- Full sections
- Half sections
- Offset sections
- Revolved sections
- Removed sections
- Broken-out sections

12.1.3 Cutting Plane

- Section views show how an object would look if a cutting plane (or saw) cut through the object and the material in front of the cutting plane was discarded.

Representation of cutting plane

According to drawing standards cutting plane is represented by chain line with alternate long dash and dot. The two ends of the line should be thick.

Full Section View

- In a full section view, the cutting plane cuts across the entire object.

  Note that hidden lines become visible in a section view.

Hatching

On sections and sectional views solid area should be hatched to indicate this fact. Hatching is drawn with a thin continuous line, equally spaced (preferably about 4 mm apart, though never less than 1mm) and preferably at an angle of 45 degrees.

(i) Hatching a single object

When you are hatching an object, but the objects has areas that are separated, all areas of the object should be hatched in the same direction and with the same spacing.

(ii) Hatching Adjacent objects

When hatching assembled parts, the direction of the hatching should ideally be reversed on adjacent parts. If more than two parts are adjacent, then the hatching should be staggered to emphasise the fact that these parts are separate.

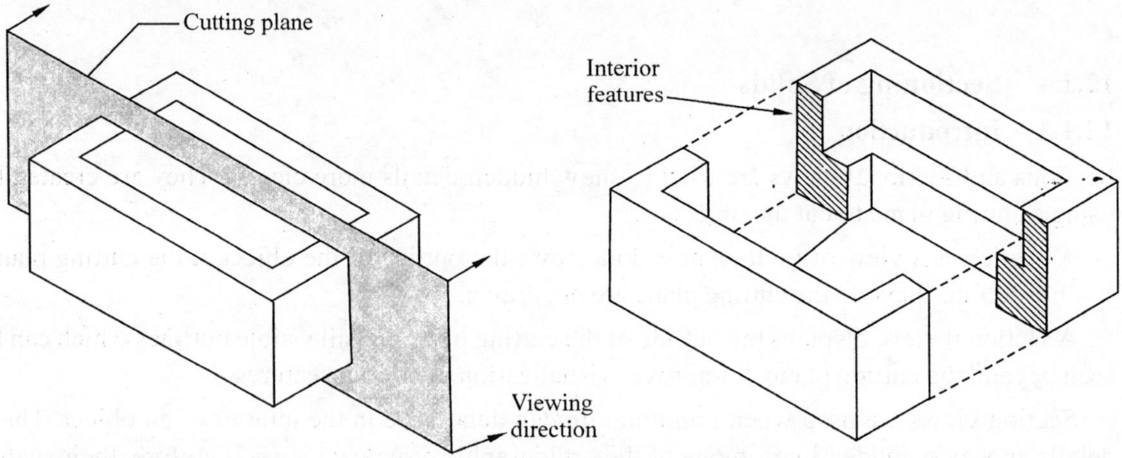

Fig. 12.1

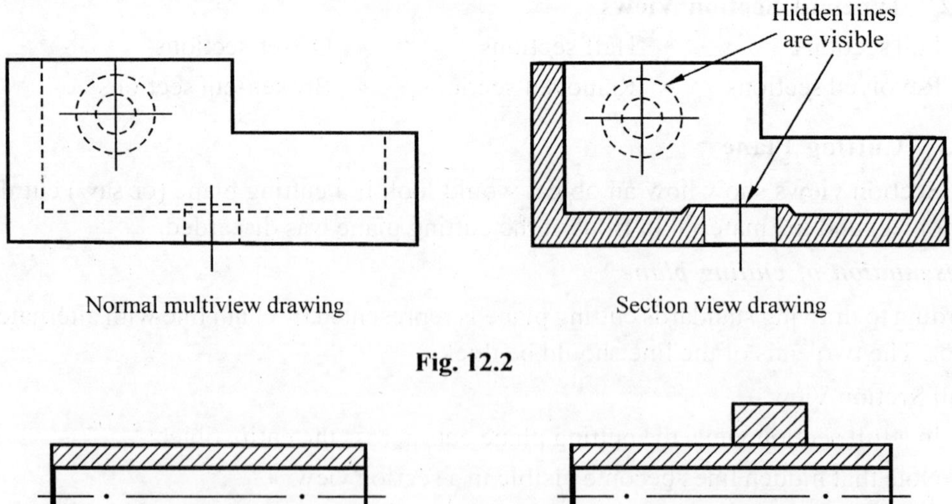

Fig. 12.2

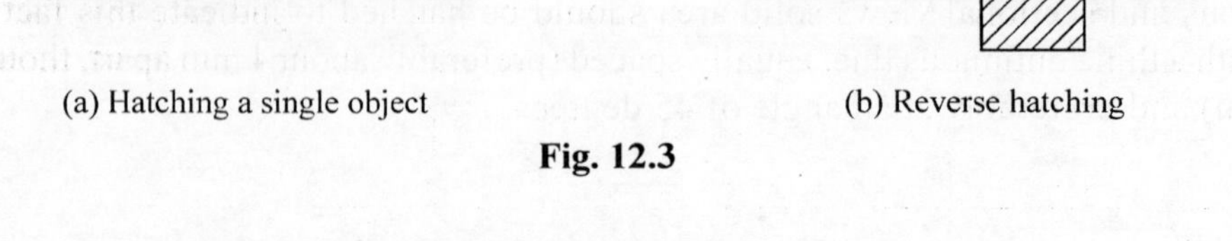

Fig. 12.3

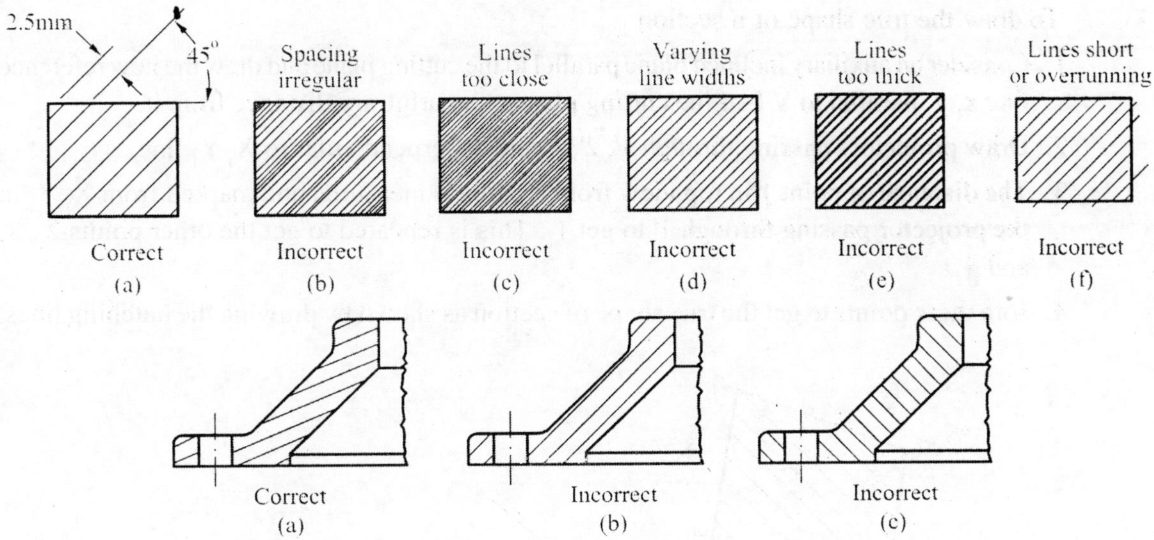

Fig. 12.4

EXAMPLES

Problem : A square prism of base side on 30 mm and axis length 60 mm is resting on HP on one of its bases, with a base side inclined at 30° to VP. It is cut by a plane inclined at 40° to HP and perpendicular to VP and is bisecting the axis of the prism. Draw its front view, sectional top view and true shape of section.

Solution : **(Fig. 12.5)** Draw the projections of the prism in the given position. The top view is drawn and the front view is projected.

To draw the cutting plane, front view and sectional top view

1. Draw the Vertical Trace (*VT*) of the cutting plane inclined at 40° to *XY* line and passing through the mid point of the axis.
2. As a result of cutting, longer edge a' p' is cut, the end a' has been removed and the new corner l' is obtained.
3. Similarly 2' is obtained on longer edge b' q', 3' on c' r' and 4' on d' s',
4. Show the remaining portion in front view by drawing dark lines.
5. Project the new points 1', 2', 3' and 4' to. get 1, 2, 3 and 4 in the top view of the prism, which are coinciding with the bottom end of the longer edges p, q, r and s respectively.
6. Show the sectional top view or apparent section by joining 1, 2, 3 and 4 by drawing hatching lines.

To draw the true shape of a section

1. Consider an auxiliary inclined plane parallel to the cutting plane and draw the new reference line $x_1 y_1$ parallel to VT of the cutting plane at an arbitrary distance from it.
2. Draw projectors passing through 1', 2', 3' and 4' perpendicular to $X_1 Y_1$ line.
3. The distance of point 1 in top view from *XY* line is measured and marked from $X_1 Y_1$ in the projector passing through 1' to get 1_1. This is repeated to get the other points 2_1, 3_1 and 4_1.
4. Join these points to get the true shape of section as shown by drawing the hatching lines.

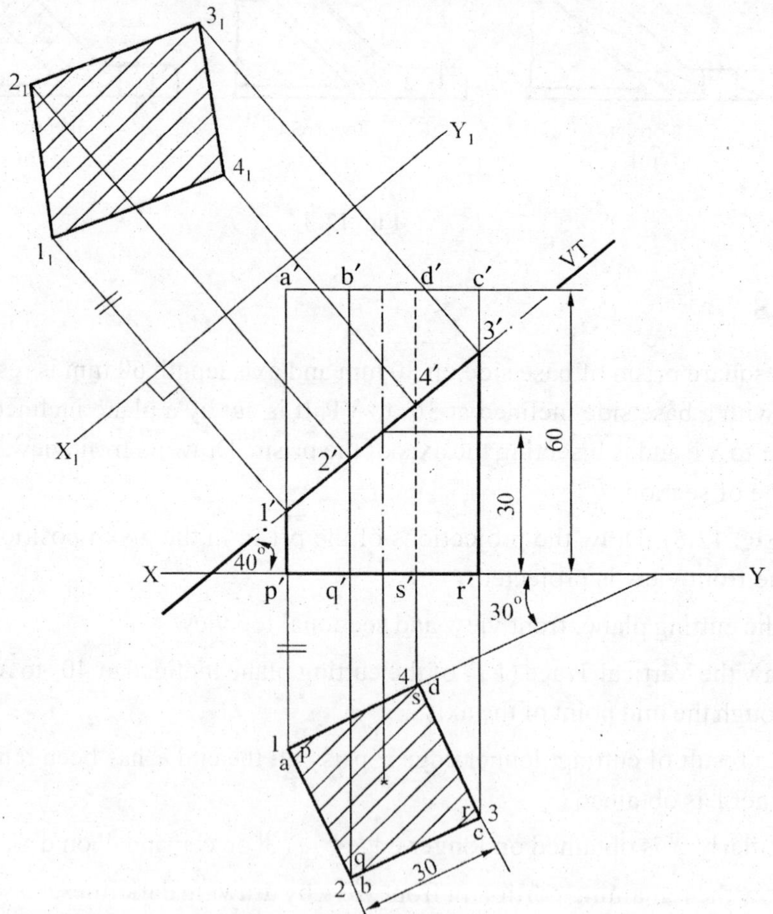

Fig. 12.5

Problem : A cube of 45 mm side rests with a face on HP such that one of its vertical face is inclined at 30° to VP. A section plane, parallel to VP cuts the cube at a distance of 15 mm from the vertical edge nearer to the observer. Draw its top and sectional front view.

Solution **: (Fig. 12.6)**

1. Draw the projections of the cube and the Horizontal Trace (*HT*) of the cutting plane parallel to *XY* and 15 mm from the vertical edge nearer to the observer.
2. Mark the new points 1,2 in the top face edge as *ab* and *bc* and similarly 3, 4 in the bottom face edge as *qr* and *pq* which are invisible in top view.
3. Project these new points to the front view to get 1', 2' in top face and 3', 4' in bottom face.
4. Join them and draw hatching lines to show the sectional front view which also shows the true shape of section.

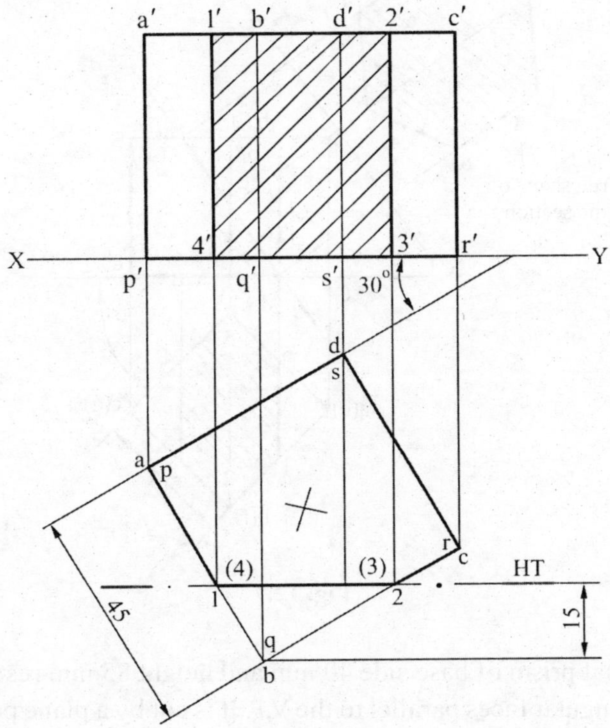

Fig. 12.6

Problem : A cube of 50 mm sides has its base edges equally inclined to the V.P. It is cut by a section plane perpendicular to the V.P., so that the true shape of the cut section is regular hexagon. Locate the plane and determine the angle of inclination of the V.T. with the reference line XY. Draw the sectional top view.

Solution **:** (Fig. 12.7)

Step 1: Draw the top and front view of the cube of base 50 mm and its base edges equally inclined to V.P.

Step 2: Draw the cutting plane of which is perpendicular to V.P. and it should pass through six edges and als should meet the edges at their mid heights. Since the true shape of a section will be a regular hexagon.

Step 3: The mid points **c'f'**, **b'f'** and **b'c'** are marked as **1'**, **2'** and **3'** respectively. The mid points **c'b'**, **d'h'** and **h'e'** are marked as **4'**, **5'** and **6'** respectively.

Step 4: Draw the projectors from cutting points to top view and mark the points **1, 2, 3, 4, 5** and **6** which is the sectional view.

Step 5: To draw the true shape of section follow the same procedure as in the previous example.

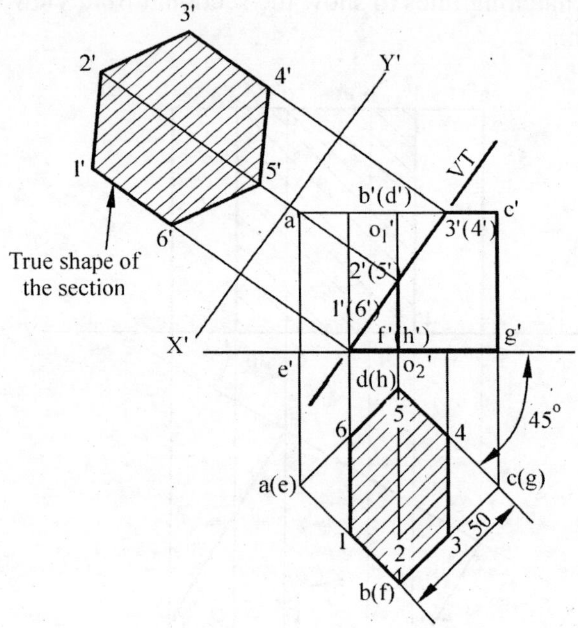

Fig. 12.7

Problem : A pentagonal prism of base side 40 mm and height 85 mm rests on the H.P. on one of its ends with two rectangular faces parallel to the V.P. It is cut by a plane perpendicular to the V.P. and inclined as 45° to the H,P. The cutting plane meets the axis at 30 mm from the top. Draw the front view and the sectional plan. And also draw an auxiliary plan on an AIP parallel to the cutting plane showing the section also.

Solution : **(Fig. 12.8)**

Step 1: Draw the top, front view of pentagonal prism of base 40 mm and 85 mm height rests on H.P. Draw the cutting plane which is inclined at 45° to H.P. and it meets with the axis at 30 mm from the top.

Step 2: Then mark the cutting points and project it to top view to obtain points **1, 2,...7** which is the section plane.

Step 3: Draw the auxiliary plan of the pentagon with a new reference line **X'Y'** parallel to the cutting plane. Project the points **1'**, **2'** etc. to auxiliary plan as shown in the figure.

Step 4: Join these points and hatch the section in both the top view and auxiliary view.

Step 5: The cutting plane in the auxiliary plan represents the true shape of the section.

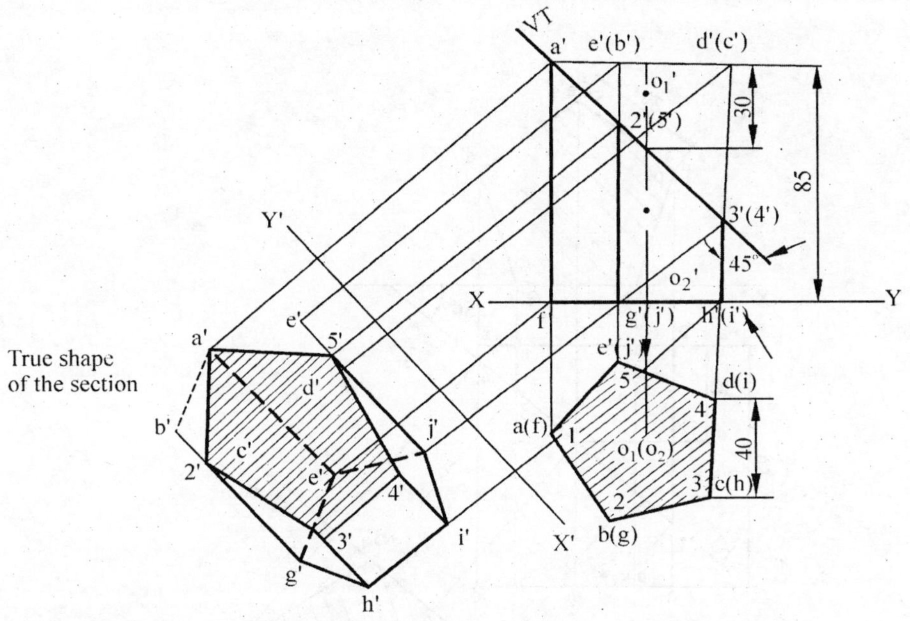

Fig. 12.8

Problem : A pentagonal prism of base side 30 mm and axis length 60 mm is resting on HP on one of its rectangular faces, with its axis perpendicular to VP. It is cut by a plane inclined at 50° to VP and perpendicular to HP and passing through a point 25 mm from rear base of the prism. Draw its top view, sectional front view and true shape of section.

Solution : **(Fig. 12.9)** To draw the cutting plane, top view and sectional front view

1. Draw the projections of the prism. Draw the *HT* of the cutting plane at 50° to *XY* and passing through the point on the axis at a distance of 25 mm from the rear base.
2. Mark the new points 1 on *ap*, 2 on *bq* etc.
3. Show the remaining portion in top view by drawing dark lines.
4. Project the new point 1, 2, etc. to the front view to get 1', 2' etc. which are coinciding with the rear end of the longer edges *p*', *q*' etc.
5. Show the sectional front view by joining 1', 2' etc. and draw hatching lines.

To draw the true shape of section

1. Consider an AVP and draw $X_1 Y_1$ line parallel to *HT* of the cutting plane.
2. Draw projectors through 1,2 etc. perpendicular to $X_1 Y_1$ line.

3. The distance of l' in front view from XY line is measured and marked from X_1Y_1 in the projector passing through 1 to get $1'_1$, and this is repeated to get $2'_1$, $3_1'$, etc.
4. Join them and show the true shape of section by drawing hatching lines.

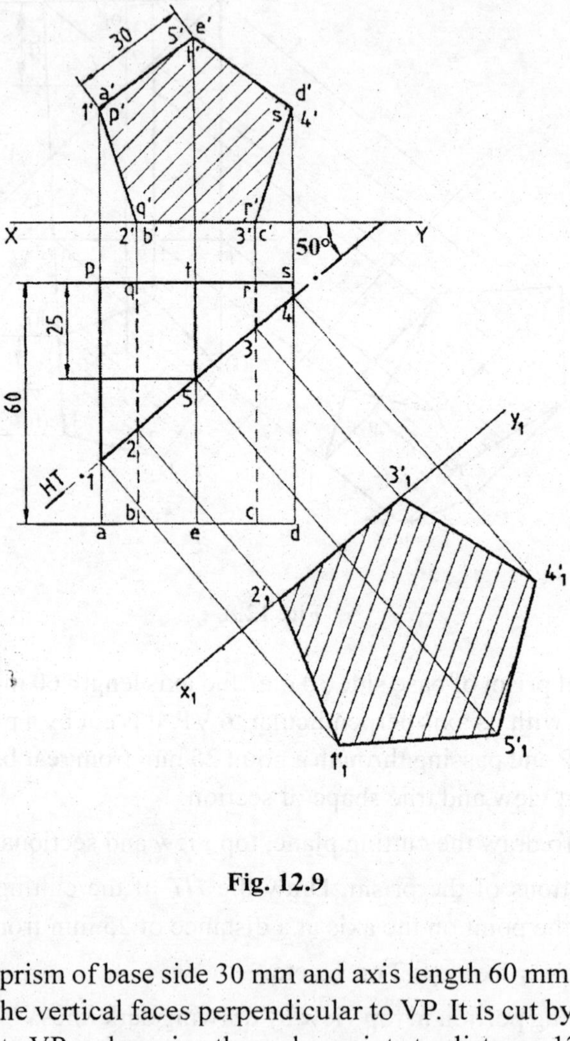

Fig. 12.9

Problem : A hexagonal prism of base side 30 mm and axis length 60 mm is resting on HP on one of its bases with two of the vertical faces perpendicular to VP. It is cut by a plane inclined at 60° to HP and perpendicular to VP and passing through a point at a distance 12 mm from the top base. Draw its front view, sectional top view and true shape of section.

Solution : **(Fig. 12.10)** Draw the projections of the prism in the given position. The top view is drawn and the front view is projected.

To draw the cutting plane, front view and sectional top view

1. Draw the *VT* of the cutting plane inclined at 60° to *XY* and passing through a point in the axis, at a distance 12 mm from the top base.

2. New points 1', 2', etc. are marked as mentioned earlier. Note that the cutting plane cuts the top base, the new point 3' is marked on base side b' c' and 4' marked on (d') (e') which is invisible.
3. Project the new points I', 2', etc. to get 1, 2, etc. in the top view.
4. Join these points and draw the hatching lines to show the sectional top view.

To draw true shape of section

1. Draw new reference line $X_1 Y_1$ parallel to the *VT* of the cutting plane.
2. Draw the projectors passing through 1', 2', etc. perpendicular to $X_1 Y_1$ line.
3. The distance of point 1 in top view from XY line is measured and marked from $X_1 Y_1$ in the projector passing through 1' to get 1_1. This is repeated to get other points 2_1, 3_1 etc.
4. Join these points to get the true shape of section and this is shown by hatching lines.

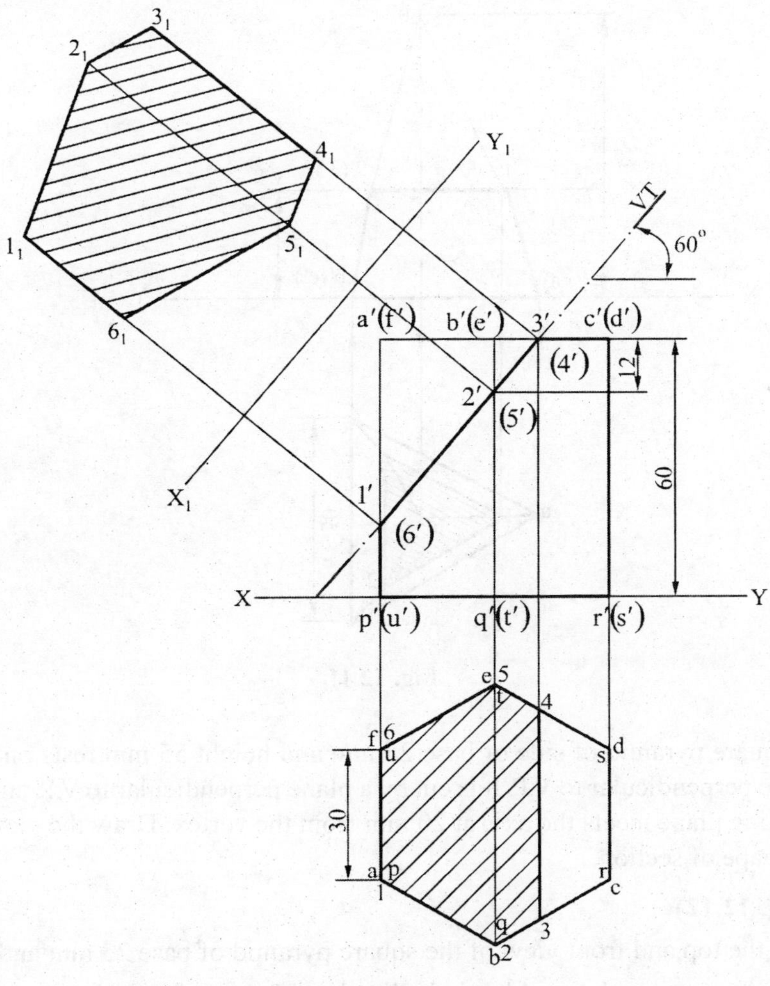

Fig. 12.10

Problem : A triangular pyramid of base side 50 mm and height 65 mm rests on H.P. on its base with a base edge perpendicular to V.P. A plane cuts it parallel to H.P. and perpendicular to V.P. meeting the axis 30 mm above the base. Draw the sectional plan.

Solution : **(Fig. 12.11)**

Step 1: Draw the top and front view of traingle pyramid of base side 50 mm and height 65 mm rests on H.P.

Step 2: Draw the cutting plane is being prallel to H.P. which meets with the axis 30 mm above the base and mark the cutting poitns **1'**, **2'** and **3'**.

Step 3: From the cutting points, draw the projectors to top view and locate the points as **1**, **2**, and **3**.

Step 4: Join these points and hatching it, which is the required sectional plan view.

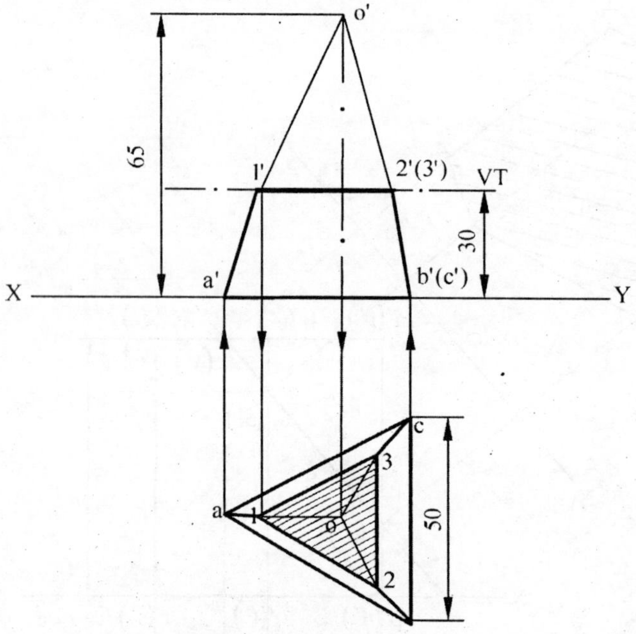

Fig. 12.11

Problem : A square pyramid of side of base 25 mm and height 55 mm rests on H.P. on its base with a base edge perpendicular to V.P. It is cut by a plane perpendicular to V.P. and inclined at 30° to H.P. The cutting plane meets the axis at 20 mm from the vertex. Draw the elevation, sectional plan and true shape of section.

Solutoin : **(Fig. 12.12)**

Step 1: Draw the top and front view of the square pyramid of base 25 mm and height 60 mm.

Step 2: Draw the cutting plane which is inclined at 30° to H.P. and it meets with the axis at 20 mm from the vertex, mark the cutting points **1'**, **2'**, **3'** and **4'**.

Step 3: Draw the projectors from cutting points to top view and mark the points **1**, **2**, **3**, and **4**.

Step 4: To draw the true shape of the section, follow the same procedure as in the previous example.

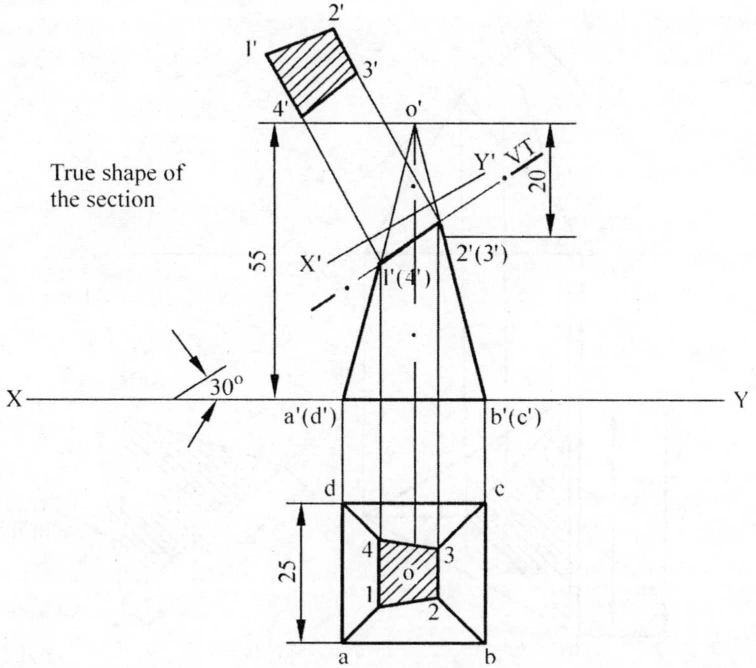

Fig. 12.12

Problem : A tetrahedron of 60 mm long edge rests with one of its faces on H.P. and an edge perpendicular to V.P. A section plane perpendicular to V.P. cuts the tetrahedron such that the true shape of section is an isosceles triangle of base 50 mm and altituve 36 mm. Draw the front view, sectional top view and the true shape of section. Also find the inclination of the section plane with H.P.

Solution : **(Fig. 12.13)**

1. Draw the top and front views of the tetrahedron of 60 mm long edges with one of its faces on H.P. and an edge is perpendicular to V.P.
2. Draw the cutting plane perpendicular to V.P. which should meet the three points, length should be 36 mm and angle of $\tan^{-1}\left(\frac{36}{25}\right)$ such that the true shape of the section is an isosceles triangle of base 50 mm and altitudes 36 mm can be maintained amd mark the cutting points **1'**, **2'** and **3'**.
3. From the cutting points, draw the projectors to top view and mark points as **1**, **2** and **3**.
4. To get the true shape of the section, draw the reference line **X'Y'** parallel to the cutting plane and project the lines from cutting points as in the drawing.

5. Measure the distance from XY plane to sectional top view and transfer it to reference line **X'Y'**. Thus mark the points **1' 2' 3'** and join these points to get the true shape of the section.

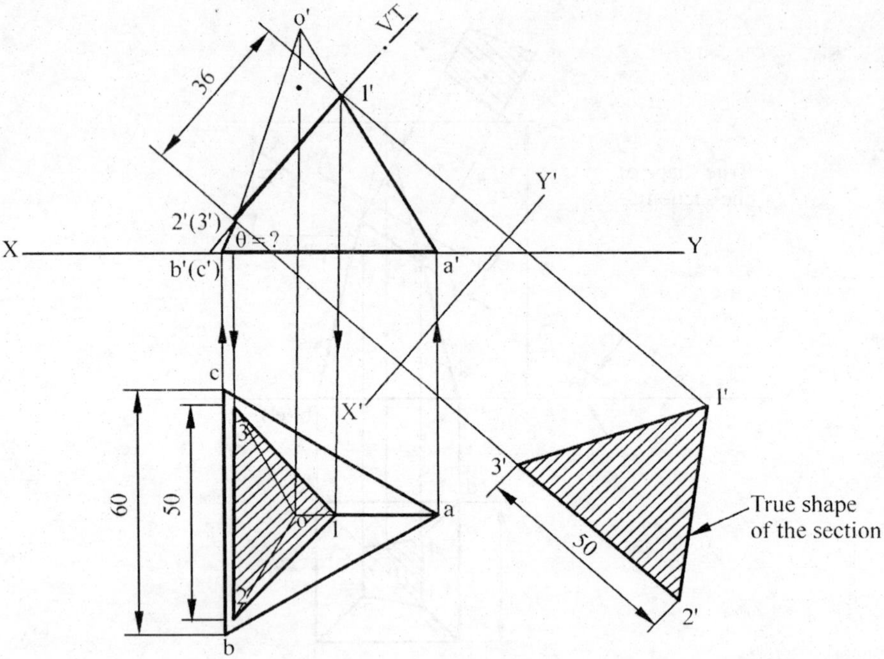

Fig. 12.13

Problem : A pentagonal pyramid of base side 40 mm and axis length 80 mm is resting on HP on its base with one of its base side parallel to VP. It is cut by a plane inclined at 30° to HP and perpendicular to VP and is bisecting the axis. Draw its front view, sectional top view, and the true shape of section.

Solution : **(Fig. 12.14)** Draw the projection of the pyramid in the given position. The top view is drawn and the front view is projected.

To draw the cutting plane, front view and sectional top view

1. Draw the *VT* of the cutting plane inclined at 30° to *XY* line and passing through the midpoint of the axis.
2. As a result of cutting, new corners 1', 2', 3', 4' and 5' are obtained on slant edges *a'o'*, *b'o'*, *c'o'*, *d'o'* and *e'o'* respectively.
3. Show the remaining portion in front view by drawing dark lines.
4. Project the new points to get 1, 2, 3, 4 and 5 in the top view on the respective slant edges .
5. Note that 2' is extended horizontally to meet the extreme slant edge *a' o' at m'*, it is projected to meet *ao* in top view at m. Considering *o* as centre, *om* as radius, draw an arc to get 2 on *bo*.

6. Join these points and show the sectional top view by drawing hatching lines.

To draw true shape of section.

1. Draw the new reference line $X_1 Y_1$ parallel to *VT* of the cutting plane.
2. Projectors from 1',2' etc. are drawn perpendicular to $X_1 Y_1$ line .
3. The distance of point 1 in top view from *XY* line is measured and marked from $X_1 Y_1$ in the projector passing through l' to get 1_1. This is repeated to get 2_1,3_1 etc.
4. Join these points and draw hatching lines to show the true shape of section.

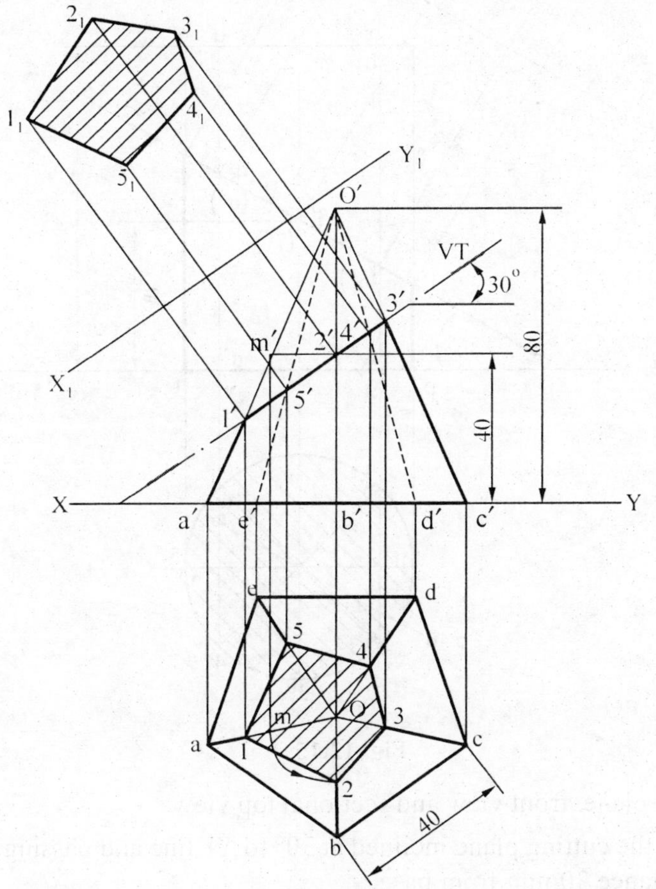

Fig. 12.14

Problem : A cylinder of base diameter 40 mm and height 60 mm rests on its base on HP. It is cut by a plane perpendicular to VP and inclined at 30° to HP and meets the axis at a distance 30 mm from base. Draw the front view, sectional top view, and the true shape of section.

Solution : **(Fig. 12.15)** Draw the projections of the cylinder. The top view is drawn and the front view is projected. Consider generators by dividing the circle into equal number of parts and project them to the front view.

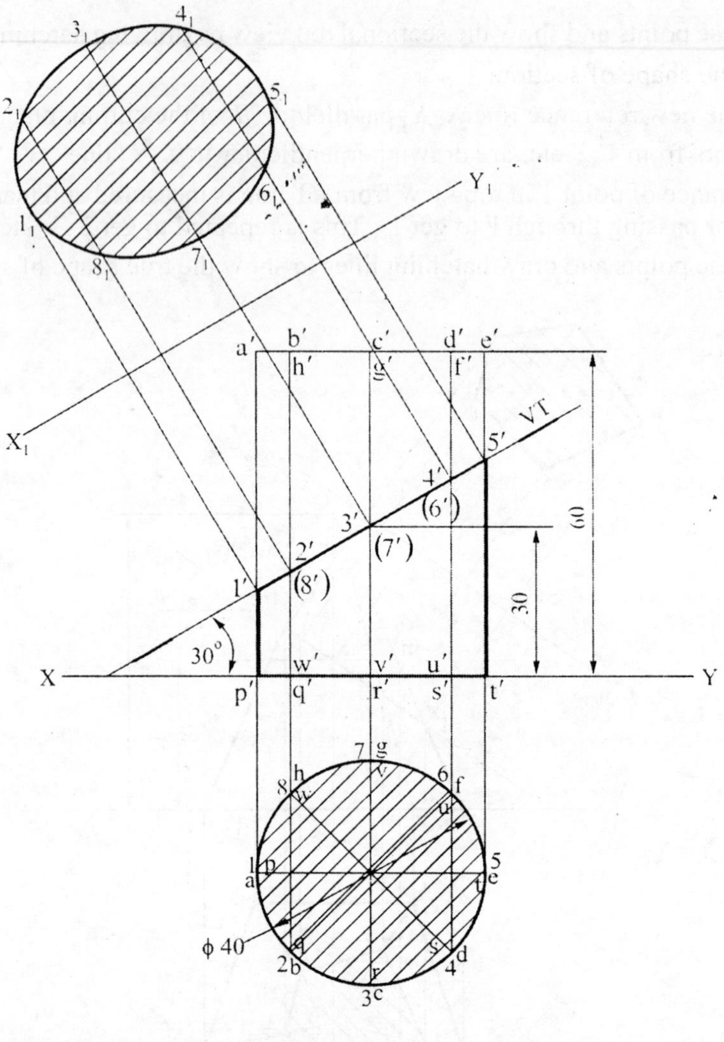

Fig. 12.15

To draw the cutting plane, front view and sectional top view

1. Draw the VT of the cutting plane inclined at 30° to *XY* line and passing through a point on the axis at a distance 30 mm from base.
2. The new point 1', 2' etc. are marked on the generators *a' p'*, *h' q'* etc.
3. Project the new points to the top view to get 1, 2, etc. which are coinciding with *p, q,* etc. on the base circle.
4. Join these points and draw the hatching lines to show the sectional top view.

To draw true shape of section.

1. Draw $X_1 Y_1$ line parallel to *VT* of the cutting plane.
2. Draw the projectors through 1', 2', etc. perpendicular to $X_1 Y_1$ line.

3. The distance of point 1 in top view from *XY* line is measured and marked from $X_1 Y_1$ in the projector passing through 1' to get 1_1. This is repeated to get other points $2_1, 3_1$ etc.
4. Join these points by drawing smooth curve to get the true shape of section and this is shown by hatching lines.

Problem : A cylinder of base diameter 45 and axis length 60 mm is resting on HP on one its generators with its axis perpendicular to VP. It is cut by a plane inclined 30° to VP and perpendicular to HP and is bisecting the axis of the cylinder. Draw its top view, sectional front view and true shape of section.

Solution : **(Fig. 12.16)** Draw the projections of the cylinder. Consider generators by dividing the circle into equal number of parts and project them to the top view.

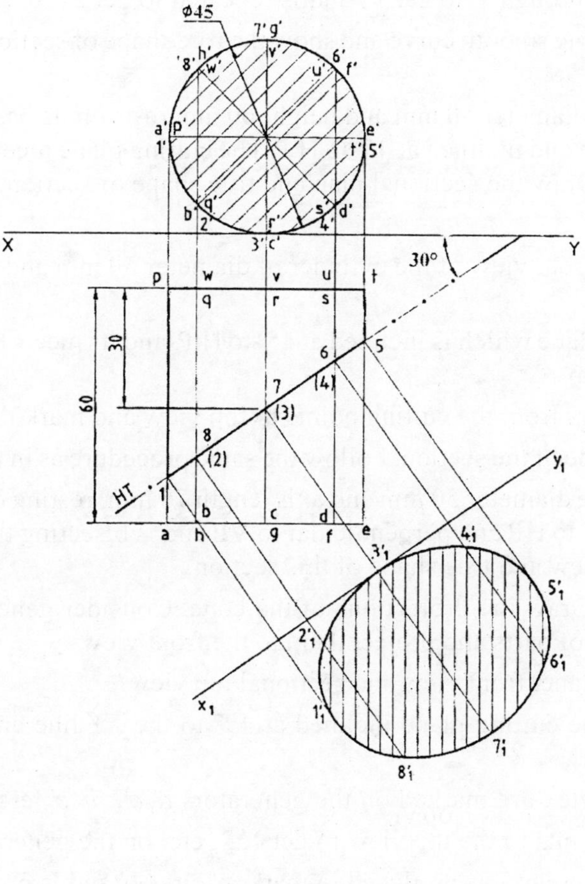

Fig. 12.16

To draw the cutting plane, top view and sectional, front view

1. Draw the *HT* of the cutting plane inclined at 30° to *XY* and passing through the midpoint of the axis.
2. The new points 1, 2, etc. are marked on generators *ap*, *bq*, etc.
3. Project the new points to the front view to get 1', 2' etc. which are coinciding with *p*, *q*, etc. on the base circle.
4. Join them and draw hatching lines to show the sectional front view.

To draw the true shape of section

1. Draw X_1Y_1 line parallel to *HT* of the cutting plane.
2. Draw projectors through 1, 2, etc. perpendicular to X_1Y_1 line.
3. The distance of 1' in front view from *XY* line is measured and marked from X_1Y_1 in the projector passing through 1' to get $1'_1$ and is repeated to get $2'_1$, $3'_1$ etc.
4. Join them by drawing smooth curve and show the true shape of section by drawing hatching lines.

Problem : A cylinder of diameter 50 mm and height 60 mm rests on its base on H.P. It is but by a plane pependicular to V.P. and inclined at 45° to H.P. The cutting plane meets the axis at a distance of 15 mm from the top. Draw the sectional plan and true shape of section.

Solution : **(Fig. 12.17)**

1. Draw the top and front view of the cylinder of diameter 50 mm and height 60 mm rests on its base on H.P.
2. Draw the cutting plane which is inclined at 45° to H.P. and its meets the axis at a distance of 15 mm from the top.
3. Draw the projectors from the cutting points to top view and mark the points (**1, 2...7**).
4. To get the true shape of the section. Follow the same procedure as in the previous examples.

Problem : A cone of base diameter 50 mm and axis length 75 mm, resting on HP on its base is cut by a plane inclined at 45° to HP and perpendicular to VP and is bisecting the axis. Draw the front view and sectional top view and true shape of this section.

Solution : **(Fig. 12.18)** Draw the projections of the cone. Consider generators by dividing the circle into equal number of parts and project them to the front view.

To draw the cutting plane, front view and sectional top view

1. Draw the *VT* of the cutting plane inclined at 45° to the *XY* line and passing through the midpoint of the axis.
2. New points 1', 2' etc., are marked on the generators *a' o'*, *h' o'*, etc.
3. Project the new points to the top view to get 1, 2, etc. on the generators *ao*, *bo* etc.
4. Note that the new point 3' is produced to mark m' on *a' o'* and is projected to get *m* on *ao*. Considering *o* as centre and *om* as radius, draw an arc to get 3 on *co* in the top view. The same method is repeated to get 7 on *go*.
5. Join these points by drawing smooth curve and draw the hatching lines to show the sectional top view.

To draw true shape of section

1. Draw $X_1 Y_1$ line parallel to *VT* of the cutting plane.
2. Draw the projectors through 1', 2' etc. perpendicular to $X_1 Y_1$ line.
3. The distance of point 1 in top view from *XY* line is measured and marked from $X_1 Y_1$ in the projector passing through 1' to get 1_1 and is repeated to get 2_1, 3_1 etc.
4. Join these points by drawing smooth curve to get the true shape of section and is shown by hatching lines.

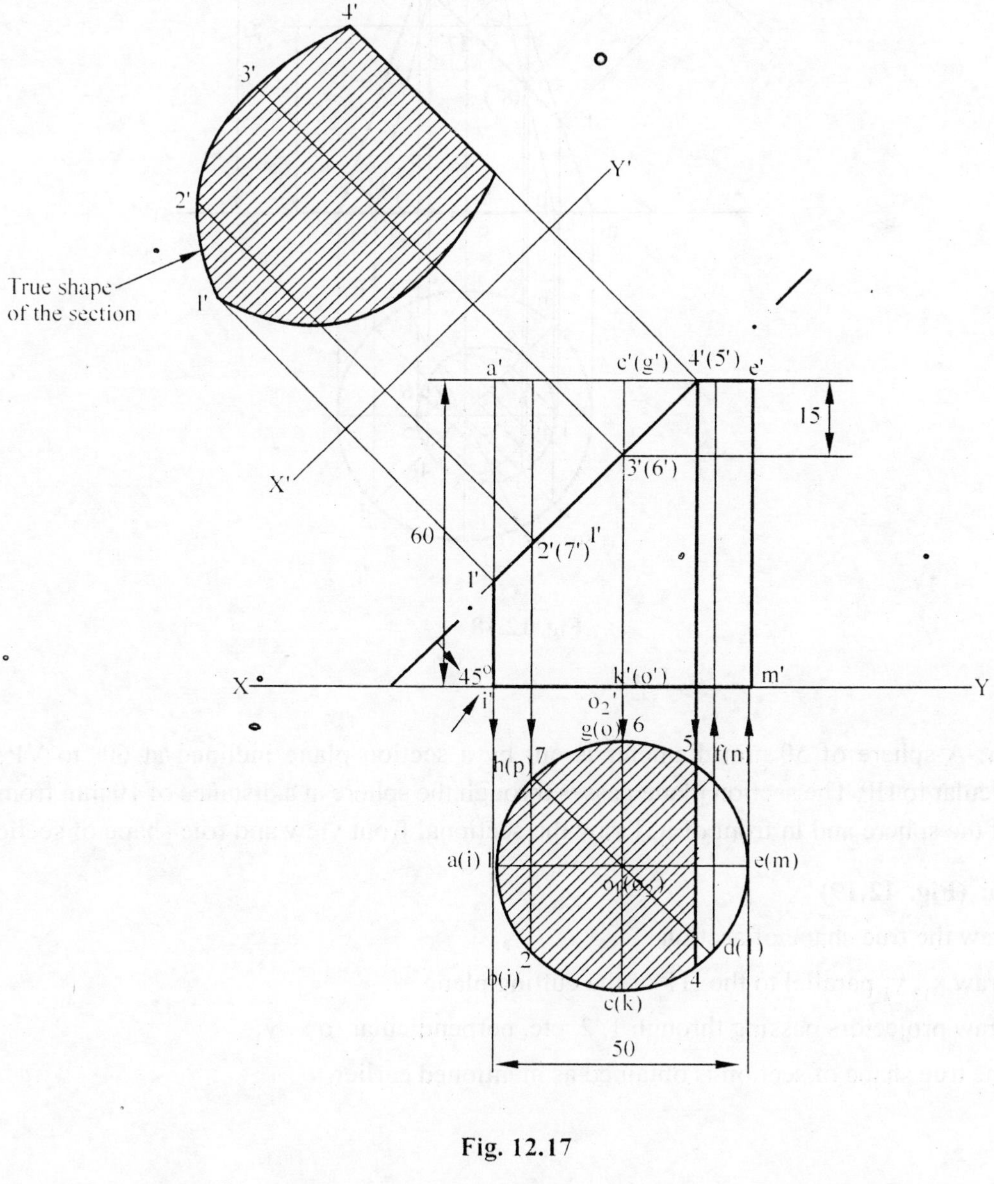

Fig. 12.17

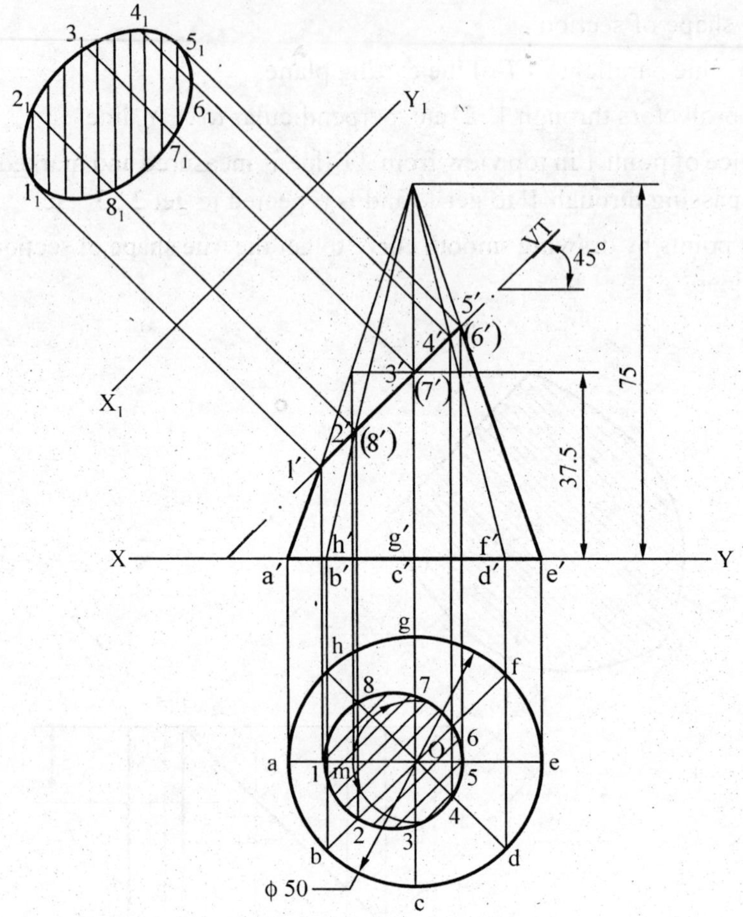

Fig. 12.18

Problem: A sphere of 50 mm diameter is cut by a section plane inclined at 60° to VP and perpendicular to HP. The section plane passes through the sphere at a distance of 10 mm from the centre of the sphere and in front of it. Draw the sectional front view and true shape of section.

Solution: (Fig. 12.19)

To draw the true shape of section

1. Draw x_1, y_1 parallel to the HT of the cutting plane.
2. Draw projectors passing through 1, 2, etc. perpendicular to x_1 y_1.
3. The true shape of section is obtained as mentioned earlier.

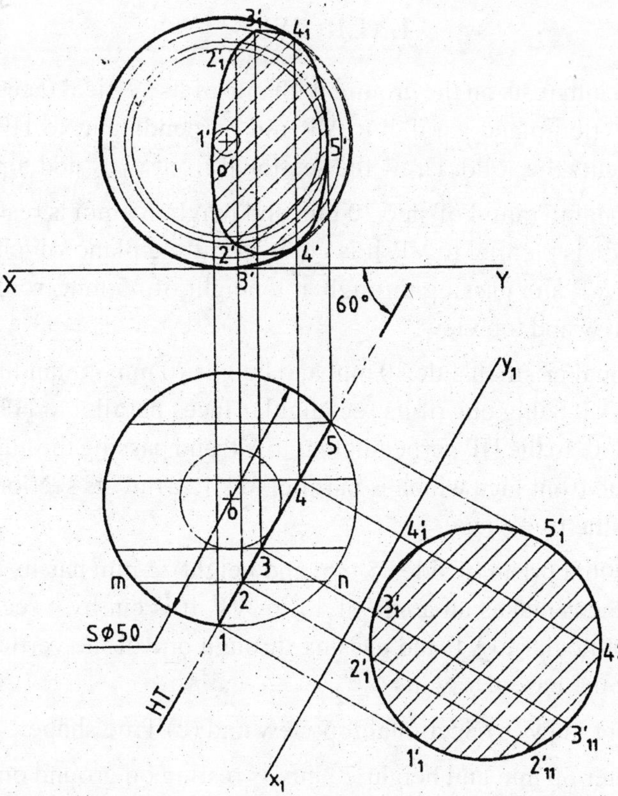

Fig. 12.19

EXERCISES

1. A cube of side 35 mm rests on the ground with one of its vertical faces inclined at 30° to the VP. A vertical section plane parallel to VP and perpendicular to HP and at a distance of 35 mm from VP cuts the solid. Draw the sectional front view and top view.
2. A regular hexagonal pyramid of side 30 mm and height 65 mm is resting on its base on HP one of its base side is parallel to VP. It is cut by a cutting plane which is parallel to HP and perpendicular to VP and passing through at a height of 45 mm from its bottom. Draw its sectional front view and top view.
3. A regular hexagonal prism of side 30 mm and height 70 mm is standing on VP with its axis perpendicular to VP being one of its rectangular faces parallel to HP. It is cut by a section plane inclined at 60° to the HP perpendicular to VP and passing through the mid-point of the bottom side on the front face which is parallel to HP. Draw its sectional front view and top view. Also draw the true shape.
4. A regular pentagonal prism of side 35 mm and height 75 mm has its base in HP and one of the rectangular faces makes an angle of 45° to VP. It is cut by a section plane inclined at 60° to HP, perpendicular to VP and passing through one of the vertical edges at a distance of 25 mm above the base. Draw its

   (a) Sectional front view (b) Sectional top view and (c) True shape.
5. A cone of diameter 60 mm and height 70 mm is resting on ground on its base. It is cut by a section plane perpendicular to VP inclined at 45° to HP and cutting the axis at a point 40 mm from the bottom. Draw the front view, sectional top view and true shape.
6. A right circular cylinder of diameter 60 mm and height 75 mm rests on its base such that its axis is inclined at 45° to HP and parallel to VP. A cutting plane parallel to HP and perpendicular to VP cuts the axis at a distance of 50 mm from the bottom face. Draw the front vi w and sectional top view.
7. A regular pentagonal pyramid of side 30 mm and height 60 mm is lying on the HP on one of its triangular faces in such a way that its base edge is at right angles to VP. It is cut by a plane at 30° to the VP and at right angle to the HP bisecting its axis. Draw the sectional view from the front, the top view from above and the true shape of the section.
8. A square pyramid base 50 mm side and axis 75 mm long is resting on the ground with its axis vertical and side of the base equally inclined to the vertical plane. It is cut by a section plane perpendicular to VP inclined at 45° to the HP and bisecting the axis. Draw its sectional top view and true shape of the section.
9. A hexagonal pyramid of base side 30 mm and height 75 mm is resting on the ground with its axis vertical. It is cut by plane inclined at 30° to the HP and passing through a point on the axis at 20 mm form the vertex. Draw the elevation and sectional plane.

10. A cube of 40 mm side rests on the HP on one of its faces with a vertical face inclined on 30° to VP. A plane perpendicular to the HP and inclined at 60° to the VP cuts the cube 5 mm away from the axis. Draw the top view and the sectional front view.
11. A cylinder 40 mm dia and 60 mm long is lying on the HP with the axis parallel to both the planes. It is cut by a vertical section plane inclined at 30° to VP so that the axis is cut a point 20 mm from one of its ends. Draw top view, sectional front view and true shape of section.

Chapter 13

Freehand Sketching

13.1 Introduction

Freehand sketching is one of the effective methods to communicate ideas irrespective of the branch of study. The basic principles of drawing used in freehand sketching are similar to those used in drawings made with instruments. The sketches are self explanatory in making them in the sequence shown (Fig. 13.1 to 13.14).

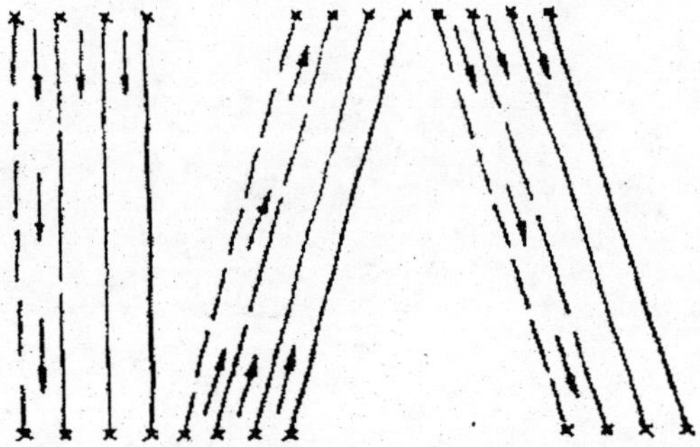

Fig. 13.1 Sketching Straight Lines.

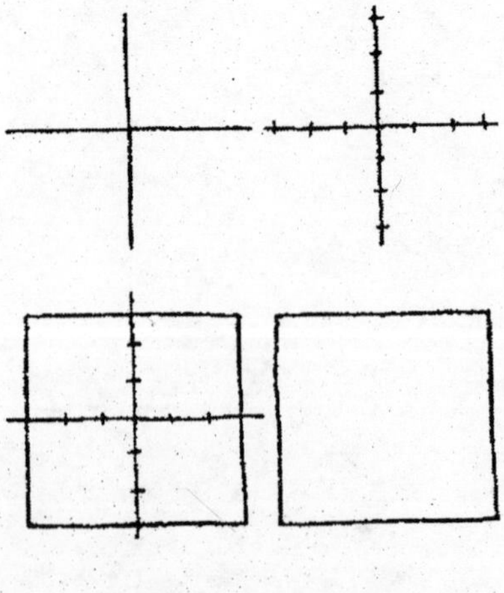

Fig. 13.2 Sketching a Square.

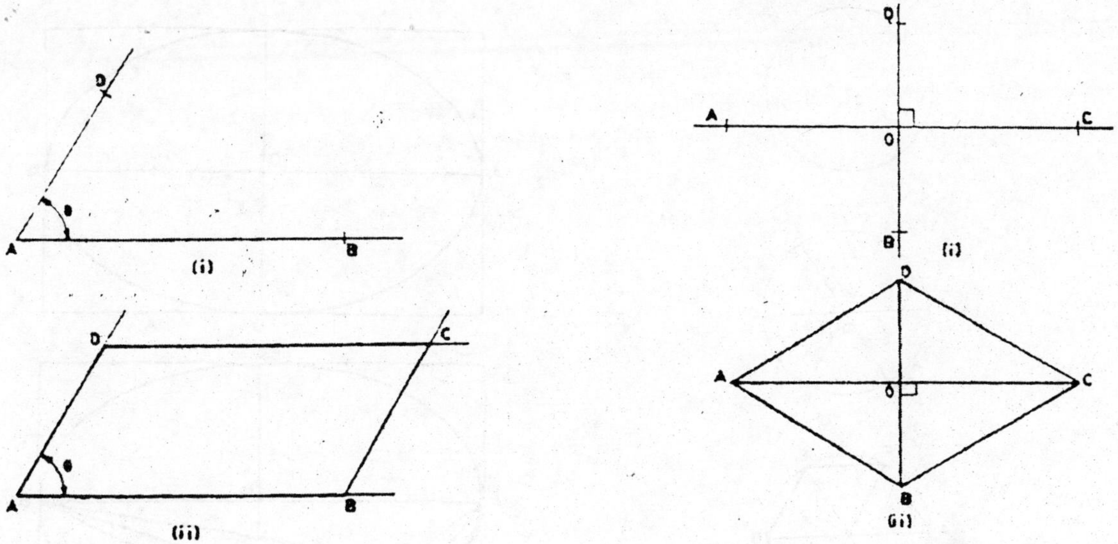

Fig. 13.3 Skectching a parallogram.

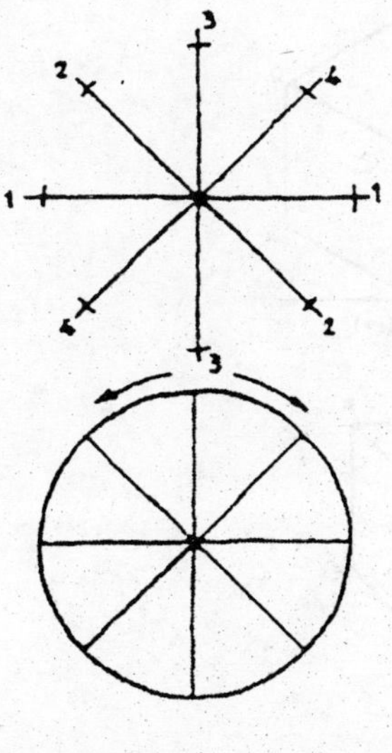

Fig. 13.4 Sketching a Circle.

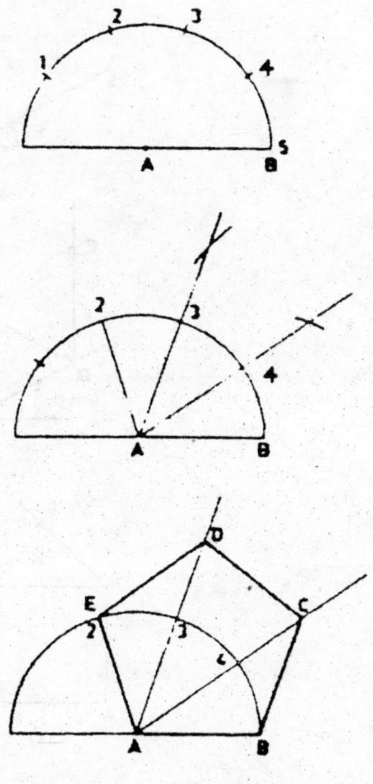

Fig. 13.5 Sketching a Pentagon.

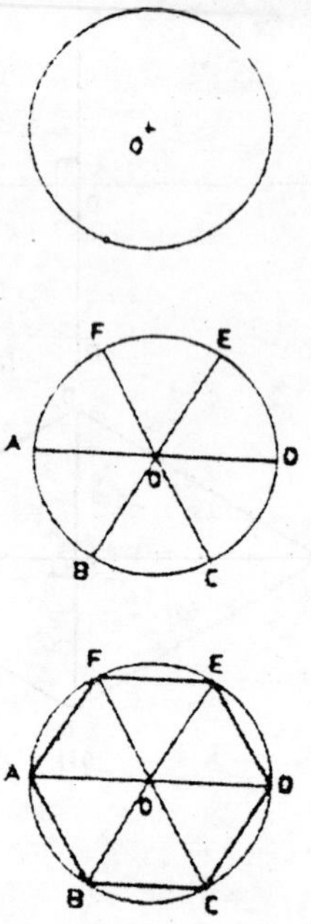

Fig. 13.6 Sketching a Hexagon.

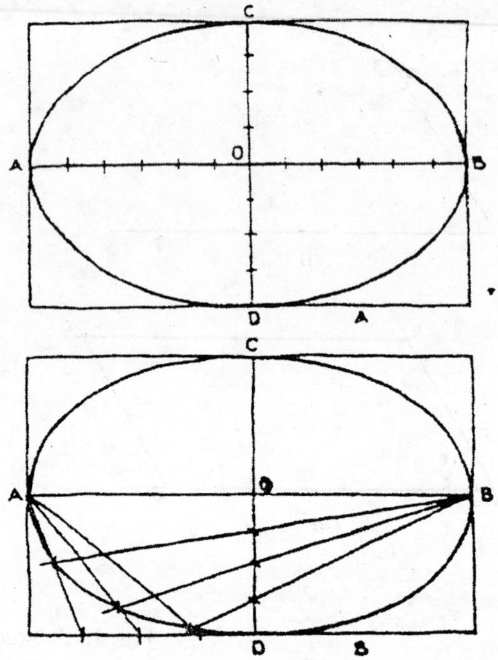

Fig. 13.7 Sketching an Ellipse.

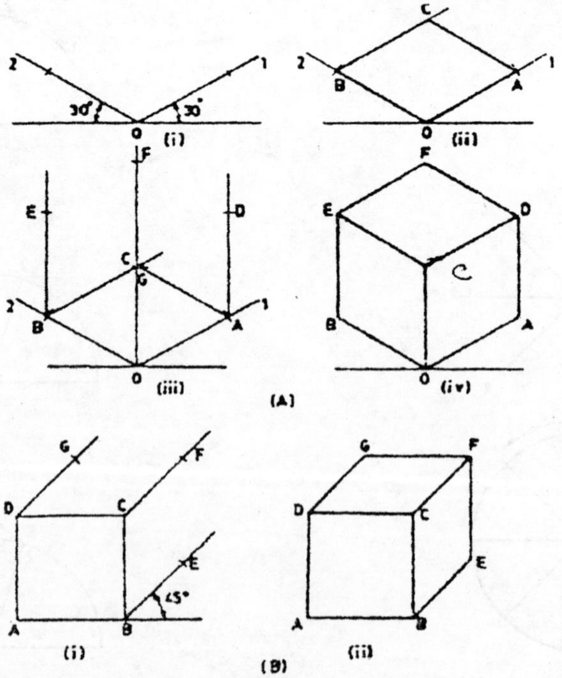

Fig. 13.8 Sketching a Cube.

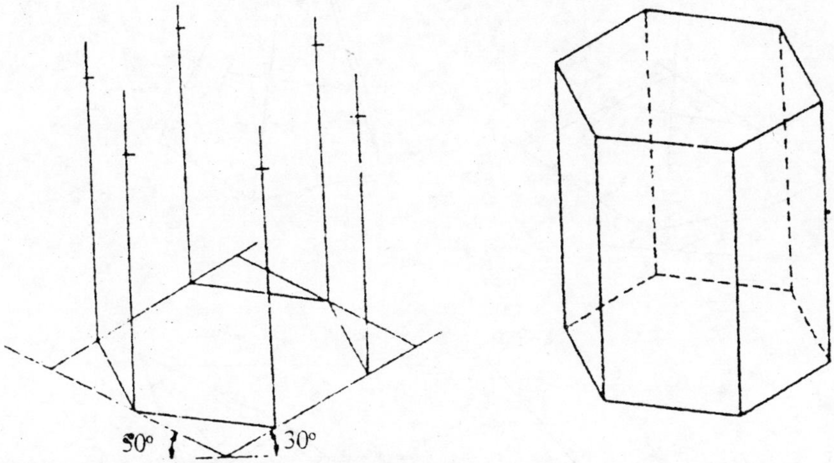

(a)

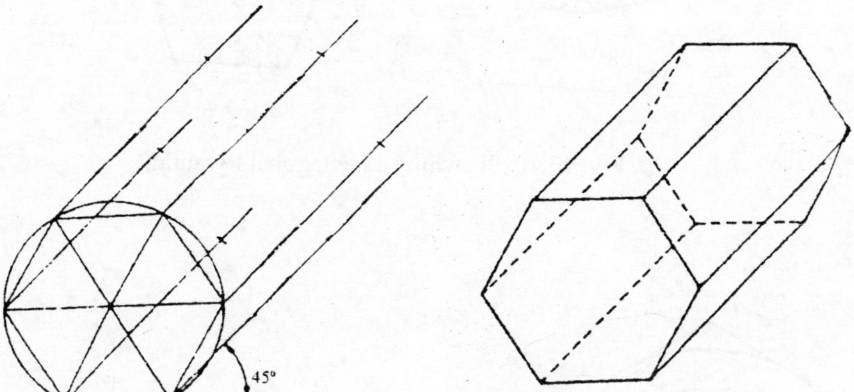

(b)

Fig. 13.9 Sketching a Hexagonal Prism.

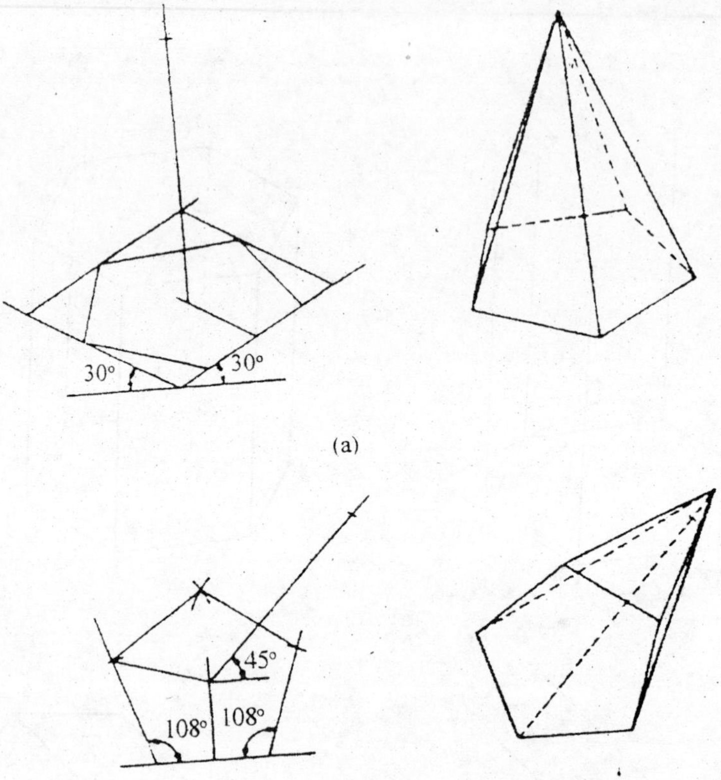

(a)

(b)

Fig. 13.10 Sketching a Pentagonal Pyramid.

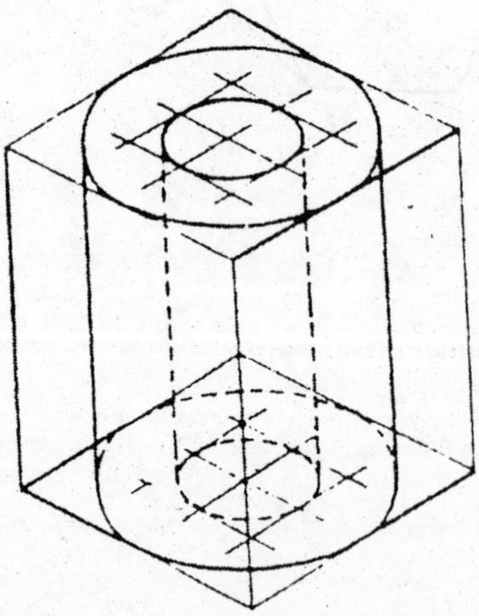

Fig. 13.11 Sketching a Hollow Cylinder.

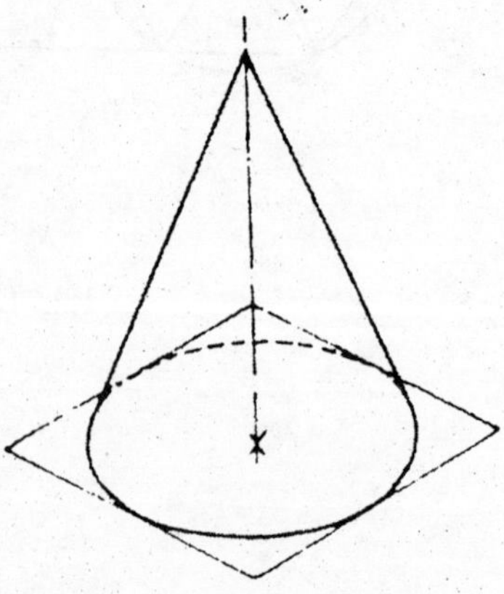

Fig. 13.12 Sketching a Cone.

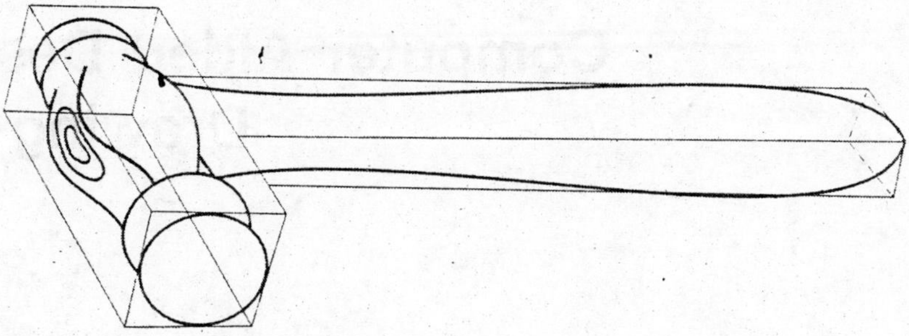

Fig. 13.13 Sketching a Ball Peen Hammer.

Fig. 13.14 Sketching a Cutting Plier.

CHAPTER 14

Computer Aided Design and Drawing (CADD)

14.1 Introduction

In previous chapters we dealt with traditional drawings in which we use essentially drawing board tools such as paper, pencils, drafter, compasses, eraser, scale etc., which will take more time and tough in complex drawings. The most drawback with traditional drawing is INFORMATION SHARING i.e. if an engineer is drawing design of machine component and suddenly the manufacturer want to modify dimension of innermost part of the component; in such situations one cannot modify the drawing already drawn, he should redraw the component.

CADD is an electronic tool that enables us to make quick and accurate drawings with the use of a computer. Drawings created with CADD have a number of advantages over drawings created on a drawing board. CADD drawings are neat, clean and highly presentable. Electronic drawings can be modified quite easily and can be presented in a variety of formats. There are hundreds of CADD programs available in the CADD industry today. Some are intended for general drawing work while others are focused on specific engineering applications. There are programs that enable you to do 2D drawings, 3D drawings, renderings, shadings, engineering calculations, space planning, structural design, piping layouts, plant design, project management, etc.

Examples of CAD software

AutoCAD, PRO/Engineer, IDEAS, UNIGRAPHICS, CATIA, Solid Works, etc.

14.2 History of CAD

In 1883 Charles Barbage developed idea for computer. First CAD demonstration is given by Ivan Sutherland (1963). A year later IBM produced the first commercial CAD system. Many changes have taken place since then, with the advancement of powerful computers, it is now possible to do all the designs using CAD including two-dimensional drawings, solid modeling, complex engineering analysis, production and manufacturing. New technologies are constantly invented which make this process quicker, more versatile and more powerful.

14.3 Advantages of CAD

(i) Detail drawings may be created more quickly and making changes is more efficient than correcting drawings drawn manually.

(ii) It allows different views of the same object and 3D pictorial view, which gives better visualization of drawings

(iii) Designs and symbols can be stored for easy recall and reuse.

(iv) By using the computer, the drawing can be produced with more accurately.

(v) Drawings can be more conveniently stored, retrieved and transmitted on disks and tapes.

(vi) Quick design analysis, also simulation and testing is possible.

14.4 Applications of Computer Graphics

There is virtually no area in which graphical displays cannot be used for some advantage, and so it is not surprising to find the use of computer graphics so widespread.

14.4.1 Computer-Aided-Design (CAD)

The biggest use of computer graphics has been as an aid to design since two decades. This is generally referred to as CAD, and it is powerful tool to design and draft the parts or figures interactively.

When an objectís dimensions have been specified to the computer system, the designer can view any side of the object, to see how it will look after construction.

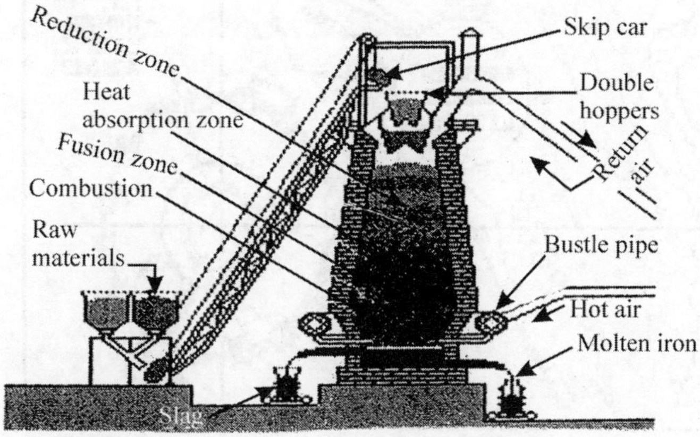

Fig. 14.1 Blast Furnace Schematic Drawing with CAD System.

Civil, Mechanical, Eletrical and Electronics Engineers rely heavily on CAD methods. For e.g., Eletronic circuits are typically desgined with computer graphics. Using pictorial symbols to represent various components, a designer can build up a circuit on monitor, by successively adding components to the circuit layout.

Automobile, Aircraft, Aerospace and Ship designers used CAD techniques for the design of various types of vehicles. Wire-frame drawings are used to model individual components and **plan** surface for Automobiles, Airplanes, Spacecrafts and Ships. Individual section of vehicle components can be designed separately and fitted together to display the total object.

Building design are also created with computer graphics by the Architects for the design of floor plans, arrangement of doors, windows, and three dimensional building models permit Architects to study the appearance of single building or a group of buildings.

14.4.2 Graphs and Chart Methods

A number of commercially graphics programs are designed specifically for generation of graphics and charts. A graph plotting program will have the capability of generating a variety of graph types, such as Bar charts, Line graphs, Surface graphs and Pie charts etc.

Business graphics are one of the most rapidly growing areas of application, that makes extensive use of visual displays as a means for rapidly communicating the vast amounts of information that are compiled for managers and other individuals in an organization.

Graphs and charts are typically used to summarize, Financial, Statistical, Mathematical, Scientific and Economical data and several graphs are often combined in one presentation. These graphs are generated for Research Reports, Managerial Report, Consumer Information bulletins and Visual aids that are used during presentation.

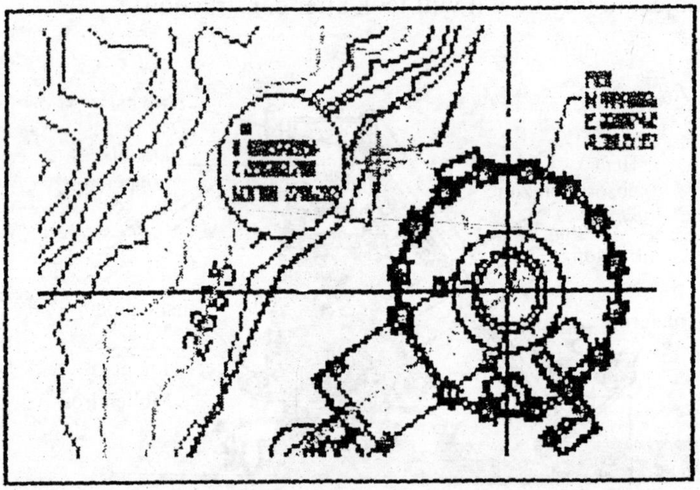

Fig. 14.2 Diagram used to Explain the Weather Report.

14.4.3 Computer Art

The creative and commercial art applications make extensive use of computer graphics. Paint brush programs allow artists to create pictures on the screen. Actually, the artist might draw the picture on a 'Graphic Tablet' using a stylus as input.

Computer generated art is widely used in commercial applications, such as Logos and Advertisement designs for TV messages are commnly produced with graphics systems. In addition, graphics programs have been developed for applications in publishing and word processing, which allows graphics and text editing operations to be combined.

Fig. 14.3 Computer Art Produced on a Pen Plotter.

Fig. 14.4 Cartoon Drawing with Paintbrush Program.

14.4.4 Computer Animation

Animation is a frame of sequence moved slightly from one frame to the next to simulate motion.

Animation methods are used in Education, Training, Entertainment and Research Applications etc. For some training applications, special systems are developed.

Fig. 14.5 Graphics Simulation.

14.4.5 Image Processing

The graphics technique used for producing visaul displays from photographs or TV scans is called Image Processing.

In computer graphics, a computer is used to create the picture. On the other hand Image processing technique uses a computer to digitilize the shadings and color patterns from already existing picture. This digitalized information is then transfered to the screen.

These methods are useful for viewing many objects or systems that we cannot see directly, such as TV scans from Spacecraft or view from the eye of an Industrial Robot.

Once, a picture has been digitalized, additional processing techniques can be applied to rearrange picture parts to enahance color separation, or to improve the quality of shading.

Fig. 14.6 A Drawing Becomes Legible After the Image Processing.

Image processing is extensively used in commercial applications involving the retouching and rearranging of section of photographs and other artwork, and medical applications make use of image processing techniques both for picture enhancements and in tomographs. (Tomography is a technique of x-ray photography that allows cross sectional views of physiological system to the displayed).

14.4.6 Wireframe Model

A wireframe model is the simplest geometric modelling type, where an object is described by points, lines, circles and curves in 3D representations.

14.4.7 Surface Model

A surface model is similar to wireframe model where an object is described by surface entities.

14.4.8 Solid Model

A solid model is a complete representation of a surface model, where the object is described by the solid entities such as a block, cylinder, cone, sphere, wedge, torus etc. It always appear as a solid to the viewer.

A solid model will have physical material properties such as mass, density and other design data rotated with an object. A solid model is also compared with an actual object. It can be rotated easily to visualize the object as it is on the computers monitor screen.

The commonly used solid modelling softwares are PRO/ENGINEER, IDEAS, CATIA, UNIGRAPHICS, SOLID WORKS, etc.

14.5 Auto Cad Main Window

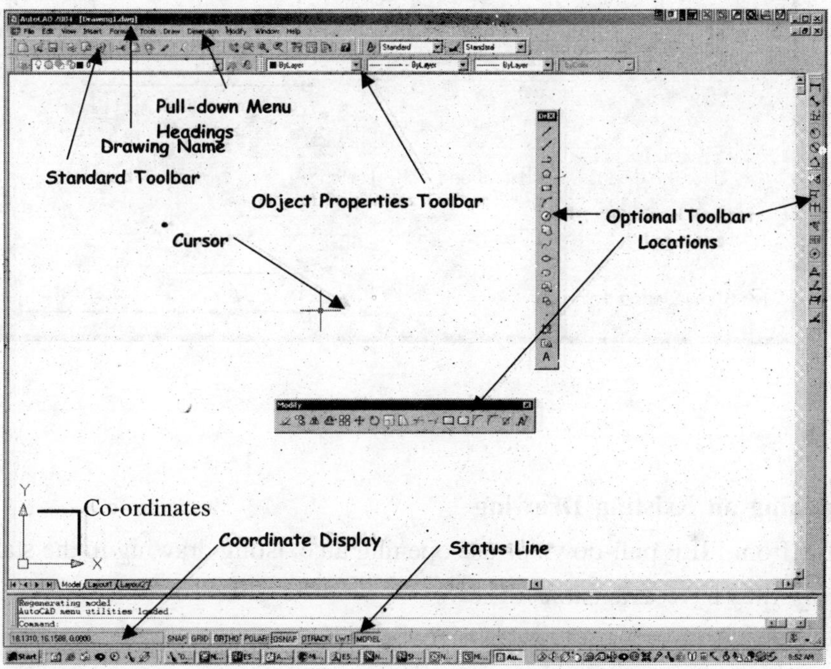

Fig. 14.1

14.5.1 Starting a New Drawing

Select NEW file from pull-down or Toolbar

File>New

Startup dialog box will be opened, with four Options

- Open an existing drawing
- Start from scratch
- Use a template
- Use a Wizard

Select *Start from Scratch* , Click on *Metric(millimetres).*

Fig. 14.18

14.5.2 Opening an Existing Drawing

Choose OPEN from FILE pull-down or use opening an existing drawing in the start-up dialogue

Note: Drawing files have extensions of .dwg

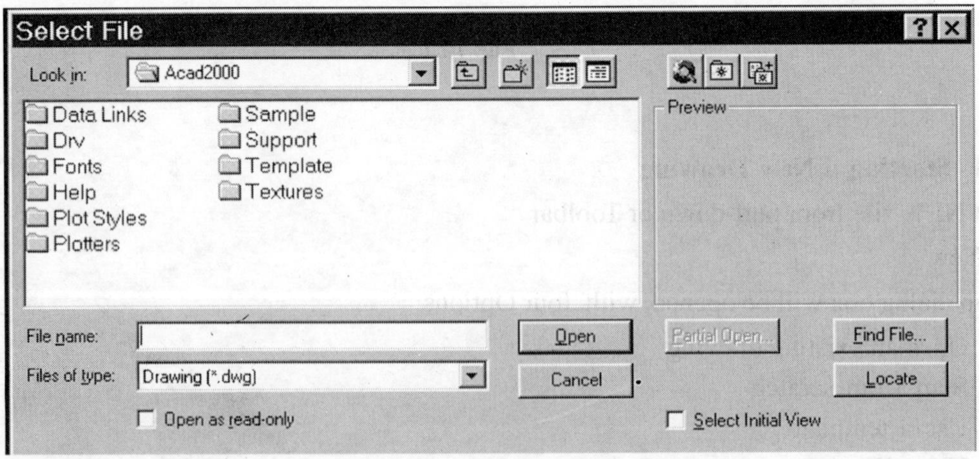

Fig. 14.19

14.5.3 Setting Drawing Limits

It is normal when using AutoCAD to draw objects full size, so it is usually necessary to reset the drawing limits to (about) the size of the object being drawn. Move the cursor to the bottom left of the screen, you can notice Command box. We can fix required paper size like A0, A1, A2, A3, A4 etc. from the Command box.

```
Command:
Command: '_limits
Reset Model space limits:
Specify lower left corner or [ON/OFF] <0.0000,0.0000>:
Specify upper right corner <420.0000,297.0000>: 297,210
Command:
```

Fig. 14.10

14.5.4 Erasing Objects

Removes objects from the drawing: Activate from Modify pull-down

Prompt will appear to.... select objects

Cursor changes to selection box

Ways to selecı objects for erasure:

Pick with selection box

Create a window to select multiple objects

Type ALL to select everything visible on screen

14.5.5 Saving a Drawing File

Save

Saves drawing to current name (Quick save)

Allows user to input name if drawing has never been saved

Saveas

Allows input of a drawing name or location every time

Provides the ability to change file saving version

14.5.6 Exiting an AutoCAD Session

Close

- Closes the drawing but does not leave the software

Exit

- Closes the drawing AND leaves the software

Note: Both will give an extra chance to save the changes in your work

14.6 The Coordinate System

The coordinate system is another method of locating points in the drawing area. It enables us to locate points by specifying distances from a fixed reference point. One can locate a point by giving its distance in the horizontal direction, vertical direction, measuring along an angle, etc.

The coordinate system is available when a function requires data input in the form of point locations. You may use it while drawing, editing or any time you need to locate a point. The most common coordinate systems are as follows:

- Cartesian coordinates
- Polar coordinates

14.6.1 Cartesian Coordinates

Cartesian coordinates is a rectangular system of measurement that enables you to locate points with the help of horizontal and vertical coordinates. The horizontal values, called X-coordinates, are measured along the X-axis. The vertical values, called Y-coordinates, are measured along the Y-axis. The intersection of the X- and Y-axes is called the origin point, which represents the 0,0 location of the coordinate system.

The positive X values are measured to the right and the positive Y values are measured above the origin point. The negative X and Y values are measured to the left and below. To enter a coordinate, you need to enter both the X and Y values separated by a comma (X, Y).

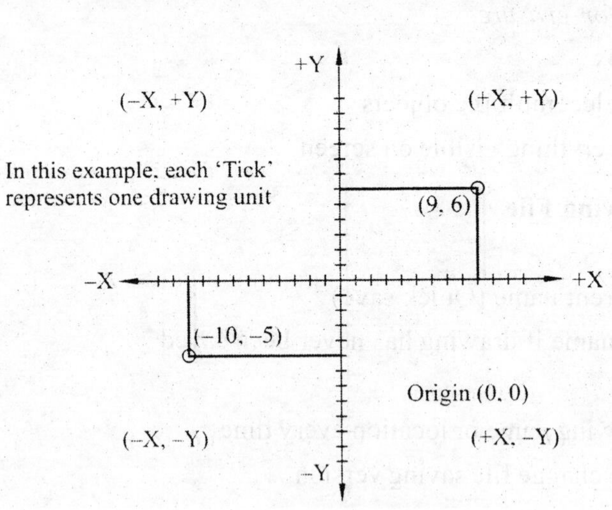

Fig. 14.11

14.6.2 Polar Coordinates

Polar coordinates allow you to define a point by specifying the distance and the direction from a given point. This mode of measurement is quite helpful in working with angles. To draw a line at an angle, you need to specify how long a line you want to draw and specify the angle.

14.7 The Formats to Enter Coordinates

Cartesian or polar coordinate values can be entered in two formats:

- Absolute
- Relative

Absolute format is a way of measuring distances from a fixed reference location (origin point). which is the 0,0 location of the coordinate system. Consider this point to be stationary at all times. In some CADD programs this point remains visible at the left bottom corner of the drawing area, while in others it is invisible.

You can use this point as a reference to measure any distance in the drawing. Absolute coordinates are primarily used to adjust the alignment of diagrams in a drawing, to align one drawing with another or to make plotting adjustments.

Relative format is a way of measuring distances from the last point entered. All measurements are taken the same way as the absolute coordinates, with the only difference being that the relative coordinates are measured from the last point entered instead of the origin point. When a point is entered, it becomes the reference for entering the next point and so on. This mode of measurement is frequently used for drawing because it is always convenient to place the drawing components relative to each other rather than a fixed reference point.

Examples

Cartesian Coordinates

- Sounds like math, and it is exactly the same as in math
- Input as either Absolute or Relative Coordinates
  - *Absolute X,Y*
  - *Relative @X,Y*

Polar Coordinates (Vector Coordinates)

- Used to input a distance and the direction angle
- Format: *@Distance<Angle*

14.7.1 User-Defined Coordinate System

CADD allows you to create a user-defined coordinate system that can help simplify drawing. When you need to work with a complex drawing that has many odd angles this mode of measurement is very useful.

Let us say you need to draw or modify an odd-shaped diagram, it is very difficult to use Cartesian or polar coordinates because they would involve extensive calculations. In this case, you can create a custom coordinate system that aligns with the odd angles of the diagram.

To define a new coordinate system, you need to specify where you want the origin point and the direction of the X and Y-axis. The computer keeps working according to this customized coordinate system unless you set it back to normal

The user-defined coordinate system is especially helpful when you are working with 3D. In a 3D drawing, you need to define each point with three coordinates and work with various surfaces of a 3D model. The user-defined coordinate system allows you to align coordinates with a specific surface.

Note: At the bottom left of the AutoCAD is the coordinate display.

Move your cursor around the drawing area and watch the coordinate change

Function Keys

F1 - AutoCAD Help Screens

F2 - Toggle Text/Graphics Screen

F3 - OSNAP On/Off

F4 - Toggle Tablet Modes On/Off

F5 - Toggle Isoplanes Modes On/Off

F6 - Toggle Coordinates Modes On/Off

- Coordinates has 2,modes when in a drawing command
- *X,Y coordinates*
- *Polar coordinates (Distance<Angle)*

F7 - Toggle Grid Modes On/Off

F8 - Toggle Ortho Modes On/Off

F9 - Toggle Snap Modes On/Off

4F10 - Toggle Polar Modes On/Off

F11 - Toggle Object Snap Tracking Modes On/Off

Definitions :

– Click
- Press once and release
- Also commonly used to refer to left-click

– Left-click
- Press left-mouse button (LMB) once and release
- Commonly used to pick or choose an item

– Right-click
- Press right-mouse button (RMB) once and release
- Commonly used to access pop-up menu

– Double-click
- Commonly referred to clicking left mouse button twice

– Click-and-drag
- Commonly referred to pressing left mouse button (and *not* releasing it) and move the mouse as required
- The left mouse button is released to finish the command

14.8 Choosing Commands in AutoCAD

Methods of choosing commands or to execute the command

- Pick from pull-down menu
- Select from Toolbar
- Type command on Command Prompt Line

14.8.1 Pull-down Menus [pd menu] (Fig 14.12)

Select pull-down through left mouse button

Move mouse to command and click left button to select command

ñ Example:

- To draw straight line,
- [pd menu] >Design > Line

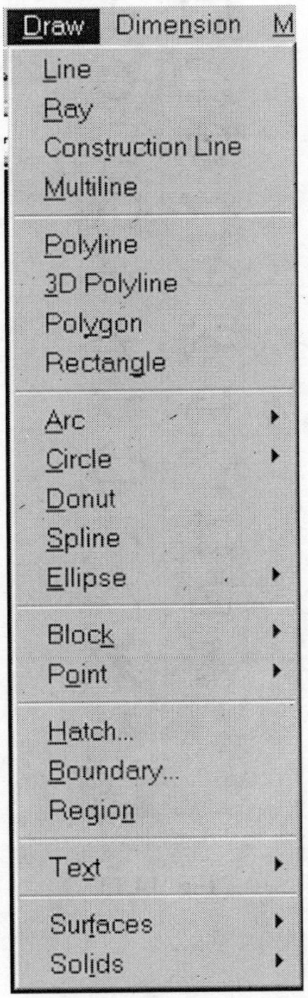

Fig. 14.12

14.8.2 Tool Bar Selection

Use mouse to track over Toolbar image for button detection

Left mouse button click will select command

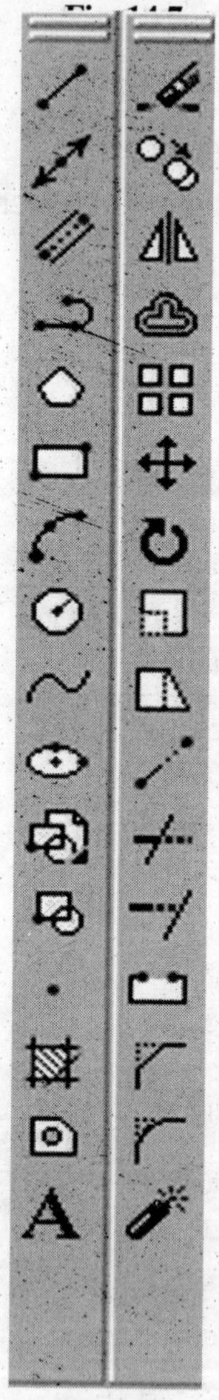

Fig. 14.13

14.8.3 Activating Tool Bars

Right mouse click on any toolbar button will bring up the toolbar activation menu

Pick toolbar menu to activate it

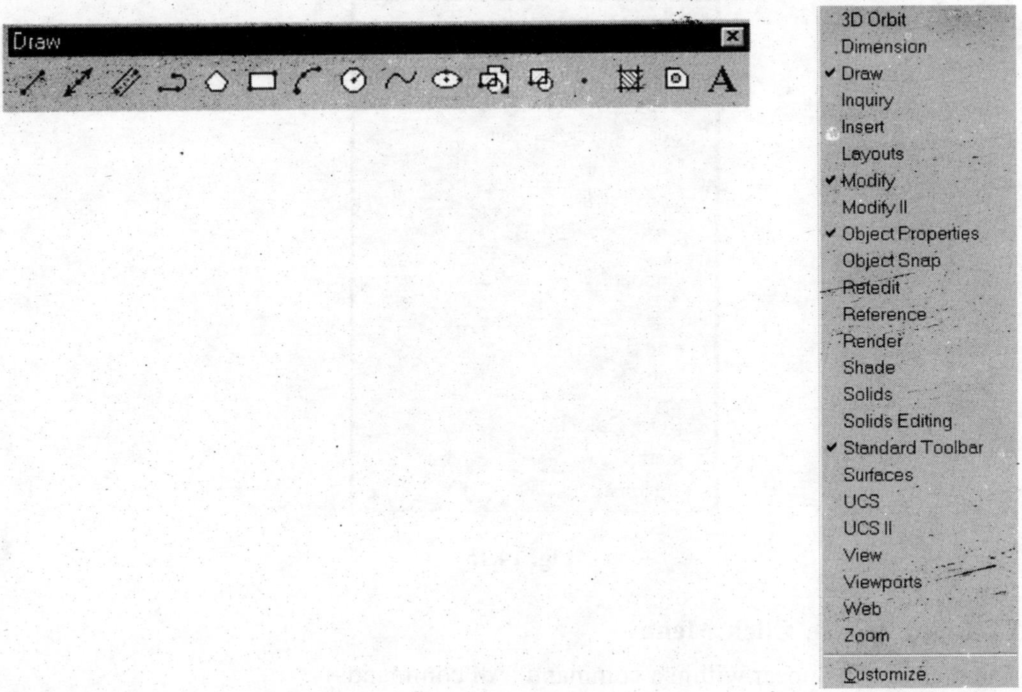

Fig. 14.14

Command Prompt Line (Fig. 14.14)

Located at bottom of screen

Left click in command prompt area and type in command

Press Enter key to input command

To draw a straight line type in

- Command: line {enter}

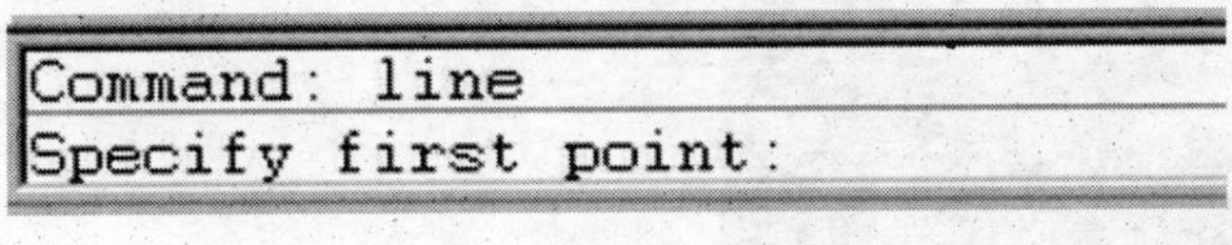

- You just have to type in the word line and press the key Enter (on the keyboard)
- The command is not case sensitive

14.9 Right Mouse Clicking

- It will get you everywhere!

Command and area specific menus will appear upon right mouse clicking

The default Right mouse click menu is shown.

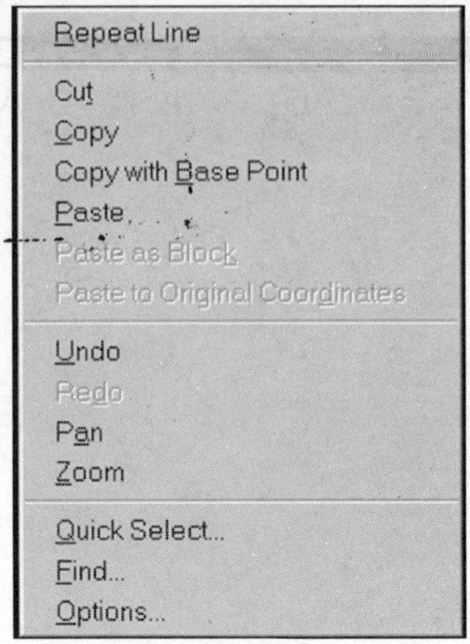

Fig. 14.15

14.9.1 Right Mouse Click Menu

Short-Cut menus will appear within a command (All commands)

The Default editing menu appears on the right

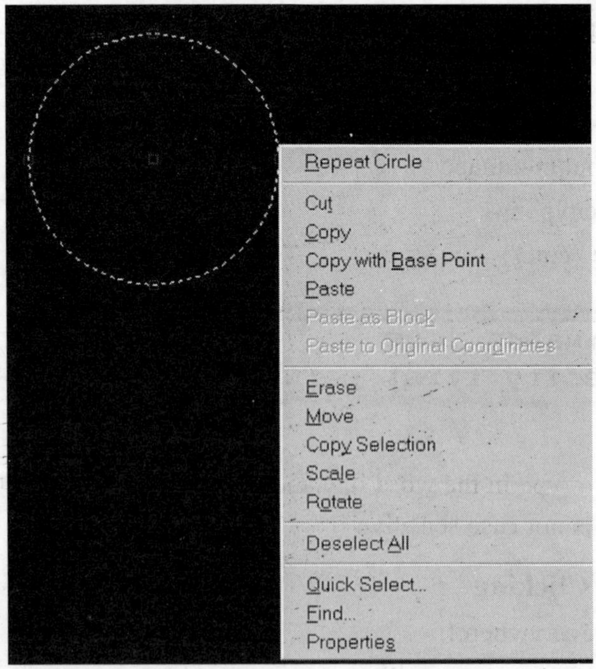

Fig. 14.16

14.10 Object Snaps

Uses geometry to locate specific positions

- Activation of Object Snaps (one time use only)
  - *Typing the first 3 letters of the object snap name*
  - *Holding down the shift key while clicking the right mouse button*

14.10.1 Types of Object Snaps

Center

Center of circle or arc

Endpoint

End of line or arc

Extension

Extends lines & arcs by a temporary path

From

Must be used with another Object snap to establish a reference point

Insert

Locates an insertion point of an object

Intersection

Finds the common intersection point between two objects

Extended Intersection

Locates the intersection between objects that do not touch

Midpoint

Locates the middle of arcs & lines

Node

Snaps to a point

Parallel

Assists in constructing a line parallel to another

Perpendicular

Snaps to an angle of 90° to the selected object

Quadrant

Finds the 0°, 90°, 180°, and 270° locations about a circle

Tangent

Assists in creating lines, arcs or circles tangent to another object

Deferred Tangent

Occurs when multiple tangent selections are needed to complete a task

[ex. Drawing a line tangent to two circles requires two tangent picks, one for each circle, the first tangent selection is a deferred selection (line does not appear) until both tangents have been selected]

Object Snap Tracking

- *Allows the user to select more than one object snap location to determine a specific position. [ex. Use for finding the center of a rectangle in conjunction with the midpoint object snap]*

14.10.2 Running Object Snaps

The object snaps that never stop working for you

Activate from Tools pull-down, drafting settings

Right click on OSNAP button on status bar

On/Off by OSNAP button

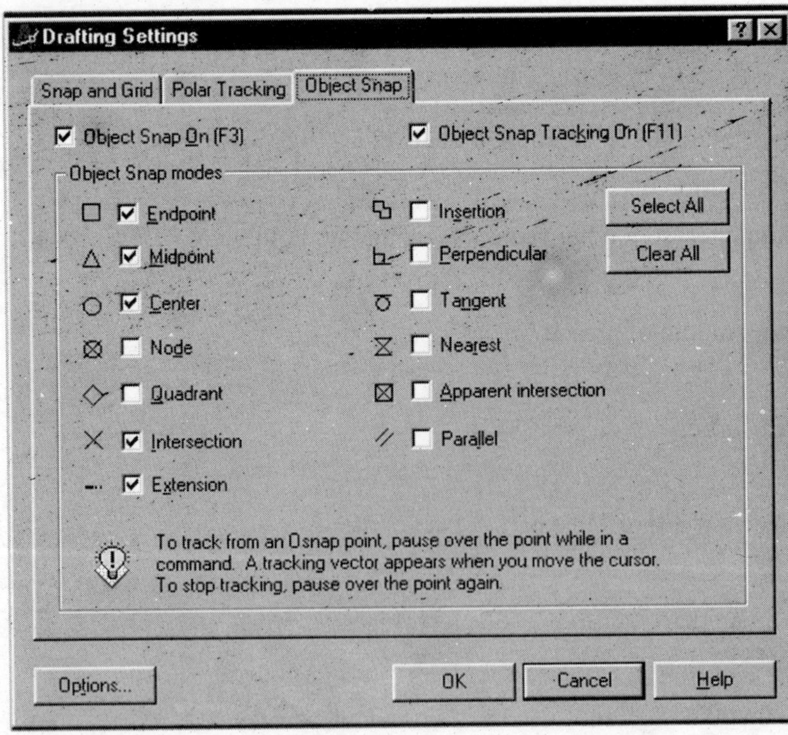

Fig. 14.17

14.10.3 Dividing an Object into Equal Segments

Divide Command

- Divides an entity into equal segments
    - *Select entity to divide*
    - *Enter the number of equal segments desired*
    - *Set Point style to another style other than default dot*
    - *Use OSNAP node to select points*

14.10.4 Setting off Equal Distances

Measure command

- Locates points based on the input distance
    - *Points are started from the closest endpoint that is used to select the line*
    - *Set Point style to another style other than default dot*
    - *Use OSNAP node to select points*

14.10.5 Polyline Command

Creates a multisided closed shape

- Located by the center and radius

  or
- Located by edges of polygon

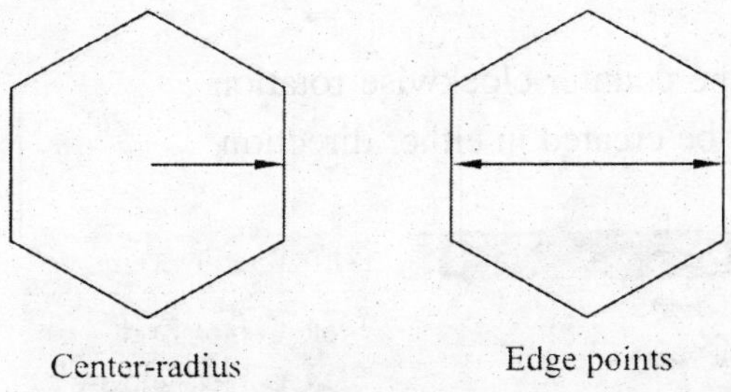

Fig. 14.18

- Polygon may be inscribed or circumscribed about a circle (Remember inscribed is inside!)

(a) Circumscribed Polygon is outside like a bolt head radius is to the middle of segment

(b) Inscribed Polygon is inside radius is to the corner

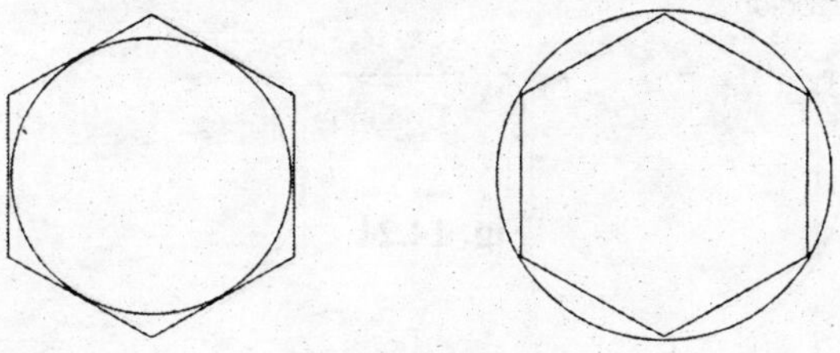

Fig. 14.19

14.10.6 Ray Command

- Creates a line that is fixed on one end and infinite in one direction
    - *When a ray is broken, one segment turns into a line and the other remains a Ray, which is infinite in one direction*

14.10.7 Rectangle Command

- Creates a rectangular object based on two opposite corner points
- Use relative coordinates: @dist<ang or @X,Y for second point
    - *Rectangle options*
    - Fillet - fillets the corners based on input radius
    - Chamfer - chamfers the ends based on chamfer distances
    - Width - changes width of lines

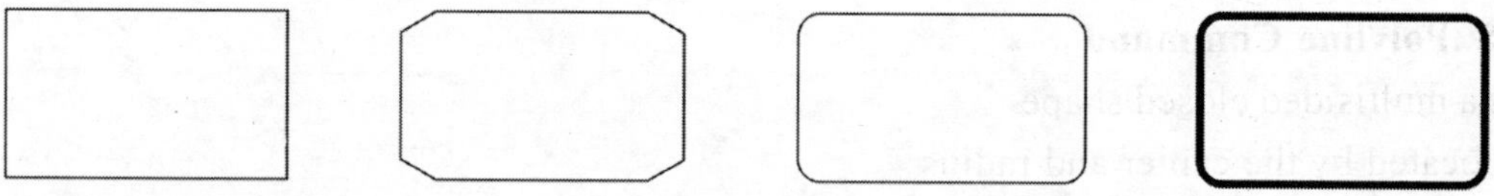

Fig. 14.20

14.10.8 Arc Command

- Arcs are created in the counter-clockwise rotation
- The 3 point Arc may be created in either direction

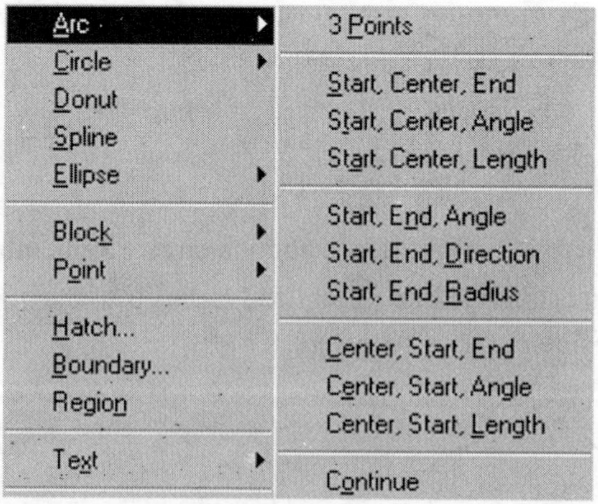

Fig. 14.21

3 point Arc

- 3 points are selected on the circumference of the circle

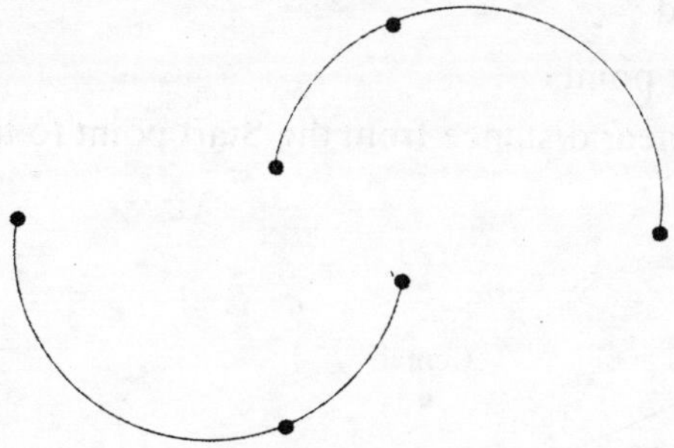

Fig. 14.22

Start, Center, End

Selection begins with a Start point on the circumference.

- Then select the center point of the arc
- Finally, select the End point on the circumference counter-clockwise from the start point

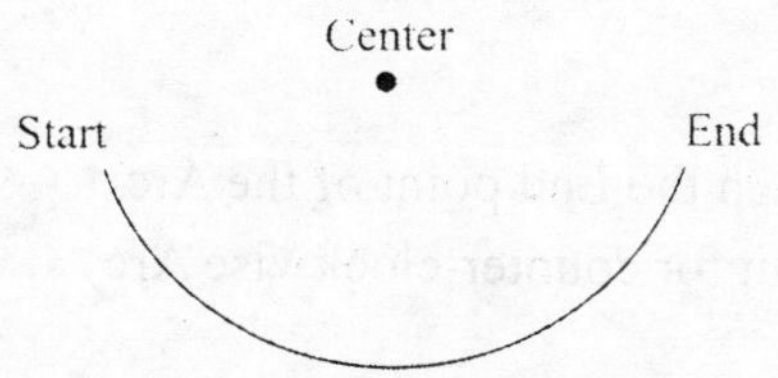

Fig. 14.23

Start, Center, Angle

- Select the Start point and Center locations.
- Type the value for the Angle in for counter-clockwise arc creation

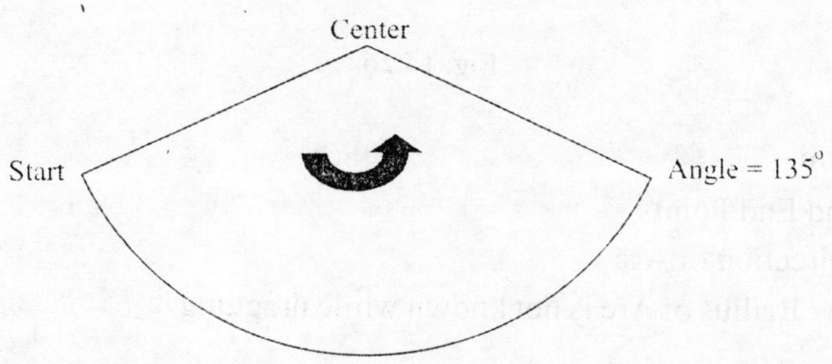

Fig. 14.24

Start, Center, Length of Chord

- Select the Start and Center points
- The length of chord is a linear distance from the Start point to the End point based on the arc's radius

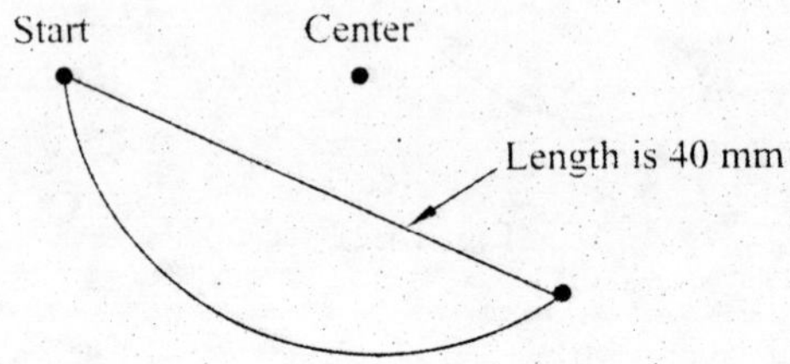

Fig. 14.25

Arc Command

Start, End, Angle

- Select the Start Point and then the End point of the Arc
- The Angle value is then input for counter-clockwise Arc

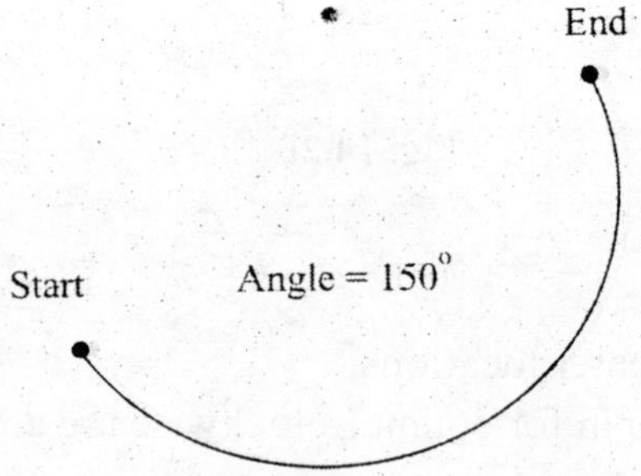

Fig. 14.26

Start, End, Direction

- Select Start and End Points
- Drag for the direction of Arc
    - Caution : Radius of Arc is not known while dragging

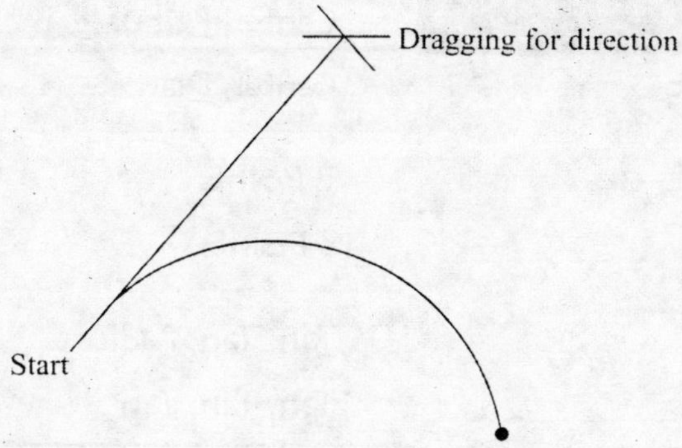

Fig. 14.27

Start, End, Radius

- Select Start point and End point of Arc
- Radius is then input
  - + *radius* = *Minor Arc*
  - – *radius* = *Major Arc*

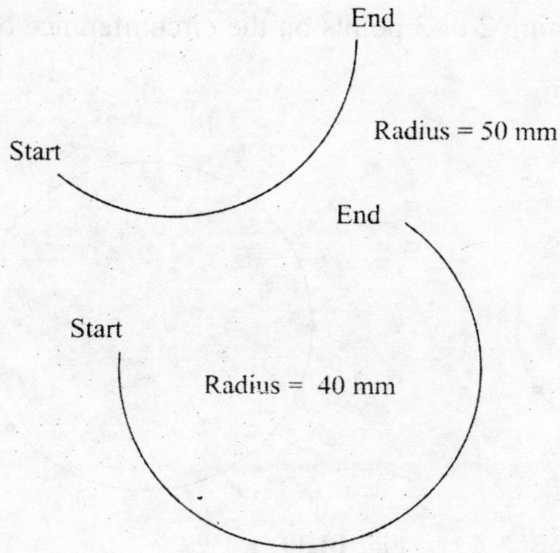

Fig. 14.28

14.10.9 Circle Command

Activate from Draw pull-down

Multiple options to create a circle

Center, Radius

Center, Diameter

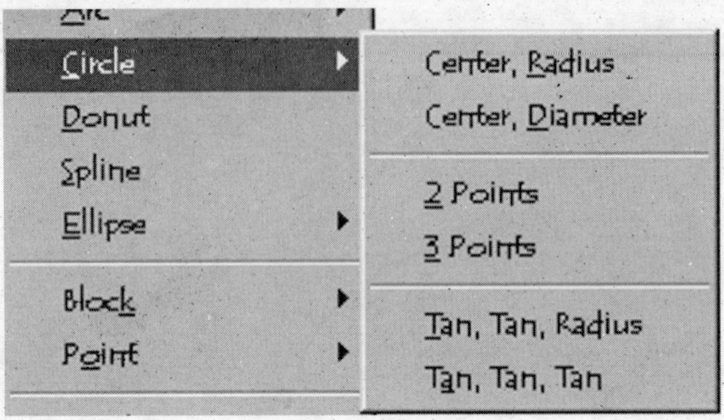

Fig. 14.29

Other options

(2 point, 3 point, Tangent) covered in Geometric Constructions Unit

A minimum of 2 points needed to create circle.

Select the center point

Type the Radius or Diameter distance at the next prompt

2 point & 3 point **(Fig. 14.30)**

- Create a circle by selecting 2 or 3 points on the circumference of the circle.

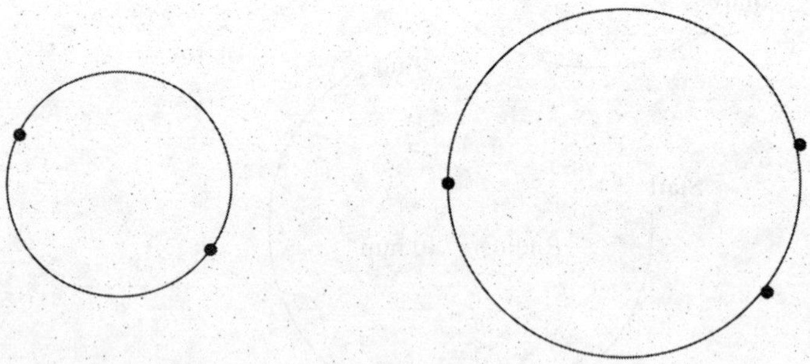

Fig. 14.30

14.10.10 Ellipse Command

- Center and Radius
- Create by selecting the center, then the 2 radius distances for the major and minor Axis
- Axis Endpoints
- Select the endpoints of the major axis first, and then select the radius distance for the minor axis

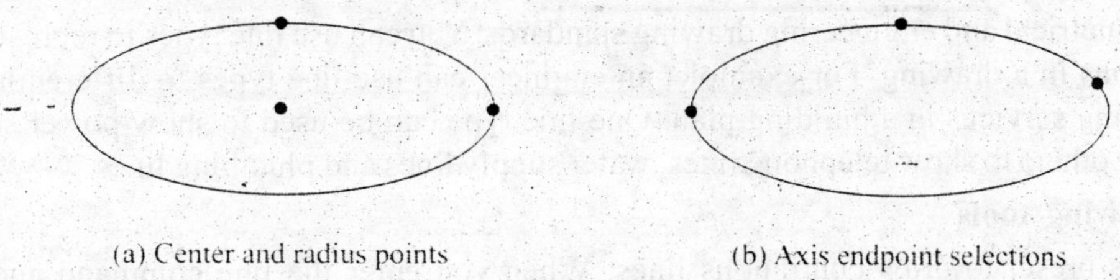

Fig. 14.31

Rotated Ellipse (Fig. 14.32)

- Created by selecting endpoints
- The first axis defined is now used as an axis of rotation that rotates the ellipse into a third dimension

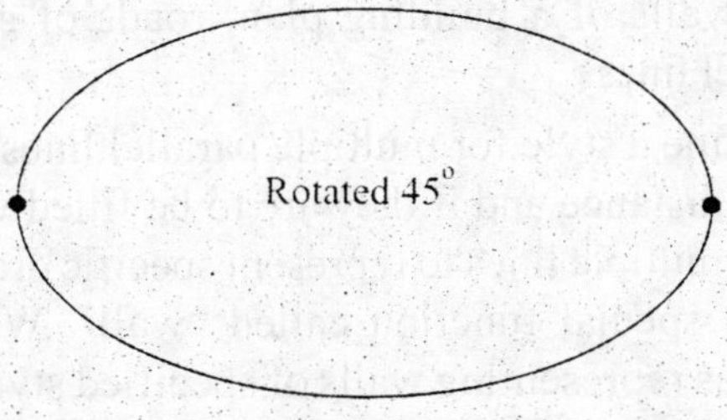

Fig. 14.32

14.11 The Drawing Tools of CADD

The following are the basic drawing tools found in a CADD program:

- Line types
- Multiple parallel lines
- Flexible curves
- Arcs and circles
- Ellipses and elliptical arcs
- Text
- Dimensions
- Hatch patterns
- Polygons
- Arrows

14.11.1 Using Line Types

There are a number of line types available in CADD that can be used to enhance drawings. There are continuous lines, dotted lines, center lines, construction lines, etc. CADD enables you to follow both geometrical and engineering drawing standards. You can use line types to represent different annotations in a drawing. For example, an engineer can use line types to differentiate between engineering services in a building plan. One line type can be used to show power supply lines, while the others to show telephone lines, water supply lines and plumbing lines.

The drawing tools

CADD is preset to draw continuous lines. When you enter the line command and indicate a starting point and end point, a continuous line is drawn. If you want to draw with another line type, you need to set that line type as the current line type. Thereafter, all the lines are drawn with the newly selected line type.

14.11.2 Drawing Multiple Parallel Lines

CADD allows you draw parallel lines simultaneously just by indicating a starting point and an end point. These lines can be used to draw something with heavy lines or double lines. For example, they can be used to draw the walls of a building plan, roads of a site map, or for any other presentation that requires parallel lines.

Most programs allow you to define a style for multiple parallel lines. You can specify how many parallel lines you need, at what distance and if they are to be filled with a pattern or solid fill. A number of add-on programs use multiple lines to represent specific drawing features. For example, an architectural program has a special function called “wall’. When you use this option, it automatically draws parallel lines representing walls of specified style and thickness.

Note:

Multiple lines are a unified entity. Even though double lines are drawn, they are treated as one line. You cannot erase or edit one line separately. However, there are functions available that can break the entities apart.

14.11.3 Drawing Flexible Curves

CADD allows you to draw flexible curves (often called splines) that can be used to draw almost any shape. They can be used to create the smooth curves of a sculpture, contours of a landscape plan or roads and boundaries of a map. To draw a flexible curve, you need to indicate the points through which the curve will pass. A uniform curve is drawn passing through the indicated points. The sharpness of the curves, the roughness of the lines and the thickness can be controlled through the use of related commands.

Drawing Arcs and Circles

CADD provides many ways to draw arcs and circles. There are a number of advanced techniques available for drawing arcs and circles, which can simplify many geometrical drawing problems. You can draw an arc by specifying circumference and radius, radius and rotation angle, chord length and radius, etc.

Arcs are drawn so accurately that a number of engineering problems can be solved graphically rather than mathematically. Suppose you need to measure the circumference of an arc, just select that arc and the exact value is displayed.

The following are basic methods for drawing arcs and circles:
(These are essentially the same methods you learn in a geometry class.
However, when drawing with CADD the approach is a little different.)

- Center point and radius
- 3 points
- Angle and radius
- 2 points
- 2 tangents and a point
- 3 tangents

14.11.4 Drawing Ellipses and Elliptical Arcs

Ellipses are much easier to draw with CADD than on a drawing board. On a drawing board, you need to find the right size template or draw a series of arcs individually to draw an ellipse. With CADD, all you need to do is specify the size of the ellipse.

The following are two basic methods for drawing ellipses:

- Length and width
- Axis and rotation angle

Adding Text to Drawings

CADD allows you to add fine lettering to your drawings. You can use text to write notes, specifications and to describe the components of a drawing. Text created with CADD is neat, stylish and can be easily edited. Typing skills are helpful if you intend to write a lot of text.

Writing text with CADD is as simple as typing it on the keyboard. You can locate it anywhere on the drawing, write it as big or as small as you like and choose from a number of available fonts.

Defining a Text Style

As discussed, there are a number of factors that control the appearance of text. It is time-consuming to specify every parameter each time you need to write text. CADD allows you to define text styles that contain all the text information such as size, justification and font. When you need to write text, simply select a particular style and all the text thereafter is written with that style. CADD offers a number of ready-made text styles as well.

Important Tip

There are a number of add-on programs available that can make working with text faster and easier. These programs provide basic word-processing capabilities that can be used to write reports and make charts. They provide access to a dictionary and thesaurus database that can be used to check spelling and to search for alternative words.

Drawing Dimensions

CADD's dimensioning functions provide a fast and accurate means for drawing dimensions. To draw a dimension, all you need to do is to indicate the points that need to be dimensioned. CADD automatically calculates the dimension value and draws all the necessary annotations.

The annotations that form a dimension are: dimension line, dimension text, dimension terminators and extension lines. You can control the appearance of each of these elements by changing the dimensioning defaults.

The following are the common methods for drawing dimensions:

- Drawing horizontal and vertical dimensions
- Dimensioning from a base line
- Dimensioning arcs and circles
- Drawing dimensions parallel to an object
- Dimensioning angles

Adding Hatch Patterns to Drawings

The look of CADD drawings can be enhanced with the hatch patterns available in CADD. The patterns can be used to emphasize portions of the drawing and to represent various materials, finishes, and spaces. Several ready-made patterns are available in CADD that can be instantly added to drawings.

Hatch patterns are quite easy to draw. You don't need to draw each element of a pattern one by one. You just need to specify an area where the pattern is to be drawn by selecting all the drawing objects that surround the area. The selected objects must enclose the area completely, like a closed polygon. When the area is enclosed, a list of available patterns is displayed. Select a pattern, and the specified area is filled.

Drawing Symbols

Symbols provide a convenient way to draw geometrical shapes. You may compare this function with the multi-purpose templates commonly used on a drawing board. To draw a geometrical shape, such as a pentagon or hexagon, select an appropriate symbol from the menu, specify the size of the symbol, and it is drawn at the indicated point.

Drawing Arrows

Arrows (or pointers) in a drawing are commonly used to indicate which note or specification relates to which portion of the drawing, or to specify a direction for any reason. There are several arrow styles available in CADD programs. You can choose from simple two-point arrows to arrows passing through a number of points, and from simple to fancy arrow styles. To draw an arrow, you need to indicate the points through which the arrow will pass.

The Command Line Box

- There are two basic ways to input a command:
  - The command line
  - Clicking on a command icon. The command icons will execute the appropriate text based command on the command line
- Additionally some commands have a keyboard shortcut option that normally involves the "cntrl" or "alt" keys on the keyboard
- The command line box size can be changed to show more or fewer command lines

Basic Commands

- The "Draw" Toolbar
  - Lines
  - Polylines
  - Circles
  - Arcs
- The "Modify" Toolbar
  - Erase
  - Copy
  - Move
  - Offset
  - Fillet
  - Array
  - Trim
  - Extend

Using the Help function

- AutoCAD has a good command reference in it's help function.
- This presentation will not duplicate that reference. You should frequently refer to the command reference as you learn the various commands.
- Some practical pointers are added here that may not be easily encountered in the command reference.

Drawing Lines

- Either type "line" on the command line or click on the line icon in the draw toolbar.
- Lines can be drawn by point and click.
  - Can keep an eye on the coordinates display to make sure that you get what you want.
- Lines can be specified by their end point coordinates.
  - Type in the coordinates on the command line
- Lines can be specified by their first point coordinates, then by a distance and angle.
  - Select starting point, type in "@distance<angle"
  - Example: @5<45 would go 5 units at a 45 degree angle
- Click the line icon (or enter "line" on the command line)
- Draw a horizontal line whose left end coordinate is 0.5 and is 5 inches long.
- Continue the line so that the second segment is starts at the end of the first line and goes vertically up 1.5 inches

The Copy Command

- The copy command is used to make a single duplicate of an entity or group of entities.
- Click on copy icon in the modify menu and follow the instructions on the command line.
- Note that copy offsets may be independent of the actual line

Selecting Objects

- AutoCAD has several ways to select objects.
  - Click on each object that you want to select.
  - Make a window that encloses all the objects that you want to select.
- Click on the lower or upper LEFT corner of desired window area
- Click on the opposite corner of the window area
  - Make a boundary that selects every thing that is within the boundary and that CROSSES the boundary.
- Click on the lower or upper RIGHT corner of desired window area
- Click on the opposite corner of the window area

EXAMPLES

FILLET command

The FILLET command is used to draw chamfering arcs connecting two lines of specificed radius.

Command	:	**FILLET**
Current settings	:	Mode = TRIM, Radius = 3,0000.
Select first object or [polyline/Radius/Trim]	:	*Specify first object using mouse.*
Select second object	:	*Specify second object using mouse.*

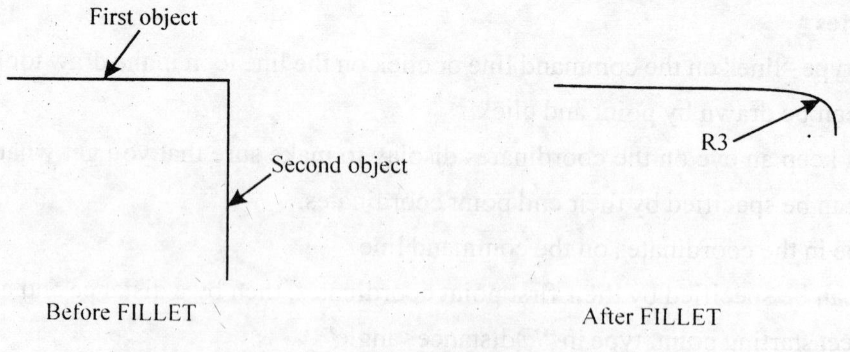

Fig. 14.33 FILLET.

CHAMFER Command

The chamfer command is used to draw bevelled lines connecting two lines at specified distances from the corner of two lines.

Command	:	**CHAMFER**
(TRIM mode) current chanfer Dist 1 = 3.000, Dist 2 = 3.0000		
Select first line or [Polyline/Distance/Angle/Trim/Method]	:	*Specify the first line using mouse.*
Select second line	:	*Specify the second line using mouse.*

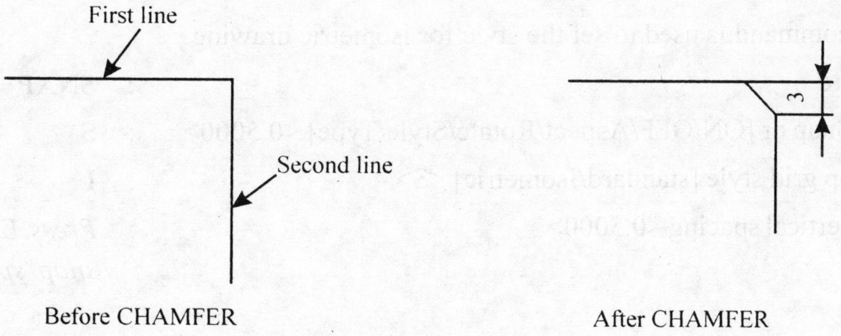

Fig. 14.34 Chamfer.

TRIM Command

The TRIM command is used to trim or cut the lines projecting beyond the specified boundary or cutting lines.

Command	:	**TRIM**
Current settings	:	Projection = UCS Edge = None
Select cutting Edges....		
Select objects	:	*Specify the cutting edges using mouse.*
Select objects	:	*Press ENTER.*
Select object to trim of [Project/Edge/Undo]	:	*Select edges.*
Select object to trim or [Project/Edge/Undo]	:	*Press ENTER.*

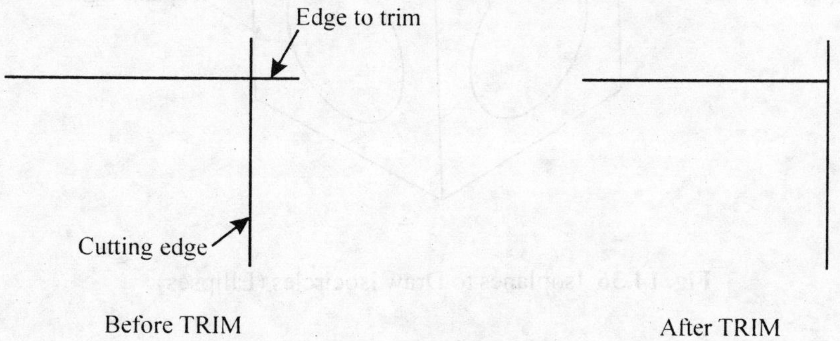

Fig. 14.35 Trim.

Isometric Drawing

Isometric drawings are used for better visualization of an object. The isometric axes as discussed in isometric projections are used to draw isometric drawings. But curves and ellipses in isometric drawings are drawn in isometric planes.

SNAP Command

The SNAP command is used to set the style for isometric drawing.

Command	:	**SNAP**
Specify Snap or [ON/OFF/Aspect/Rotate/Style/Type] <0.5000>	:	**S**
Enter snap grid style [standard/Isometric] <S>	:	**I**
Specify vertical spacing <0.5000>	:	*Press ENTER or new snap spacing*

Note:

1. The SNAP setting is activated by pressing F9 key on keyboard.
2. The GRID used to show a grid in drawing area is activated by pressing F7 key on key-board.

Drawing Isometric Circle

Circles are seen as ellipses in isometric drawings. An isoplane is set to draw the isometric circle (ellipse) using ELLIPSE command with Isocircle option.

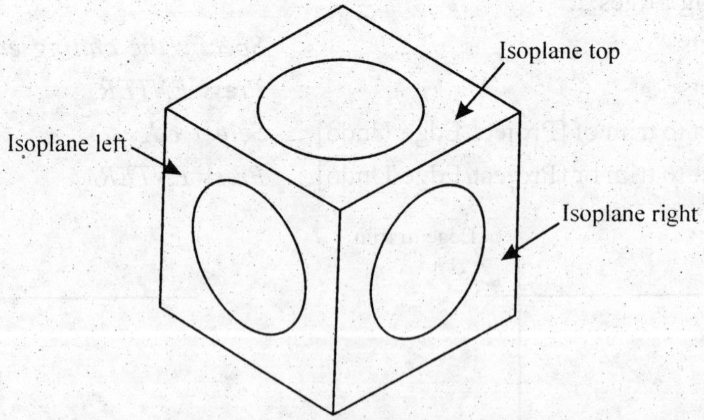

Fig. 14.36 Isoplanes to Draw Isocircles (Ellipses)

ISOPLANE Command

The ISOPLANE commnd is used to get the curve ellipse in isoplane (Top/Right/Left) for isometric drawing.

Command	:	**ISOPLANE**
Enter isometric plane setting [Left/Top/Right] <Top>	:	*Press* L *or* R *or* T *to select isoplane.*
Current isoplane	:	Top

Note that to toggle among different isoplanes is also obtanied by using F5 key on the keyboard.

HATCH (or) BHATCH

The HATCH command is used to crosshatch or pattern-fill an area.

Format : HATCH Pattern (? or name/U.style) <default>

?	-	Lists the standard hatch patterns in "acad.pat".
name	-	Name of a hatch pattern. You are prompted for scale and an angle for the pattern.
U	-	Allows you to draw a simple pattern on the fly. You are prompted for an angle, the spacing between the lines, and a single or double hatch area.
style	-	Defines what areas of the selected items are to be filled with the specified pattern.

Style codes	Example
N - Normal	BRICK,N or U,N
O - Outermost area only	BRICK,O or U.O
I - Ignore internal structure	BRICK,I or U.I

The specified parameters are remembered and are displayed as the defaults for subsequent HATCH commands

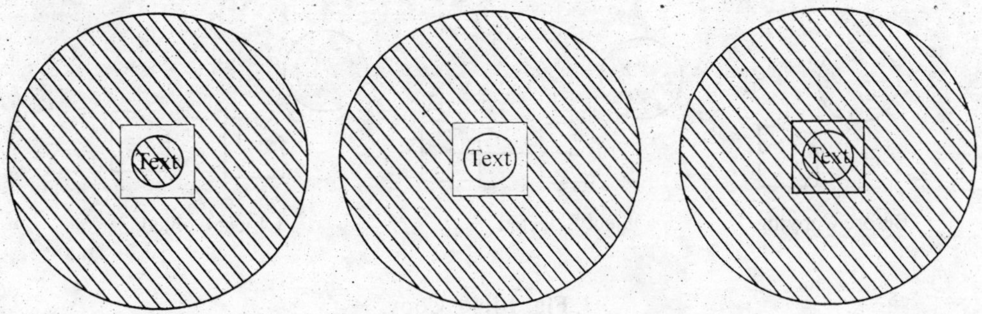

Fig. 14.38 Hatch.

ARRAY

The **ARRAY** command makes multiple copies of selected objects, in a rectangular or circular pattern.

Format : ARRAY select objects : (Show what to copy)

Rectangular or polar array (R/P):

For a rectangular array, you are asked for the number of columns and rows, and the spacing between them. The array is built along a baseline defined by the current snap rotation angle set by the "SNAP Rotate" command.

For a polar, or circular array, you must first supply a center point. Following this, you must supply two of the following three parameters:

- the number of items in the array
- the number of degrees to fill
- the angle between items in the array

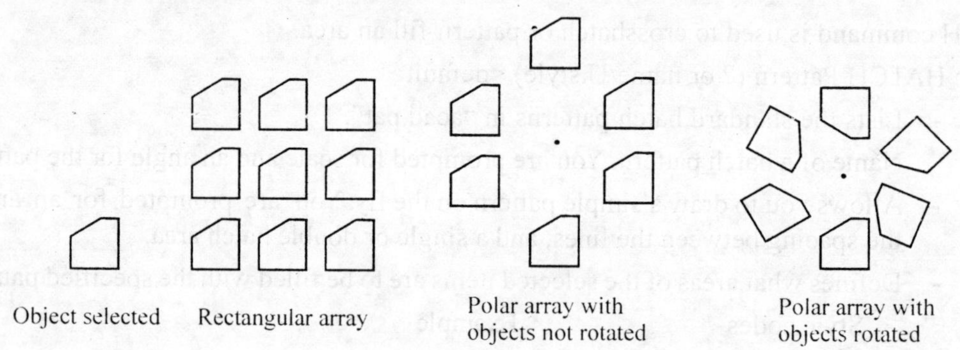

Fig. 14.39 Array

Copy

The COPY command is used to duplicate one or more existing drawing entities at another location (or locaations) without erasing the original.

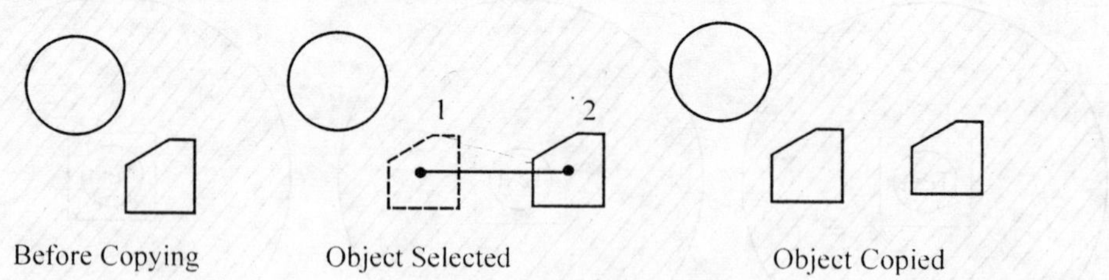

Fig. 14.40 Copy.

Extend

The EXTEND command allows to lengthen existing objects in a drawing so that they end precisely at a boundary defined by one or more other objects in the drawing.

You many use any form of entity selection to define the boundary objects. Lines, Arcs, Circles, and 2D Polylines may serve as boundary objects. When using a 2D Polyline as a boundary, its width information is ignored so that objects are extended to its centre line.

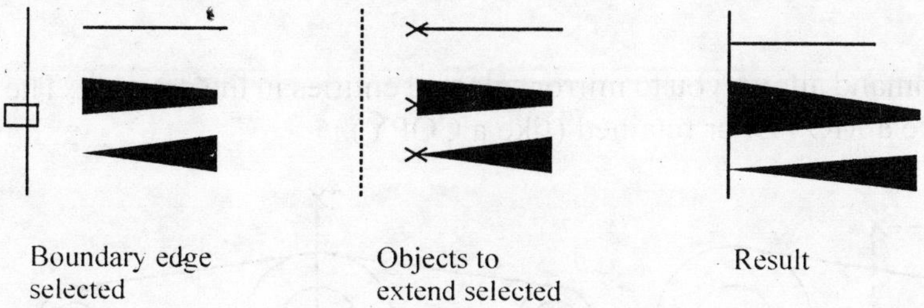

Fig. 14.41 Extend.

MOVE

The MOVE command is used to move one or more existing drawing entitles from one location in the drawing to another.

Format : MOVE select objects : (select)

Base point or displacement:

Second point of displacement : (if base selected above)

You can "drag" the object into position on the screen. To do this, designate a reference point on the object in response to the "Base point...". prompt, and then reply "DRAG" to the "second point". prompt. The selected objects will follow the movement of the screen crosshairs. Move the objects into position and then press the pointers "pick" button.

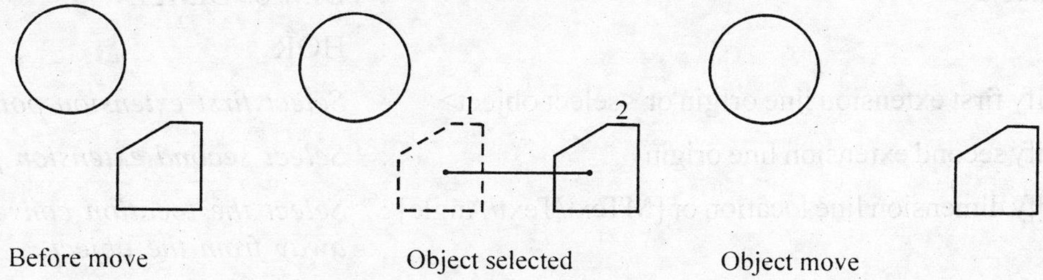

Fig. 14.42 Move.

TRIM

The TRIM command allows to trim objects in a drawing so that they end precisely at a "cutting edge" defined by one or more other objects in the drawing.

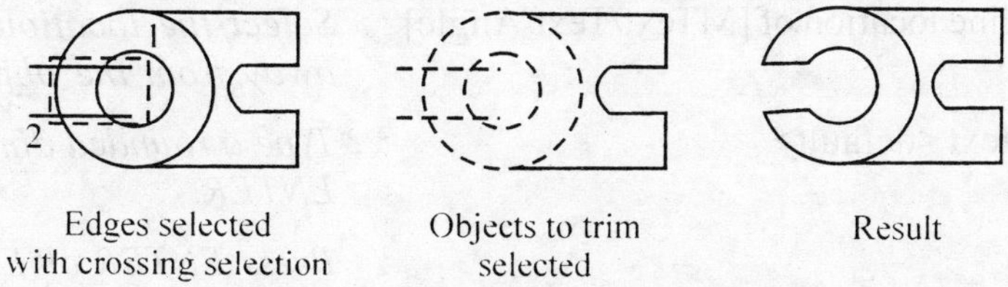

Fig. 14.43 Trim.

MIRROR

The MIRROR command allows you to mirror selected entities in the drawing. The original objects can be deleted (like a MOVE) or retained (like a COPY).

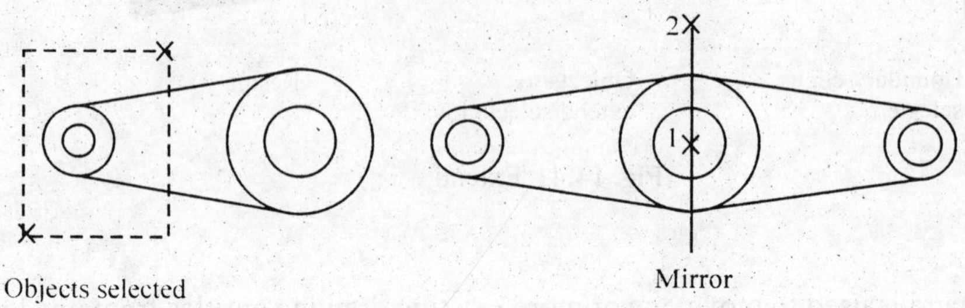

Fig. 14.44 Mirror.

Basic Dimensioning

Any engineering drawing prepared for an application purpose, should accompany with the dimensions of the object. The method of dimensioning is similar to the normal method followed in engineering drawing. Following dimensioning types are commonly used in engineering drawing practice.

Linear Dimensioning

The horizontal and vertical dimensions of an object are marked by using linear dimensioning method.

Command	:	DIM or DIMLIN
DIM	:	HOR
Specify first extension line origin or <select object>	:	*Select first extension point.*
Specify second extension line origin	:	*Select second extension point.*
Specify dimension line location or [MText/Text/Angle]	:	*Select the location conveniently away from the object.*
Enter dimension text <default>	:	Type a rounded dimension or press *ENTER.*
DIM	:	VER
Specify first extension line origin or <select object>	:	*Select first extension point.*
Specify second extension line origin	:	*Select second extension point.*
Specify dimension line location of [MText/Text/Angle]	:	*Select the location conveniently away from the object.*
Specify dimension text <default>	:	*Type a rounded dimension or press ENTER.*
DIM	:	*Press ENTER or Esc to complete dimensioning.*

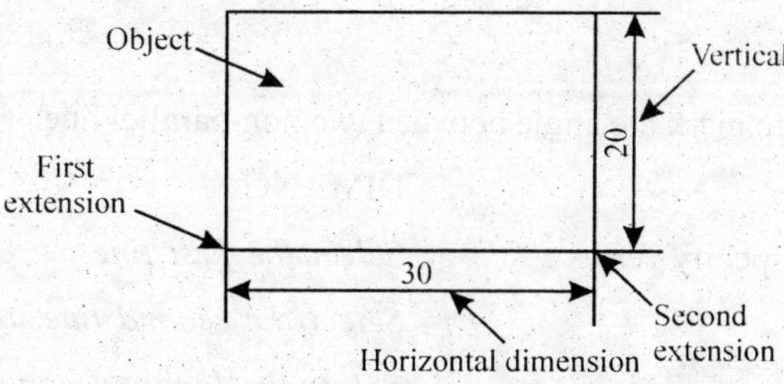

Fig. 14.45 Linear Dimensioning.

Aligned Dimensioning

Aligned dimensioning is similar to linear dimensioning but the dimension line is prallel to the edge of the object which is inclined at any angle.

Command	:	**DIM or DIMALIGNED**
DIM	:	Aligned
Specify first extension line origin or <select object>	:	*Select first extension point.*
Specify second extension line origin	:	*Select second extension point.*
Specify dimension line location or [MText/Text/Angle]	:	*Select the location conveniently away from the object.*
Enter the dimension text <default>	:	*Type a rounded dimension or press ENTER.*
DIM	:	*Press ENTER or ESc to complete dimensioning.*

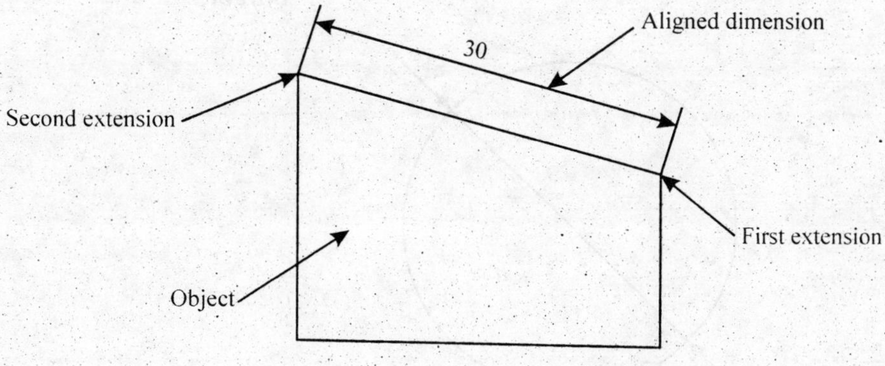

Fig. 14.45 Aligned dimensioning.

Angular Dimensioning

The angular dimensioning is used to mark the angle between two non-parallel line.

Command : DIMANG

Select arc, circle, line, or < specify vertex> : *Select the first line.*

Select second line : *Select the second line using mouse.*

Specify dimension arc line location or [MText/Text/Angle] : *Select the location conveninently away from the object.*

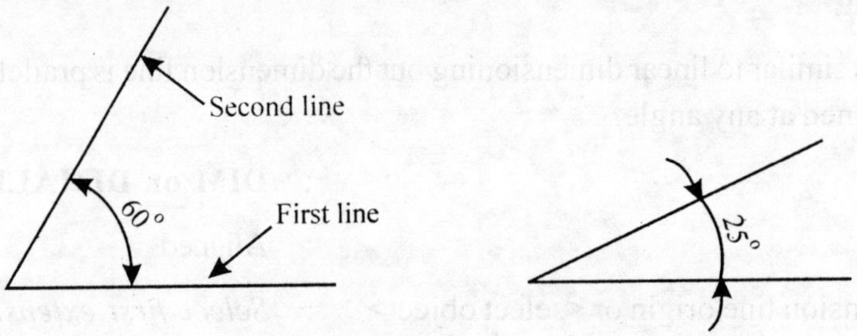

Fig. 14.47 Angular Dimensioning.

Diameter Dimensioning

The diameter dimensioning is used to mark diameter of a circle.

Command : DIMDIA

Select arc or circle : *Select a circle using mouse.*
Dimension text = current

Specify dimension line location of [MText/Text/Angle] : *Select the location conveniently away from the object.*

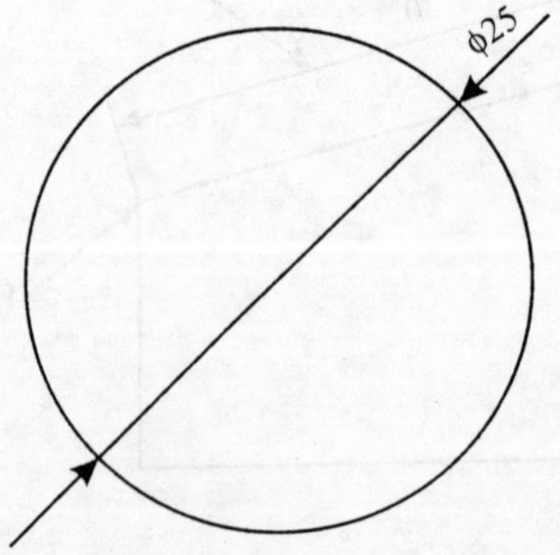

Fig. 14.48 Diameter Dimensioning.

Radius Dimensioning

The radius dimensioning is used to mark radius of a circle or an arc.

Command	:	DIMRAD
Select arc or circle	:	*Select an arc using mouse.*
		Dimension text = Current.
Specify dimension line location or [MText/Text/Angle]	:	*Select the location conveniently away from the object.*

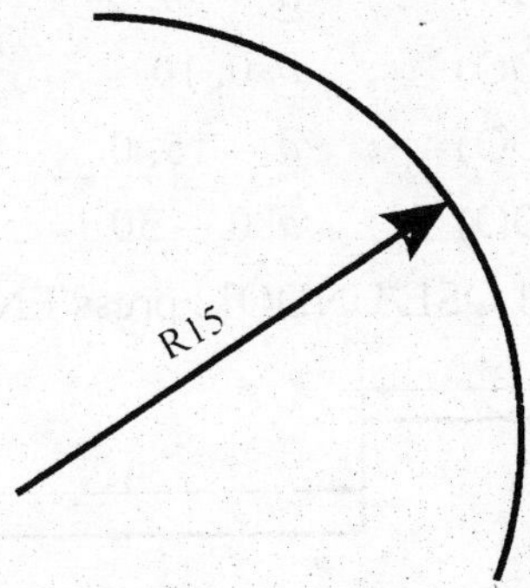

Fig. 14.49 Radius Dimensioning.

Draw the figure given below by different methods

(a) Absolute Co-ordinates method

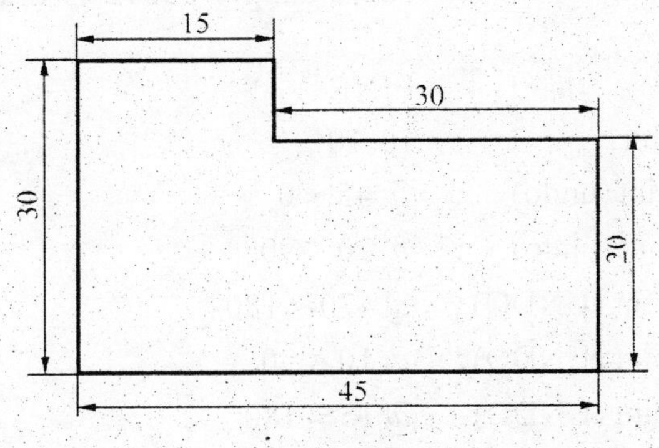

Fig. 14.50

(b) Relative Coordinates

- In this method the points are located to draw a line with reference to the previous point. The above example is drawn in this method.

Command	:	LINE
Specify the first point	:	10, 10
Specify the next point (UNDO)	:	@ 45, 0
Specify the next point (UNDO)	:	@ 0, 20
Specify the next point (UNDO)	:	@ – 30, 0
Specify the next point (UNDO)	:	@ 0, 10
Specify the next point (UNDO)	:	@ – 15, 0
Specify the next point (UNDO)	:	@ 0, – 30

Specify the next point or [CLOSE/UNDO] : press ENTER to complete the drawing.

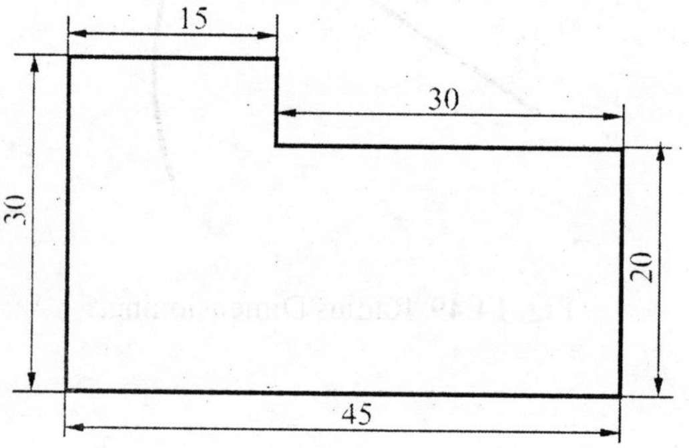

Fig. 14.52

(c) Polar Coordinates

In this method the points are located to draw the line by defining the distance of the point from the current position and the angle made to that line. The angle referred in AU to CAD is given in Fig. 14.53(a). The example above is drawn by this method in Fig. 14.53(b)

Command	:	LINE
Specify the first point	:	10, 10
Specify the next point (undo)	:	@ 45 < 0
Specify the next point (undo)	:	@ 20 < 90
Specify the next point (UNDO)	:	@ 430 < 180
Specify the next point (UNDO)	:	@ 10 < 90
Specify the next point (UNDO)	:	@ 15 < 180
Specify the next point (UNDO)	:	@ 30 < 270

Specify next point or [CLOSE/ UNDo] press ENTER to complete the drawing.

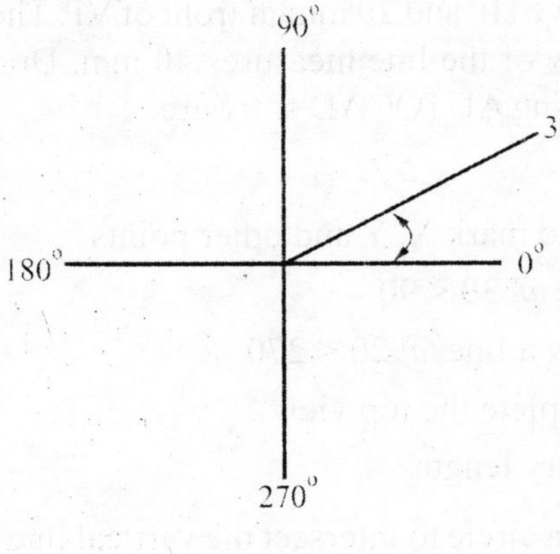

(a)

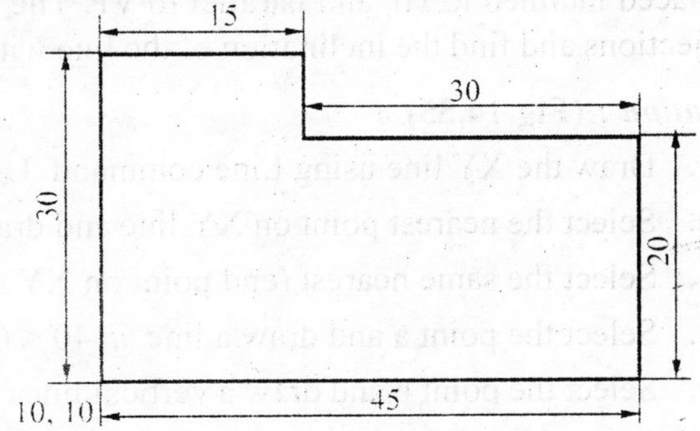

(b)

Fig. 14.53

By direct distance entry

In this method the points are located to draw a line using the distance entry in the direction of the cursor

Command	:	LINE
Specify the first point	:	LINE [ORTHOM]
Specify the next point (UNDO)	:	10, 10 move mouse horizontally right
Specify the next point (UNDO)	:	45 move mouse vertically up
Specify the next point (UNDO)	:	20 move mouse horizontally left
Specify the next point (UNDO)	:	30 move mouse vertically up
Specify the next point (UNDO)	:	15 move mouse horizontally left
Specify the next point (UNDO)	:	10 move mouse vertically down
Specify the next point (UNDO/CLOSE)	:	press ENTER to complete the drawing

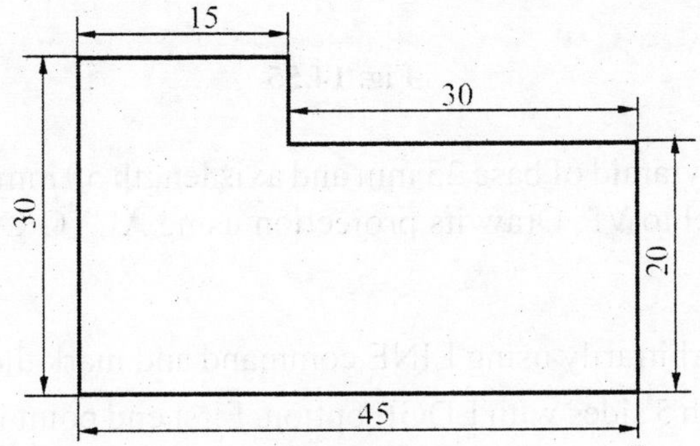

Fig. 14.54

Problem : A line AB 60 mm long has its end A 30 mm above HP and 20 mm in front of VP. The line is placed inclined to HP and parallel to VP. The top view of the line measures 40 mm. Draw its projections and find the inclination of the line with HP using AUTOCAD software.

Solution : **(Fig 14.55)**

1. Draw the XY line using Line command. Use Text to mark X, Y and other points
2. Select the nearest point on XY line and draw a line @ 30 < 90
3. Select the same nearest (end point on XY and draw a line @ 20 < 270
4. Select the point a and draw a line @ 40 < 0 to complete the top view
5. Select the point b and draw a vertical line of arbitrary length
6. Select the point a′ as centre and radius 60mm draw a circle to intersect the vertical line at b′
7. Join a′ and intersection point b′ to complete the line in the front view
8. Use Erase to delete the circle or Break the circle to return a small arc
9. Use DIMANG to mark and measure the inclination θ of the line with HP
10. Use DIM LIN/ DIMALINGNED to mark other dimentions on the drawing

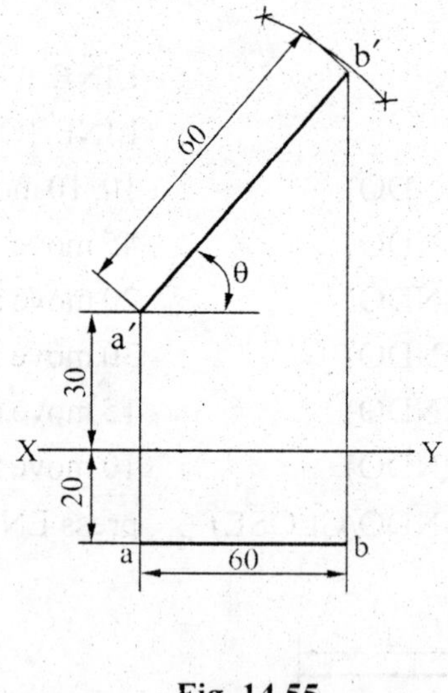

Fig. 14.55

Problem : A pentagonal pyramid of base 25 mm and axis length 50 mm is resting on HP on its base with a side of base parallel to VP. Draw its projection using AUTO CAD software.

Solution : **(Fig 14.56)**

1. Draw the XY line arbitrarily using LINE command and mark the TEXT points
2. Draw a polygon with 5 sides with EDGE option. First end point is selected arbitrarily below XY line and second end point @ 25 < 0

3. Draw line from a, b, c, d and e to the MID point of opposite sides
4. Mark the vertex O at the intersection point of there lines. TRIM all the unnecessary lines from O
5. Draw vertical line from a, b, etc. PERpendicular to XY
6. Draw a LINE from the intersection of the vertical line from O with XY @ 50 < 90 to get O′
7. Draw line joining a′, b′, etc with O'
8. Use CHPROP to change the line e′O′ L type hidden line
9. Use DIM LIN/ DIM ALIGNED to mark base side and axis length

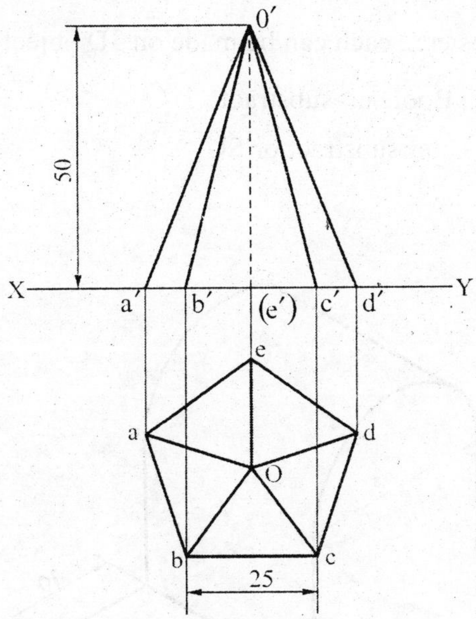

Fig. 14.56

Example : To draw the isometric view of the figure given below

Solution : **(Fig. 14.57)**

1. Select the point a arbitrarily and draw a LINE @ 40 < 30, @ 15 < 90, @ 30 < 150, @ 40 < 210, @ 30 < 330, @ 40 < 30
2. Again select the point a and draw a LINE @ 60 < 150, @ 40 < 90
3. Draw vertical LINE from b and c @ 25 < 90, from a @ 15 < 90
4. Draw LINE joining e and f and use CHPROP to change L type to CENTER line and if SCALE is used, say 7 to view the centre line
5. Set SNAP to isometric style
6. Set ISOPLANE to left

7. Use ELLIPSE command and select Isocircle, select MID point of ef to draw the ellipse with a radius 15
8. TRIM the portion of the ellipse below ef
9. Copy the portion of the ellipse with base point and the second of displacement d
10. Draw LINE above the curved portion using QU Adrant option
11. TRIM the lower portion
12. Mark the dimentions using DIMALINGED command
13. Edit the dimentions to isometric by using DIM OBLIQUE options

SUBSTRACT

Using the command slots, holes etc., each can be made on 3D object.

- from modify menu, select Boolean>substract
- at the command prompt enter substract or SU.

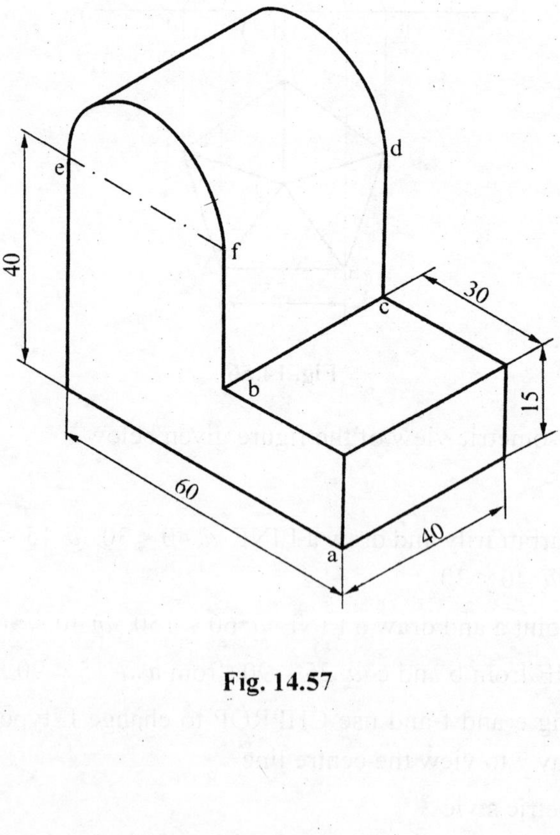

Fig. 14.57

First select solids and re-substract from goins to substract from by selecting parent objects and select solids and regions to substract by selecting them the solid shown in Fig. 14.58a&b is drawn.

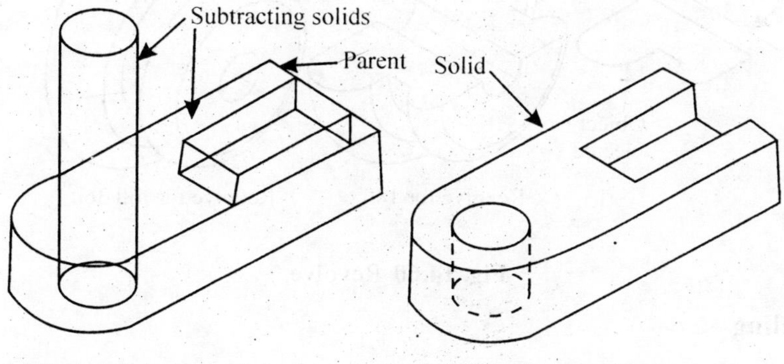

(a) Solid before subtract (b) Solid after subtract

Fig. 14.58 Solid Substraction.

SURF

The TABSURF command creates a polygon mesh representing a general tabulated starting at the point of the path curve closest to your pick point. The direction vector may be a line, 2D polyline, or 3D polyline. It is determinated by substrating the endpoint of the entry clost to your pick point from the entity's other endpoint.

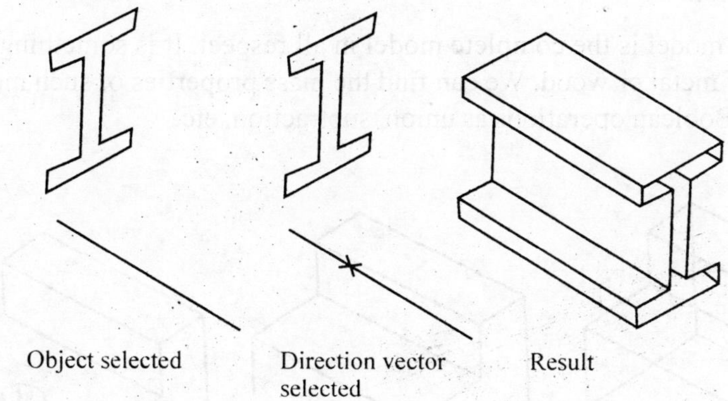

Fig. 14.59 Tabsurf.

REVOLVE

It is to get 3D solid object from a 2D polyline entity by revolving with the axis.

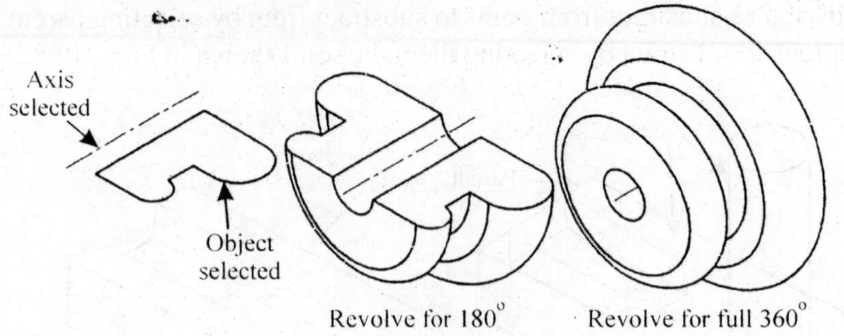

Fig. 14.60 Revolve.

Types of Modeling

With Auto CAD one can construct 3 different types of 3D (i) wire frame models. (ii) Surface models and (iii) Solid models.

Wire frame model : A wirw frame model is a Skeletal representation of the 3d models, built out mostly of line and arcs. This has just the edge information only. From the wire frame model it is difficult to understand the 3d model completely, for example, from Fig. 14.61 shown it is difficult to understand which edge is a front which edge is at the back side of the model.

Surface model : A surface model is made of surface objects such as 3D faces. The model is made of opaque surfaces which can hide any thing that is lying at the back of these surfaces. When the inner portion of a surface model is hallow you can cut in to slices (or) do such Boolean operations as union and substraction.

Solid model : A solid model is the complete model in all respect. It is something like a real solid made out of a piece of metal or wood. We can find the mass properties of such model. We can cut into slices or do such Boolean operations as union, subtraction, etc.

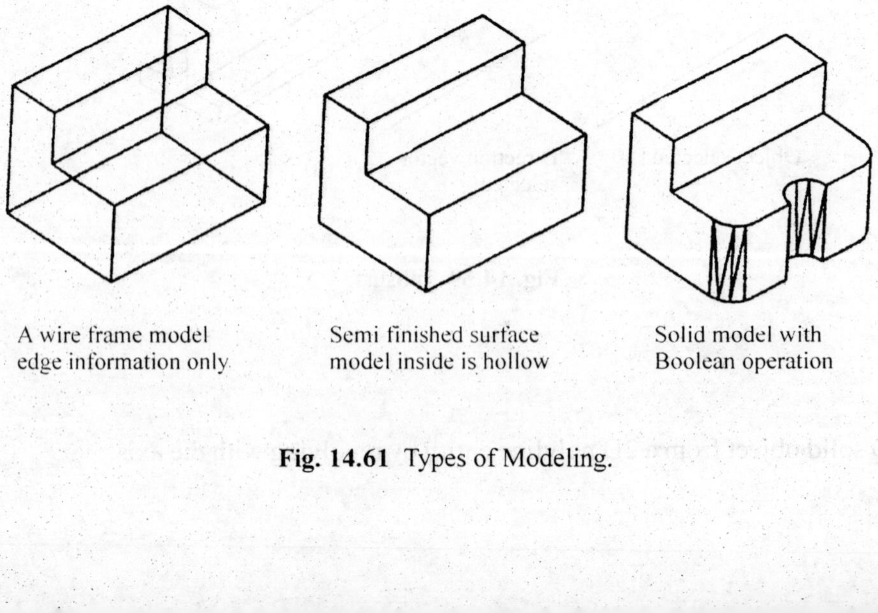

Fig. 14.61 Types of Modeling.

Program for I section solid (Fig. 14.62)

Command: limits

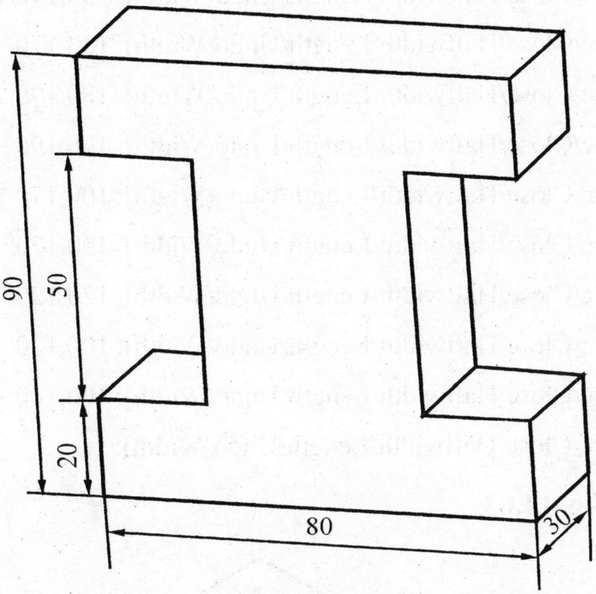

Fig. 14.62

Reset Model space limits:

Specify lower left corner or [ON/OFF] <0.0000,0.0000>:

Specify upper right corner <12.0000,9.0000>: 250,250

Command: z

ZOOM

Specify corner of window, enter a scale factor (nX or nXP), or

[All/Center/Dynamic/Extents/Previous/Scale/Window/Object] <real time>: a

Regenerating model.

Command: pl

PLINE

Specify start point: 100,100

Current line-width is 0.0000

Specify next point or [Arc/Halfwidth/Length/Undo/Width]: 180,100

Specify next point or [Arc/Close/Halfwidth/Length/Undo/Width]: 180,120

Specify next point or [Arc/Close/Halfwidth/Length/Undo/Width]: 160,120

Specify next point or [Arc/Close/Halfwidth/Length/Undo/Width]: 160,170

Specify next point or [Arc/Close/Halfwidth/Length/Undo/Width]: 180,170

Specify next point or [Arc/Close/Halfwidth/Length/Undo/Width]: 180,190

Specify next point or [Arc/Close/Halfwidth/Length/Undo/Width]: 100,190

Specify next point or [Arc/Close/Halfwidth/Length/Undo/Width]: 100,170

Specify next point or [Arc/Close/Halfwidth/Length/Undo/Width]: 120,170

Specify next point or [Arc/Close/Halfwidth/Length/Undo/Width]: 120,120

Specify next point or [Arc/Close/Halfwidth/Length/Undo/Width]: 100,120

Specify next point or [Arc/Close/Halfwidth/Length/Undo/Width]: 100,100

Specify next point or [Arc/Close/Halfwidth/Length/Undo/Width]:

Program for step in Fig. 14.63

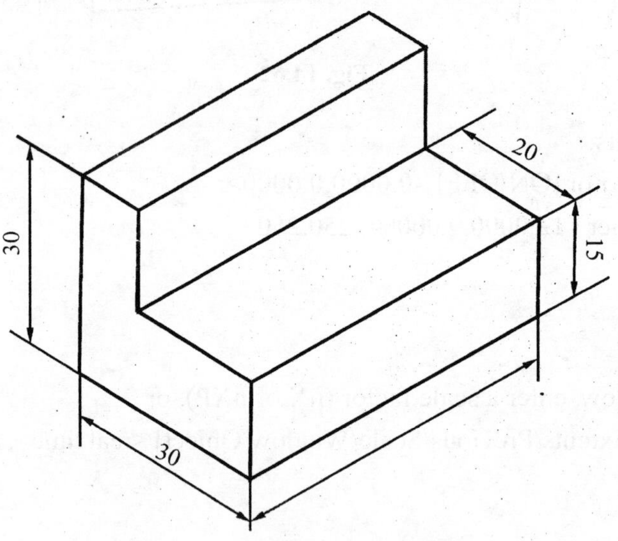

Fig. 14.63

Command; limits
Reset Model space limits:
Specify lower left corner or [ON/OFF] <0.0000,0.0000>:
Specify upper right corner <12.0000,9.0000>: 250,250

Command: z
ZOOM

Specify corner of window, enter a scale factor (nX or nXP), or
[All/Center/Dynamic/Extents/Previous/Scale/Window/Object] <real time>: a
Regenerating model.

Command: pl
PLINE
Specify start point: 100,100

Current line-width is 0.0000
Specify next point or [Arc/Halfwidth/Length/Undo/Width]: 130,100
Specify next point or [Arc/Close/Halfwidth/Length/Undo/Width]: 130,115
Specify next point or [Arc/Close/Halfwidth/Length/Undo/Width]: 110,115
Specify next point or [Arc/Close/Halfwidth/Length/Undo/Width]: 110,130
Specify next point or [Arc/Close/Halfwidth/Length/Undo/Width]: 100,130
Specify next point or [Arc/Close/Halfwidth/Length/Undo/ Width]: c

Command: extrude
Current wire frame density: ISOLINES=4
Select objects: I
1 found
Select objects:
Specify height of extrusion or [Path]: 50
Specify angle of taper for extrusion <0>:

Command: extrude
Current wire frame density: ISOLINES=4
Select objects: I
1 found

Select objects:

Specify height of extrusion or [Path]: 30

Specify angle of taper for extrusion <0>:

Program for Boolean operation (Fig. 14.64)

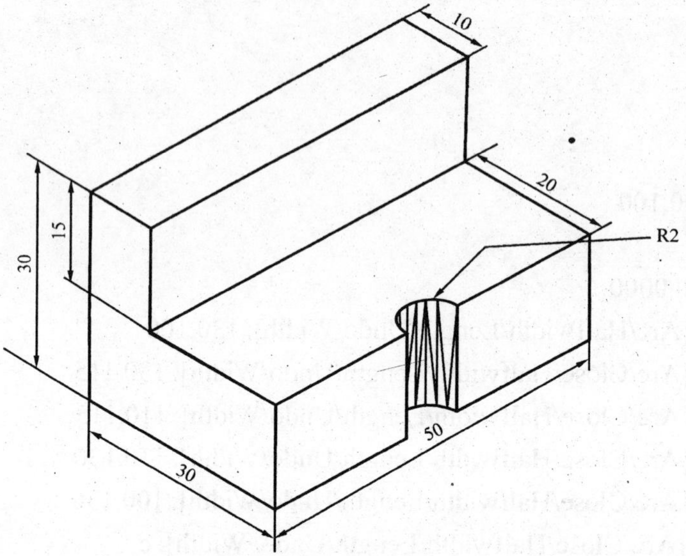

Fig. 14.64

Command: p1
PLINE
Specify start point:
Current line-width is 0.0000
Specify next point or [Arc/Halfwidth/Length/Undo/Width]: 30
Specify next point or [Arc/Close/Halfwidth/Length/Undo/Width]: 15
Specify next point or [Arc/Close/Halfwidth/Length/U nd 0/Width]: 20
Specify next point or [Arc/Close/Halfwidth/Length/Undo/Width]: 15
Specify next point or [Arc/Close/Halfwidth/Length/Undo/Width]: 10
Specify next point or [Arc/Close/Halfwidth/Length/Undo/Width]: c

Command: extrude
Current wire frame density: ISOLINES=4
Select objects: I
1 found
Select objects:
Specify height of extrusion or [Path]: 50
Specify angle of taper for extrusion <0>:

Command: hi

HIDE Regenerating model.

Command: ucs

Current ucs name: *FRONT*

Enter an option [New/Move/orthoG raphic/Prev/Restore/Save/Del/Apply/?/World]

<World>: g

Enter an option [Top/Bottom/Front/BAck/Left/Right]<Front>: t

Command: c

CIRCLE Specify center point for circle or [3P/2P/Ttr (tan tan radius)]:

Specify radius of circle or [Diameter] <2.3146>: 2

Command: extrude

Current wire frame density: ISOLINES=4

Select objects: I

1 found

Select objects:

Specify height of extrusion or [Path]: -15

Specify angle of taper for extrusion <0>:

Command: subtract

Select solids and regions to subtract from..

Select objects: 1 found

Select objects:

Select solids and regions to subtract..

Select objects: 1 found

Select objects:

Command: hi

HIDE Regenerating model.

Drawing standard 3D shapes

Tool bar for 3D shapes

Drawing a sphere

Command: SPHERE or choose from the solids toolbar.

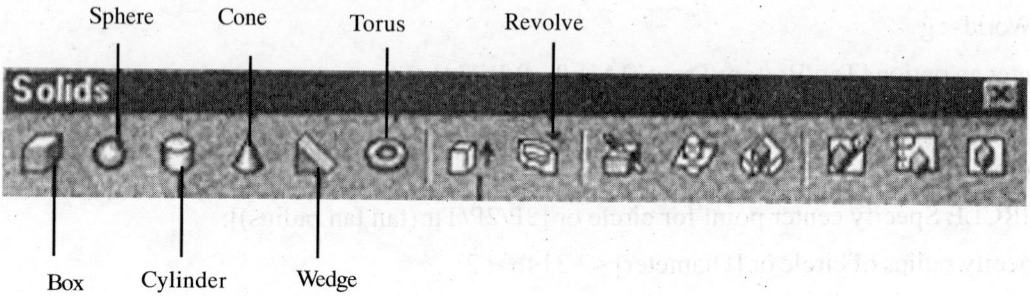

Specify centre of sphere < 0, 0, 0 >: Specify the centre of the sphere. If sphere is to lie on the xy plane, the z coordinate of the centers should be equal to the radius of the sphere

Specify radius of the sphere or [Diameters]: Type the radius or right click and choose diameters to specify the diameters

Drawing a Cylinder (Fig 14.63)

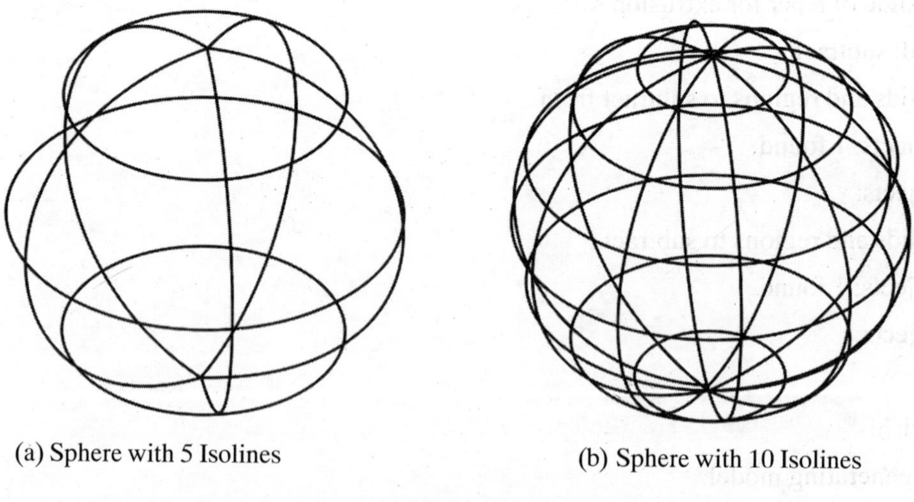

(a) Sphere with 5 Isolines

(b) Sphere with 10 Isolines

Fig. 14.63

Command: Cylinder or choose cylinder from the solids toolbar

Specify center point for the base of cylinder or [Elliptical] < 0, 0, 0 >: Specify the center point for a circular cylinder or choose the elliptical option to define an ellipse as a base

Specify radius for base of cylinder or (Diameter): Specify a radius or use the diameter option to specify the diameter

Specify height of cylinder or (center of other end): Specify the height or use the center of the other end option to specify the center of the top of the cylinder

Drawing a Torus (Fig 14.65)

Command: Torus or choose Torus from solid toolbar

Specify center of the torus < 0, 0, 0 >: Specify the center of the torus (the center of the role)

Specify the radius of just the tube or use the diameters option to specify the diameter

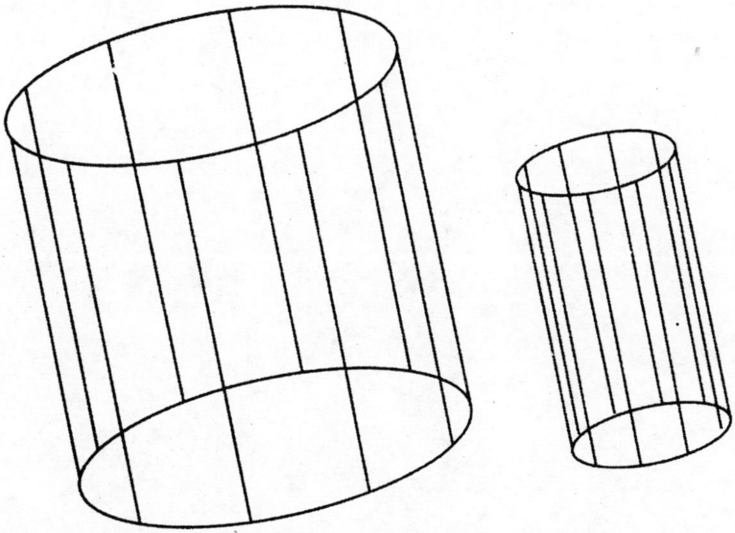

Fig. 14.64 Cylinder with Different Radius and Height.

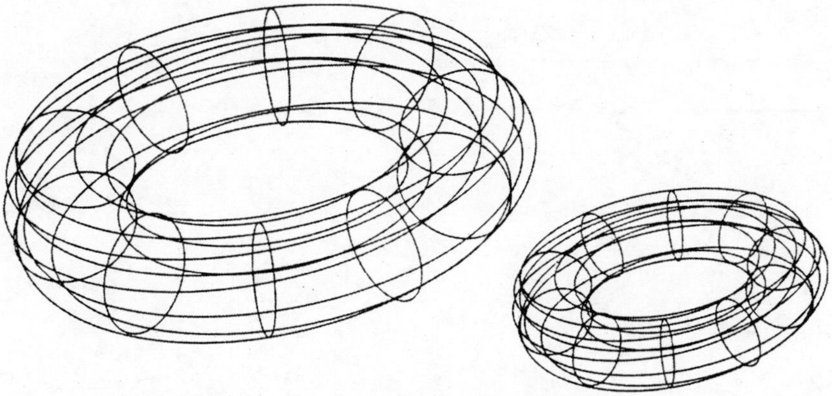

Fig. 14.65 Torus with Different Tubes and Torus Radii.

Objective Type Questions and Answers

CHAPTER - 1

1.1 Writing a letter is known as ______________ and making a drawing is known ______________.

1.2 The difference in height between the lead and leg points in compass is ______________ mm.

1.3 A mechanical pencil is specified by the ______________ of the lead.

1.4 Pencil leads are graded by the ______________ or ______________ of the lead.

1.5 An enlarged scale is to represent small dimension as big. (True/False)

1.6 Normally used scale is ______________

1.7 Templated are used for ______________ features.

1.8 As per BIS standard drawing sheets vary from ______________ to ______________

1.9 French curves help to make a smooth curve joining the points marked. (True/False)

1.10 M2 scale is used to represent ______________ scale.

1.11 ______________ is used in compass to draw large arcs or circles.

1.12 ______________ compass is used for small circles.

1.13 Mini draughter is used to make angles (True/False)

CHAPTER - 2

2.1 The principle involved in arriving at drawing sheet size is given by the relations ______________ and ______________.

2.2 For A1 size, the surface area is ______________. The length is equal to ______________ and width is equal to ______________.

2.3 The size of the title block is ______________.

2.4 The location of the title block is ______________.

2.5 The ratio of thick to thin line thicknesses is ____________.

2.6 Hidden lines are represented by ____________ lines.

2.7 A dimension with importance in the component is known as ____________ dimension.

2.8 Axes line cross at small dashes. (True/False)

2.9 The thickness of the line of letters having 10mm height is ____________ mm.

2.10 Drawing title is written in ____________ or ____________ mm size.

2.11 Hatching and sub titles are written in ____________ or ____________ mm size.

2.12 The inclination of the inclined lettering is ____________ degrees.

2.13 Lettering with adjoining stems require ____________ spacing. (More/Less)

2.14 Some letter combinations have over lapping space. (True/False)

2.15 Each feature shall be dimensioned once only on a drawing. (True/False)

2.16 Dimension should be placed on the view where the shape is best seen. (True/False)

2.17 As far as possible dimension should be placed out side the view. (True/False)

2.18 Dimension should be taken from the hidden lines. (True/False)

2.19 A gap should be left between the feature and the starting of the dimension line. (True/False)

2.20 The proportions of the arrow head of the dimension line is ____________.

2.21 Dimension should follow the symbol of the shape. (True/False)

2.22 Extension lines should meet without a gap. (True/False)

2.23 Hidden lines should meet without a gap. (True/False)

2.24 Horizontal dimension line should be to ____________ to insert the value of the dimension in both methods. (Broken/Not broken)

2.25 Dimension may be placed above or below the line. (True/False)

2.26 The terms Elevation and Plan are obsolete in drawing. (True/False)

2.27 Elevation is replaced by front view and plan by top view. (True/False)

2.28 Two methods of dimensioning are (a) ____________ (b) ____________.

CHAPTER - 3

3.1 What is RF?

3.2 Drawings are normally made to full size scale. (True/False)

3.3 The scale factor for reducing is ____________.

3.4 For drawing of small components ____________ scale is used.

3.5 To measure three units ____________ scale is used.

3.6 What is a plain scale?

3.7 What is the application of diagonal scale?

3.8 What are the differences between a vernier and diagonal scale?

3.9 Name the types of verniers.

3.10 The main scale of vernier scale is a ____________ scale.

3.11 The main scale of a vernier is divided into primary and seconary divisions. (True/False).

3.12 In a direct vernier, 9 main sclae divisions are divided into 10 equal parts on the vernier. True/False).

3.13 In a retrograde vernier, 19 main scale divisions are divided into 20 equal parts on the vernier (True/False).

3.14 What is a least count?

CHAPTER - 4

4.1 Name the solids of revolution.

4.2 What is a conic section?

4.3 Define (a) Ellipse, (b) Parabola, (c) Hyperbola

4.4 When a cone is cut by a place the intersection curve obtained is known as ____________.

4.5 A cone with an apex angle 2θ is cut by a cutting plane at an angle α.

(a) When "α is greater than θ, the intersection curves is ____________

(b) "α is equal to θ," then ____________.

(c) "less than θ," ____________.

(d) "α = 90°" ____________.

4.6 When a cone is but by a section plane that is parallel to the axis and passing through it, their intersection curve is ______________.

4.7 In a rectangular hyperbola, the asymptotes intersect each other at ______________ angle.

4.8 Ellipse is a curve traced by a point moving such that the sum of its distances from the two fixed points, foci, is constant and equal to the major axis. (True/False)

4.9 The terms transverse and conjugate axes refer to the curve ______________.

4.10 Differentiate between epicycloid and hypocycloid.

4.11 What is a generating circle?

4.12 Define directing circle.

4.13 Define an involute.

4.14 The curves generated by a fixed point on the circumference of a rolling circle is knwon as ______________.

4.15 The size of the cycloidal curve is the same irrespective of the size of the generating circle. (True/False)

4.16 The curve generated by a point on the circumference of a rolling circle, rolling along another circle, outside it is called ______________.

4.17 The curve traced by a point on a straight line, when it rolls without slipping along a circle or a polygon is called ______________.

4.18 The profile of a gear tooth is ______________.

CHAPTER - 5

5.1 Name two systems of projections.

5.2 Sketch the symbols of projection.

5.3 In orthographic projectins, the ______________ are perpendicular to the ______________ of projection.

5.4 In ______________ projection, any view is so placed that it represents the side of the object away from it.

5.5 Placing the view represented by any side of the object nearer to, it is called ______________ projection.

5.6 In first angle projection, the object is placed in between the observer and the plane of projection. (True/False).

5.7 In third angle projection, the object is placed in between the ovserver and the plane of projection. (True/False).

5.8 A surface of an object appears in its true shape, when it is ______________ to the plane of projection. (Parallel/Perpendicular).

5.9 The front view of an object is obtained as a projection on the ______________ plane, by looking the object, ______________ to its front surface.

5.10 The top view of an object is obtained as a projection on the ______________ plane, by looking the object, ______________ to its front surface.

5.11 In first angle projection (a) the top view is ______________ of the front view and (b) the right side view is to the ______________ of the front view.

5.12 In third angle projection (a) the top view is ______________ of the front view and (b) the right side view is to the ______________ of the front view.

5.13 The side view of an object is obtained as a projection on the plane, by looking the object ______________ to its ______________ surface.

5.14 In both the methods of projection, the views are indentical in shape and detail but only their location with respect to the front view is different. (True/False).

5.15 A straight line is generated as the ______________ of a moving point.

5.16 The ______________ view of a point is obtained as the intersection point between the ray of sight and VP.

5.17 The top view of a point is obtained as the intersection point between the ray of sight and ______________.

5.18 The projecting lines meet the plane of projection at an angle of 90° to it. (True/False)

5.19 The distance of a point from HP is marked from XY to ______________ ((a) top view, (b) front view, (c) side view).

5.20 When a point lies on HP its front view will lie in ______________. (XY, below XY, above XY).

5.21 When both the projections of a point lie below XY the point is situated in ______________ quadrant.

5.22 When a point lies on VP, its top view lines on XY. (True/False)

5.23 When a point is above HP its front view is ______________ XY.

5.24 When a point is ______________ VP its top view is above XY.

5.25 When a point lies on ______________ and ______________ its two views lie on XY.

5.26 A straight line is defined as the ______________ distance between two points.

5.27 The projection of a line on a plane parallel to it, appears in its true length. (True/False).

5.28 When a line is perpendicular to one of the planes, it is ______________ to the other plane.

5.29 When a line is perpendicular to HP, its front view is ____________ to XY. (Parallel/Perpendicular).

5.30 When a lne is perpendicular to VP, its ____________ is a point. (Front view/Top view).

5.31 When a line is inclined to ____________ and parallel to ____________ its front view represents the true length of the line.

5.32 When a line is contained by a plane, its projection on that plane is (Point, equal to its true length)

5.33 When a line is inclined to a plane, its projection on the plane is a lines shorter than its true length. (True/False).

5.34 Define trace of a line?

5.35 What is horizontal trace of a line.

5.36 Define vertical trace of a line.

5.37 The trace of a line is a ____________.

5.38 A line parallel to both the HP and VP has no vertical and horizontal traces. (True/False).

5.39 A line inclined to VP and parallel to HP has ____________ trace.

5.40 A line inclined to HP and parallel to VP has ____________ trace and no ____________ trace.

5.41 The HT and VT of a line will always lie on one and same projector. (True/False).

5.42 The HT of a line will always lie below XY line. (True/False).

5.43 The vertical trace of a line situated in the first quadrant will always lie above XY line (True/False).

5.44 A line is perpendicular to the HP. Its ____________ trace coincides with its ____________ view.

5.45 A line is perpendicular to the VP. Its ____________ trace coincides with its ____________ view.

5.46 A line is perpendicular to the VP. Its ____________ trace and a line perpendicular to VP as no ____________ trace.

5.47 A line perpendicular to HP has no VT because the line is ____________ to the VP. (Parallel/Perpendicular)

5.48 A line contained by a profile plane will always have both HT and VT. (True/False).

5.49 A line is said to be ____________ if it is inclined to both HP and VP.

5.50 When a plane is perpendicular to a reference plane, its projection on the plane is a ______________.

5.51 One of the projections of an oblique plane is a straight line. (True/False).

5.52 When two planes intersect each other, their intersection is a straight line. (True/False).

5.53 The topview of a circular plane, inclined to VP and perpendicular to HP is a line ___.

5.54 The true shape of the front view of a circular plane, which is parallel to HP and Perpendicular to VP is ______________.

CHAPTER - 6

6.1 What is a solid?

6.2 What is a polyhedron?

6.3 What is a regular polyhedron?

6.4 What is a tetrahedron?

6.5 How many faces does an octahedron and dedecahedron have?

6.6 Define a prism?

6.7 What is a right regular prism?

6.8 Define a pyramid.

6.9 A prism is named according to the shape of its ends. (True/False).

6.10 A prism is a regular polyhedron. (True/False).

6.11 All polyhedra are bounded by only equal equilateral triangles. (True/False).

6.12 Define the axis of a pyramid.

6.13 Pyramids are named depending on the shape on their base. (True/False).

6.14 Three examples of solids of revolution are __________ __________ __________

6.15 Define frustrum.

6.16 What is truncated solid?

6.17 When the axis of an object is perpendicular to the VP, it is __________ to the H.P.

6.18 The front view of the corners resting on the HP are on the __________ line.

6.19 The shape of a top view of a cone with its base on the VP is a __________.

6.20 A cylinder of diamter D and a high H rests on its base on the HP. Its front view is a __________ of width __________ and height __________.

ANSWERS

CHAPTER - 1

1.1	Drafting/Draughting.	1.2	1mm.	1.3	Diameter
1.4	Hardness, Softness	1.5	True	1.6	Full scale
1.7	Standard (circles, etc.,)	1.8	A0 to A4	1.9	True
1.10	1:5	1.11	Lengthening bar	1.12	Bow

CHAPTER - 2

2.1	X:Y = 1 : $\sqrt{2}$, XY = 1	2.2	594 × 841, 841 , 594	2.3	170 × 65
2.4	at RH corner	2.5	2:1	2.6	Thin dotted lines
2.7	Functional	2.8	F	2.9	1mm
2.10	7 or 10mm	2.11	3.5 or 10mm	2.12	15°
2.13	More	2.14	True	2.15	True
2.16	True	2.17	True	2.18	False
2.19	False	2.20	3:1	2.21	True
2.22	True	2.23	True	2.24	Not broken
2.25	False	2.26	True	2.27	True
2.28	(a) Aligned and (b) undirectional.				

CHAPTER - 3

3.1	Representative factor also knwn as scale factor				
3.2	False	3.3	Greater than 1:1	3.4	Enlarging
3.5	Diagonal	3.6	True		
3.9	Forward, Backward vernier scales			3.12	True
3.13	False				

CHAPTER - 4

4.4	Conic	4.5	(a) Ellipse, (b) Parabola, (c) Hyperbola, (d) Circle		
4.6	Isoscles triangle,	4.7.	90°	4.8	True
4.9	Hyperbola	4.14	Cycloids	4.15	False
4.16	Epi-cycloid	4.17	Involute	4.18	Involute

CHAPTER - 5

5.1	First and third angle	5.3	Projectors, plane	5.4	First angle
5.5	Third angle	5.6	True	5.7	False
5.8	Parallel	5.9	Vertical, Normal	5.10	Horizontal, top.
5.11	(a) Below (b) Left	5.12	(a) above (b) right		
5.13	Profile plane, normal, side	5.14	True	5.15	Locus
5.16	Front	5.17	HP	5.18	True
5.19	(b) Front view	5.20	(a) XY	5.21	Fourth
5.22	True	5.23	Above	5.24	Behind
5.25	On HP and VP	5.26	Shortest	5.27	True
5.28	Parallel	5.29	Perpendicular	5.30	Front
5.31	HP, VP	5.32	Equal to its true length	5.33	True
5.37	Point	5.38	True		
5.39	Vertical, horizontal				
5.40	Horizontal, vertical	5.41	False	5.42	False
5.43	False	5.44	Horizontal, Top	5.45	Vertical, Front
5.46	Vertical, Horizontal	5.47	Parallel	5.48	Parallel
5.49	False	5.50	Oblique.	5.51.	Line
5.52	False	5.53	True	5.54	Inclined to XY.

CHAPTER - 6

6.9	True	6.10	False	6.11	False
6.13	True	6.14	Cyclinder, cone, sphere	6.17	Parallel
6.18	xy or reference line	6.19	Triangle	6.20	Rectangle, D, H

Model Question Papers

Question Paper – 1

1. Draw a parabola passing though three vertices of a triangle of sides 30 mm, 45 mm and 60 mm. The corner of the triangle common to 45 mm and 60 mm sides lies on the axis of parabola. Draw a tangent and normal at a point on the curve 20 mm from the axis.
2. The front view of a line B, 80 mm long measures 55 mm while its top view measures 70 mm. End A is in the both HP and VP. Draw the projections of the line, find its inclinations with the reference planes. Also locate the traces.
3. A hollow cylinder of 40 mm outside diameter and 30 mm inside diameter is resting on a point on the rim in VP with axis inclined at 30° to VP and parallel to HP. The axis length of the cylinder is 60 mm. It is cut by a vertical section plane inclined at 60° to VP and bisecting the axis. Draw the sectional front view, top view and true shape of the section.
4. Two pipes of 40 mm diameter are joined in elbow shape. Their mean axis heights are 100 mm each. Draw the development of surface of the pipes.
5. A horizontal steam boiler of 3 m diameter is surmounted by a dome of the shape of a vertical cylinder of 1.4 m diameter. Draw the projections showing the curves of intersection. When their axes intersect each other at right angles.
6. Draw the isometric view of the object whose orthographic projections are given in Fig. 1. All dimensions are in mm.

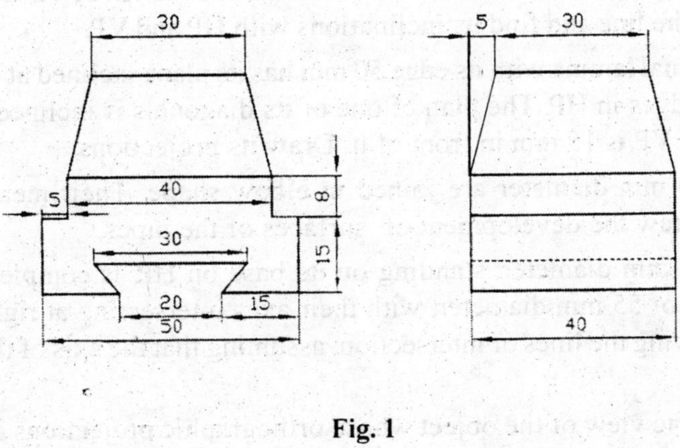

Fig. 1

7. Draw the following views of the object given in Fig. 2. All dimensions are in mm.
   (a) Front view
   (b) Top view
   (c) Both side views

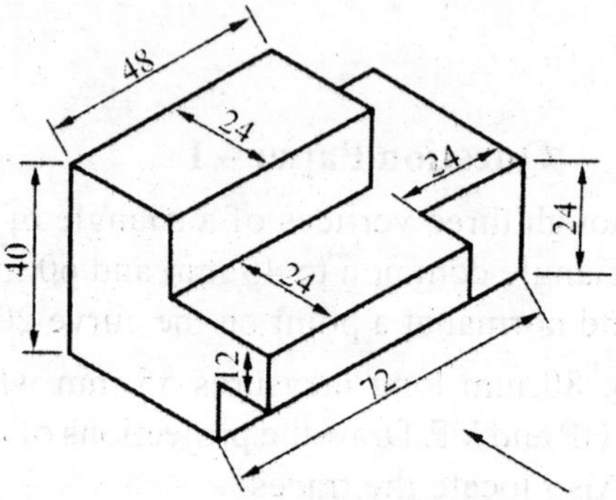

Fig. 2

8. A model of steps has three steps of 10 mm tread and 10 mm rise. The length of the steps is 60 mm. The model is placed with the vertical edge of the first step touching the PP and its longer edge inclined at 30° to PP. The station point is 70 mm in front of PP, 55 mm above the ground plane and lies in a central plane which is at 30 mm to the right of the vertical edge touching the PP. Draw the perspective view.

Question Paper - 2

1. Construct a cycloid having a rolling circle diameter as 50 mm for one revolution. Draw a normal and tangent to the curve at a point 35 mm above the directing line.
2. A 100 mm line AB measures 70 mm in top view and 80 mm in profile view. The end A 80 mm from profile plane, 90 mm above HP and 30 mm infront of VP. Draw the front view and top view of the line and find its inclinations with HP and VP.
3. A regular hexagonal lamina with its edge 50 mm has its plane inclined at 45° to HP and lying with one of its edges in HP. The plan of one of its diagonals is inclined at 45° to XY. The corner nearest to VP is 15 mm in front of it. Draw its projections.
4. Two pipes of 40 mm diameter are joined in elbow shape. Their mean axis heights are 100 mm each. Draw the development of surfaces of the pipes.
5. A cylinder of 75 mm diameter, standing on its base on HP, is completely penetrated by another cylinder of 55 mm diameter, with their axes intersecting at right angle. Draw the projections, showing the lines of intersection, assuming that the axis of the smaller cylinder is parallel to VP.
6. Draw the isometric view of the object whose orthographic projections are given in Fig. 1. All dimensions are in mm.

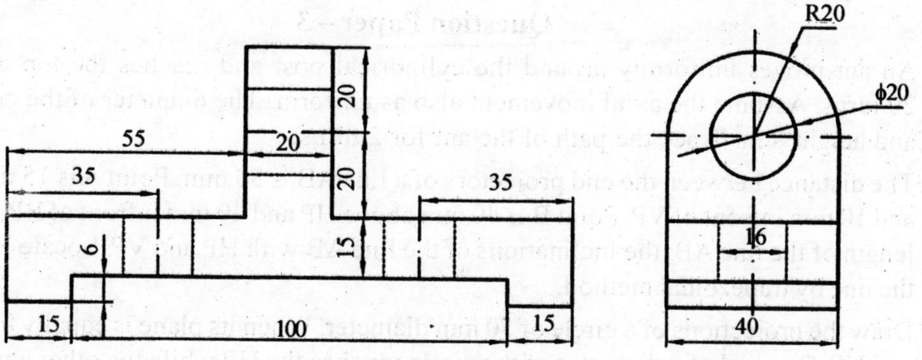

Fig. 1

7. Draw the following views of the dove tail bracket given in Fig. 2. All dimensions are in mm.

(a) Front view (b) Top view and (c) Side view

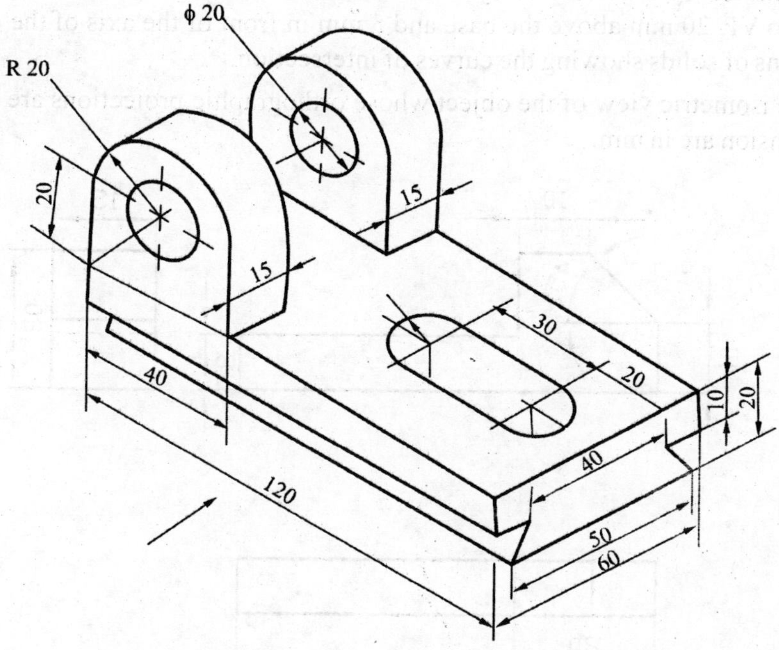

Fig. 2

8. A square plane of 35 mm sides stands vertically with one of its edges on the ground and inclined at 45° to picture plane. The vertical edge nearest to picture plane is 20 mm behind it. The station point is 30 mm in front of picture plane, 40 mm above the ground and lies in a central plane which passes through the centre of the plane. Draw the perspective view of the plane.

Question Paper – 3

1. An ant moves uniformly around the cylindrical post and reaches the top at the end of 20 turns. Assume the axial movement also as uniform. The diameter of the post is 60 mm and height 8 m. Trace the path of the ant for 2 turns.
2. The distance between the end projectors of a line AB is 50 mm. Point A is 15 mm above HP and 10 mm infront of VP. Point B is 40 mm above HP and 40 mm infront of BP. Find the true length of the line AB, the inclinations of the line AB with HP and VP. Locate HT and VT of the line by trapezoidal method.
3. Draw the projections of a circle of 50 mm diameter. When its plane is equally inclined to HP and VP. One end of a diameter of the circle touches the HP while the other end touches the VP.
4. A cone of base diameter 50 mm and axis 75 mm long has its base in VP. It is sectioned by a vertical section plane 10 mm to the right of the axis. Draw its projections and develop the surface of the truncated cone.
5. A cone of base 60 mm diameter and axis 70 mm long stands vertically with its base on HP. It is penetrated by a horizontal cylinder of 26 mm diameter. The axis of the cylinder is parallel to VP, 20 mm above the base and 5 mm in front of the axis of the cone. Draw the projections of solids showing the curves of intersection.
6. Draw the isometric view of the object whose orthographic projections are given in Fig. 1. All dimension are in mm.

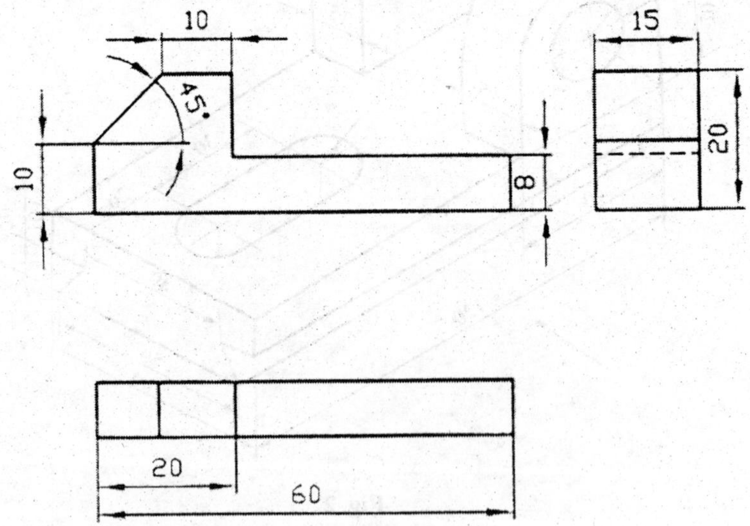

Fig. 1

7. Draw the following views of the flange given in Fig. 2. All dimensions are in mm.
   (a) Front view (b) Top view and (c) Side view

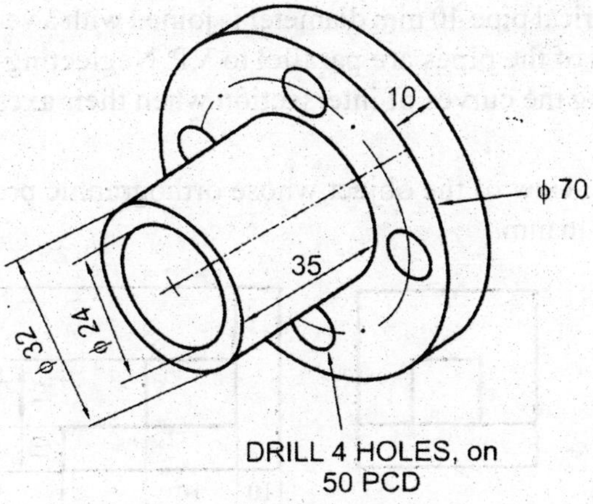

Fig. 2

8. A model of steps has three steps of 10 mm tread and 10 mm rise. The length of the steps is 60 mm. The model is placed with the vertical edge of the first step touching the PP and its longer edge inclined at 30° to PP. The station point is 70 mm in front of PP, 55 mm above the ground plane and lies in a central plane which is at 30 mm to the right of the vertical edge touching the PP. Draw the perspective view.

Question Paper – 4

1. Draw two complete coils of a conical spring made up to 20 mm round stock. The outside diameters are large 110 mm, small 60 mm, pitch 50 mm.
2. Front view of a line PQ is inclined at 30° to XY-line and measures 60 mm. The line is inclined at 45° to VP. The end P is in HP and VT of the line is 20 mm below HP. Draw the projections of the line and find its true length and inclinations with the reference planes. Also locate HT.
3. A regular pentagon of side 40 mm has its surface inclined to HP at 45°. It is resting with its base on HP and the line joining the vertex to mid-point of the base making an angle of 60° with VP. Draw its projections.
4. Draw the development of the lateral surface of the part P of the hexagonal pyramid two sides of the base parallel to the VP as shown in Fig. 1. All dimensions are in cm.

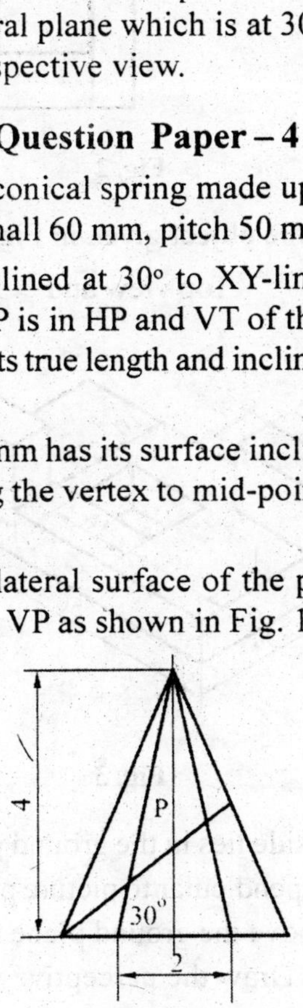

Fig. 1

5. A horizontal cylindrical pipe 40 mm diameter is joined with a vertical cylindrical pipe of same diameter. The axes of the pipes are parallel to VP. Neglecting the pipe thickness, draw the projections showing the curves of intersection when their axes intersect each other at right angles.

6. Draw the isometric view of the object whose orthographic projections are given in Fig. 2. All dimensions are in mm.

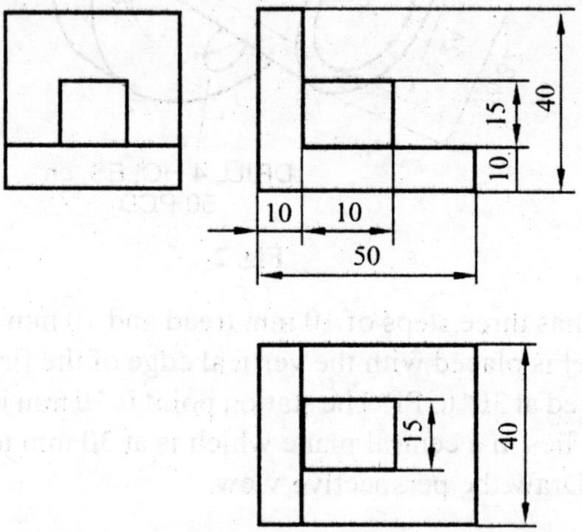

Fig. 2

7. Draw the following views of the object given in Fig. 3. All dimensions are in mm

(a) Front view (b) Top view and (c) Side view

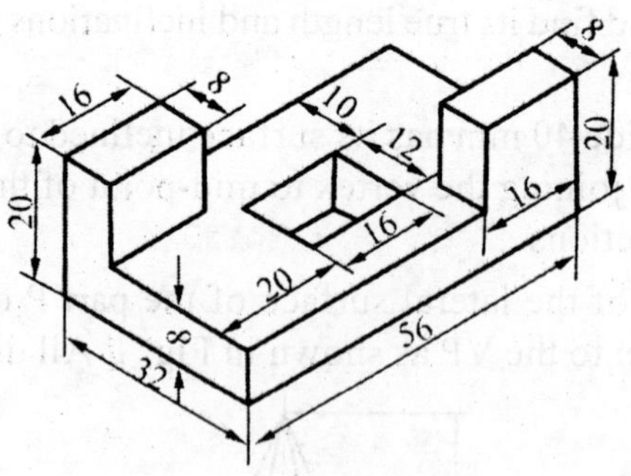

Fig. 3

8. A hexagonal plane of 30 mm side lies in the ground plane. One of its corners is touching the picture plan and an edge is perpendicular to picture plane. The station point is 30 mm in front of the picture plane 60 mm above the ground plane and lies in a central plane which passes through the centre of lamina. Draw the perceptive view.

Question Paper – 5

1. Inscribe an ellipse in a parallelogram having sides 150 mm and 100 mm long and an included angle of 120°.

2. (a) The top view of a 75 mm long line measures 55 mm. The line is in the VP, its one end being 25 mm above the HP, draw its projections.

   (b) The front view of a line, inclined at 30° to the VP is 65 mm long. Draw the projections of the line, when it is parallel to and 40 mm above the HP, its one end being 30 mm in front of the VP.

3. A semi circular plate of 80 mm diameter has its straight edge in the VP and inclined at 60° to the HP, the surface of the plate makes an angle of 30° with the VP. Draw its projections.

4. A regular square prism lies on its axis inclined at 60° to the HP and 30° to the VP. The prism is 60 mm long and has a face width of 25 mm. The nearest corner is 10 mm away from the VP and the farthest shorter edge is 100 mm from the HP. Draw the projections of the solid.

5. A right circular cylinder diameter of base 50 mm and length of axis 70 mm, rests on HP on its base rim such that its axis is inclined at 45° to HP and the top view of the axis is inclined at 60° to the VP. Draw its projections.

6. Draw the isometric projection of a hexagonal prism of side of base 35 mm and altitude 50 mm surmounting a tetrahedron of side 45 mm such that the axes of the solids are collinear and at least one of the edges of the two solids are parallel.

7. Three views of a machine part are shown in Fig. 1. Draw the isometric view of the part (al' dimensions are in mm).

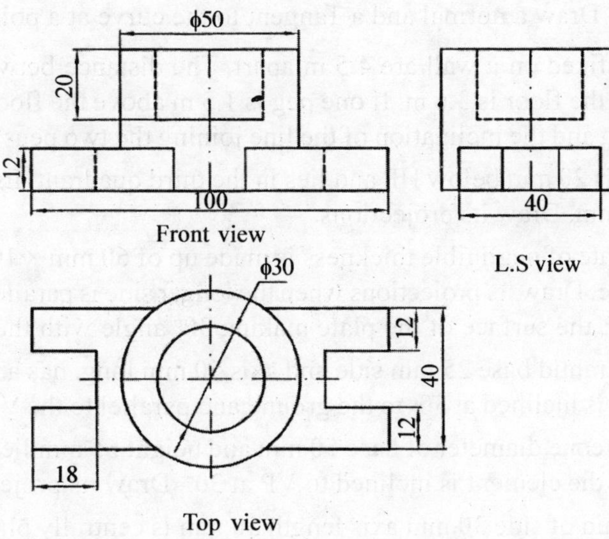

Fig. 1

8. Draw the orthographic views of the object as shown in the Fig. 2. (all dimensions are in mm).
   (a) Front view (b) Top view (c) Side view

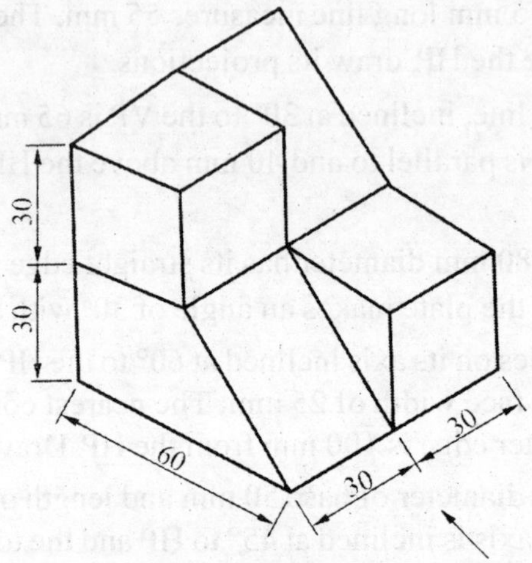

Fig. 2

Question Paper - 6

1. Draw a straight line AB of any length. Mark a point F, 65 mm from AB. Trace the paths of a point P moving in such away, that the ratio of its distance from the point F, to its distance from AB is 2:3. Draw a normal and a Tangent to the curve at a point on it, 50 mm from F.
2. (a) Two pegs fixed on a wall are 4.5 m apart. The distance between the pegs measured parallel to the floor is 3.6 m. If one peg is 1.5 m above the floor, find the height of the second peg and the inclination of the line joining the two pegs with the floor.
   (b) A point P is 20 mm below HP and lies in the third quadrant. Its shortest distance from XY is 40 mm. Draw its projections.
3. A composite plate of negligible thickness is made up of 60 mm × 10 mm, and a semi circle on its longer side. Draw its projections when the longer side is parallel to the HP and inclined at 45° to the VP, the surface of the plate making 30° angle with the HP.
4. A hexagonal pyramid base 25 mm side and axis 50 mm long, has an edge of its base on the ground. Its axis is inclined at 30° to the ground and parallel to the VP. Draw its projections.
5. A right circular cone diameter of base 50 mm and height 65 mm lies on one of its elements in HP such that the element is inclined to VP at 30°. Draw its projections.
6. A square pyramid of side 30 mm axis length 50 mm is centrally placed on top of a cube of side 50 mm. Draw the isometric projections of solids.

7. Two views of a casting are shown in Fig. 1. Draw the isometric view of the casting (all dimensions are in mm).

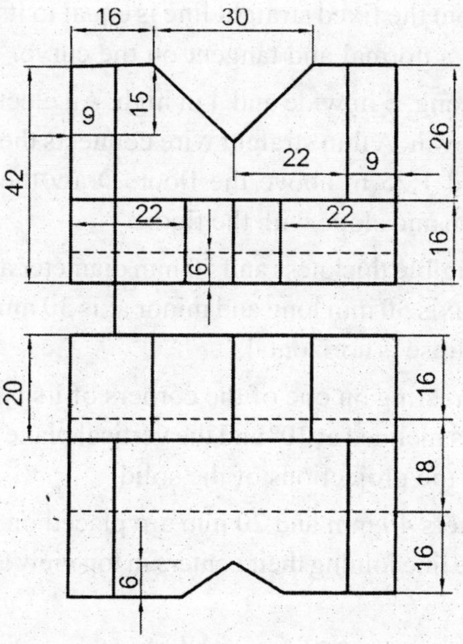

Fig. 1

8. Draw the three views of the object shown in Fig. 2. (all dimensions are in mm).

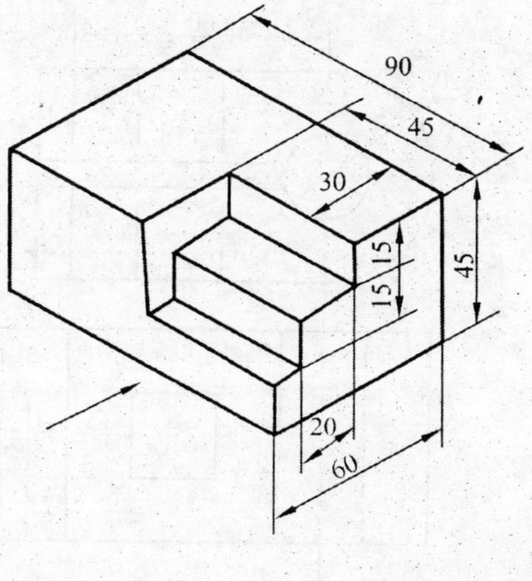

Fig. 2

Question Paper – 7

1. A fixed point is 75 mm from a fixed straight line. Draw the locus of a point P moving such a way that its distance from the fixed straight line is equal to its distance from the fixed point. Name the curve. Draw a normal and tangent on the curve.
2. A room measures 8 m long, 5 m wide and 4 m high. An electric bulb hangs in the centre of the ceiling and 1 m below it. A thin straight wire connects the bulb to a switch kept in one of corner of the room and 1.25 m above the floor. Draw the projections of the wire, also determine its true length and slope with the floor.
3. A circular plate of negligible thickness and 50 mm diameter appears as an ellipse in the front view, having its major axis 50 mm long and minor axis 30 mm long. Draw its top view when the major axis of the ellipse is horizontal.
4. A pentagonal prism is resting on one of the corners of its base on the HP. The longer edge containing that corner is inclined at 30° and the vertical plane containing that edge is inclined at 45° to the VP. Draw the projections of the solid.
5. Two spheres of diameters 40 mm and 20 mm are placed on HP touching each other. Draw its projections when the line joining their centers in top view appears to be inclined 40° to XY line.
6. A triangular prism of base edge 30 mm and height 60 mm stands on one of its corners on the ground with the axis inclined at 30° to the HP and 45° to the VP. The base of the object is nearer to VP compared to the top. Draw an isometric view of the object.
7. Two views of a casting are shown in Fig. 1. Draw the isometric projection of the casting (all dimensions are in mm).

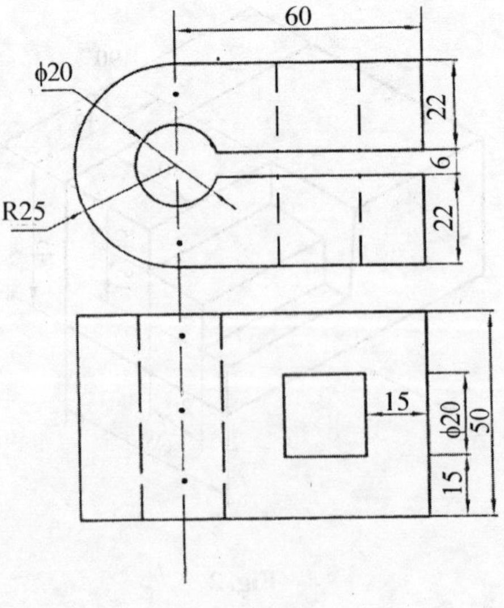

Fig. 1

8. Draw the front view, top view and right side view of the part shown in the Fig. 2. (all dimensions are in mm).

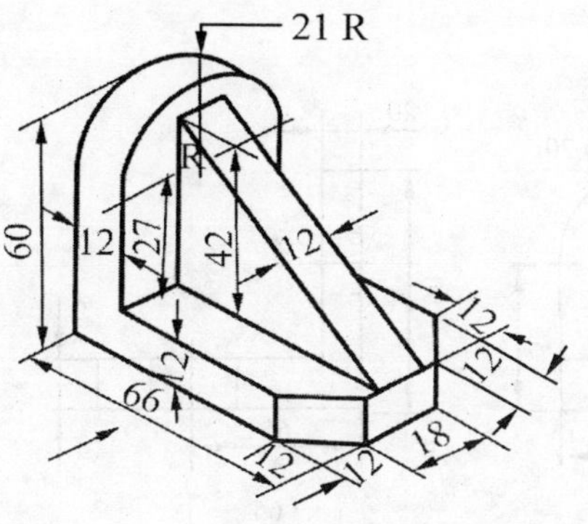

Fig. 2

Question Paper – 8

1. Draw a straight line AB of any length. Mark a point F, 65 mm from AB. Trace the paths of a point P moving in such a way, that the ratio of its distance from the point F, of its distance from AB is 1. Draw a normal and a tangent to the curve at a point on it, 50 mm from F.
2. A line AB, 90 mm long, is incline at 45° to the HP and its top view makes an angle of 60° with the VP. The end A is in the HP and 12 mm in front of the VP. Draw its front view and find its true inclination with the VP.
3. A regular hexagon of 40 mm has a corner in the HP. Its surface is inclined at 45° to the HP and the top view of the diagonal through the corner which is in the HP makes an angle of 60° with VP. Draw its projections.
4. One end of a longer edge of a regular hexagonal prism of side of base 30 mm and height 80 mm is on the VP and the other end of the same edge is on the ground. The axis makes 30° to the VP and 40° to the ground. Draw its projections.
5. An ash tray made up of a thin sheet of steel, is spherical in shape with flat, circular top o 68 mm diameter and bottom of 52 mm diameter and parallel to each other. The greates diameter of it is 100 mm. Draw the projections of the ash tray when its axis is parallel to th VP and; (a) makes an angle of 60° with the HP and (b) its base is inclined at 30 to the HI
6. A square pyramid of side 30 mm, axis length 50 mm is centrally placed on top of a cube c side 50 mm. Draw the isometric projections of solids.

7. Orthographic views of an angle plate are shown in Fig. 1. Draw the isometric projection, (all dimensions are in mm)

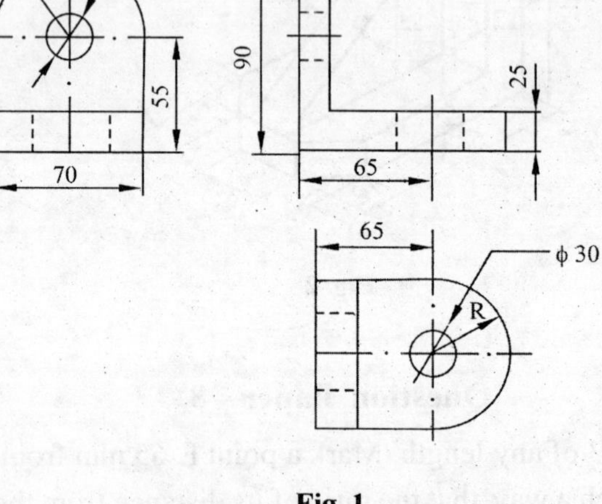

Fig. 1

8. Draw the three views of the object shown in Fig. 2 (all dimensions are in mm).

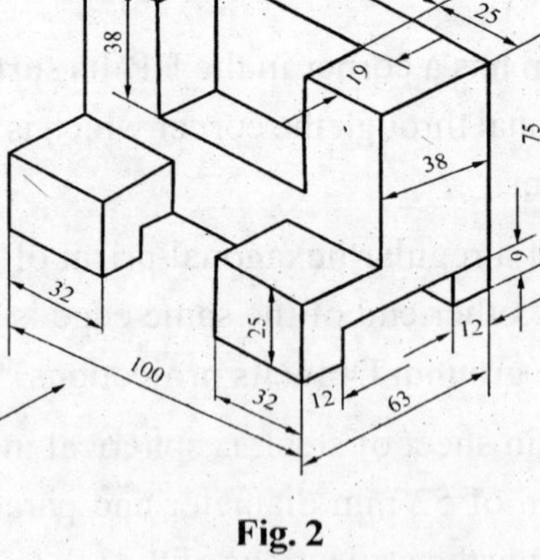

Fig. 2

Question Paper – 9

1. Construct an ellipse, with distance of the focus from the directix as 50 mm and eccentricity as 2/3. Also draw normal and tangent to the curve at a point 40 mm from the directix.
2. (a) A line AB is 75 mm long. A is 50 mm in front of VP and 15 mm above HP. B is 15 mm in front of VP and is above HP. Top view of AB is 50 mm long. Draw and measure the front view. Find the true inclinations.

(b) A line measuring 80 mm long has one of its ends 60 mm above HP and 20 mm in front of VP. The other end is 1500 above HP and in front of VP. The front view of the line is 60 mm long. Draw top view.

3. (a) A pentagonal plate of 35 mm side is perpendicular to VP and parallel to HP. One of its edges is perpendicular to VP. Draw its projections.

(b) An equilateral triangular lamina of side 30 mm is parallel to HP and perpendicular to VP. One of its sides of 20 mm in front of VP and 30 mm above HP. Draw its projections.

4. (a) A cylinder base 35 mm diameter and axis 60 mm long lies with one of its generators on HP such that its axis is parallel VP. Draw its projections.

(b) Draw the projections of cube of 40 mm side, resting with a face on HP such that one of its vertical faces is inclined at 30° to VP.

5. Hexagonal pyramid side of base 30 mm and axis 50 mm long rests with one of the corners of its base on HP. Its axis is inclined at 35° to HP and 45° to VP. Draw its projections.

6. The frustum of a hexagonal pyramid side of top and bottom 25 mm and 40 mm respectively with axis 50 mm height rests on its base in HP. Its axis is parallel to VP. Draw the orthographic projections and provide the isometric view of the solid.

7. Two views of a casting are show in Fig. 1. Draw the isometric view of the casting (all dimensions are in mm).

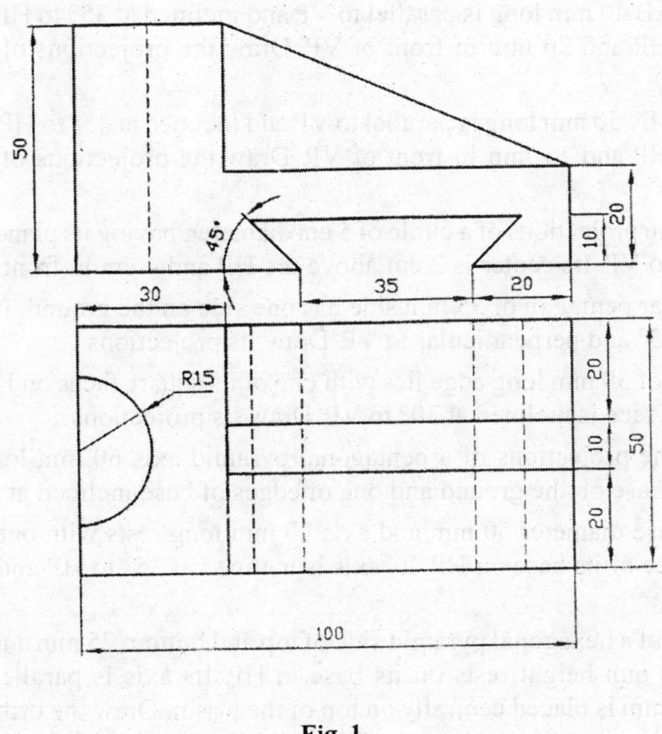

Fig. 1

8. Draw the front view, top view, and right side views of the object shown in Fig. 2. (all dimensions in mm).

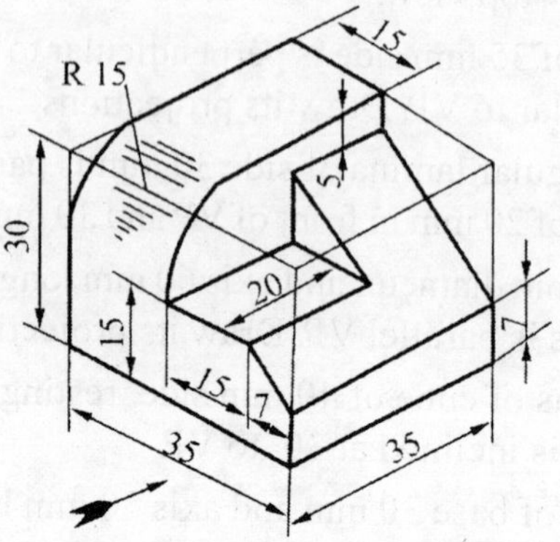

Fig. 2

Question Paper – 10

1. A fixed point F is 7.5 cm from a fixed straight line. Draw the locus of a point P moving in such a way that its distance from the fixed straight line is 2/3 times its distance from F. Name the curve. Draw normal and tangent at a point 6 cm from
2. (a) A line AB 40 mm long is parallel to VP and inclined at 35° to HP. The end A is 15 mm above HP and 20 mm in front of VP. Draw the projections of the line and find its traces.

   (b) A line MN 50 mm long is parallel to VP and inclined at 45° to HP. The end M is 20 mm above HP and 15 mm in front of VP. Draw the projections of the line and find its traces.
3. (a) Draw the projections of a circle of 5 cm diameter, having its plane vertical and inclined at 30° to VP. Its center is 3 cm above the HP and 4 cm in front of the VP.

   (b) A regular pentagon of 35 mm side has one side on the ground. Its plane is inclined to HP at 75° and perpendicular to VP. Draw its projections.
4. (a) A cube of 30 mm long edge lies with one of its square faces on HP such that one of its vertical face is inclined at 30° to VP. Draw its projections.

   (b) Draw the projections of a pentagonal pyramid axis 60 mm long, base 30 mm side having base on the ground and one of edges of base inclined at 45° to VP.
5. A cone of base diameter 50 mm and axis 70 mm long rests with one of the points on the circumference of its base on HP. Its axis is inclined at 35° to HP and 45° to VP. Draw its projections.
6. The frustum of a hexagonal pyramid side of top and bottom 25 mm and 40 mm respectively with axis 50 mm height rests on its base in HP. Its axis is parallel to VP. A sphere of diameter 40 mm is placed centrally on top of the prism. Draw the orthographic projections and provide the isometric projection of the solid.

7. Two views of a casting are shown in Fig. 1. Draw the isometric view of the casting (All dimensions are in mm).

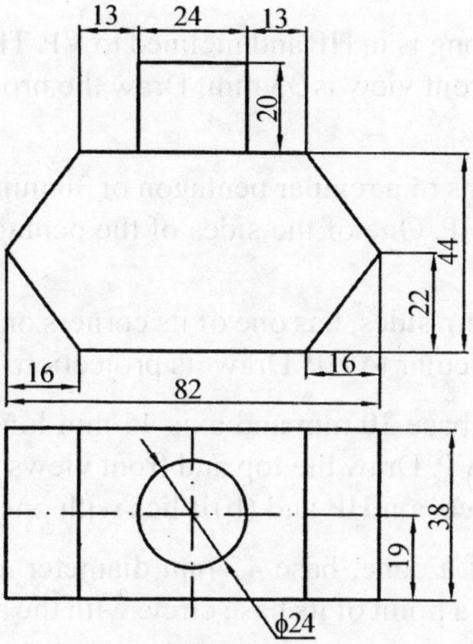

Fig. 1

8. Draw the front view, top view and and right side views of the part shown in the Fig. 2 (dimensions in mm).

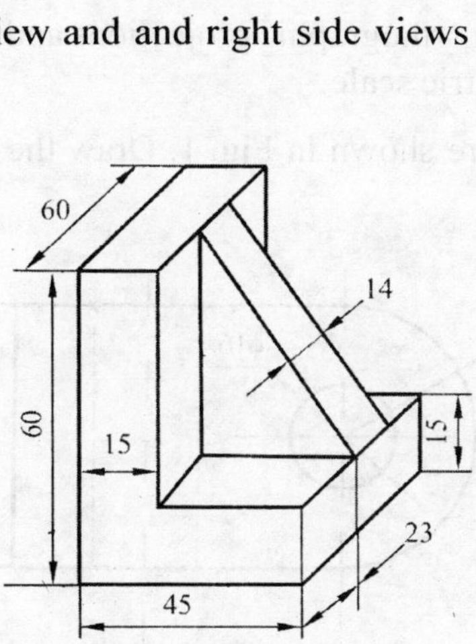

Fig. 2

Question Paper – 11

1. A fixed point F is 7.5 cm from a fixed straight line. Draw the locus of a point P moving in such a way that its distance from the fixed straight line is 2/3 times its distance from F. Name the curve. Draw normal and tangent at a point 6 cm from F.

2. (a) The line EF 60 mm long is in VP and inclined to HP. The top view measures 45 mm. The end E is 15 mm above HP. Draw the projections of the line. Find its inclination with HP.

   (b) A line GH 45 mm long is in HP and inclined to VP. The end G is 15 mm in front of VP. The length of the front view is 35 mm. Draw the projection of the line. Determine its inclination with VP.

3. (a) Draw the projections of a regular pentagon of 30 mm side, with its surface making an angle of 50° with HP. One of the sides of the pentagon is parallel to HP and 15 mm away from it.

   (b) A pentagon of 30 mm sides, has one of its corners on HP and its plane inclined at 65° to VP and perpendicular to HP. Draw its projection.

4. A square prism, side of base 30 mm and axis 45 mm long lies on HP such that its axis is parallel to both HP and VP. Draw the top and front views of the prism when (a) it lies with one of its rectangular faces on HP and (b) It lies with one of its longer edges on HP.

5. Draw the projections of a cone, base 45 mm diameter and axis 60 mm long, when it is resting on the ground on a point of its base circle with the axis making an angle 30° with the HP and 45° with the VP.

6. A hexagonal prism, side of base 25 mm and axis 50 mm long rests on its base in HP. Its axis is parallel to VP. Draw the orthographic projections and provide the isometric projection of the solid. Show the isometric scale.

7. Two views of a casting are shown in Fig. 1. Draw the isometric view of the casting (Al dimensions are in mm).

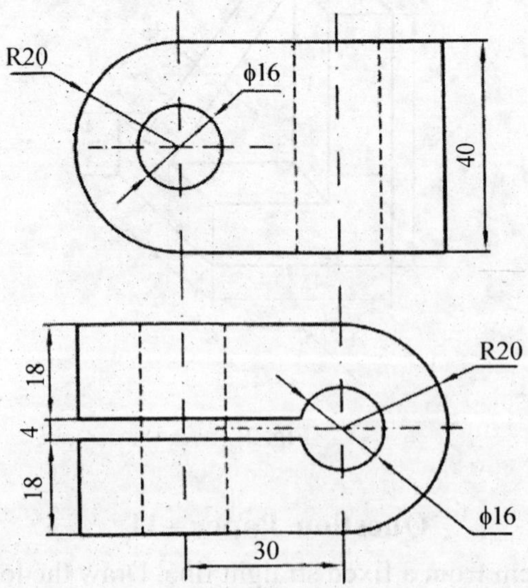

Fig. 1

8. Draw the front view, top view and right side view of the object shown in Fig. 2 (all dimensions in mm).

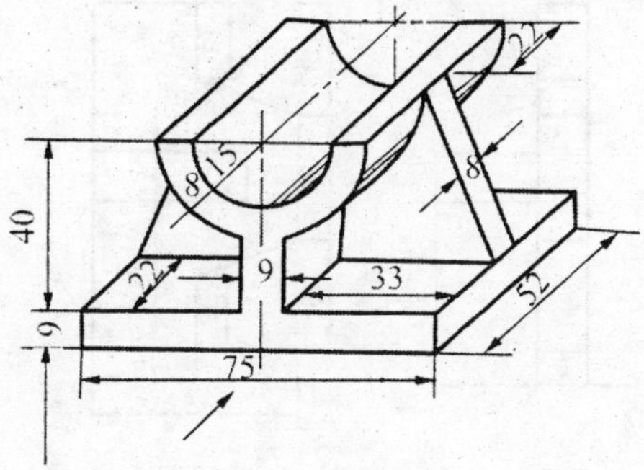

Fig. 2

Question Paper – 12

1. Two fixed points A an B are 100 mm apart. Trace the complete path of a point P moving (in the same plane as that of A and B) in such a way that the sum of its distances from A and B is always equal to 125 mm. Name the curve. Draw another curve parallel to and 25 mm away from this curve.
2. A line AB is 30 mm long inclined at 30° to VP and parallel to HP. The end A of the line is 15 mm above HP and 20 mm in front of VP. Draw the projections.
3. (a) A square lamina of 40 mm side is perpendicular to HP. One of sides is 20 mm above HP and 15 mm in front of VP. Draw its projections.

   (b) Draw projections of a square.
4. (a) Square pyramid base 40 mm side, axis 65 mm long has base in VP, one edge of base inclined to 30° to HP and corner contained by that edge is on HP. Draw its projections.

   (b) Draw the projections of a hexagonal prism having one of its rectangular faces parallel to the HP. Its axis is perpendicular to the VP and 3.5 cm above the ground.
5. A triangular prism of base side 45 mm and length of axis 75 mm has corner in the HP, the face opposite to that corner makes 50° to the HP while the axis of the solid makes 30° to the VP. Obtain the two views of the solid.
6. A cone radius of base 25 mm and axis 50 mm long rests with one of its base edges on HP. Its axis is parallel to VP. Draw the orthographic projections and provide the isometric projection of the solid showing the tip towards viewer. Show the isometric scale.

7. Three views of a model, in third angle projection are shown in Fig. 1. Draw the isometric view of the casting (all dimensions are in mm).

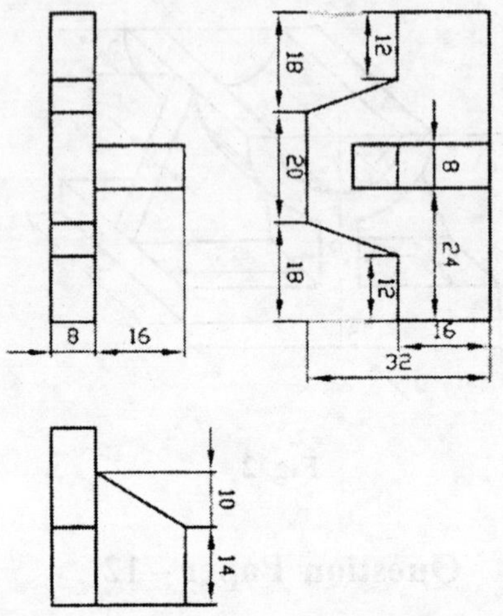

Fig. 1

8. Draw the front, top and right views of the block shown in the Fig. 2 (All dimensions are in mm).

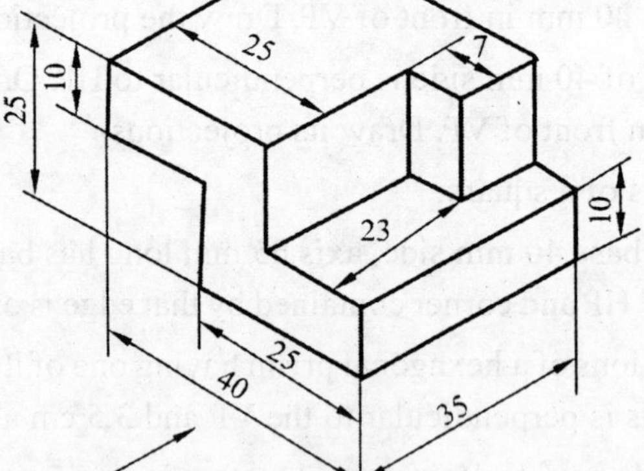

Fig. 2

Question Paper – 13

1. The vertex of a hyperbola is 65 mm from its focus. Draw the curve if the eccentricity is 3/2. Draw a normal and a tangent at a point on the curve, 75 mm from the directrix.
2. Show by means of a drawing that when the diameter of the directing circle is twice that of the generating circle, the hypocycloid is a straight line. Take the diameter of the generating circle equal to 50 mm.

3. (a) A point A is 2.5 cm above the HP and 3 cm infront of the VP. Draw its projections.

   (b) A point A is 2 cm below the HP and 4 cm behind the VP. Draw its projections.

   (c) Two points A and B are in the HP. The point A is 30 mm infront of the VP while B is behind the VP. The distance between their projections is 75 mm and the line joining their top views makes an angle of 45° with XY. Find the distance of the point B from the VP.

4. A line AB 120 mm long is inclined at 45° to the HP and 30° to the VP. Its mid point C is in VP and 20 mm above HP. The end A is in the third quadrant and B is in the first quadrant. Draw the projections of the line.

5. (a) A regular pentagon of 25 mm side has one side on the ground. Its plane is inclined at 45° to the HP and perpendicular to the VP. Draw its projections.

   (b) Draw the projections of a circle of 5 cm diameter, having its plane vertical and inclined at 30° to the VP. Its centre is 3 cm above the HP, and 2 cm in front of the VP.

6. (a) Draw the projections of a hexagonal prism of base 25 mm and axis 60 mm long, when it is resting on one of its corners of the base on HP. The axis of the solid is inclined at 45° to HP.

   (b) Draw the projections of a pentagonal prism of base 25 mm side and axis 50 mm long. When it is resting on one of its rectangular faces on HP, the axis of the solid is inclined at 45° to VP.

7. Draw the isometric view of Door-Step having three steps of 22 cm trend and 15 cm rise. The steps measure 75 cm widthwise.

8. Draw the following views of the block shown in Fig. 1. All dimensions are in mm.

   (a) Front view

   (b) Top view

   (c) Both side views

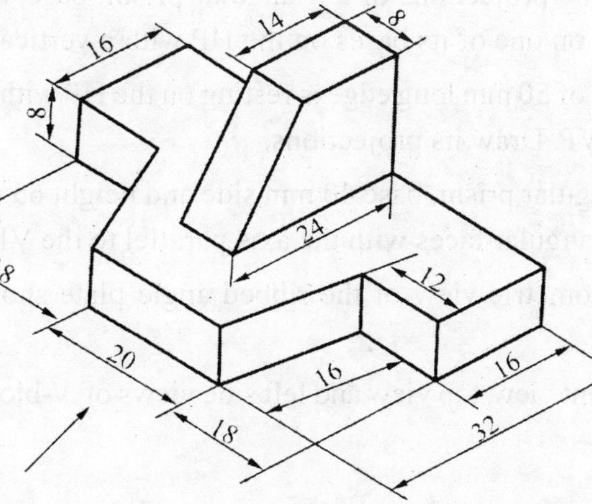

Fig. 1

Question Paper – 14

1. The foci of an ellipse are 80 mm apart and the minor axis is 55 mm long. Determine the length of the major axis and draw the ellipse by concentric-circle method. Draw a curve parallel to the ellipse and 20 mm away from it.

2. A circle of 50 mm diameter rolls on the circumference of another circle of 175 mm diameter and outside it. Trace the locus of a point on the circumference of the rolling circle for one complete revolution. Name the curve. Draw a tangent and a normal to the curve at a point 125 mm from the center of the directing circle.

3. (a) A point A is 2.5 cm above the HP and 3 cm infront of the VP. Draw its projections.

   (b) A point A is 2 cm below the HP and 4 cm behind the VP. Draw its projections.

   (c) Two points A and B are in the HP. The point A is 30 mm infront of the VP while B is behind the VP. The distance between their projections is 75 mm and the line joining their top views makes an angle of 45° with XY. Find the distance of the point B from the VP.

4. (a) A 100 mm long line is parallel to and 40 mm above the HP. Its two ends are 25 mm and 50 mm infront of the VP respectively. Draw its projections and find its inclination with the VP.

   (b) A line AB, 50 mm long, has its end A in both the HP and the VP. It is inclined at 30° to the HP and at 45° to the VP. Draw its projections.

5. A circular plane of 60 mm diameter, rests on VP on a point A on its circumference. Its plane is inclined at 45° to VP. Draw the projections of the plane when (a) the front view of the diameter AB makes 30° with HP and (b) the diameter AB itself makes 30° with HP.

6. (a) Draw the projections of a triangular prism, base 40 mm side and axis 50 mm long, resting on one of its bases on the HP with a vertical face perpendicular to the VP.

   (b) A cube of 50 mm long edge is resting on the HP with the vertical faces equally inclined to the VP. Draw its projections.

   (c) A triangular prism, base 40 mm side and height 65 mm is resting on the HP on one of its rectangular faces with the axis parallel to the VP. Draw its projections.

7. Draw the isometric view of the ribbed angle plate shown in Fig. 1. All dimensions are in mm.

8. Draw the front view, top view and left side views of V-block shown in Fig. 2. All dimensions are in mm.

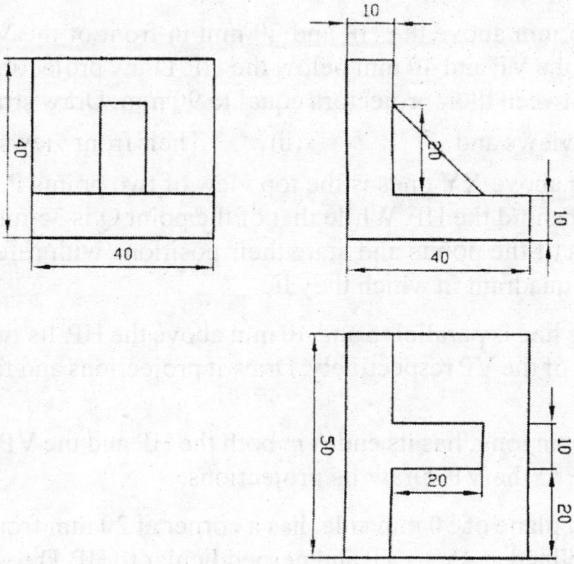

Fig. 1

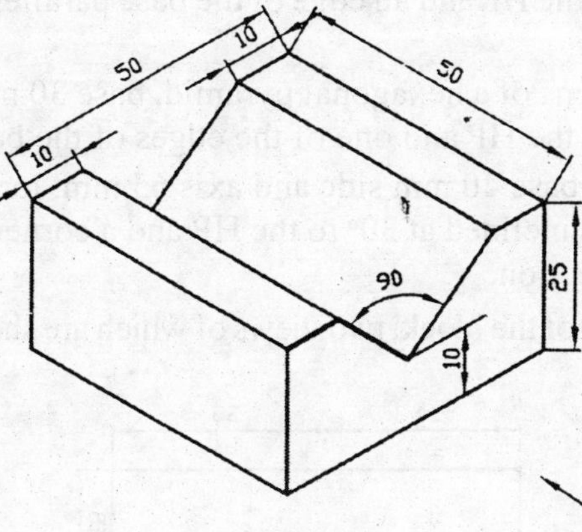

Fig. 2

Question Paper – 15

1. Two straight lines OA an OB make an angle of 75° between them. P is a point 40 mm from OA and 50 mm from OB. Draw a hyperbola through P, with OA and OB as asymptotes, marking at least ten points.

2. A circle of 35 mm rolls on a horizontal line. Draw the curve traced out by a point R on the circumference for one half revolution of the circle. For the remaining half revolution, the circle rolls on the vertical line. The point R vertically above the center of the circle in the starting position.

3. (a) A point P is 15 mm above the HP and 20 mm in front of the VP. Another point Q is 25 mm behind the VP and 40 mm below the HP. Draw projections of P and Q keeping the distance between their projectors equal to 90 mm. Draw straight lines joining
   (i) Their top views and (ii) Their front views
   (b) A point 30 mm above XY lines is the top view of two points P and Q. The front view of P is 45 mm behind the HP. While that of the point Q is 35 mm below the HP. Draw the projections of the points and state their positions with reference to the principal planes and the quadrant in which they lie.
4. (a) A 100 mm long line is parallel to and 40 mm above the HP. Its two ends are 25 mm and 50 mm in front of the VP respectively. Draw it projections and find its inclination with the VP.
   (b) A line AB, 50 mm long, has its end A in both the HP and the VP. It is inclined at 30° to the HP and 45° to the VP. Draw its projections.
5. A regular hexagonal plane of 30 mm side, has a corner at 20 mm from VP and 50 mm from HP. Its surface is inclined at 45° to VP and perpendicular to HP. Draw the projections of the plane.
6. (a) Draw the projections of a pentagonal pyramid, base 30 mm edge and axis 50 mm long, having its base on the HP and an edge of the base parallel to the VP. Also draw its side view.
   (b) Draw the projections of a hexagonal pyramid, base 30 mm side and axis 60 mm long, having its base on the HP and one of the edges of the base inclined at 45° to the VP.
   (c) A square pyramid base 40 mm side and axis 65 mm, long has its base in the VP. One edge of the base is inclined at 30° to the HP and a corner contained by that edge is on HP. Draw its projection.
7. Draw the isometric view of the block, two views of which are shown in Fig. 1. (All dimensions are in mm)

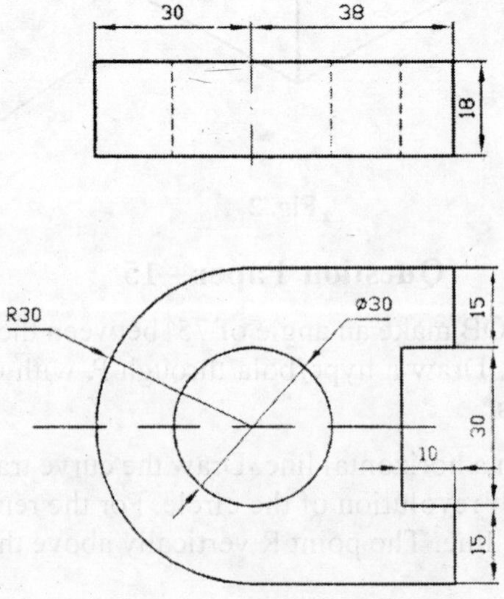

Fig. 1

8. Draw the front view, top view and right side views of the block shown in Fig. 2. (all dimensions are in mm).

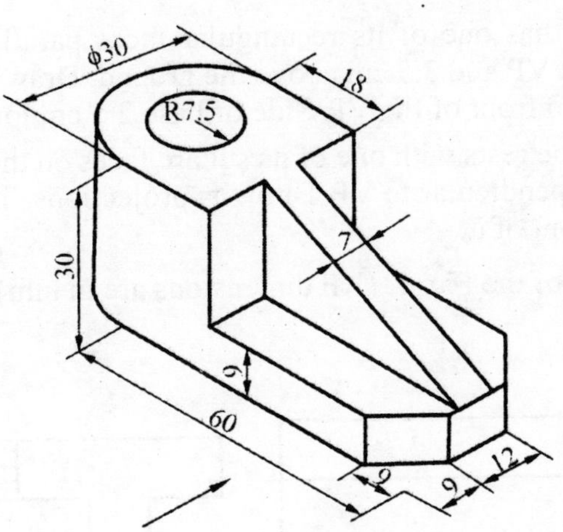

Fig. 2

Question Paper – 16

1. (a) Inscribe an ellipse in a parallelogram having sides 150 mm and 100 mm long and an inclined angle of 120°.

   (b) Draw a rectangle having its sides 125 mm and 75 mm long. Inscribe two parabolas in it with their axis bisecting each other.

2. Draw a hypo-cycloid of a circle of 30 mm diameter which rolls inside another circle of 160 mm diameter, for one revolution counter clock wise. Draw a tangent and a normal to it at a point 60 mm from the center of the directing circle.

3. Draw the projections of the following points on the same ground line, keeping the projectors 20 mm apart.

   (a) Point C, in the VP and 40 mm above the HP.

   (b) Point D, 25 mm below the HP and 25 mm behind the VP.

   (c) Point E, 15 mm above the HP and 50 mm behind the VP.

   (d) Point F, 40 mm below the HP and 25 mm infront of the VP.

4. A line AB of 70 mm long, has its end A at 10 mm above HP and 15 mm in front of VP. Its front view and top view measure 50 mm and 60 mm respectively. Draw the projections of the line and determine its inclinations with HP and VP.

5. Draw the projections of a circle of 60 mm diameter, resting on VP on a point on the circumference. The plane is inclined at 45° to VP and perpendicular to HP. The centre of the plane is 40 mm above HP.

6. (a) Draw the projections of
   (i) a cylinder, base 40 mm diameter and axis 50 mm long, and
   (ii) a cone, base 40 mm diameter and axis 50 mm long resting on the HP on their respective bases.

   (b) A hexagonal prism has one of its rectangular faces parallel to the HP. Its axis is perpendicular to the VP and 3.5 cm above the ground. Draw its projections when the nearer end is 2 cm in front of the VP. Side of base 2.5 cm long axis 5 cm long.

   (c) A cube of 40 mm side rests with one of its square faces on the HP such that one of its vertical faces is perpendicular to VP. Draw its projections. The nearest edge parallel to VP is 5 mm in front of it.

7. Draw the isometric view of the Fig. 1. (All dimensions are in mm).

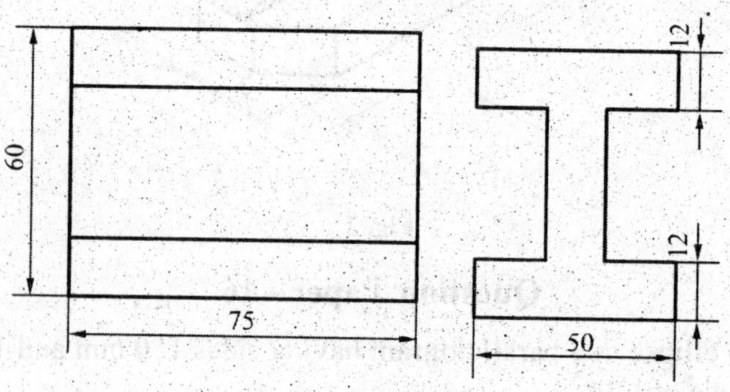

Fig. 1

8. Draw the front view, top view and left side views of the part shown in the Fig. 2. (all dimensions are in mm).

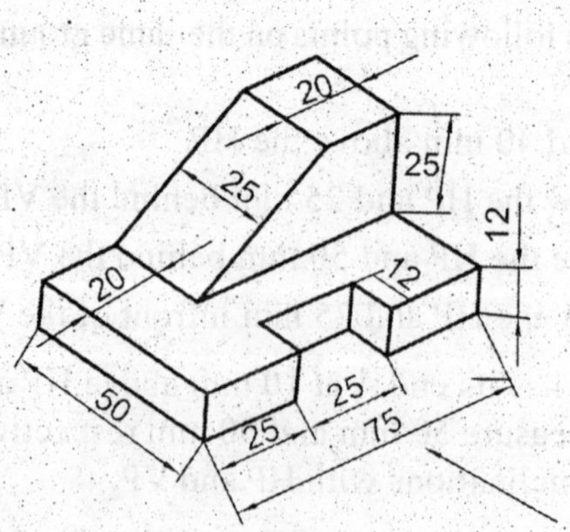

Fig. 2

Question Paper – 17

1. A point P is 30 mm and 50 mm respectively from two straight lines which are at right angles to each other. Draw a rectangular hyperbola from P within 10 mm distance from each line.

2. Show by means of a drawing that when the diameter of the directing circle is twice that of the generating circle, the hypocycloid is a straight line. Take the diameter of the generating circle equal to 60 mm.

3. A line AB, 80 mm long, makes an angle of 30° with the VP and lies in a plane perpendicular to both the HP and the VP. Its end A is in the HP and the end B is in the VP. Draw its projections.

4. A plate having shape of an isosceles triangle has base 50 mm long and altitude 70 mm. It is so placed that in the front view it is seen as an equilateral triangle of 50 mm sides, one side inclined at 45° to xy. Draw its top view.

5. Draw the projections of a cylinder 75 mm diameter and 100 mm long, lying on the ground with its axis inclined at 40° to the VP and parallel to the ground.

6. One of the body diagonals of a cube of 45 mm edge is parallel to the HP and inclined at 45° to the VP. Draw the front view and the top view of the cube.

7. Draw the isometric view of the object whose orthographic projections are shown in Fig. 1, all dimensions are in mm.

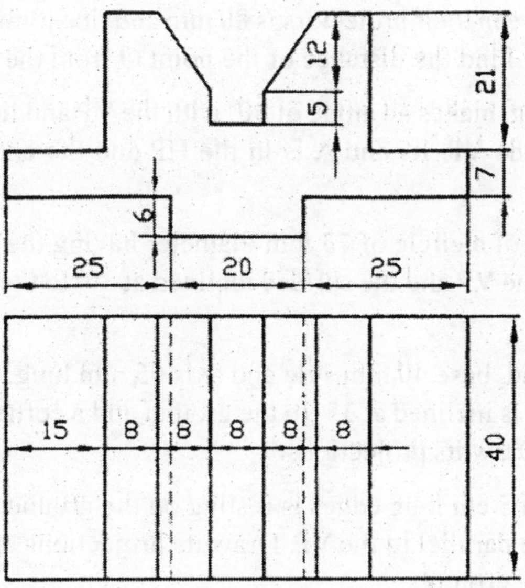

Fig. 1

8. Draw the orthographic projections for the isometric view shown in Fig. 2. All dimensions are in mm.

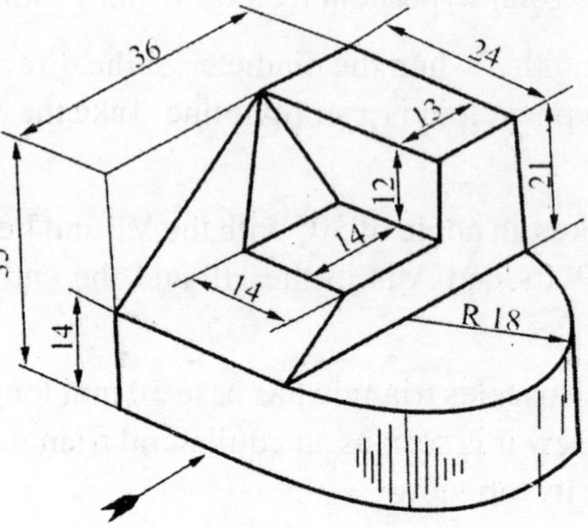

Fig. 2

Question Paper – 18

1. Two points A and B are 50 mm apart. Draw the curve traced out by a point P moving in such a way that the difference between its distances from A and B is always constant and equal to 20 mm.

2. Two points P and Q are in the HP. The point P is 30 mm in front of VP and Q is behind the VP. The distance between their projectors is 80 mm and line joining their top view makes an angle of 40° with XY. Find the distance of the point Q from the VP.

3. A line AB, 80 mm long, makes an angle of 30° with the VP and lies in a plane perpendicular to both the HP, and the VP. Its end A is in the HP and the end B is in the VP. Draw its projections.

4. Draw the projections of a circle of 75 mm diameter having the end A of a diameter AB in the HP, the end B in the VP and the surface inclined at 30° to the HP and inclined 60° to the VP.

5. (a) A square pyramid, base 40 mm side and axis 65 mm long, has its base in the VP. One edge of the base is inclined at 35° to the ground and a corner contained by that edge is on the ground. Draw its projections.

   (b) A tetrahedron of 5 cm long edges is resting on the ground on one of its faces with an edge of that face parallel to the VP. Draw its projections and measure the distance of its apex from the ground.

6. A pentagonal pyramid, base 25 mm side and axis 75 mm long has one of its triangular faces in the VP and the edge of the base contained by that face makes an angle of 30° with the HP. Draw its projections.

7. Draw the isometric view of the object whose orthographic projections are shown in Fig. 1. All dimensions are in mm.

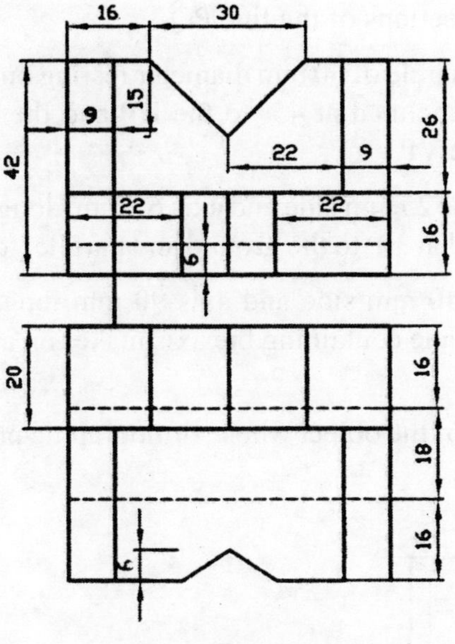

Fig. 1

8. Draw the orthographic projections for the given isometric view Fig. 2. All dimensions are in mm.

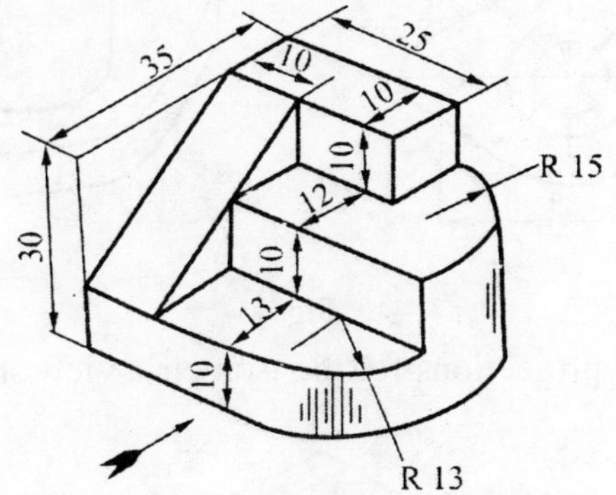

Fig. 2

Question Paper – 19

1. Draw a straight line AB of any length. Mark a point F, 75 mm from AB. Trace the path of a point P moving in such a way that the ratio of its distance from the point F, to its distance from AB is 3:2. Plot at least 8 points. Name the curve. Draw a normal and a tangent to the curve at a point on it, 50 mm from F.
2. Show by means of a drawing that when the diameter of the directing circle is twice that of the generating circle, the hypocycloid is a straight line. Take the diameter of the generating circle equal to 60 mm.

3. The front view of a 125 mm long line PQ measures 80 mm and its top view measures 100 mm. Its end Q and the mid-point M are in the first quadrant. M being 20 mm from both the planes. Draw the projections of the line PQ.

4. Draw the projections of a circle of 60 mm diameter resting on the ground on a point A on the circumference, its plane inclined at 45° to the HP and the top view of the diameter AB making 30° angle with the VP.

5. A hexagonal pyramid, base 25 mm side and axis 65 mm along, has an edge of its base on the ground. Its axis is inclined at 30° to the ground and parallel to the VP. Draw its projections.

6. A square pyramid, base 40 mm side and axis 90 mm long, has a triangular face on the ground and the vertical plane containing the axis makes an angle of 45° with the VP. Draw its projections.

7. Draw the isometric view of the object whose orthographic projections are shown in Fig. 1. All dimensions are in mm.

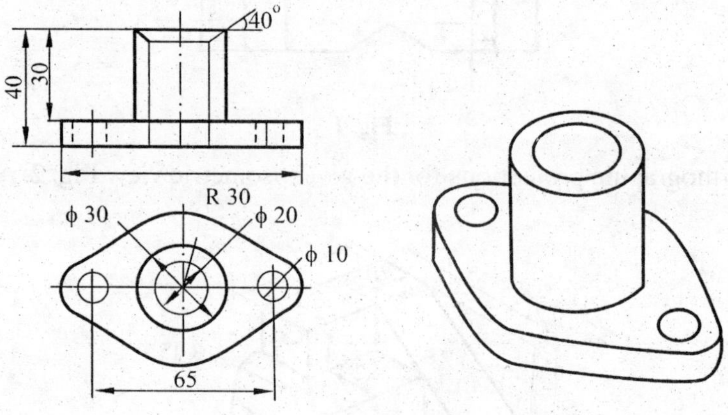

Fig. 1

8. Draw the orthographic projections for the isometric view shown in Fig. 2. All dimensions are in mm.

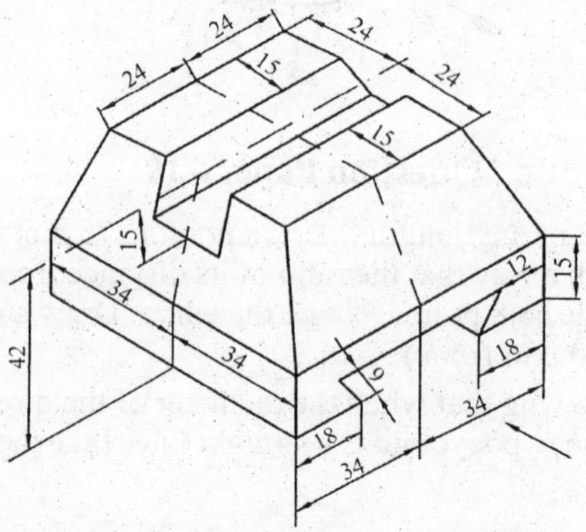

Fig. 2

WORKSHEETS

Study the Isometric Views in Figs. 1 to 24 and draw the Orthographic Views required.

Arrow indicates the direction of view.

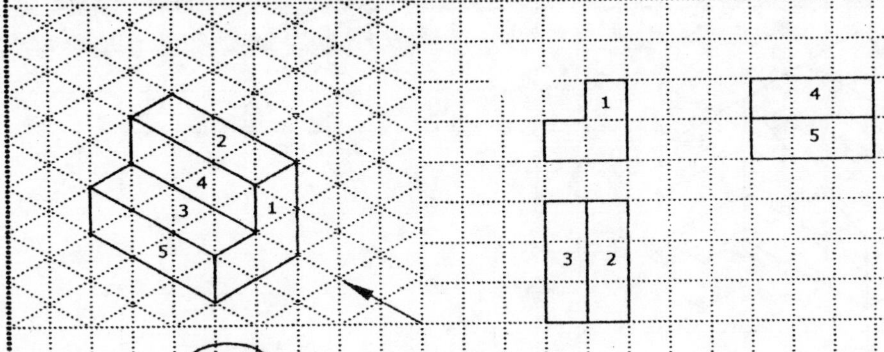

Fig.1

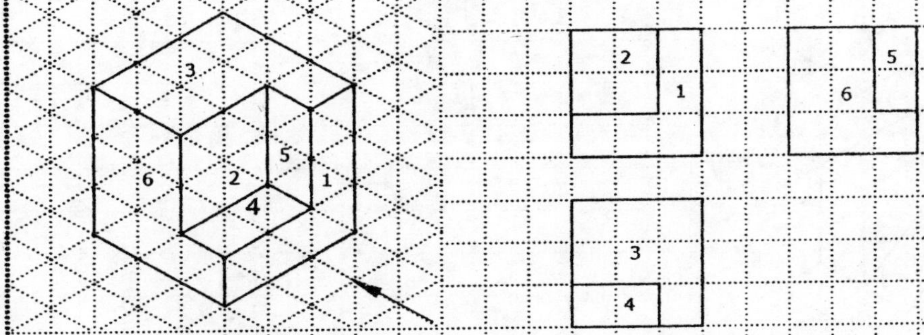

Fig.2

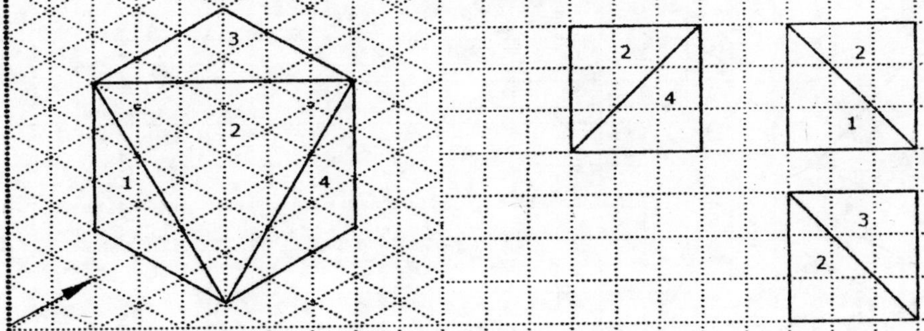

Fig.3

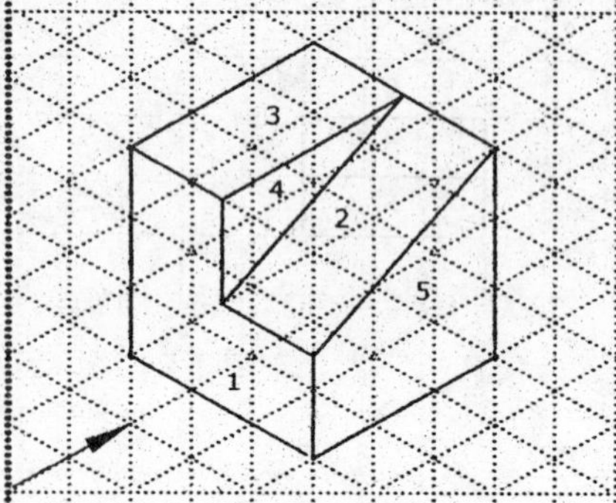

Fig.4

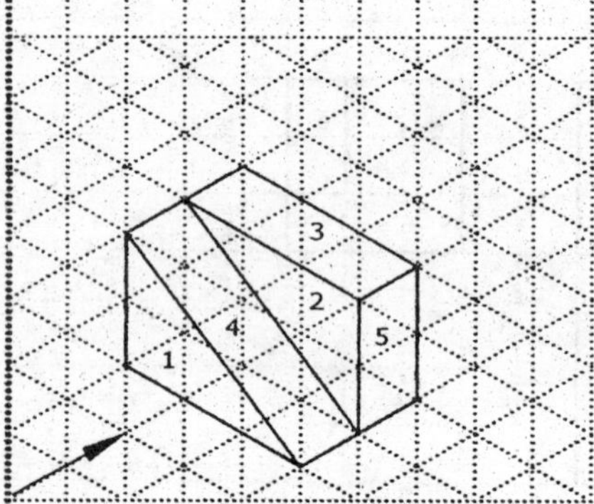

Fig.5

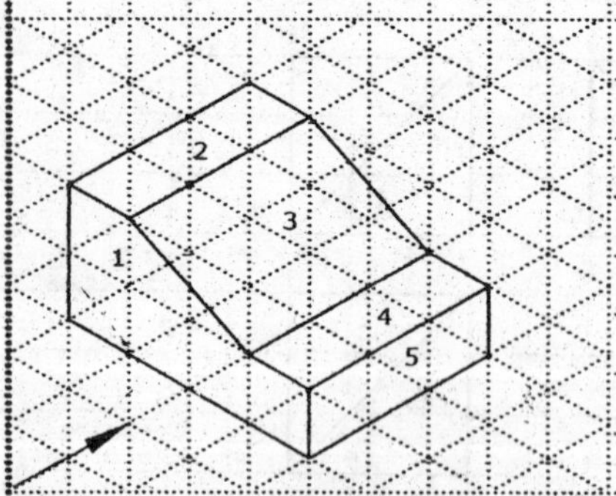

Fig.6

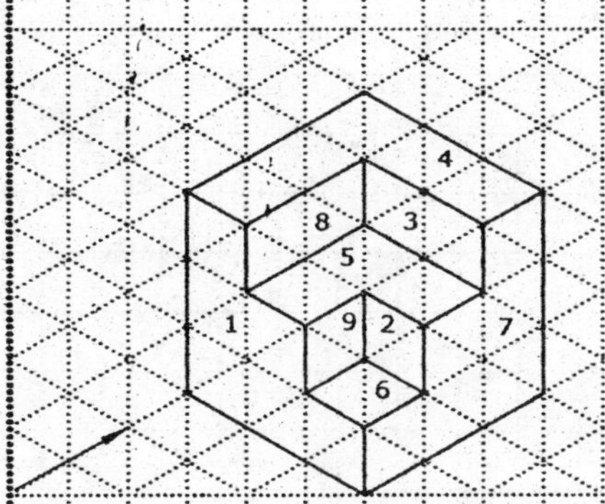

Fig.7

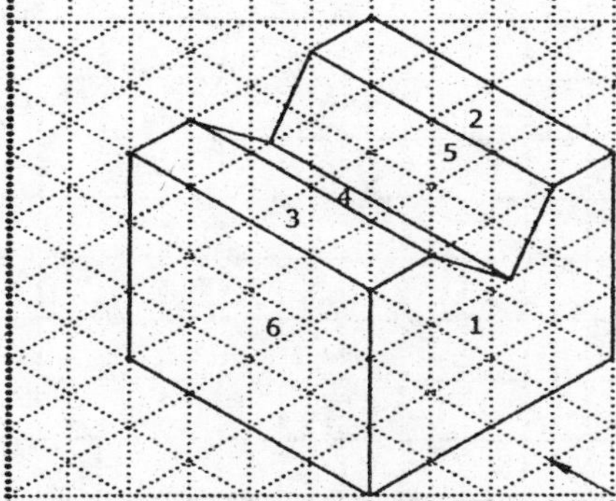

Fig.8

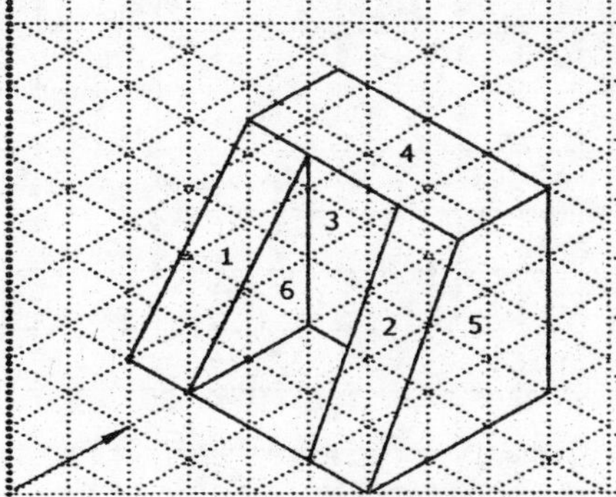

Fig.9

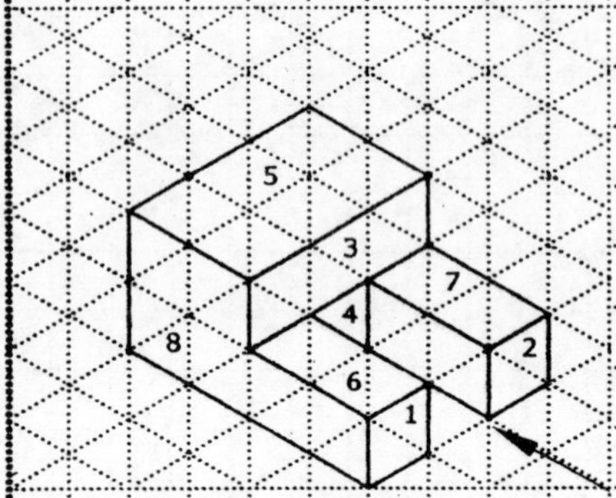

Fig.10

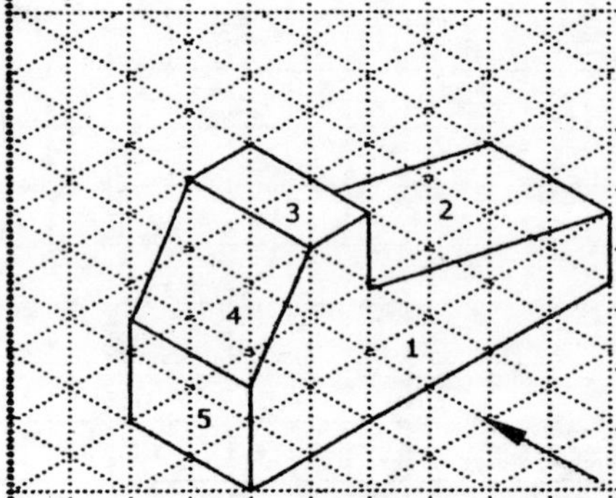

Fig.11

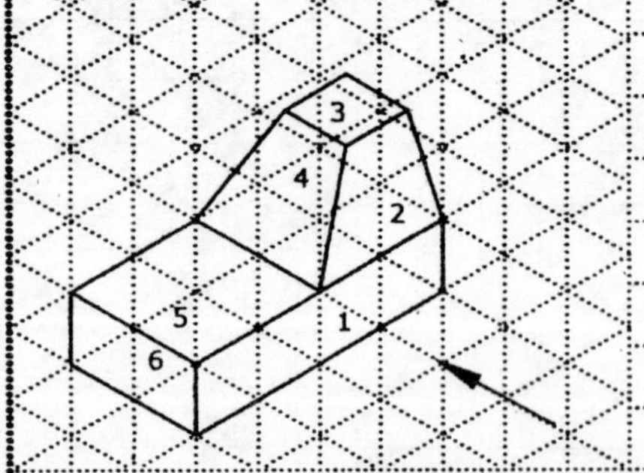

Fig.12

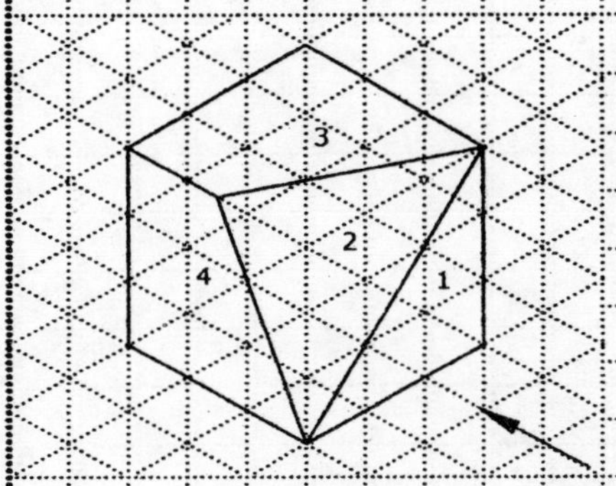

Fig.13

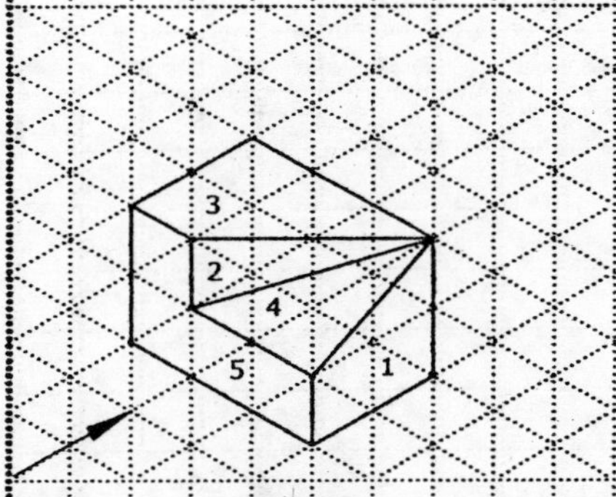

Fig.14

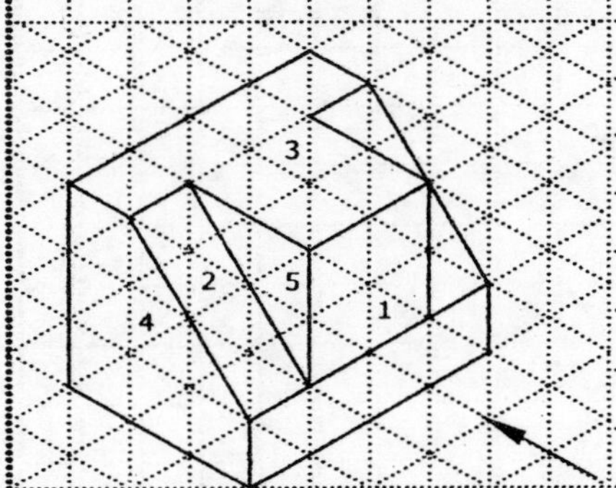

Fig.15

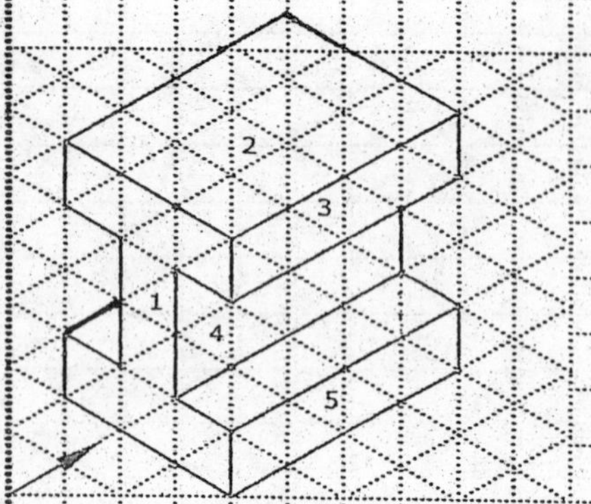

Fig.16

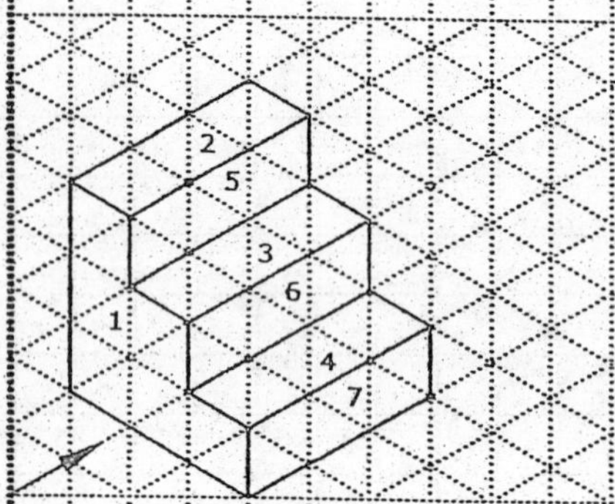

Fig.17

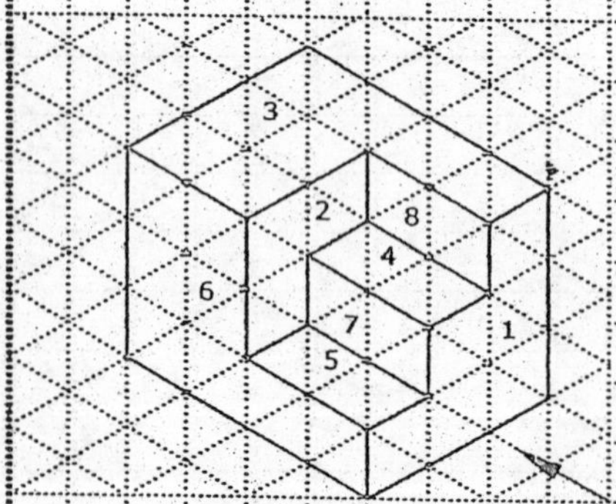

Fig.18

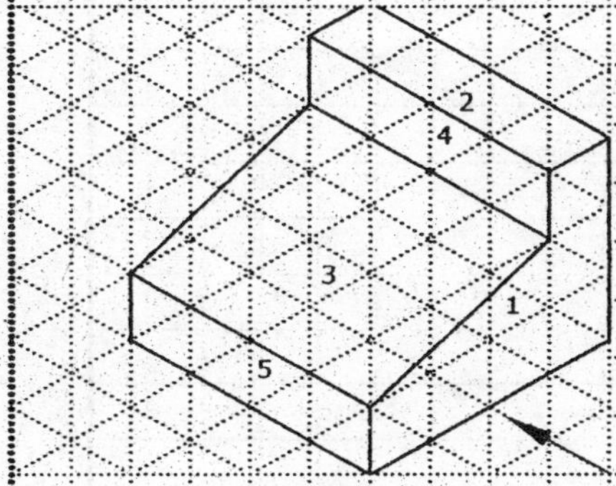

Fig.19

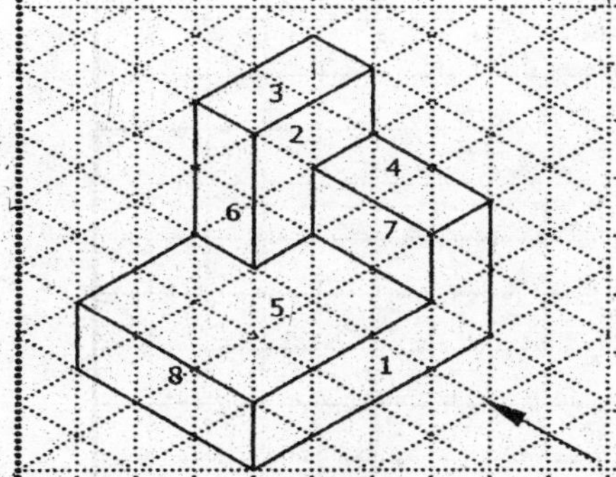

Fig.20

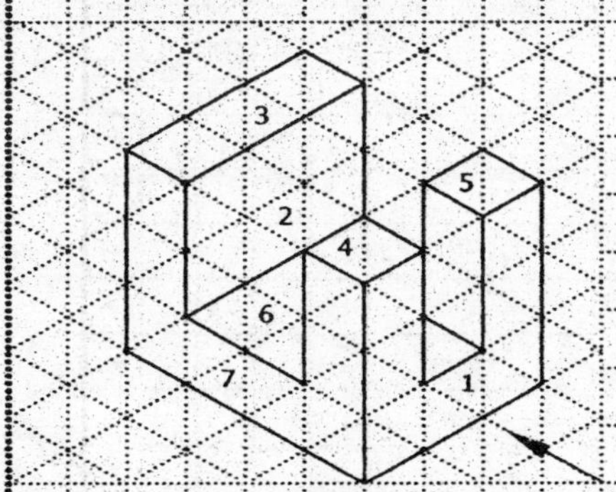

Fig.21

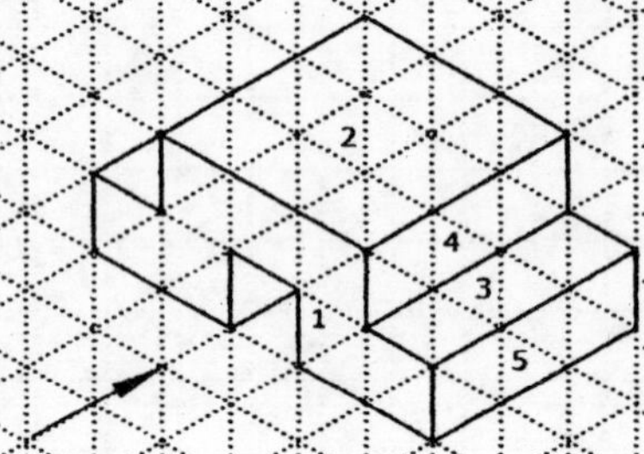

Fig.22

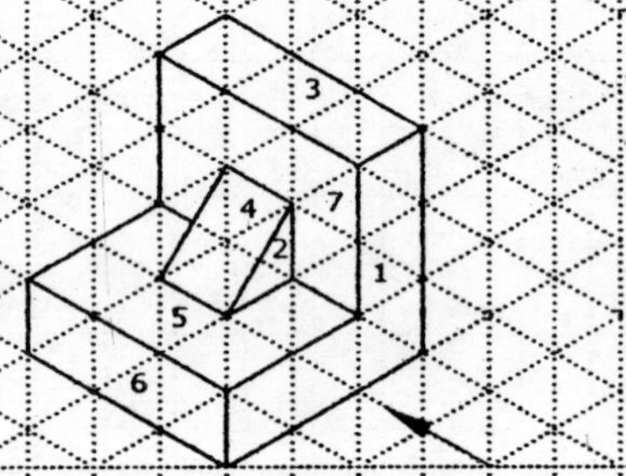

Fig.23

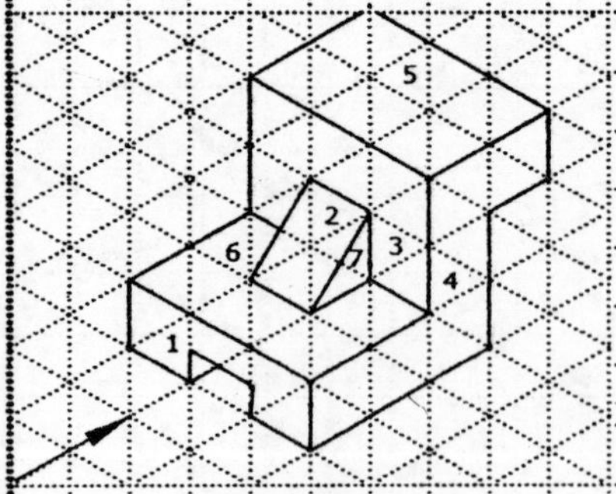

Fig.24

MISSING LINE(S)

Study the Isometric and Orthographic Views in Figs. 25 to 33.

Complete the views Marked 'X' by drawing the Missing Lines.

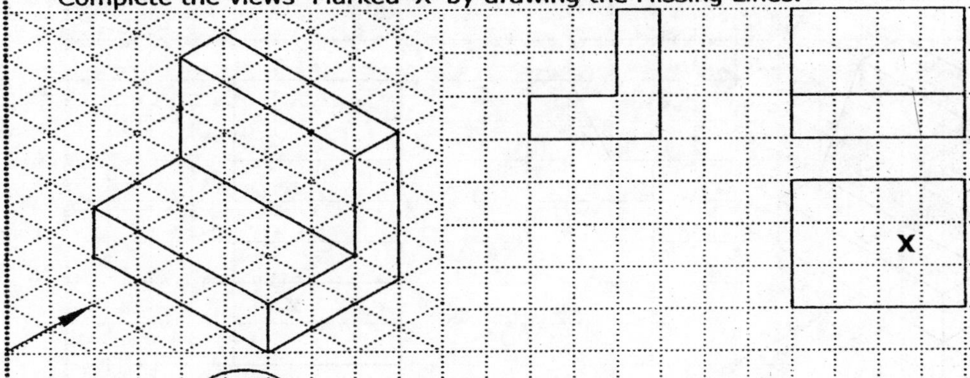

Fig.25

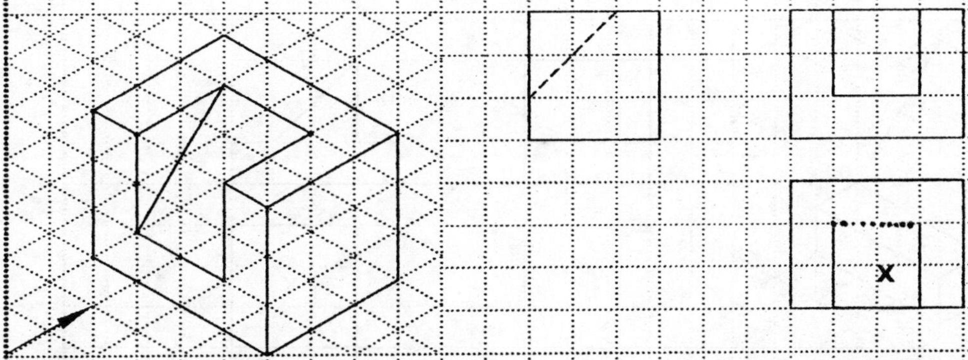

Fig.26

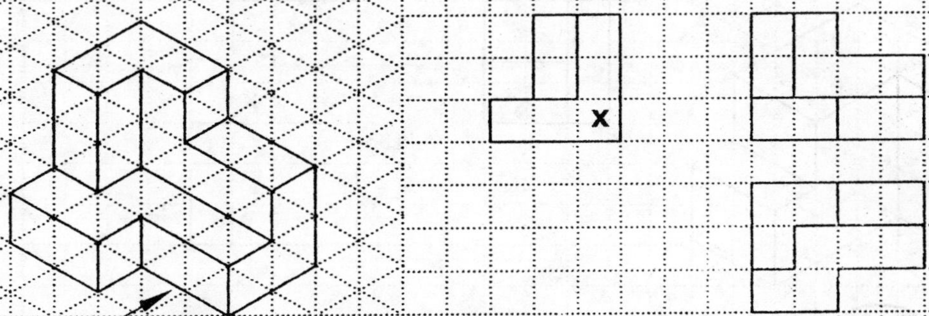

Fig.27

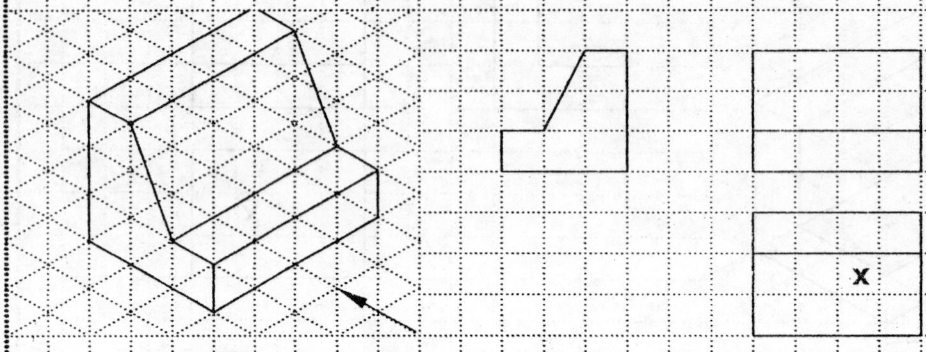

Fig.28

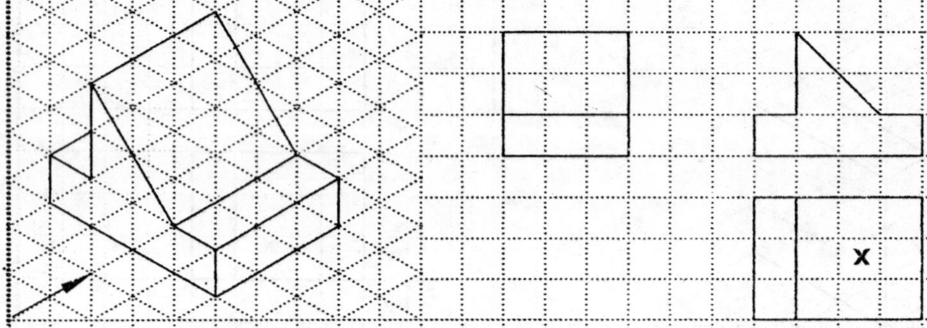

Fig.29

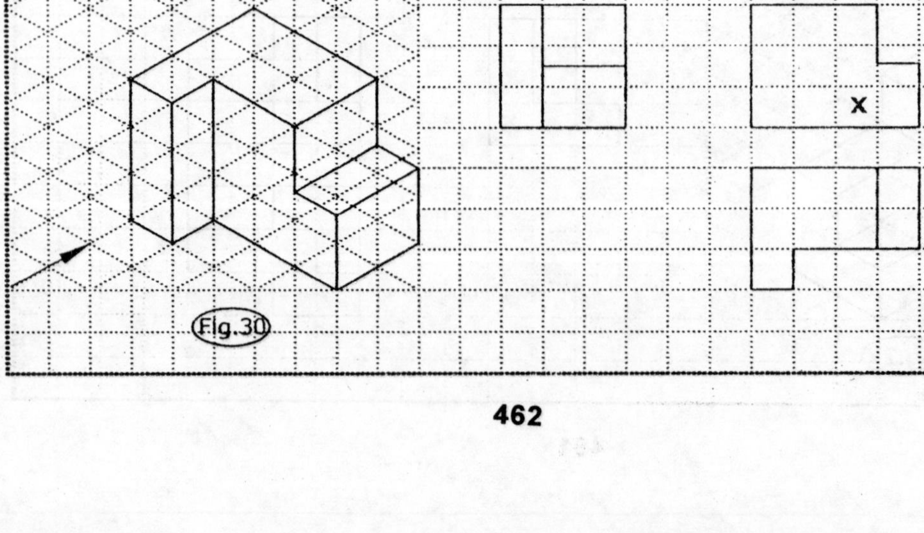

Fig.30

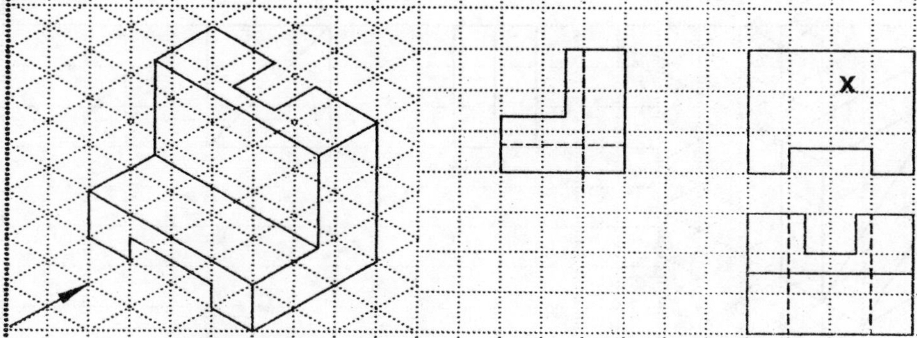

Fig.31

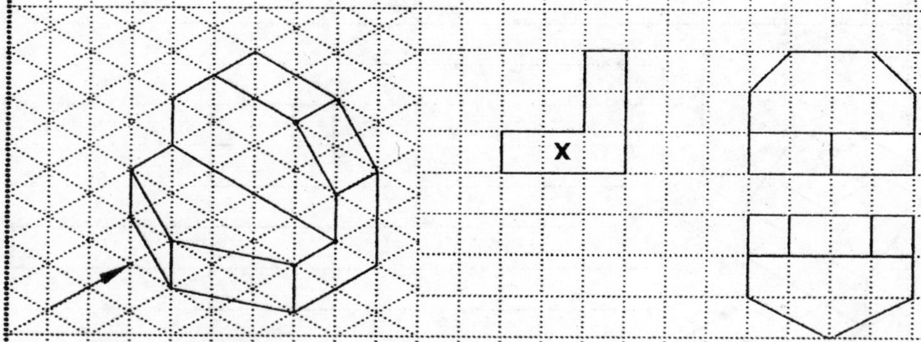

Fig.32

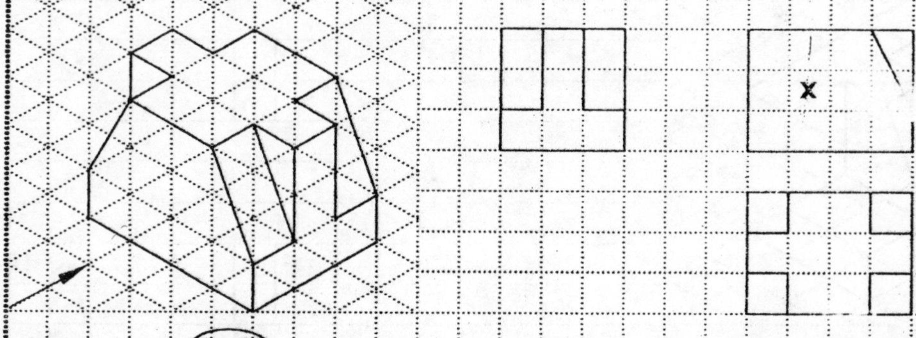

Fig.33

MISSING VIEWS

Study the Isometric and Orthographic Views in Figs.34 to 54 and draw the Missing views.

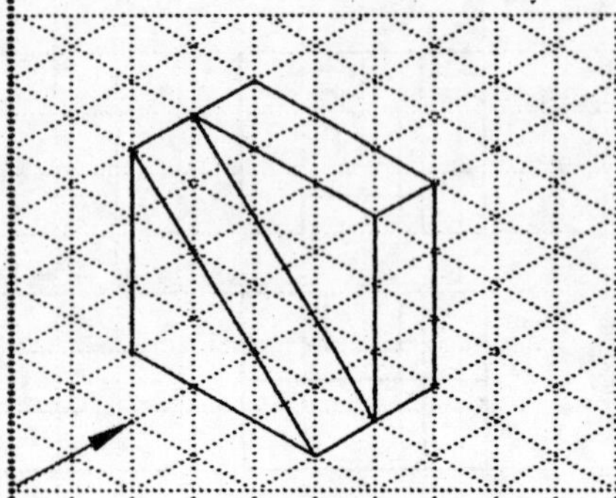

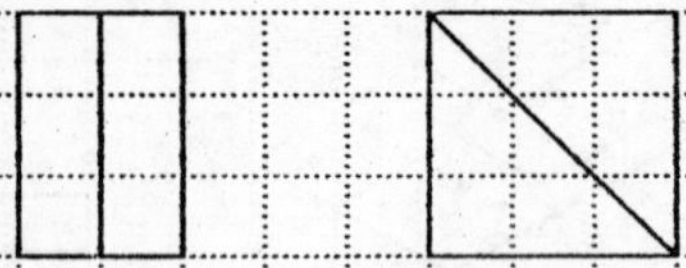

Fig.34

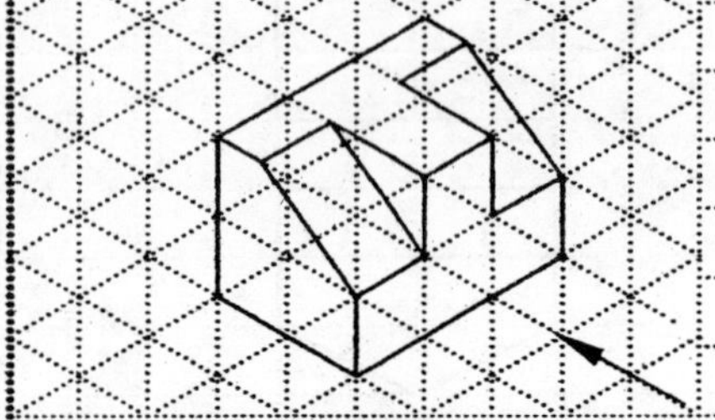

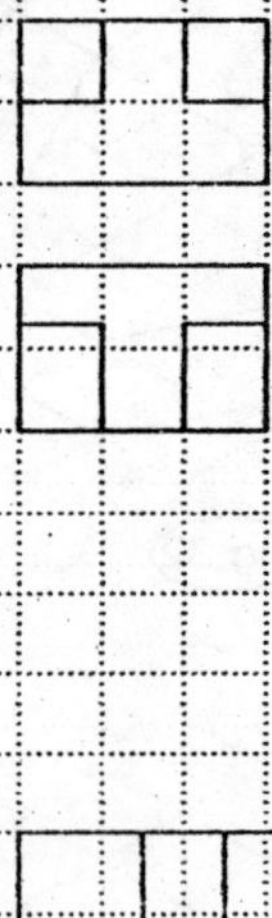

Fig.35

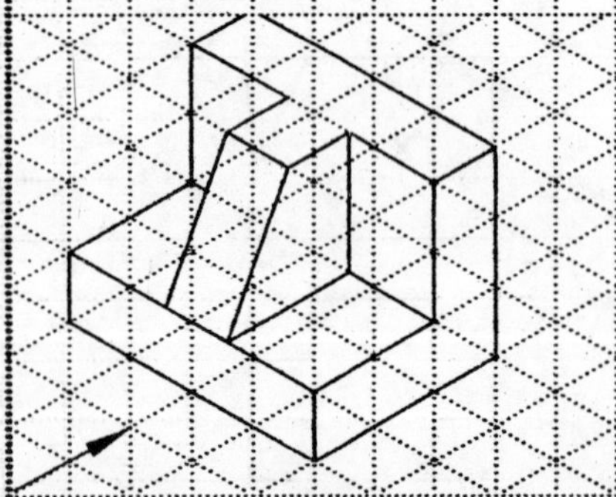

Fig.36

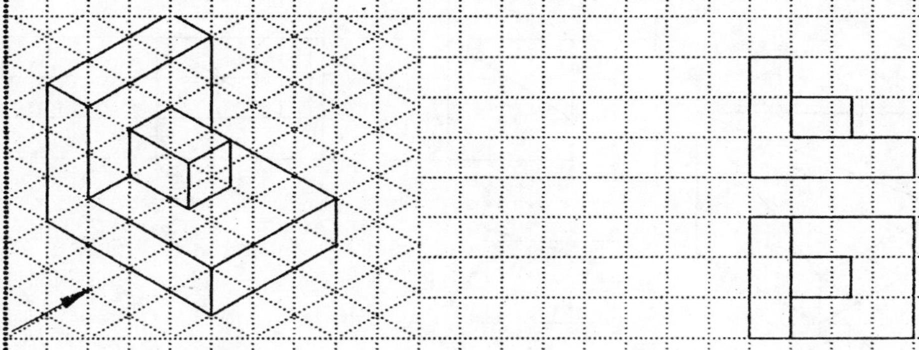

Fig.37

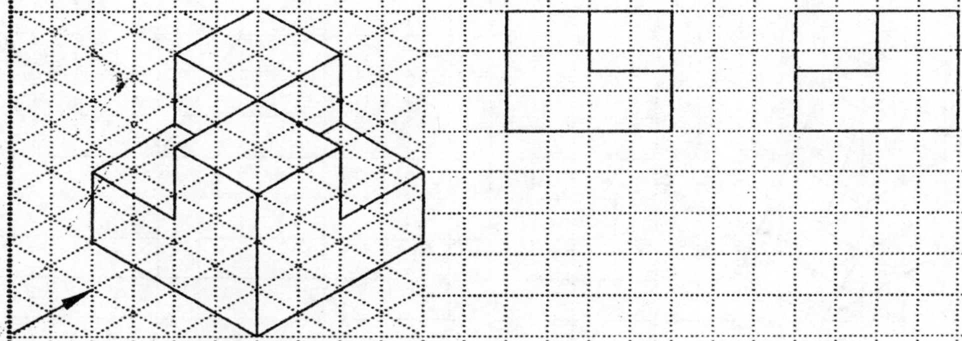

Fig.38

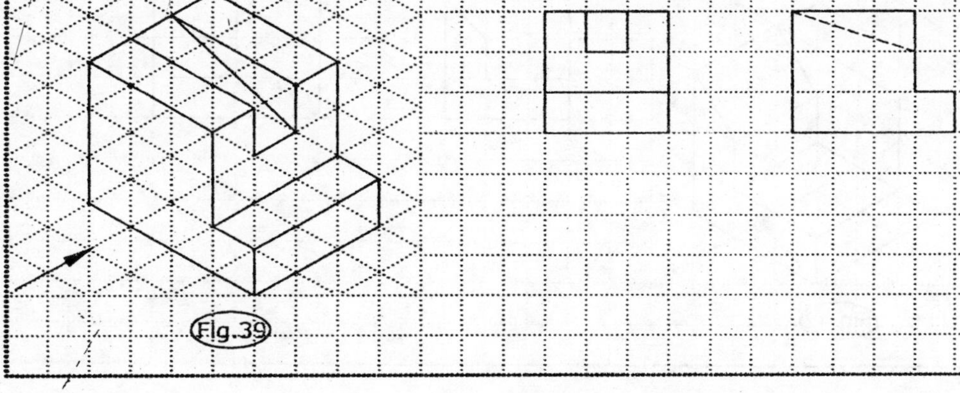

Fig.39

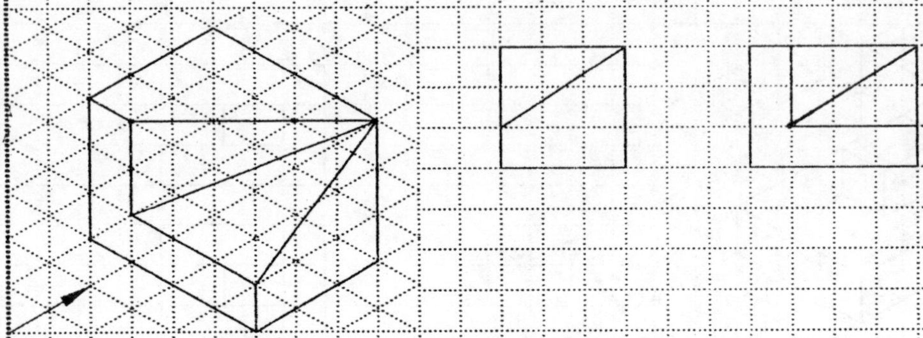

Fig.40

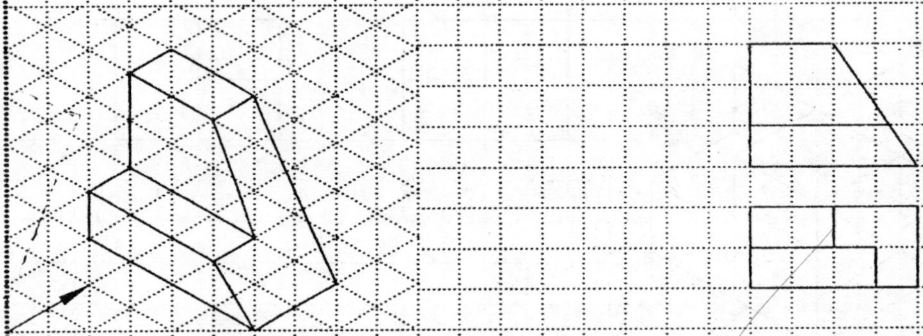

Fig.41

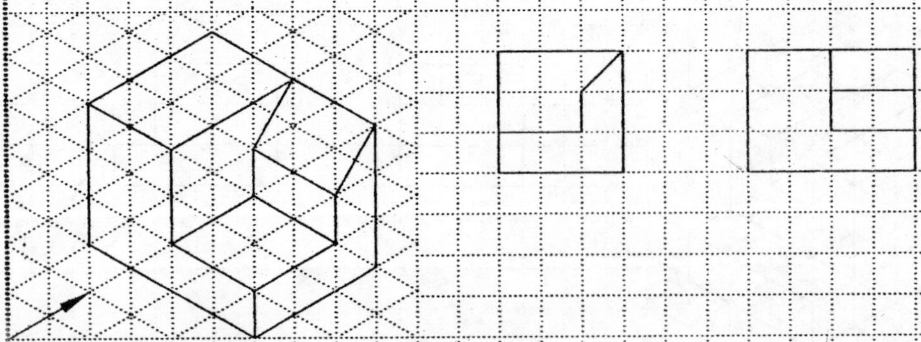

Fig.42

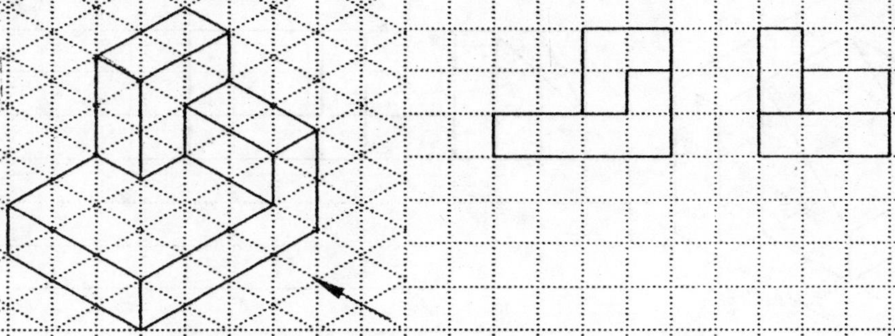

Fig.43

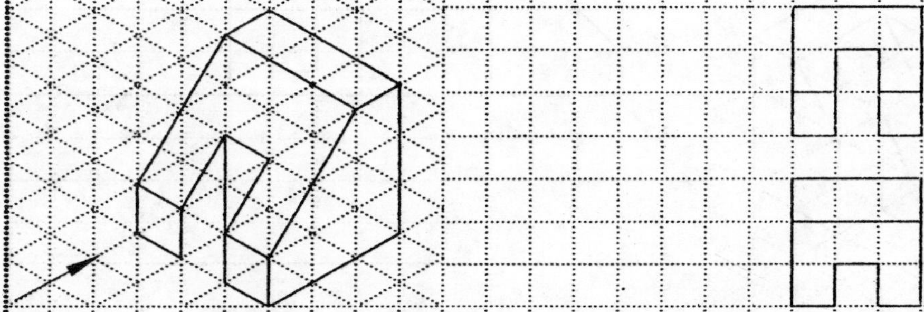

Fig.44

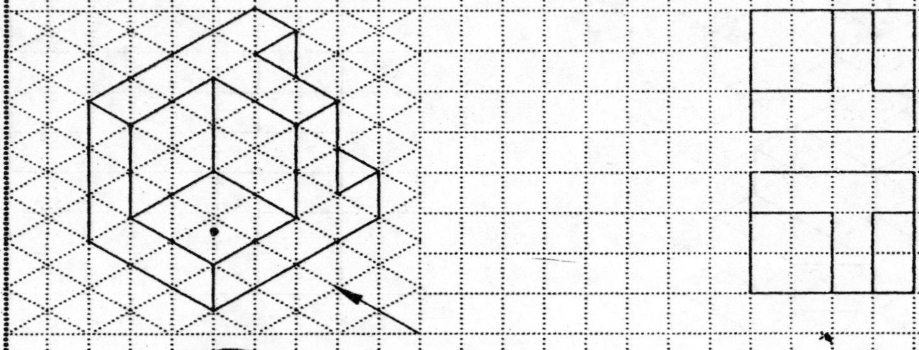

Fig.45

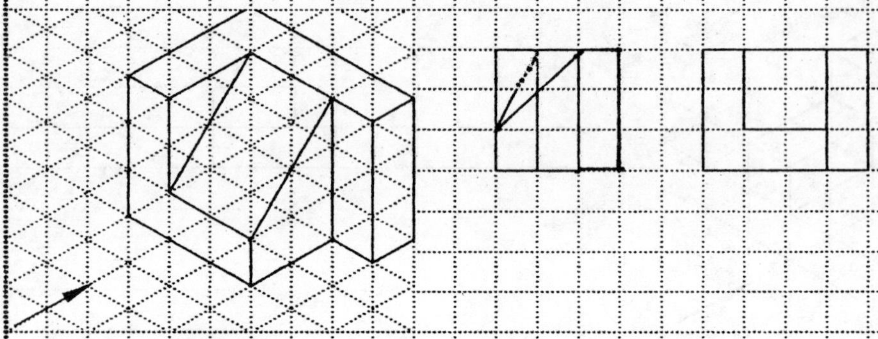

Fig.46

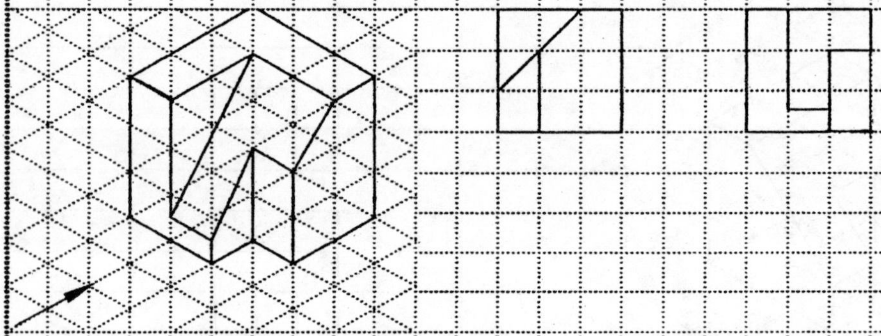

Fig.47

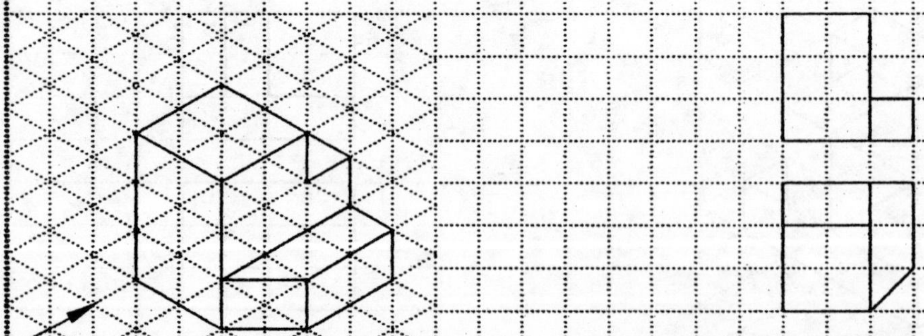

Fig.48

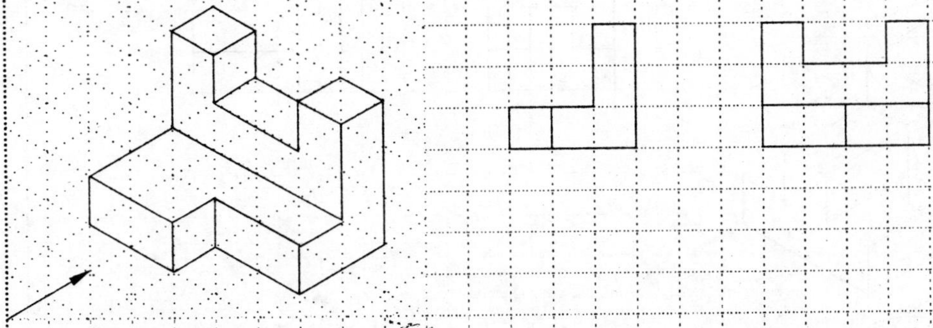

Fig. 49

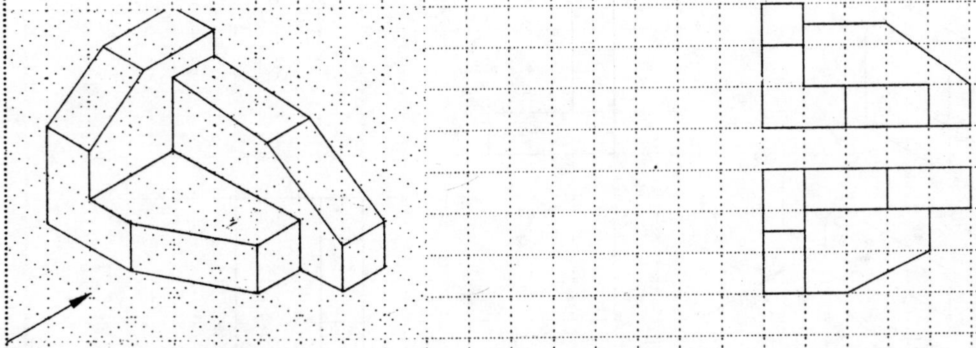

Fig. 50

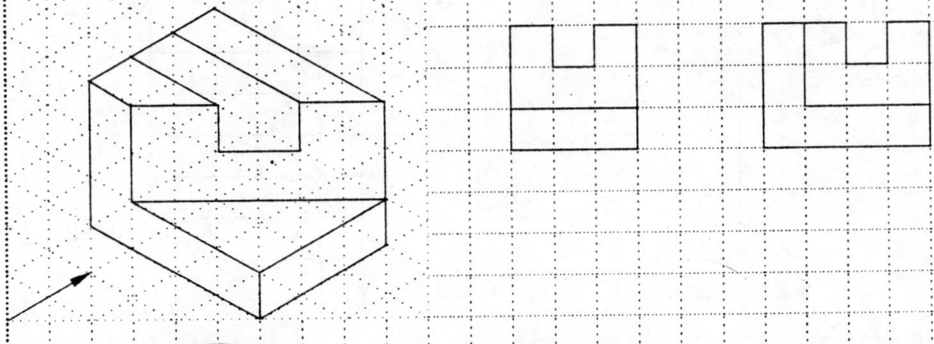

Fig. 51

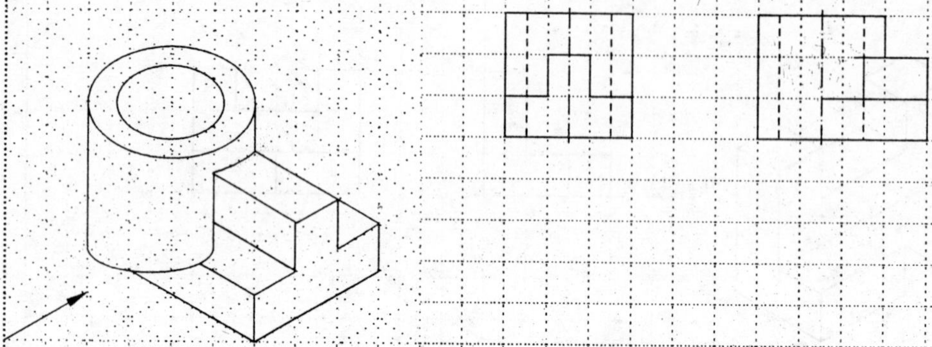

Fig. 52

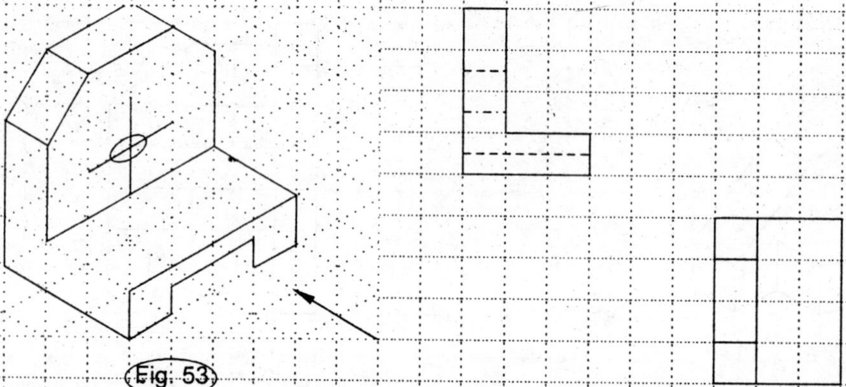

Fig. 53

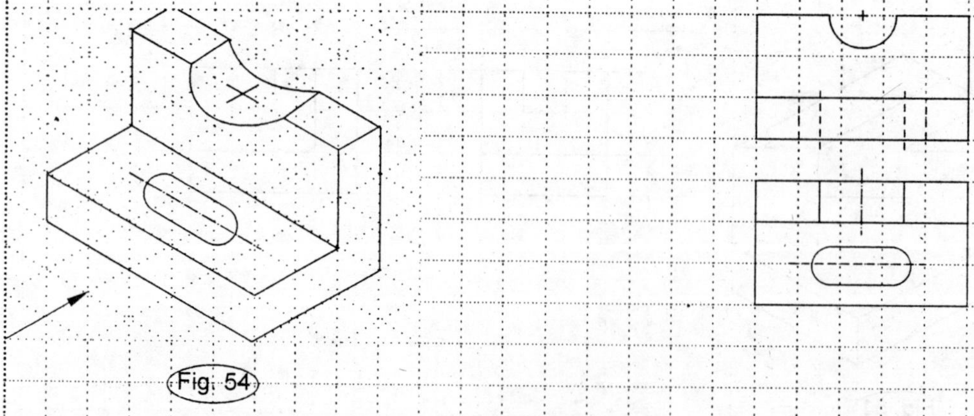

Fig. 54

Study the Isometric views in Figs. 55 to 75 and draw their orthographic views.

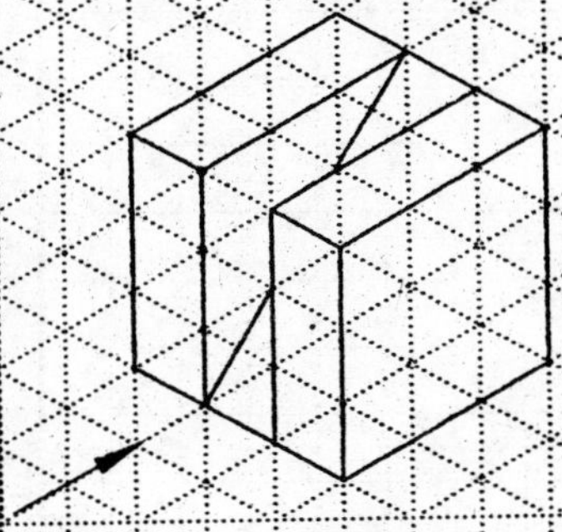

Fig.55

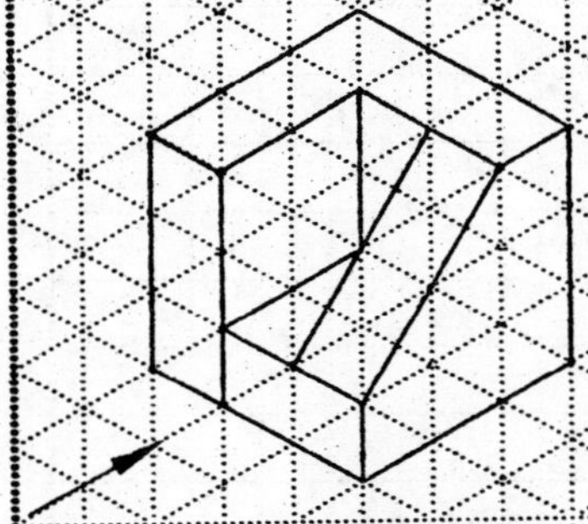

Fig.56

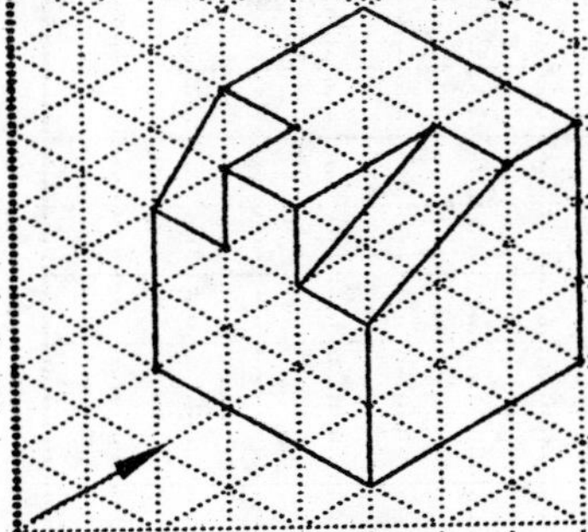

Fig.57

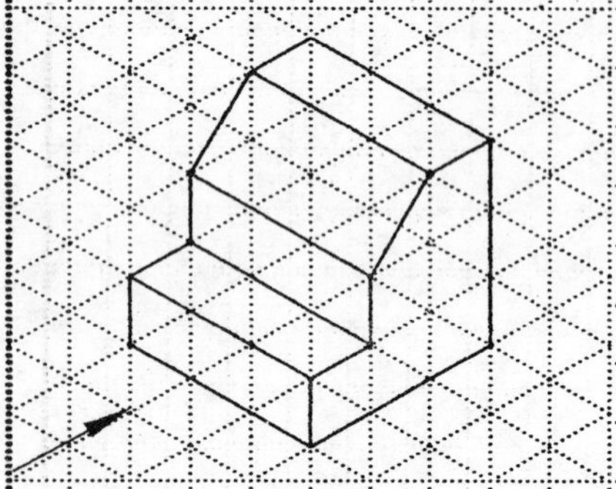

Fig.58

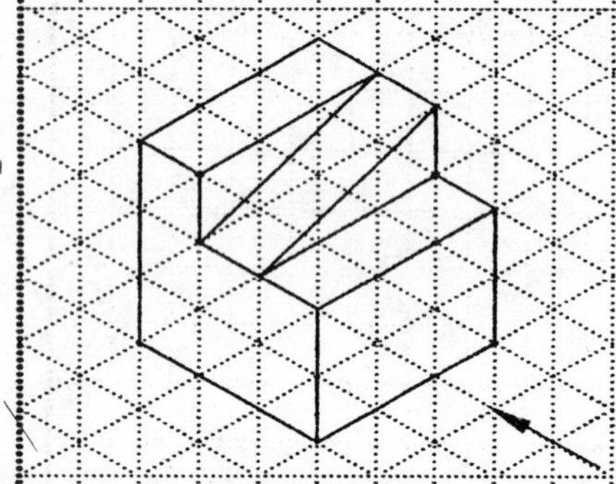

Fig.59

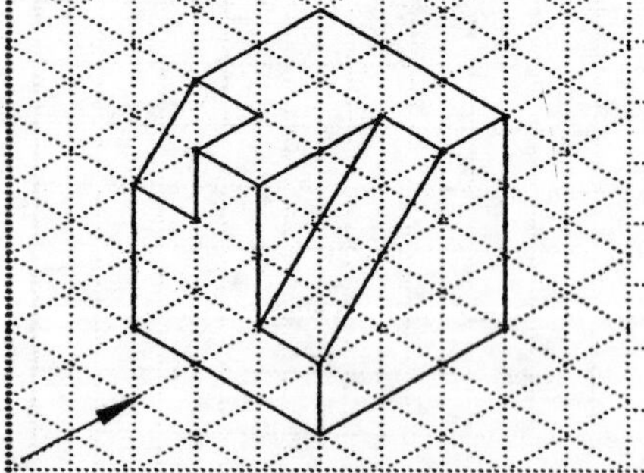

Fig.60

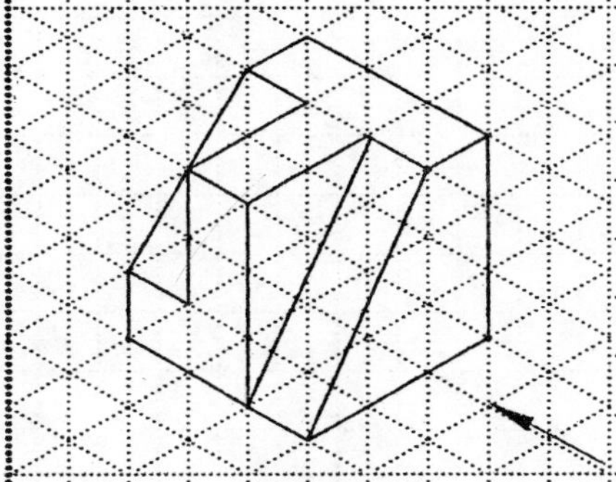

Fig.61

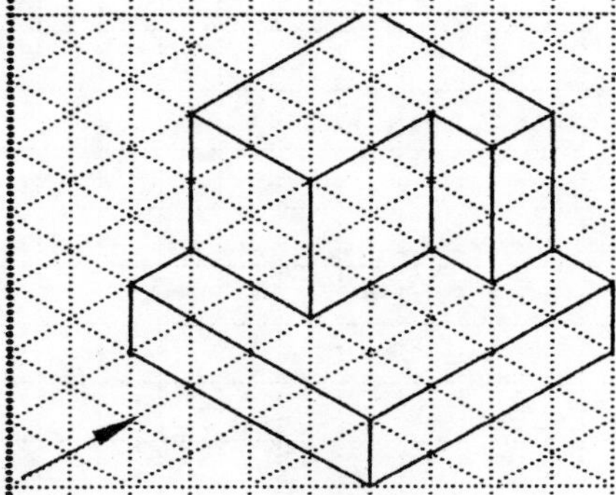

Fig.62

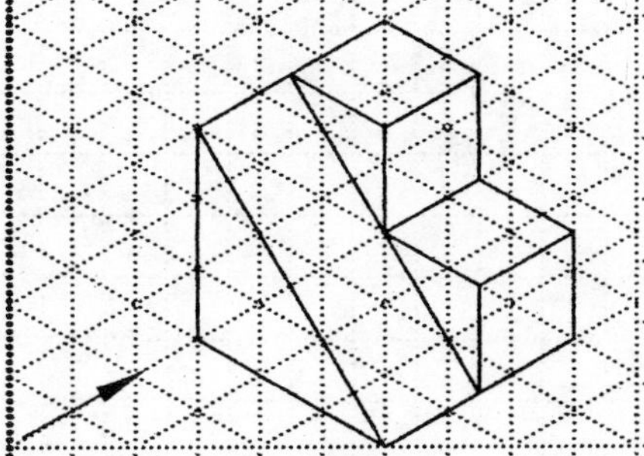

Fig.63

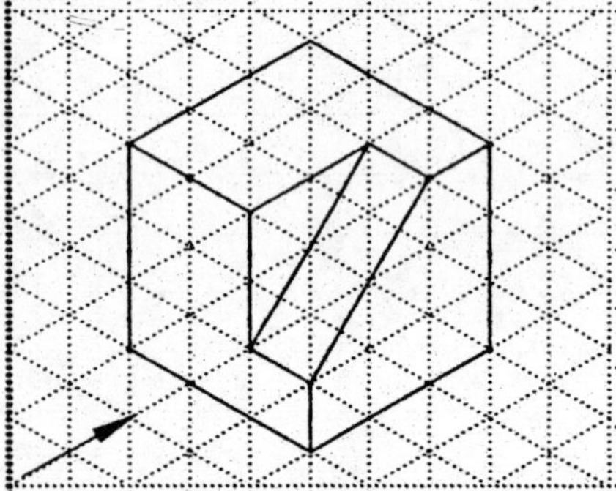

Fig.64

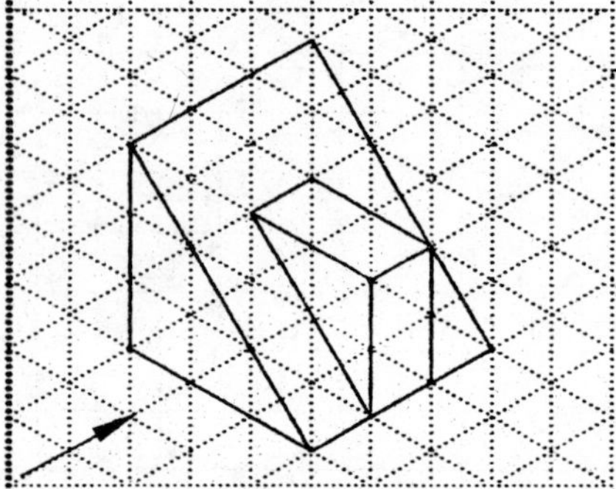

Fig.65

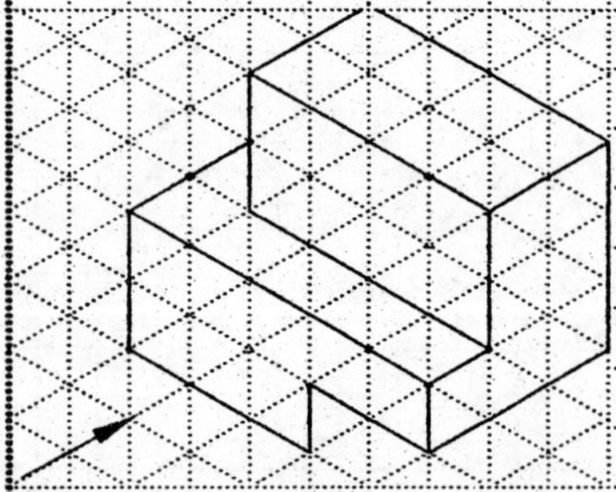

Fig.66

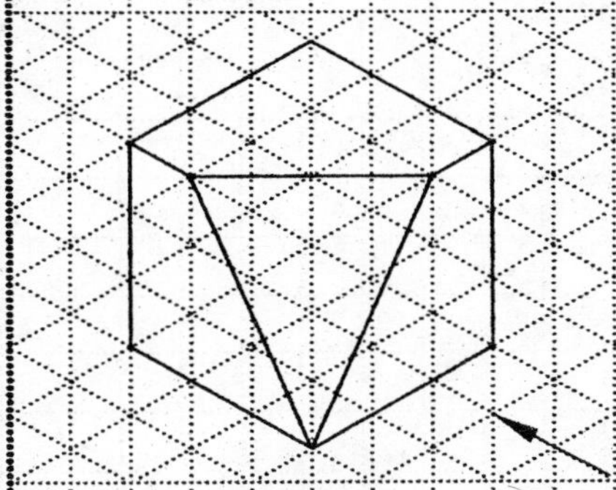

Fig.67

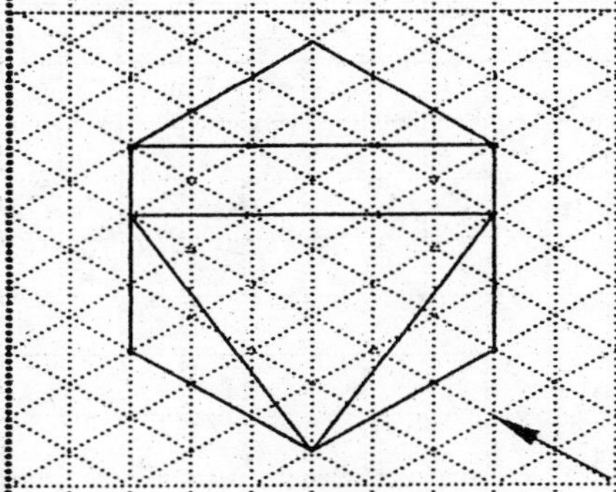

Fig.68

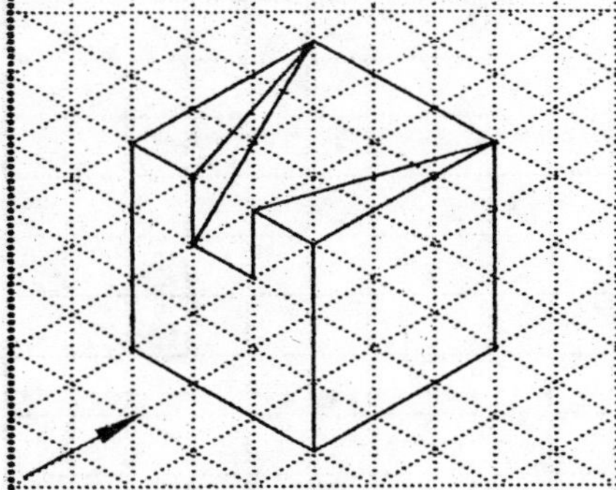

Fig.69

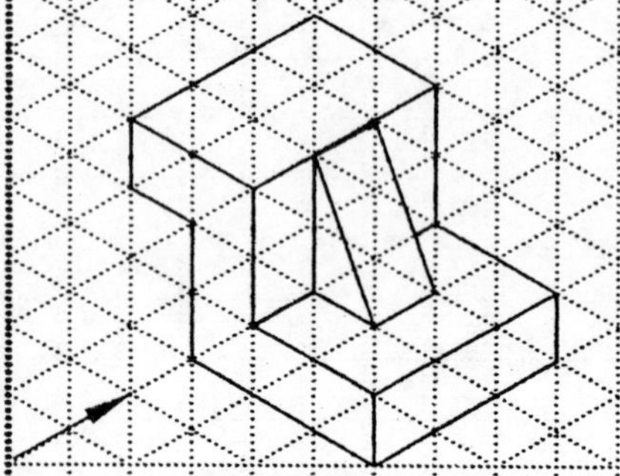

Fig.70

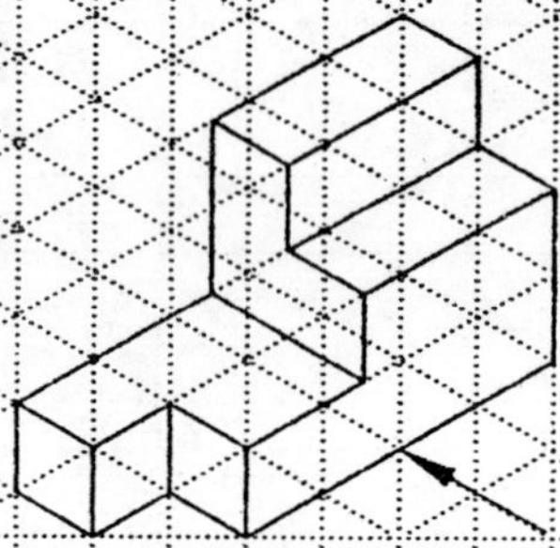

Fig.71

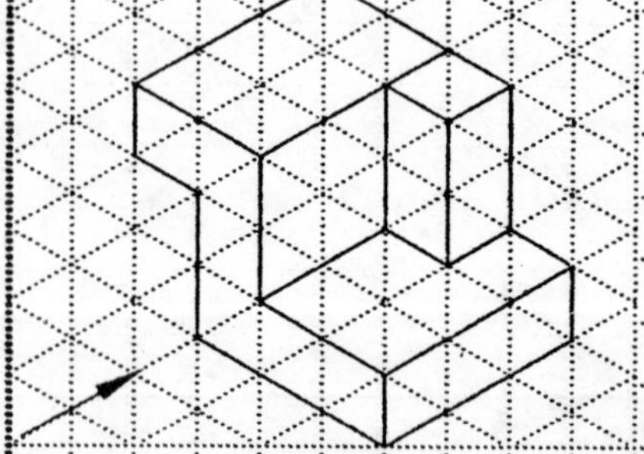

Fig.72

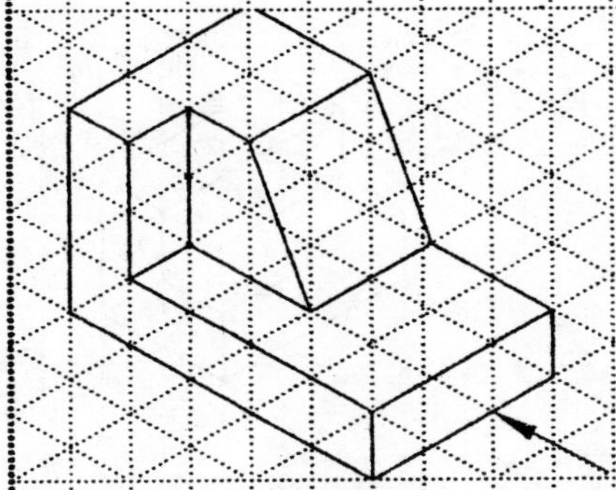

Fig.73

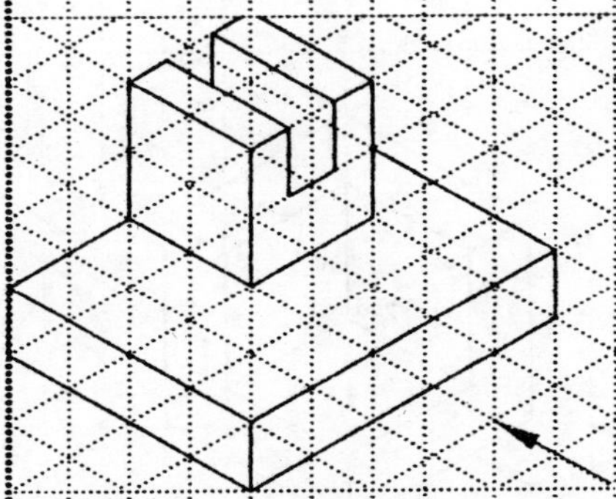

Fig.74

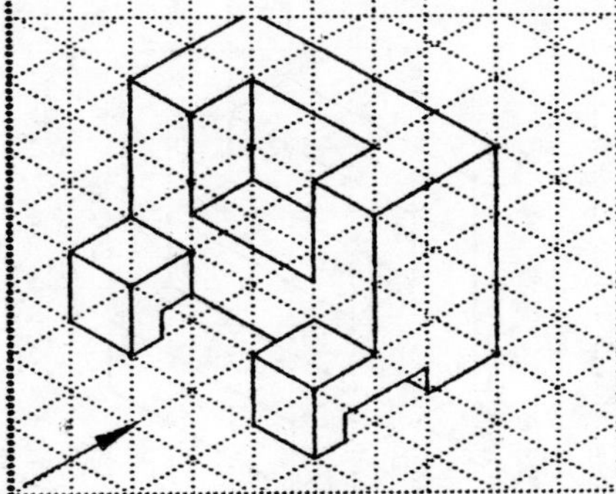

Fig.75

Study the orthographic views in Figs. 76 to 96 and draw their Isometric views.

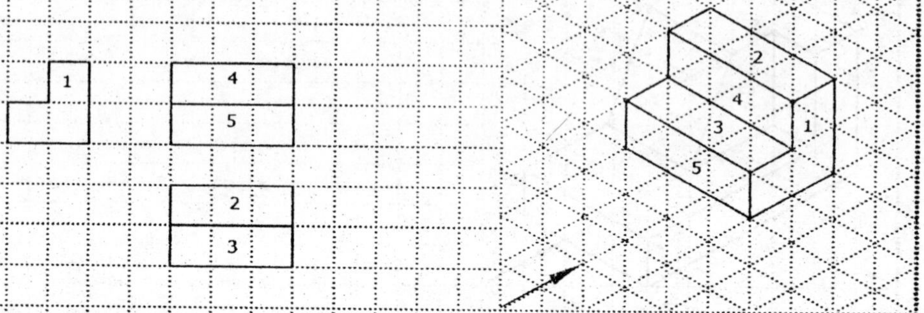

Fig.76

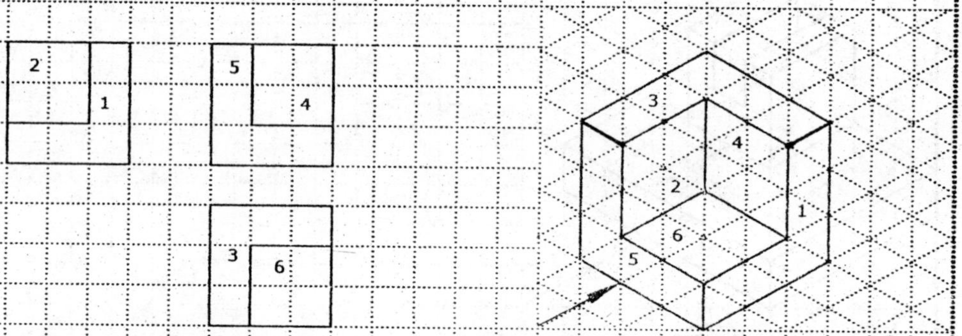

Fig.77

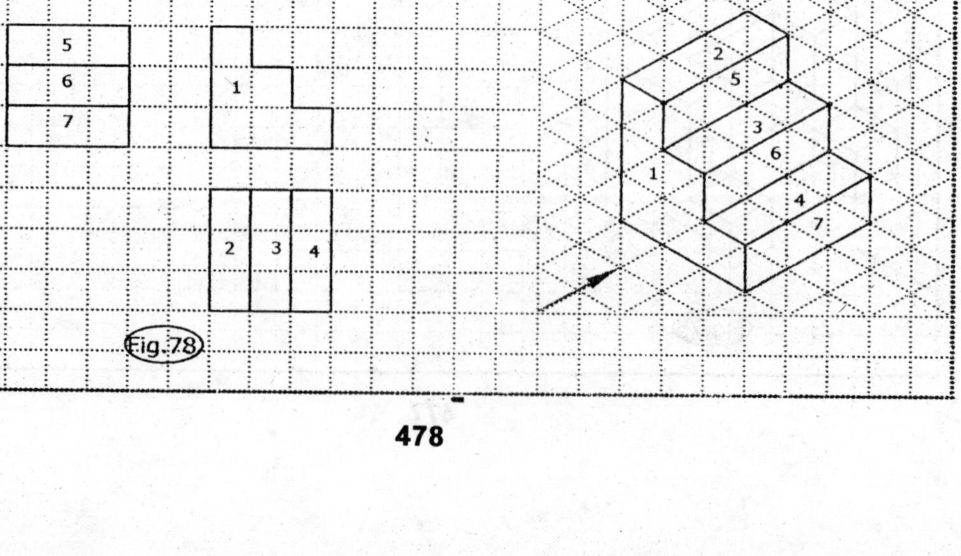

Fig.78

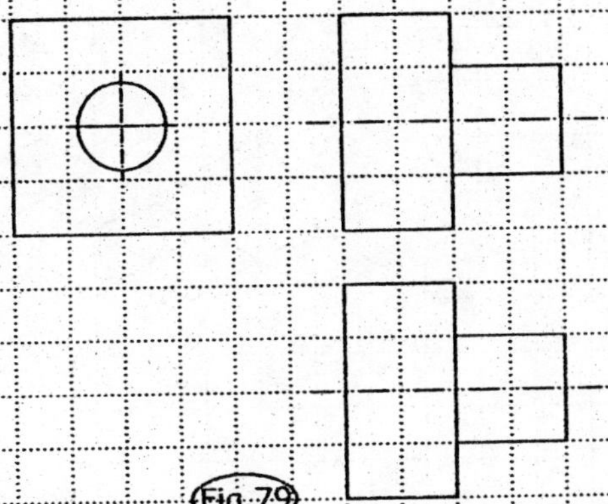

Fig.79

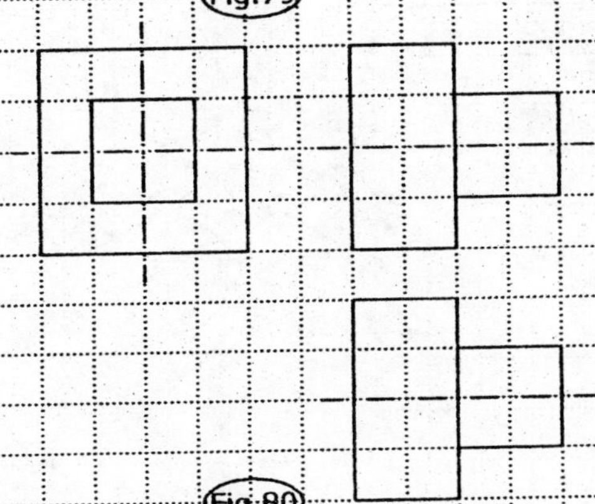

Fig.80

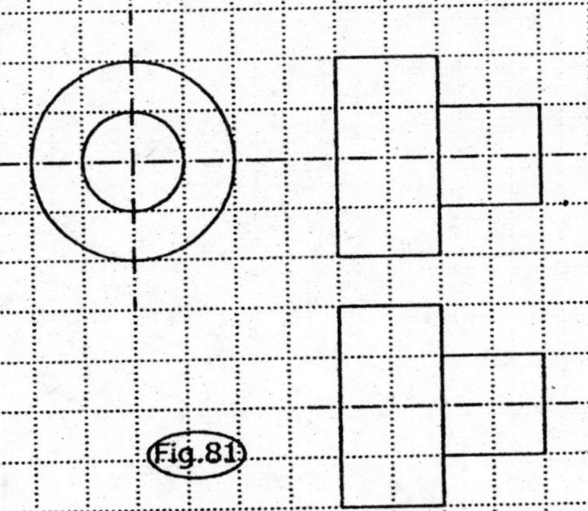

Fig.81

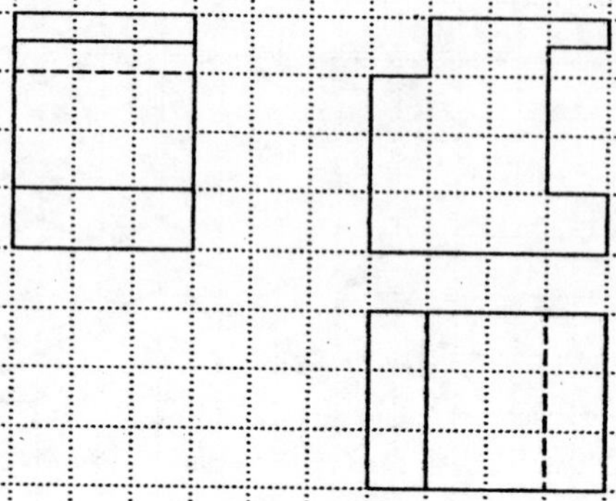

Fig.82

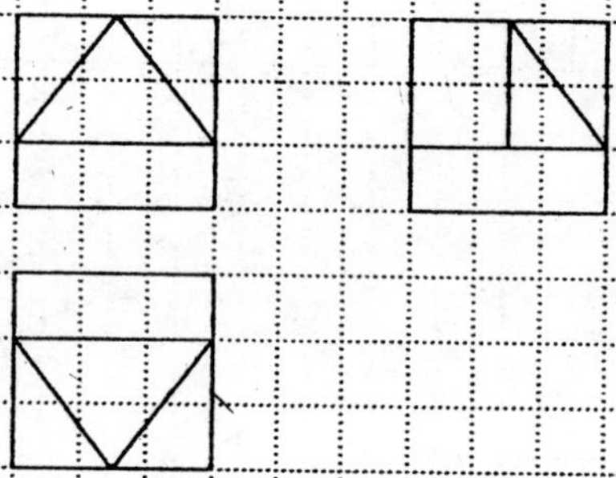

Fig.83

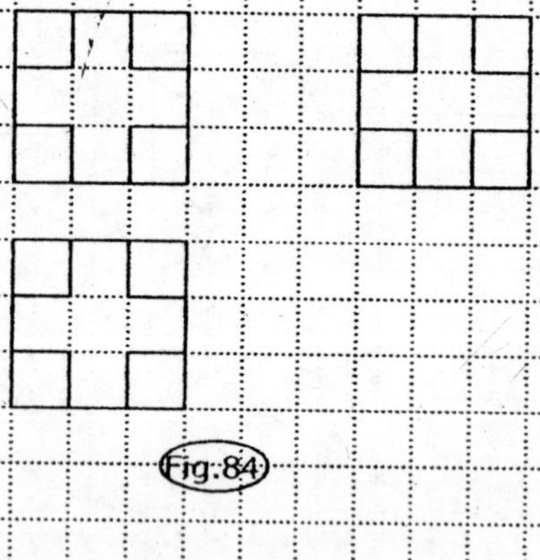

Fig.84

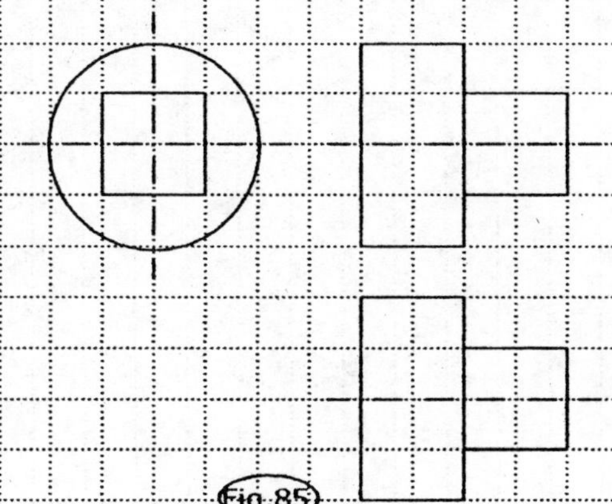

Fig.85

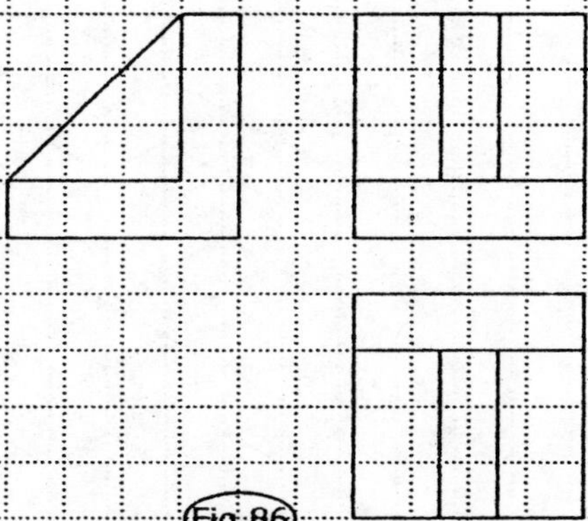

Fig.86

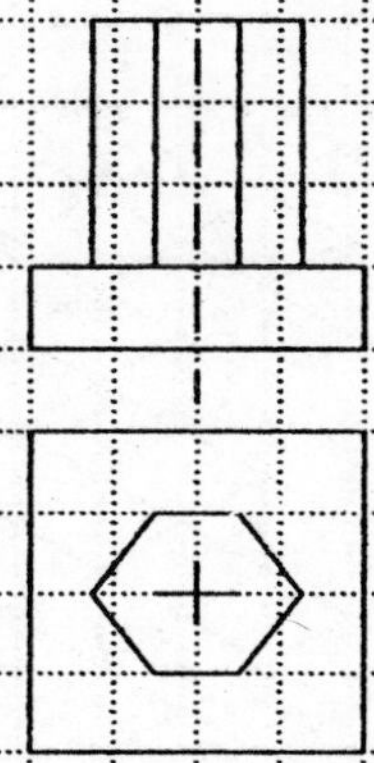

Fig.87

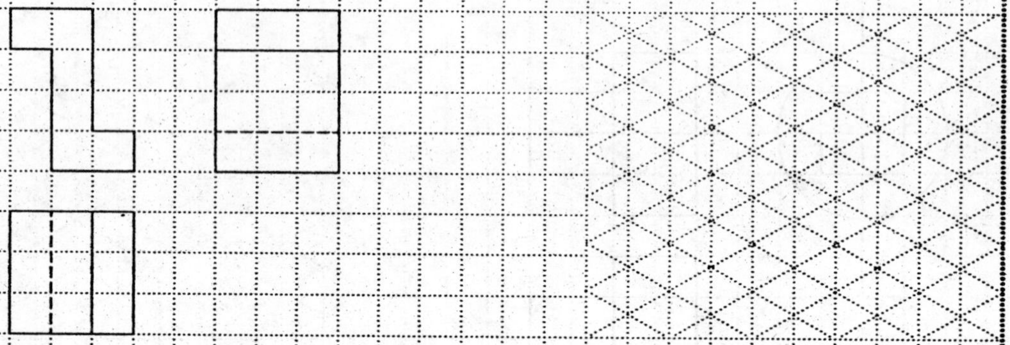

Fig.88

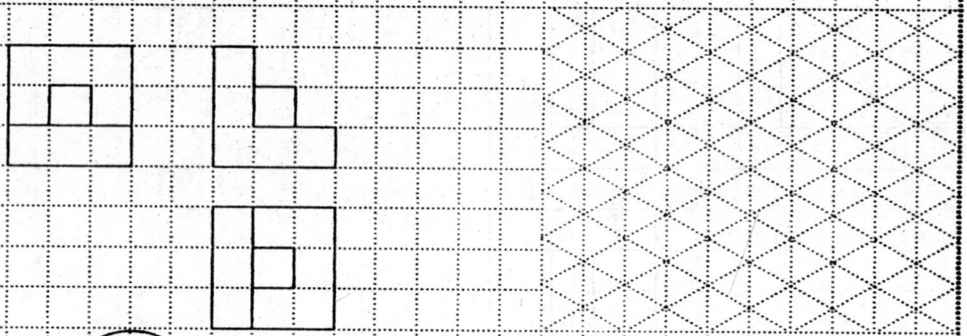

Fig.89

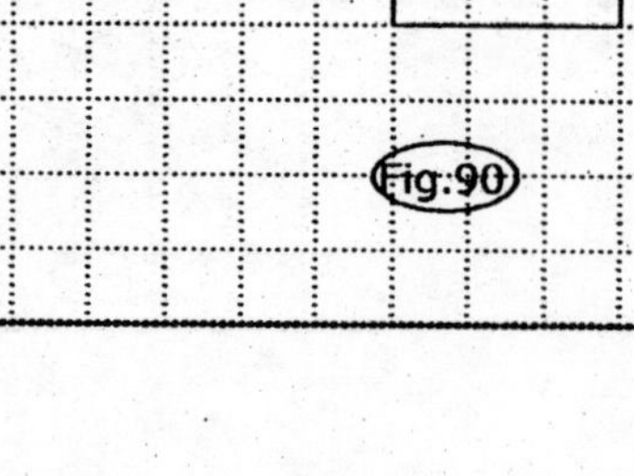
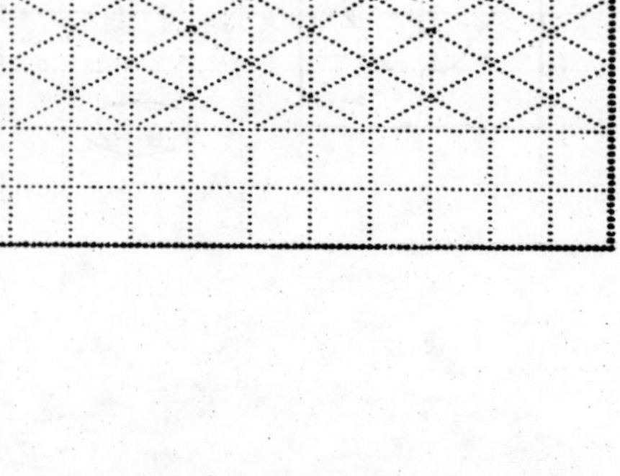

Fig.90

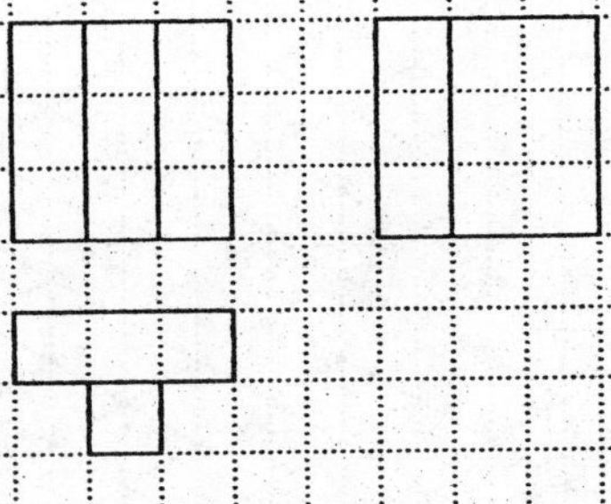

Fig.91

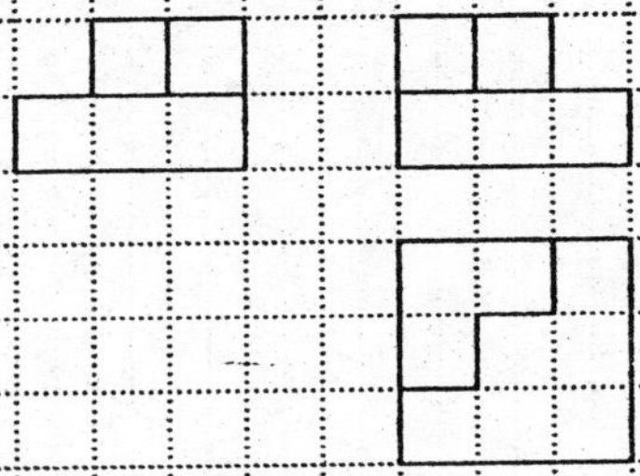

Fig.92

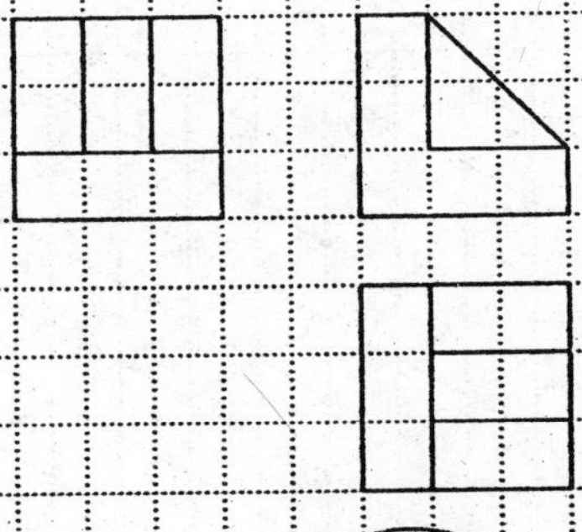

Fig.93

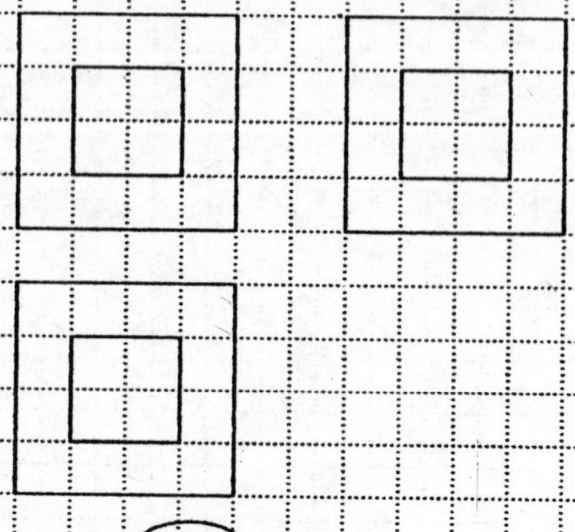

Fig.94

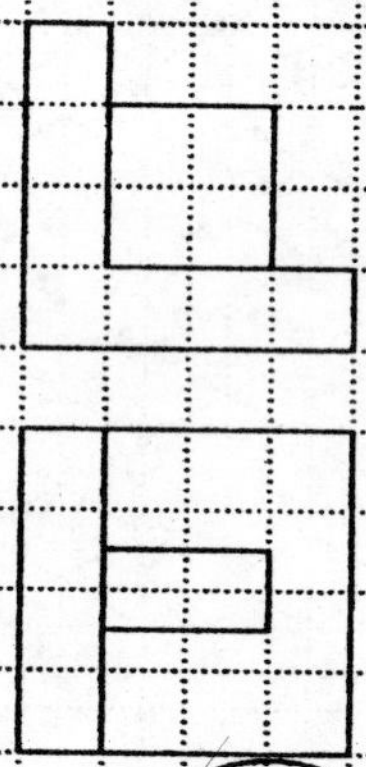

Fig.95

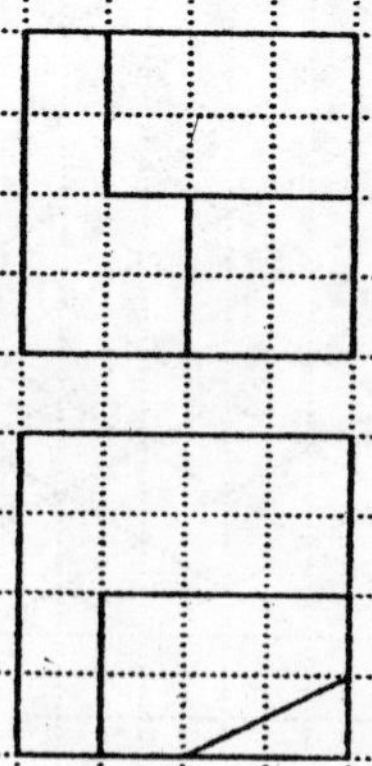

Fig.96